Quantity	Symbol	
Speed of light	c	
Diameter of hydrogen atom	r_0	5.29×10^{-11} m
Atomic mass unit	amu	1.66×10^{-27} kg
Absolute zero	O K	$-273°C, -460°F$
Unit of electric charge	e	1.60×10^{-19} C
Planck's constant	h	6.63×10^{-34} J·s
Coulomb's law constant	K	8.99×10^9 N-m^2/C^2
Gravitation constant	G	6.67×10^{-11} N-m^2/kg^2
Surface gravity of earth	g	$9.81 \times$ m/sec^2
1 gram/cubic centimeter	g/cm^3	1000 kg/m^3
1 year	yr	3.16×10^7 sec
1 atmosphere	atm	1.01×10^5 N/m^2
1 kilowatt hour	kWh	3.60×10^6 J
1 megaton	Mton	4.18×10^{15} J

FACETS OF PHYSICS:
A CONCEPTUAL APPROACH

FACETS OF PHYSICS:
A CONCEPTUAL APPROACH

Roger B. Culver
Colorado State University

WEST PUBLISHING COMPANY

MINNEAPOLIS/ST. PAUL   NEW YORK   LOS ANGELES   SAN FRANCISCO

....
PRODUCTION CREDITS

Composition: Parkwood Composition, Inc.
Copyediting: Lorretta Palagi
Indexing: Sonsie Conroy, Catalyst Communication Arts
Interior Design: Roslyn M. Stendahl
Illustrations: J/B Woolsey Associates
Page Layout: Roslyn M. Stendahl
Chapter Opening Art: © 1993 Roslyn M. Stendahl
Cover Photos:

Space shuttle Eric Meola/The Image Bank
Darts Terje Rakke/The Image Bank
Fiber optics Dominique Sarraute/The Image Bank
Gymnast Co Rentmeester/The Image Bank

WEST'S COMMITMENT TO THE ENVIRONMENT

In 1906, West Publishing Company began recycling materials left over from the production of books. This began a tradition of efficient and responsible use of resources. Today, up to 95 percent of our legal books and 70 percent of our college and school texts are printed on recycled, acid-free stock. West also recycles nearly 22 million pounds of scrap paper annually—the equivalent of 181,717 trees. Since the 1960s, West has devised ways to capture and recycle waste inks, solvents, oils, and vapors created in the printing process. We also recycle plastics of all kinds, wood, glass, corrugated cardboard, and batteries, and have eliminated the use of styrofoam book packaging. We at West are proud of the longevity and the scope of our commitment to the environment.

Production, Prepress, Printing and Binding by West Publishing Company.

PHOTO CREDITS

1.1, page 2: Neg# 317640. Courtesy Department of Library Services, American Museum of Natural History.
1.2(a), page 3: Lawrence Migdale/Stock Boston.
1.2(c), page 3: ©Dallas & John Heaton/Stock Boston.
1.4(top), page 5: Rouxaime/Jacana/Photo Researchers, Inc.
1.4(bottom), page 5: J. M. Pasachoff/Visuals Unlimited.

Photo Credits continue following index

Library of Congress Cataloging-in-Publication Data

Culver, Roger B.
    Facets of physics / Roger Culver.
       p. cm.
    Includes index.
    ISBN 0-314-00969-8
    1. Physics. I. Title.
QC21.2.C85 1993   92-34465
530—dc20   CIP

To

Rosalie and Gail,
with love

AND

In Loving Memory of
My Parents
Louise K. Culver   and   Theodore G. Culver
who gave so much for so long to so many

Brief Contents

Contents

Preface

The creation of a physics text at the conceptual level presents both an opportunity and a challenge. On one hand, physics has supplied much of the driving force behind the science that, over the past four centuries, has permitted human understanding of the physical world to progress further and faster than at any time in recorded history. In the process, we have uncovered a universe that ranges from subatomic to extragalactic and is incredibly fascinating everywhere throughout that range.

The opportunity is thus to convey that sense of interest, wonder, and fascination to the introductory student whose concerns and interests often lie elsewhere. The challenge arises from the fact that this sense of excitement must be conveyed against the backdrop of what British author C.P. Snow has referred to as the ''two-culture'' society in which science and technology form one culture and everything else the other.

One of the familiar manifestations of the two cultures can be found in the fact that physics is routinely ranked high on the list of most-feared classes in the course curriculum by the non-science student population. A physics text written at this level must therefore not only convey the wonder of the world of physics, but also do so in such a way that the apprensive, science-anxious student is not intimidated by the presentation.

The first chapter of this text reflects this author's view that any student who emerges from an elementary science course should have a reasonably clear-cut exposure to the philosophy by which that science operates, and the historical forces that have shaped that philosophy. While this text emphasizes the descriptive or qualitative elegance of physics, homage is also necessarily paid to the mathematical and quantitative elegance of physics as well. In paying that homage it is assumed that any student at the college level should be able to perform basic arithmetic operations. On the other hand, it is also recognized that there exists a large measure of ''math anxiety'' in a significant fraction of the student clientele for this course. As a result, word equations are employed throughout the main text in lieu of the more traditional equations using mathematical symbols. Basic mathematical concepts which will be of use to the student are described in some detail in Chapter 2, and worked example involving numerical calculations appear at the end of each section, where appropriate.

More advanced mathematical concepts, derivations, and discussions relating to the material in a given chapter are deferred to the end of each chapter via short presentations entitled Physics Formulations. The Physics Formulations are essentially more detailed explorations of results presented in the main text. Their placement at the end of each chapter emphasizes the optional nature of this material to both the instructor and the student.

Chapters 3 through 13 and Chapter 15 are concerned with the standard physics topics discussed and covered in an elementary physics course. In presenting this material, an attempt has been made to provide at least a beginning historical background in order that the student gain some insight and appreciation of the ways in which physics has made its progress.

Students very often express concern over the relevance of a given field of knowledge. Chapter 14 on physics and modern technology describes how the principles of physics have been employed in a number of devices that have had a profound impact on human society. Such discussions are generally dispersed throughout a given text, and in some cases this has been done here. Devoting an entire chapter to this topic serves to starkly emphasize the point in a much better fashion.

Chapter 16 on astrophysics represents something of a departure from the material contained in a conventional physics text. Of all the areas of physics, however, it is the physics-generated descriptions of the celestial objects above us that perhaps catches the interest and imagination of the non-science student most of all. Moreover, our ongoing attempts to understand the more exotic and mysterious of these objects represent an important frontier of physics that ought not be ignored or glossed over in a text of this type.

The final chapter on energy and the environment provides a discussion of what is in effect the ultimate relevance of physics. It is hoped that it will at least generate an increased awareness of some of the issues and questions raised therein.

Sprinkled throughout the text are a number of descriptive vignettes called Physics Facets. The Physics Facets cover a spectrum of topics as diverse as bioelectricity, Marie Sklodowska Curie, and nuclear war. All are topics that the student should find interesting, but do not qualify as mainstream text items. Unlike the Physics Formulations, the Physics Facets are non-mathematical in nature and hence have been included as features of the main text.

There are associated with this text a number of supplementary materials that have been developed as aids to both teacher and student alike.

Supplements include:

Instructor's Solutions Manual, by Jerry O'Connor of San Antonio College, with answers and solutions for each chapter;

Test Bank, written by the text author, with 50 questions per chapter including multiple choice and problem questions;

West's 3.0 Computerized Testing, available for IBM PCs and compatible computers and MacIntosh computers;

Transparency Acetate Overheads that show approximately 100 figures from the text;

West's Physics/Astronomy Video Library; and

Great Ideas in Teaching Physics, which includes demonstrations, analogies, and diagrams for lecture enrichment, written by teachers throughout the country.

Acknowledgments

Like most of us, I often become rather annoyed during award shows or ceremonies when a smiling recipient of some kind of award or other proceeds to thank everyone in sight in a highly time-consuming fashion. One cannot spend the considerable amount of time required for a project of this nature, however, without piling up a tremendous number of individuals who, in varying degrees, have contributed to the completion of this text, and who deserve to be so mentioned. And thus, like the grinning award recipient, I will express my gratitude to the cadre of individuals that have made it all possible.

First of all, my thanks goes to Peter Marshall at West Publishing Company who has steadfastly stuck with me throughout this endeavor. There have been a number of times when he has exhibited a patience well above and beyond the call of duty.

Thanks and gratitude, along with much love go to my sons Ken and Larry who have been two of the great joys of my life as they march from their teens into adulthood; to the magnificent Gail Matulewicz who has been loving, patient, and supportive almost from the beginning of this work; and to Rosalie Huther, a grand and ageless aunt who has encouraged my interest in the world around us for as long as I can remember.

Preface acknowledgements are also due to Drs. Eric Craine and Philip Ianna, who at the risk of their own careers have been my collegues, coauthors, and valued friends for over two decades. Their unending encouragement, cajoling, and outright teasing have helped immensely in the completion of this project. Similar comments can be made on a lesser scale for a myriad of friends and family who have contributed to this effort in one way or another.

In putting a text of this type together, one is beholding to a large infrastructure of competent and skilled individuals. No less than six secretaries at Colorado State University worked on the manuscript at one time or other, including "Linda-Linda" Balcomb; Sandy Demlow; Bonnie Gillmore, one of the "Linchpins of Western Civilization"; Karen, "The Big K" Knorr; Heidi Meyers; and Jackie Tunberg. At the CSU end of the process, I must also acknowledge the personnel of Preston Davis' Office of Instructional Services. Their fine efforts produced many of the photographs and graphics which appear in this text.

In addition to Pete Marshall, I would like to thank the individuals at West Publishing for their competent professionalism, including Becky Tollerson for her help in editing the manuscript, and to Deanna Quinn for firmly but gently keeping an otherwise entropy-laden professor on a tight production schedule.

Thanks are also due to John Woolsey and his crew of talented artists who have translated my ideas for the artwork in this text into superb reality. Ken Culver also gets kudos for his work on some of the preliminary artwork as well. The chapter opening illustrations are the result of the creative genius of Ms. Roslyn Stendahl.

A large number of reviewers were also involved at various stages of the development of this manuscript, the two most prominent of which are Dr. Allen Miller of Syracuse University who checked the equations and numerical calculations in the text and Dr. Jerry O'Connor of San Antonio College who has done a splendid job on the *Instructor's Solutions Manual*. In addition, each of the following at one time or other have offered their valuable comments on the manuscript as it evolved and developed, and to all of them I express my thanks.

Peter Bartel
Wichita State University

Claire Chapin
Chabot College

Lowell Christensen
American River College

Dawn Dressler
Portland State University

John Erdei
University of Dayton

Sherman Frye
Northern Virginia Community College

David Haase
North Carolina State University

Gary Layton
Northern Arizona University

David Lichtman
University of Wisconsin—Milwaukee

Bob Martin
Tarrant Jr. College—Northeast

James Merkel
University of Wisconsin—Eau Claire

Victor Michalk
Southwest Texas State University

Allen Miller
Syracuse University

David Murray
University of Wisconsin—Rocky
County Center

Van Neie
Purdue University

Jerry O'Connor
San Antonio College

Kenneth O'Dell
Northern Arizona University

Hans Plendl
Florida State University

Dan Quisenberry
Northern Arizona University

Raymond Robinson
Colorado State University

Elwood Schapansky
Santa Barbara City College

Charles Shirkey
Bowling Green State University

John Stanford
Iowa State University

Paul Varlashkin
East Carolina University

Robert Zelenka
Anne Arundel Community College

Although I dearly wish otherwise, two people who were important in my life while this text was in preparation are no longer with us. Dr. Ian Spain of the CSU Physics Department was a close friend and one of the most multitalented individuals I have ever known. His passing only weeks after his fiftieth birthday in 1990 was a difficult loss. Sadly, Ian's death was followed some six months later by that of my mother, Louise K. Culver, who shone as a supernova to me for nearly fifty-two years and who can never be replaced.

Finally, I wish to express my deepest gratitude to the students with whom I have worked, played, argued, and otherwise interacted in a number of ways down through the years. It is they who have helped shape this text and have made teaching as much fun as it has been for me over the past quarter-century.

Roger B. Culver

FACETS OF PHYSICS:
A CONCEPTUAL APPROACH

1 DEVELOPMENT OF SCIENTIFIC PHILOSOPHY

FIGURE 1.1 ··········

A Stone Age Rendition of a Wild Boar The two sets of legs in this Altamira, Spain, Stone Age cave painting are most likely an attempt by the artist to represent the physical reality of the boar's running motion.

Ancient time line

Pythagoras
500— B.C.
Anaxagoras
Socrates
Democritus Peloponnesian war
Plato

Aristotle Alexander the Great

Euclid Library at Alexandria built

Archimedes

Julius Caesar

⊕ Birth of Christ

Ptolemy

Roman empire at greatest extent

St. Ambrose

Hypatia
St. Augustine Fall of Rome
500— A.D.

Physics is perhaps the most pervasive of those reservoirs of human knowledge that we call the modern sciences. Presiding over a dominion that includes topics as seemingly diverse as light, sound, motion, and matter, physics has come to play an extraordinary and prominent role in shaping much of our current society. As a result of this influence, a significant number of the problems and questions that confront and challenge contemporary civilization possess a distinctly scientific and technological flavor. An almost obligatory prerequisite for informed citizenship in the twenty-first century, therefore, is at least a basic understanding of the principles by which the physical world operates.

Central to the discovery and development of these principles are the philosophical assumptions that have come to us from centuries past. These assumptions were more or less classified into a mode of operation known as the scientific method during the intellectual fervor of the Western European Renaissance. These scientific axioms, postulates and "rules of conduct," however, were not a sudden inspiration to such as Copernicus, Galileo, or Newton, but rather are the product of an arduous, eons-old human struggle to gain an understanding of—and perhaps even a mastery over—a physical universe that does not readily yield its secrets. In this opening chapter we will examine some of the details of that struggle and the philosophical product it has fashioned.

1.1 The Physics of Antiquity

Like almost all human achievements and endeavors, the beginnings of our perceptions and understanding of the laws of nature lie deeply embedded in the murky mire of prehistory. Although our knowledge of these long-ago millennia is quite limited, the few tantalizing glimpses of "prehistoric physics" that we have gleaned from cave paintings, rock carvings, and the like portray a "physics" that was both descriptive and practical, (Figure 1.1). A prehistoric hunter may not have been able to write the mathematical equations of motion describing the trajectory of an arrow or spear, but that individual had certainly developed from experience an excellent sense of the angle and speed at which the arrow or spear should be launched in order to strike the desired target. For that hunter, a failure to develop such a sense of projectile motion meant quite simply and bluntly a failure to eat. A similar practicality fueled a beginning understanding of a number of other basic physical principles of nature, including the principles of heat transfer involved in keeping a dwelling warm in cold weather or cool in hot weather, those of fluid flow involved in the irrigation of crops and the supply of water to a primitive village, and those of static and rotational mechanics utilized in the construction of the wheel and various simple machines. This practical approach to "prehistoric physics" was maintained throughout recorded history, and is manifested in such feats as the pyramids of Egypt, the roads and aqueducts of Rome, and the medieval cathedrals of Western Europe (Figure 1.2). Currently, we see it in the more contemporary forms of engineering and technology that are so important to our modern society. Although this chapter is concerned primarily with the flow of Western culture, we want to emphasize that other cultures from around the world, while not directly involved in the development of the scientific method that emerged from Renaissance Europe, nevertheless generated impressive achievements of their own. The Chinese, for example,

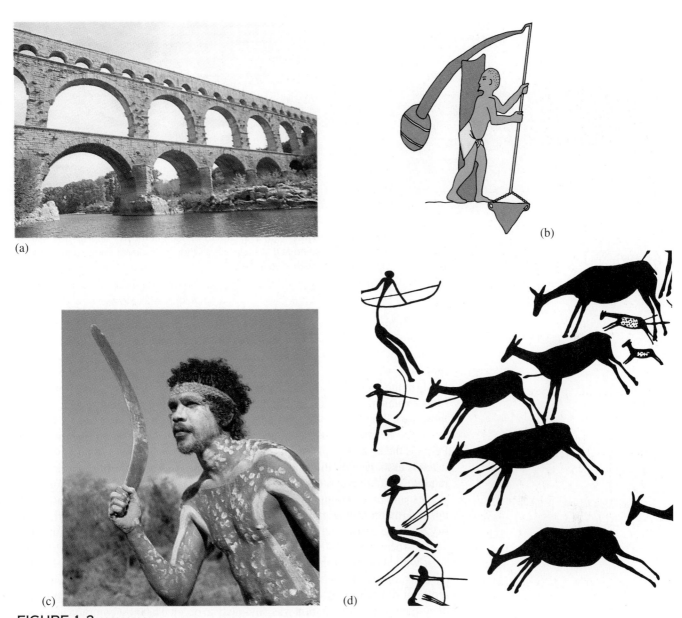

FIGURE 1.2 ··········
Practical Physics in Antiquity   (a) Water flow in a Roman aqueduct; (b) torque in shaduf, a simple lifting machine of ancient Egypt; (c) aerodynamics in the boomerang of an Australian aborigine; (d) projectile mechanics of Stone Age hunters.

developed the magnetic compass, were the first to make use of rockets, and built the Great Wall, a structure so large and extensive that it could be telescopically recognized from the Moon and the nearby planets Venus and Mars as the work of an intelligent civilization. The blowguns of the Amazon tribes make use of the gas laws, and the amazing aerodynamics of the boomerangs of the Australian aborigines are legendary. Even the universal art form of music is based in large measure on the physics of vibrations and waves.

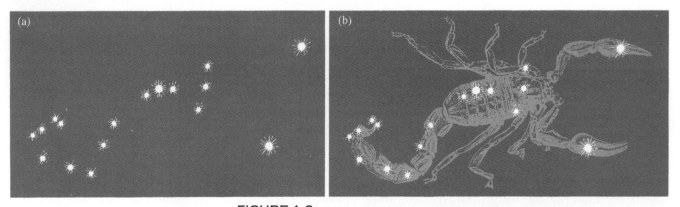

FIGURE 1.3 ··········
The Constellation of Scorpius   (a) The pattern of stars in the sky appears to have the rough outline of a scorpion and (b) hence was thought by most cultures to be a heavenly manifestation of such a creature.

As impressive as they were however, such efforts and achievements lacked a unifying theme. Connections that are now known to us still lay hidden to these cultures. In the absence of a more complete view, the phenomenon of a falling rock bore no relationship to the movement of the moon through the heavens. To be sure, the ''physics'' of prehistory and of antiquity saw the start of efforts to offer explanations for the phenomena observed in the physical world. For the most part, however, these explanations took the form of myths and legends involving heroes, heroines, deities, and other assorted characters from folklore. Perhaps the best example of this type of approach can be found in the myths and legends associated with the astronomical constellations. Out of the more or less random sprinkling of stars in the night sky, one can imagine a variety of groups, shapes, and patterns, just as one can lie down on a nice soft stretch of grass, look up, and imagine a variety of faces and forms in the puffy clouds of a springtime sky. In some cases, such as Orion the Hunter and Scorpius the Scorpion, the star patterns bear a reasonable resemblance to their constellation namesakes (Figure 1.3), while others, such as Canes Venatici, the Hunting Dogs, do not. For the ancients, however, the resemblance of a pattern of stars to a great warrior or a scorpion did not seem fortuitous. If such star patterns were indeed significant, then some sort of explanation was obviously in order. The most common approach was to regard the sky as a kind of ''Celestial Hall of Fame'' into which various legendary characters from a culture's folklore had been placed for various reasons. Thus Orion, the mighty hunter of Greek mythology, and the scorpion that killed him are both to be found in the heavens, but placed in such a way that Orion and Scorpius, in the northern hemisphere at least, are never in the sky at the same time. This is presumably so that in the afterlife Orion need not ever view the creature that handed him the only loss of his career. The vain queen Cassiopeia, also from Greek mythology, eternally circles the north celestial pole spending half of the time in a very unqueenly upside-down position as eternal punishment for her boast that she was more beautiful than the sea nymphs.

In many cases the explanations for a given phenomenon involved the idea of a god or goddess possessed of considerable power (Figure 1.4). Thus the sun, moon, and certain ''wandering stars'' or planets were able to execute their

FIGURE 1.4 ··········
Mythological Explanations for Natural Events A lightning bolt was depicted by the ancient Greeks as being hurled from heaven by the god Zeus. In Chinese lore a dragon munches on the sun to produce a solar eclipse.

observed motions relative to the background of "fixed" stars and constellations because of their powers as gods and goddesses. Although we have long since given up such a view of the planets, traces of that ancient concept are still to be found in the deistic names of the planets, such as Venus and Mars.

Inevitably, myths and legends that have highly religious components tend to become non-negotiable dogma in which the given culture fiercely resists any alternate explanations. Thus when the Ionian philosopher Anaxagoras proclaimed to the Athenians of the fifth century B.C. that the sun was not the sun-god Helios but was a red-hot stone the equivalent of 65 kilometers (40 miles) across and 6500 kilometers (4000 miles) distant, he left town one jump ahead of a very angry mob. Ironically Anaxagoras's views could be challenged quite rightly on nonreligious grounds. Aristotle observed, for example, that Anaxagoras's red-hot stone would be expected to cool off very quickly over a matter of hours or at most days. Obviously the sun showed no signs of such an effect, and therefore Anaxagoras's explanation for the sun's shining could not possibly be correct. Religious dogmatism of the type that led to the exile of Anaxagoras served to channel the sciences of antiquity into more practical modes. In a very real sense, it was far safer in most cultures to proclaim the practical "blessings" from the various aspects of the physical world rather than to speculate on their intrinsic nature or ultimate cause.

The treatment of Anaxogoras and his more famous contemporary Socrates notwithstanding, the most profound thinking of early civilization came from what we now refer to as the Greek and Hellenistic cultures. Centered on the Greek peninsula, the islands of the Aegean Sea, and the western coast of Asia Minor, the Greek culture flourished from roughly the sixth to the fourth centuries B.C. As a result of the meteoric career of Alexander the Great in the fourth century B.C., the Greek culture was expanded far beyond the shores of the Aegean Sea and blended with the Near Eastern cultures of Egypt, Babylonia, and Persia into a mixture called the Hellenistic civilization. The Hellenistic era lasted some five centuries after Alexander. During this period the intellectual and cultural center of action shifted from the storied but shopworn Athens to a newer and more vibrant city founded on the Nile River delta by Alexander himself—the famed city of Alexandria.

The contributions made by the Greek and Hellenistic culture to our western heritage are myriad, and to do proper justice to them is well outside the purpose of this chapter. Suffice to say, however, that in light of the immense impact of these cultures on western thought, it is not the least bit surprising to find that several of the basic concepts that underlie our modern scientific method originated during this extraordinary era of human history.

One of the earliest of these ideas was that proposed in the sixth century B.C. by Pythagoras, a man perhaps most familiar for his Pythagorean theorem involving right triangles in plane geometry. Pythagoras firmly believed that a fundamental relationship existed between numbers and the physical world. In fact, the school founded by Pythagoras in the Greek city of Crotona in southern Italy had as its motto ''All is Number.'' Most of Pythagoras's ideas and concepts have faded into history, but his belief that nature could be described by numbers and, through logical extension, by mathematics in general remains one of the cornerstones of our present scientific endeavors.

Mathematics, however, often involves concepts that are difficult to reproduce physically in a way that is consistent with their abstract definitions. For example, in the perception of Greek geometry put together in the third century B.C. by the Alexandrian mathematician, Euclid, a straight line has length but no width and a point defined by the intersection of two straight lines has no dimensions at all! Thus when we draw a point or a straight line, such a drawing can only be a *representation* of that point or line. In other words, a drawing of a point or line can only approximate the ''true'' nature of that point or line. Moreover, representing nature with numbers often turned out to be ''messy.'' The ratio pi (π) of the circumference of a circle to its diameter, for example, cannot be expressed as a ''nice'' ratio of two integers, or ''whole'' numbers, but is rather an ''irrational'' number whose true value can only be approximated by such ratios as 22/7, 377/120, etc.

Many of the Greek thinkers, most notably Plato, felt that such representations were totally inadequate in determining the ''true'' nature of the universe. Such a determination could only be properly done at the ''higher plane'' of pure thought and reason. Conclusions drawn from observations and measurements made by the ''eye of the body'' were imperfect and not to be trusted, while conclusions drawn from the thought and reason of the ''eye of the soul'' were regarded as the essence of truth. Other Greek thinkers vehemently disagreed with Plato. The most notable of these was one of Plato's very own students, a young man by the name of Aristotle. Aristotle regarded the conclusions drawn from observations and measurements made on the physical world as having validity to the extent

Observed path of planet

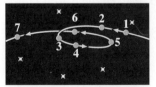

Apparent path of planet
predicted by heliocentric theory

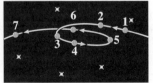

Apparent path of planet
predicted by geocentric theory

Heliocentric explanation

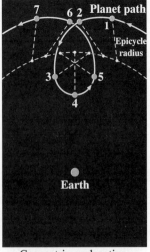

Geocentric explanation

FIGURE 1.5 ··········

**"Saving the Phenomenon": Heliocentric Versus
Geocentric** As the planets Mars, Jupiter, and Saturn orbit
about the sun, they appear from earth to move "backward" in
the sky. This retrograde motion is explained in the heliocentric
view as being caused by the earth catching up to and passing
the given planet in its orbit. The geocentric view requires the
planet to move in an epicycle about a point that in turn orbits
the earth. Both views account for the planet's observed motion,
but the heliocentric theory is simpler.

that any explanation proposed for a phenomenon in nature had to "save the
phenomenon." That is, the explanation had to satisfy as many of the observed
characteristics associated with the phenomenon as possible.

On one hand, such a view concedes in effect that any such explanations
offered were in essence approximations, representations, or models of the "true"
nature of a given physical phenomenon. Thus we are left with the disconcerting
fact that an ultimately "true" nature of the physical world may well turn out to
be an impossible quarry to capture. On the other hand, such a view permits our
ideas concerning natural phenomena to change and evolve in light of new information. Thus the sun-centered or *heliocentric* theory proposed by Aristarchus
of Samos in the third century B.C. was rejected in favor of the earth-centered or
geocentric view of Claudius Ptolemy about 140 A.D. The geocentric theory was
accepted primarily because the Ptolemaic view with its system of orbits and
"suborbits" or "epicycles" did a better job of "saving the phenomenon" of
the observed motions of the sun, moon, and planets relative to the background
stars and constellations than did Aristarchus's system in light of the observational
data available at the time (Figure 1.5). In like fashion, the atomic theory proposed
around 530 B.C. by Democritus, in which he envisioned matter as being composed
of an infinitude of tiny indivisible spherical particles he called atoms, was

superseded by Aristotle's view that matter was continuous and composed of four basic elements: earth, water, air, and fire.

Interestingly, as new experimental evidence has become available in more recent times, these two basic ideas, a heliocentric system of planets and matter composed of fundamental particles, have been resurrected and now enjoy a preeminence far beyond what they were able to achieve in antiquity. For all intents and purposes, the death of Ptolemy around 150 A.D. marked the end of Greek and Hellenistic scientific achievement, and for the next 14 centuries scientific advance was at a virtual standstill, as discussed in Section 1.2.

REVIEW
1.1
QUESTIONS

1. Think of some examples of ''practical physics'' in antiquity.
2. Why was there an emphasis on practical applications in the ''sciences'' of antiquity?
3. Give an example in which a myth or legend has been proposed to explain a natural phenomenon.
4. What contributions did the Greek and Hellenistic cultures make to the philosophy of science?

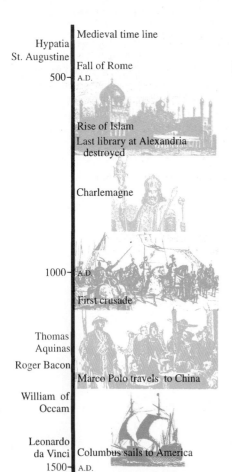

Medieval time line

Hypatia
St. Augustine
Fall of Rome
500— A.D.

Rise of Islam
Last library at Alexandria destroyed

Charlemagne

1000— A.D.

First crusade

Thomas Aquinas

Roger Bacon

Marco Polo travels to China

William of Occam

Leonardo da Vinci   Columbus sails to America
1500— A.D.

1.2 The Medieval "Interlude"

The slowing of progress in science that characterized the 1400 years after Ptolemy was the result of several historical factors that, at first glance, appear to be unrelated to the province of science. The death of Ptolemy about 150 A.D., however, corresponds to within a few decades the various dates assigned by historians as the high water mark of the Roman Empire. By 180 A.D. nearly all of the centers of Greek and Hellenistic culture including Athens and Alexandria had fallen under the domination of the Romans. The Roman culture was the epitome of practicality. The Romans were superb warriors, political organizers, and engineers. At the same time, they showed little interest or enthusiasm for the Greeks' abstract speculations concerning the nature of the physical world. Instead, they concerned themselves with practical morality and the philosophy of how one should conduct one's affairs so as to achieve the highest levels of order, peace, prosperity, and pursuit of pleasure. Indeed, the Romans' grudging admiration for the famed Greek mathematician Archimedes of Syracuse centered not on his abstract mathematical ideas, but rather on the fact that the machines of war he designed during the Roman siege of Syracuse in 212 B.C. almost single-handedly kept the vaunted Roman army at bay for over two and one-half years (Figure 1.6). In addition, the Romans were highly adept at ''Romanizing'' the cultures they conquered. As a result, Roman cultural attitudes, concerns, and priorities permeated the populace of their empire. Thus the first attack on the advances of Greek and Hellenistic scientific thought came about from what was essentially neglect on the part of a Roman Empire preoccupied with more practical and materialistic concerns.

An even more profound impact on scientific progress was generated by the rise of Christianity. Christianity first appeared in the region of the Near East that

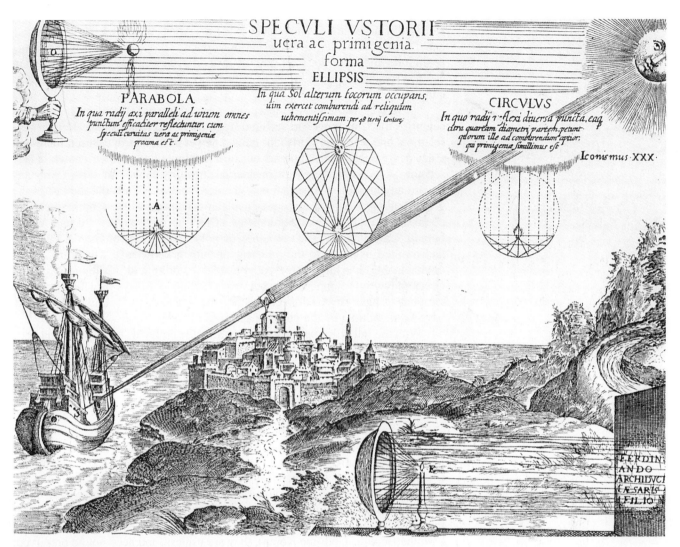

FIGURE 1.6

The Principles of Optics   17th century represetation of Archimedes' destruction of the Roman fleet by means of a system of mirrors at the seige of Syracuse in 212 B.C.

is now modern Israel barely a century before the time of Ptolemy. Despite persecution by Roman authorities, the influence of Christianity steadily increased until it was finally installed as the state religion of the Roman Empire in the fourth century A.D. The success of the Christians during this period was due in no small part to the well-known material excesses and moral bankruptcy that had gripped the Roman Empire, particularly in its latter stages. The spiritually oriented Christians simply stepped into the spiritual vacuum left by the Romans. For the Christian, life on earth was a temporary prelude to an eternal hereafter. Thus for Christians the primary purpose of one's existence here on earth was to save one's soul, and in this context, therefore, the physical world represented a dangerous distraction from that task. In such a scheme, any field of human endeavor, whether science or attendance at the chariot races, that did not have as its primary concern the salvation of souls was at the very least to be deemphasized if not shunned altogether. When St. Ambrose, the impressive and talented bishop of Milan, wrote in 400 A.D. that ''To discuss the nature and the position of the earth does not help us in the hope of life to come,'' he was in essence speaking against the basic motivation of scientific thought, i.e., human curiosity.

There was also a more sinister side to the Christian ascendancy. For three centuries the Christians had suffered mightily at the hands of the Roman persecution, and when the Christians gained power in the fourth century, there was a decidedly unchristian ''now its our turn'' attitude toward their erstwhile persecutors. As a result, there was considerable hostility on the part of the Christians toward anything or anyone that was deemed ''pagan.'' Thus did the last of the great Hellenistic astronomer-mathematicians, Hypatia of Alexandria, meet her doom when in 415 she was murdered by a Christian mob and her ''pagan'' works burned. Unfortunately, such mindless activity was all too common at this time and resulted in losses to the reservoir of human knowledge which are almost unimaginable. It is known, for example, that in addition to ''Oedipus Rex,'' the Greek playwright Sophocles, wrote over one hundred plays of which less than ten percent have survived. We can only speculate on what similar losses may have been endured by the physical sciences as well.

With the fall of Rome in the fifth century, the political stability that had been a hallmark of the Roman Empire and had made possible the continued existence of learning centers such as Alexandria was now swept away. Roman order in Western Europe was now shattered into a multitude of small, often warring feudal states. Such conditions were not conducive to learning of any kind, let alone scientific endeavor.

The final blow to scientific endeavor was provided in the seventh century A.D. by the rise of a third major religion out of the Near East, the religion called Islam. From its origin in the Arabian peninsula, the influence of Islam was expanded via military conquest over the next two centuries throughout most of the Near East and along the southern Mediterranean coast. In the course of his conquest of Alexandria in 642 A.D., the Caliph Omar burned what was left of the great library there with the words, ''If these writings of the Greeks agree with the Book of God (the Koran), they are useless, and need not be preserved; if they disagree, they are pernicious, and ought to be destroyed.'' These were tough centuries indeed for the library at Alexandria!

The Islamic expansion continued for another century before finally being stopped in the west at the battle of Tours in southern France and in the Near East by the power of the Byzantine Empire. When a revised political-religious equilibrium was reached in the eighth century, the Mediterranean was polarized between mutually antagonistic Christian and Islamic areas of control, with most of the great Hellenistic centers of culture and learning, most notably Alexandria, ending up in Islamic hands. Despite their early excesses, the Islamic civilization did an admirable job of preserving the remnants of the Greek and Hellenistic heritage, but the animosity between Islam and Christianity precluded the accessibility of that knowledge to western scholars. Thus by the end of the eighth century, scientific endeavor in Western Europe had come to a complete and total standstill.

History, however, is not a static entity and the conditions that had combined so effectively to create the Dark Ages eventually began to dissolve. Perhaps the first manifestations of this dissipation came in the form of renewed contact between Christian West and Islamic East. Late in the eleventh century the Western

Europeans launched the first of the crusades against the Islamic East. The purpose of the crusades was to regain Christian control of the holy places around Jerusalem, Bethlehem, Nazareth, etc. The first crusade was far and away the most successful and resulted in the establishment of the Latin kingdom of Jerusalem, which stretched from the Red Sea northward along the eastern Mediterranean coast into Asia Minor. By the end of the thirteenth century, despite additional crusades and the support of the Italian mercantile states such as Venice and Genoa, the Latin possessions in the Near East were destroyed.

Nevertheless, the close contact between Christianity and Islam not only in the Latin kingdom of the Near East but also in Spain and on the island of Sicily served to reintroduce the West to Greek and Hellenistic ideas and triggered what historians have called the "Twelfth Century Renaissance." However, the return of the Greek heritage to Western Europe in the twelfth century did not come without changes. A definite attempt was made, for example, to "Christianize" the Greek ideas so as to make them compatible with the Christian faith. Thus Thomas Aquinas's thirteenth century work *Summa Theologica* is in essence a combination of Christian doctrine and the rational common sense morality of Aristotle. To this day it stands as an important work in Christian philosophy.

There was also a tendency on the part of twelfth and thirteenth century scholars to impart to the rediscovered Greek works a level of authority that the original authors neither enjoyed nor even desired in their own time. Although such a view was totally consistent with an "Age of Faith," Aristotle and the others would have been totally dismayed at the thought of their works being relegated to the realm of dogma. The rediscovery of the Greek and Hellenistic heritage led to the development in Western Europe of a curious blend of faith and reason called Scholasticism, in which new "knowledge" was generated by rational appeals to revered sources of authority, most notably the Bible, Aristotle, and the like.

Although most of the Scholasticism of the twelfth, thirteenth, and fourteenth centuries dealt with trivial matters, such as how many angels can dance on the head of a pin, etc., several important concepts arose out of the Middle Ages that have become incorporated into our modern science. The thirteenth century English Franciscan Roger Bacon, for example, advanced the idea that the only viable way to investigate the physical world was through observation and measurement. A century later, another Englishman, William of Occam, claimed that because faith and reason were fundamentally different entities, it was impossible to rationally demonstrate the truth of many of the "revealed" articles of Christian belief. Conversely, the truth of the physical world was not necessarily to be fathomed by appeals to a religious faith whose "mysteries" were in direct contradiction to the workings of that world. Occam also proposed a philosophical "razor" that he believed would "cut out" theories that were too cumbersome. "Occam's razor," as it is called, states that if two competing theories are equally able to "save the phenomena" then the least complicated theory should be the preferred theory. This idea, which is sometimes called the principle of simplicity, tells us that the most acceptable explanation for a natural phenomenon must be the simplest explanation that is the most consistent with the available facts.

Other changes were occurring in Western Europe that also contributed to the eventual demise of the Dark Ages. The rise of the modern nation states and the domestic tranquility that their central governments were able to provide was instrumental in the restoration of a climate conducive to creative thinking. It was not just historical coincidence that the rise of more extensive and centralized governments, particularly in France and England, was attended by the appearance of the first universities in Western Europe. Also, the Church of the latter Middle Ages was not the spiritually oriented church it had been. The spiritual leader of Christendom, the Pope, had also become the temporal ruler of vast tracts of land in central Italy which were called the Papal States. In their capacity as the temporal rulers of the Papal States, popes thus participated in wars, alliances, and a wide variety of political intrigues which bore little relationship to the spiritual emphasis of the Christianity of earlier centuries. The increasingly materialistic and temporal orientation on the part of the Church, particularly the hierarchy, as manifested by tithing, payment for indulgences, and so forth, disturbed a significant part of the Church membership and would eventually be one of the principal factors which triggered the Protestant Reformation in 1517. In short, the preoccupation of the early Christians with spiritual matters at the expense of concern about the material and physical world had, by the turn of the fifteenth century, become irrevocably compromised, thereby setting the stage for a renewed interest in that material and physical world.

1. Describe the factors that contributed to the decline of Greek ''science'' after the time of Ptolemy.
2. What is scholasticism?
3. Discuss the contributions made to scientific philosophy in the Middle Ages.
4. What is Occam's razor? Of what value is it in scientific philosophy?

1.3 Renaissance and Revolution

The fourteenth century in Western Europe saw the onset of a most extraordinary era of human history called the Renaissance. The Renaissance was characterized by a rebirth of interest in a wide variety of intellectual and cultural activities. Contrary to the medieval disdain for worldly concerns, the Renaissance saw a renewed appreciation for all of the wonder and mystery of the physical world. It was an era in which one could still be a ''jack of all trades'' and master of them all. Thus the man who epitomizes the Renaissance, Leonardo da Vinci, could play the roles of architect, artist, biologist, engineer, philosopher, and physicist and play each of them as well as any human being of his time.

As the enthusiastic explorations of the physical world began to gain momentum, it was not long before many of the Greek and Hellenistic views of nature were found to be severely lacking in their ability to ''save the phenomena.'' In 1572, for example, less than 60 years after the death of da Vinci, a ''new star'' or nova appeared in the constellation of Cassiopeia. This star was carefully observed by a Danish astronomer named Tycho Brahe (Figure 1.7). ''Tycho's star'' attained an apparent brightness equal to that of the planet Venus and could be seen for well over a year before fading from naked-eye visibility. Here was spectacular observational evidence that was totally at odds with the Greek view of an eternal and unchanging star sphere.

FIGURE 1.7 ···········
Tycho's Star   Danish postage stamps commemorating the astronomer Tycho Brahe and his 1573 work "De Stella Nova" describing the supernova explosion of 1572 in the constellation of Cassiopeia.

Tycho also made extensive observations of the wispy celestial objects called comets. Aristotle had taken the view that comets were a rare form of atmospheric phenomenon similar to the thin, high-altitude ice crystal cirrus clouds that are sometimes visible in the sky. Tycho's attempts to determine the altitudes of several of these objects demonstrated conclusively that the comets were located at distances beyond the moon, well outside of any reasonable extension of the earth's atmosphere. Thus the comets were not atmospheric phenomena as Aristotle had claimed, but in fact were interplanetary phenomena.

The Aristotelian view of nature was further called into question by a young Florentine named Galileo Galilei. According to Aristotle, larger, heavier objects were supposed to fall to the ground faster than smaller, lighter objects. While seemingly reasonable, such a view could not be verified experimentally. Galileo performed a number of experiments on falling bodies and found, contrary to Aristotle's pronouncements, that two objects of different weight dropped from the same height fall to the ground at exactly the same rate.

Throughout the sixteenth century, the list of the failures of Greek and Hellenistic views of nature to "save the phenomena" grew at such a rate that Galileo and others began to take the view that facts about nature can *only* be deduced through experimentation and observation. Moreover, these experimentally deduced facts should not have to be derivable from or even consistent with "authoritative" sources. Up to this point the Church authorities had by and large looked on the activities of the Renaissance experimenters with a certain amount of bemusement. After all, what difference did it really make if the stars weren't eternal and unchanging, comets weren't atmospheric phenomena, and objects of unequal size fell to the ground at equal rates? As long as these newly discovered discrepancies did not impinge on matters of faith and morals, the Church hierarchy, already bogged down in the religious conflicts of the Protestant Reformation, was largely content to pursue a policy of noninterference when it came to the new discoveries about nature. It was only a matter of time, however, before one or more of the pronouncements of the Renaissance rationalists would be perceived as a threat to the articles of faith and morality, thereby resulting in a

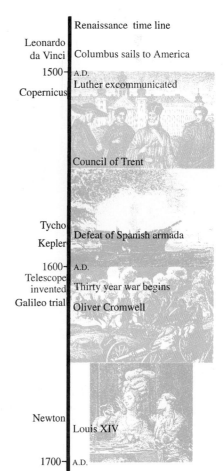

Renaissance time line

Leonardo
da Vinci   Columbus sails to America

1500   A.D.
Luther excommunicated

Copernicus

Council of Trent

Tycho
Kepler   Defeat of Spanish armada

1600   A.D.
Telescope
invented   Thirty year war begins
Galileo trial   Oliver Cromwell

Newton   Louis XIV

1700   A.D.

fundamental confrontation. That collision between the Church and the emerging empirical science finally came to pass early in the seventeenth century. The confrontation occurred at the very doorstep of the Church hierarchy in Rome and was over an astronomical issue—the ''true nature'' of the planetary system. Although the nature of the planetary system is an astronomical question rather than one of physics, the effects of the philosophical and political drama that was about to be played out over this astronomical issue were to have a profound impact on all of the science of the day, physics included.

As we have seen, the geocentric view of the system of planets emerged triumphant from the Hellenistic age. The victory had been won quite simply because the system of orbits, epicycles, and other geometric gyrations, developed by the Greek astronomers Hipparchus and Ptolemy, best ''saved the phenomena'' of the perceived celestial motions of the Sun, Moon, and planets (refer to Figure 1.5). The triumph had thus been one of observation and rationality rather than one of faith and authority. As the Greek and Hellenistic ideas filtered back into Western thought, however, the concept of an Earth-centered system of planets came to be particularly revered and cherished by the Church. After all, what more appropriate place for human beings, the greatest of God's creatures, to be located than at the center of the physical universe? Moreover, the scholars of the period noted that such a view had been developed by pagan thinkers operating only out of rational considerations, thus lending support to one of the central themes of scholasticism—that faith and reason were two different roads to the same eternal truths.

The beginning of the end for this ''neat and clean'' world view came with the advent of the Polish mathematician-astronomer Nicholas Copernicus. In the early years of the sixteenth century, Copernicus became interested in the problem of predicting the motions of the Sun, Moon, and five naked-eye planets: Mercury, Venus, Mars, Jupiter, and Saturn. He found to his dismay that the Ptolemic scheme of orbits and epicycles had become extremely complicated. More than 75 ''wheels'' were required by the Ptolemic system to describe, and imperfectly at that, the motions of seven objects in the sky. For Copernicus this was an intolerable situation and he set about the task of developing a less tedious method for his calculations. Soon Copernicus was able to demonstrate that a somewhat simpler system was to be had by assuming that the Sun and not the Earth marked the center of the planetary system. Copernicus published his findings in 1543, the same year as his death, in a text entitled *De Revolutionibus Orbium Celestium (On the Revolutions of the Celestial Spheres)*. Copernicus, however, clung to the Greek idea of circular orbits and epicycles and, as a result, his Sun-centered system was not much simpler than that of Ptolemy; in fact, it was observationally indistinguishable from Ptolemy's system. Through the persuasiveness of his writings, however, Copernicus had managed to resurrect the idea of a heliocentric planetary system and place it on an equal intellectual footing with the geocentric system of Ptolemy. After the death of Copernicus, matters languished for the next half-century, but it was only the lull before the storm.

Galileo Galilei was born in Pisa in 1564 and, as we have already seen, had performed a number of basic experiments on falling bodies whose results were contrary to Aristotelian mechanics. In addition, he also studied the properties of the motion of a pendulum and developed one of the first thermometers. Most importantly, however, was the fact that sometime in the 1590s Galileo read *De Revolutionibus* and became intuitively convinced that the Copernican system was correct despite a lack of supporting evidence at that time. In 1609 Galileo heard of the invention of the telescope in Holland. He then constructed several similar instruments on his own and turned them toward the heavens. The results spelled the demise of virtually all astronomical thought to that point in time. Instead of being a smooth and perfect celestial orb, the Sun was found to be blemished with spots and stains. The Moon possessed a topography of craters, mountains, and valleys as complex as anywhere on the Earth. Even the very shapes of the planets Mercury and Venus changed in a series of phases not unlike those of the Moon. Galileo was quick to seize on the significance of the latter observation. The phases of the Moon had long been known to be the result of our viewing the Moon's sunlit hemisphere from various angles as the Moon orbits the Earth. For Mercury and Venus to exhibit this same effect and at the same time display their observed closeness to the Sun,* Galileo correctly deduced that these planets must be moving about the Sun in orbits whose radii were less than the Earth-Sun distance (Figure 1.8).

Galileo's observations of Jupiter revealed that planet to be orbited by four smaller bodies, which have become known as the ''Galilean'' satellites. In Galileo's mind, here was a solar system in miniature. His observations of the stars provided strong indications that these objects were at enormous distances from the Earth and hence could not be expected to reveal to mere naked-eye observations the apparent shifts in position among themselves that an orbiting Earth would require. All of these observations were collected by Galileo into a short but impressive work entitled *Sidereus Nuncius (The Starry Messenger).* With its publication in 1610, Galileo became an instant celebrity.

Galileo's telescopic observations made him more convinced than ever of the correctness of Copernicus' heliocentric theory. If Mercury and Venus were orbiting the Sun, it was logical to assume that the rest of the planets did likewise. The satellite system of Jupiter dramatically demonstrated that centers of motion could exist that were outside of the earth and were themselves in motion, thus invalidating the centuries-old argument that if the Earth were orbiting the Sun, the Moon would be quickly left behind. Galileo had also found that in the Jupiter system small objects orbited a larger one, just as a smaller moon orbited a larger earth. Moreover, Mercury and Venus, which were both considerably smaller than the Sun, were now found to be orbiting that larger Sun. The fact that the Sun was known to be larger than the Earth dictated to Galileo that the smaller Earth should therefore orbit the larger Sun (Figure 1.9). While all of this provides strong circumstantial evidence in favor of the Copernican theory, none of it proves unambiguously that the Earth orbits the Sun. A possible response from a Ptolemaic astronomer might be, for example, to note that since the Moon orbits the Earth, it is ''logical'' to assume that the rest of the planets do likewise.

Galileo nevertheless continued to gather and organize his evidence. In 1616 he went to Rome to plead the cause of the Copernican system, but ended up being forbidden by formal decree to ''hold or defend'' a doctrine that the Church regarded as ''false and absurd, formally heretical, and contrary to scripture.'' A decade later, Galileo, badly misreading the political and religious climate, began work on his fateful text *Diologo dei Due Massimi Sistemi (Dialogue on the Two Chief World Systems),* which was finally published in the northern Italian city

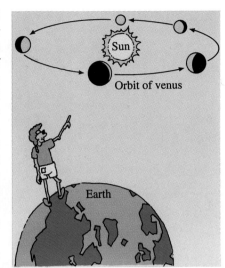

FIGURE 1.8 ···········

The Phases of Venus   As the planet Venus moves about the Sun, an observer on Earth will see the illuminated half of Venus from various angles, thus giving rise to a series of apparent shape changes or phases similar to those exhibited by the Moon. Since Venus, unlike the Moon, is never seen at an angle of more than 47 degrees from the Sun, Galileo concluded that for Venus to exhibit such phases, it must orbit the Sun and not the Earth.

---

*Mercury is never observed to be more than 27 degrees from the sun and Venus never more than 47 degrees.

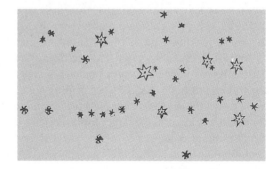

FIGURE 1.9 ··········

Galileo, His Telescope, and His Discoveries   Galileo's discovery that the planets
Mercury and Venus exhibited phases similar to those of the Moon, that Jupiter
possessed moons of its own, and that the shapes of stars could not be resolved even in
his telescope convinced the Florentine in 1610 that the sun was the center of the
planetary system.

of Florence in 1632. The *Dialogue* is a discussion of the relative merits of the
Ptolemaic and Copernican theories and is presented in the form of a conversation
among three individuals, one a Copernican, the second a supporter of the Ptolemaic theory, and the third a sort of neutral moderator. Galileo leaves no doubt

about where he stands. The supporter of the heliocentric system of Aristarchus and Copernicus is called "Salviati" or "Brains" while the supporter of Ptolemy's geocentric view is handed the name of "Simplicio." Moreover, the *Dialogue* was published in Italian rather than Latin and sold out in Florence almost immediately. Church authorities in Rome were outraged. Having endured a century of Protestant Reformation and decades of religious wars, the Catholic hierarchy in 1632 was in no mood for anything that remotely smacked of defiance or heresy. In the spring of 1633, at the age of 70 Galileo was ordered to stand trial before the Roman Inquisition on charges of violating the decree of 1616 and otherwise holding beliefs that were "false and contrary" to the Bible.

What followed was one of the more fascinating episodes in the history of Western Civilization, an event that is commonly called the "Galileo trial." Under severe pressure from the inquisitors, an old and relatively frail Galileo was forced to recant his belief in the Copernican system and was also placed under house arrest in his villa near Florence for the remaining 10 years of his life. His *Dialogue* was placed on the Index of Prohibited Books where it remained for nearly two centuries. Galileo was not exonerated by the Church until 1980, some three and one-half centuries after the fact.

The results of the Galileo trial reverberated throughout intellectual Europe. Although no direct experimental evidence had been advanced in favor of the Earth's orbital motion about the Sun, by 1633 it was obvious to anyone familiar with the situation that it was only a matter of time before such evidence would be forthcoming. Tycho Brahe had spent a quarter of a century from 1572 to 1597 systematically charting planetary positions with a painstaking naked-eye precision that was unparalleled in human history and also allowed for an observational distinction to be made finally between the two competing cosmologies. That distinction was to be made by a German mathematician, Johann Kepler, who spent the first two decades of the seventeenth century forging Tycho's observations into three ingenious laws of planetary motion that now bear his name (see Chapter 5). *All* of them required the Sun to be located at the center of the planetary system. The clear-cut observational evidence, however, came later in the century when in 1675 Danish astronomer Olaus Roemer discovered the so-called "light-time" effect (see p. 350), which could be accounted for only if the Earth orbited about the Sun. The Church's attempt to extend its authority over faith and morals to matters of "natural law,"—matters pertaining to the physical world—had in the end failed. Its dogmatic views of nature simply did not agree with the facts garnered by precise observations. The Galileo trial thus marks a parting of paths between rational, empirical scientific endeavor on one hand and authoritarian, dogmatic religious endeavor on the other. To this day, there remains a legacy of uneasy truce, apprehension, and opposition between these two great areas of human thought. Galileo himself did not anticipate such a division. To him the God of his religious beliefs was the same God who created the universe, the workings of which are better understood by empirical methods.

After the Galileo trial, scientific thought took on a distinctly amoral quality. The laws of nature were no longer regarded as "good" or "evil," but rather existed as morally neutral entities, and scientific investigation of those laws would henceforth proceed with little homage being paid to religious doctrine, morals, or authority. As a result, the "stereotype" scientist has been portrayed, as far back as Mary Shelley's Frankenstein and Goethe's Faust, as a person whose hunger for scientific knowledge far outweighs any other consideration, including moral scruples. Such stereotyping is, of course, both unfair and inaccurate. Many scientists throughout history have been deeply religious men and women, and almost all have scrupulously upheld the integrity of the empirical method. Moreover, it is quite often the scientist, particularly in the twentieth century, who has been among the first to warn that as science and technology play an increasingly dominant role in modern society, the concept of an amoral scientist doing amoral science without regard to its potential impact on society could lead to disastrous consequences.

For all intents and purposes the Galilean trial had a distinctly chilling effect on scientific endeavor in the countries of the Mediterranean. As a result, with few exceptions after 1633, scientific achievement in Europe took place north and west, in Scandinavia, England, France, Germany, and the Netherlands. The most important legacy of Galileo's seventeenth century, however, was the emergence of a philosophy of gathering knowledge that has thus far enjoyed three and one-half centuries of favor. During this time, human beings have advanced their understanding of the physical world more rapidly and more profoundly than has been possible during all of the other eras of human history combined. This philosophy of knowledge gathering came to be known as natural philosophy or more commonly as the empirical or scientific method, which we discuss in the next section.

1. Describe how the proponents of the heliocentric and geocentric systems explained the motions observed for the Sun, Moon, and planets.
2. Why did the Church hierarchy find the idea of a geocentric system of planets so attractive?
3. Of what significance was the Galileo trial?

1.4 The Scientific Method

Although the seventeenth century in Europe is often referred to by historians as the "Scientific Revolution," the scientific method that emerged from that collage of political, religious, and philosophical forces was in fact a product forged from many prior centuries of human history. The scientific method is basically a philosophy of acquiring knowledge. Like any other philosophy, it begins with a statement or statements that are the equivalent of "we hold these truths to be self-evident. . . ." A considerable amount of heated debate can arise when one person's set of truths, postulates, or axioms are not "self-evident" to someone else. Such is the driving force, for example, behind the diversity of political, religious, and other viewpoints that exist in our society.

The scientist's most important self-evident truth is the fundamental claim that the only viable way to gain significant information about the nature of the physical world is through observation and experimentation. Such an empirical view, of course, implies some basic assumptions regarding the ultimate nature of reality. The scientific method says in effect that the world perceived by our senses constitutes a "reality" within which universal truths can be found.

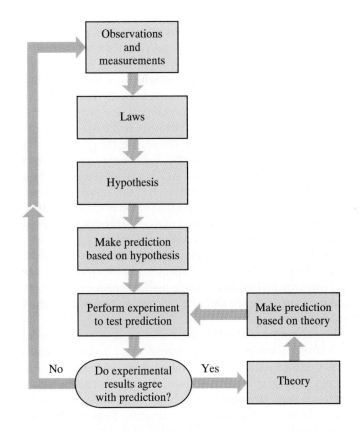

FIGURE 1.10 · · · · · · · · · ·
A Flow Chart for the Scientific Method

Reliance on this sort of perceived reality has not enjoyed a universal acceptance, and in fact has been rejected by many of the world's philosophies.

Implied in the assumption of a physical world that can be observed and experimented on is the additional inference that the results of those experiments and observations can be expressed quantitatively, i.e., represented by numerical quantities. Once in numerical form, the experimental results can then be described in terms of mathematical equations and formulas. Thus the scientific method is based in part on a revised and more extensive version of Pythagoras's "All is Number" theme.

The actual scientific method proceeds along the general outline shown in Figure 1.10. Experiments, observations, and measurements are performed on a given object or phenomenon. After a sufficient number of observations have been made, behavior patterns or "laws" begin to emerge in the data that ideally can be expressed and described in terms of mathematical equations and relationships involving the measured quantities. Once a sufficient number of laws have been established, they are collected into a general overall description of the phenomenon under investigation, which is called a *hypothesis*. Thus, in the last half of the seventeenth century, the Englishman Isaac Newton wove the various laws pertaining to the phenomenon of motion, which had been derived by Galileo, Kepler, and others, into a majestic hypothesis called "classical" or "Newtonian" mechanics (Chapter 3).

When such a hypothesis is proposed, it then serves as the basis on which predictions are made regarding the behavior of the given phenomenon, with each prediction tested by an appropriate observation or experiment. If the predictions of a given hypothesis agree with experimental and observational results, then the hypothesis becomes a theory, as did Newton's ideas on motion. Once a theory is in place, additional predictions are made and empirically tested until the theory reaches the limits of its application.

When the predictions of a given theory do not agree with experimental results, the theory must be expanded on, amended, or changed, but only in accordance with a rule of the scientific method called the *correspondence principle*. The correspondence principle states that any new or revised theory must not only account for the discrepancy between experimental results and the predictions of the old theory, but also must be successful where the old theory was successful as well. Thus the theory of quantum mechanics, which was developed only a few decades ago to describe the behavior of phenomena at the atomic or microscopic level of nature, must also match the success at the everyday or macroscopic level of nature enjoyed by the older, more limited classical theories of physics such as Newtonian mechanics. In like fashion, competing theories can be distinguished by their success or lack thereof in being able to predict experimental and observational results. Should two theories of unequal complexity be equally successful in predicting experimental results, then the modern version of Occam's razor tells us to discard the more complex theory. Thus has the scientific method operated for over three centuries.

During this period the scientific method has enjoyed a most spectacular level of success for several reasons. First of all, there appears to be a sort of "commonality" about the way nature operates. This commonality is sometimes expressed as the "principle of universality." The principle of universality states that the basic principles and laws governing the behavior of the physical world do not change with time or one's location in the universe. Thus the laws of aerodynamics that keep a jetliner flying over the United States are the same as those keeping that aircraft flying over Eurasia, Africa, Australia, or anywhere else on the earth (Figure 1.11). The principle of universality holds that experimental results can be reproduced at any time and at any location. Thus many of the important and fundamental experiments of scientific history are often repeated in elementary laboratory science courses.

In a similar vein, the important and fundamental experiments performed at the frontiers of human scientific knowledge must be replicable or they will be cast aside as spurious results. The principle of universality also comes to the rescue of the astronomer whose scientific investigation of celestial objects is severely hampered by the size and distances of those objects. Using the principle of universality, astronomers have been able to glean enormous amounts of information by interpreting the observations made on celestial objects from afar in terms of the laws of physics, chemistry, and geology that have been derived on Earth.

The theories of science are only good as long as their predictive prowess holds up. Therefore, we must regard such theories as successive representations or approximations to the "true" workings of the physical world, which can and must be altered or even overthrown in favor of different theories that better account for new experimental results. Such a view does not provide us with the psychological security that eternal and unchanging "truths" can bring, and many nonscientists are very much troubled by what they perceive as an ever-changing scientific view of the "truth" of nature. It is, however, this very capacity for change within the scientific method that has permitted great progress to be achieved in our understanding of the physical world. Differences over articles of religious faith have been debated for centuries, and the debate promises to

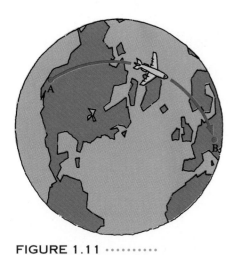

FIGURE 1.11 ···········

The Principle of Universality When an aircraft flies from point A to point B on the surface of the earth, the same laws of aerodynamics apply everywhere throughout the flight. The idea that these and other laws of nature do not change with time or one's location is referred to as the principle of universality.

1.1 Scientific Serendipity

The scientific method rightly projects itself as a systematic and rational approach to our investigations of nature. However, some great discoveries and creative theories have come to us as a result of fortuitous circumstances or unexpected inspiration.

The stories of scientific discoveries made by accident are legion. In 1820 a Danish high school physics teacher named Hans Øersted set out to prove to his class via a lecture demonstration that the phenomena of electricity and magnetism were completely unrelated to each other. To his amazement, he discovered during the course of his demonstration that quite the opposite was true! Thus began the recognition of the existence of the much more encompassing phenomenon of electromagnetism.

At the end of the nineteenth century, French physicist Henri Becquerel was initially quite perturbed when the uranium compounds he was studying were somehow able to fog photographic plates, even though the latter were in light-proof containers. His investigations of this ''annoyance'' led to the discovery of the phenomenon of radioactivity.

In more recent times, two radio astronomers at Bell Telephone Laboratory named Robert Wilson and Arno Penzias set out in 1964 to detect sources of radio waves in outer space. In the course of their study they were continually plagued by a level of background radio static. Initially this radio ''noise'' was regarded as a nuisance and the two scientists tried mightily to eliminate it somehow. Despite their best efforts, the static persisted and eventually came to be recognized as being the quite real vestiges of a primordial ''Big Bang'' in which our universe literally burst onto the scene.

The scientific inspirations visited on human beings have been many, varied, and generally unexpected. Newton developed his theory of motion while on his family farm waiting out the plague that was ravaging Cambridge. Einstein's inspiration for his theory of relativity came while he was working at a nondescript job in the Swiss patent office, and Archimedes' principle of buoyancy came to him in his bathtub!

The greatness of the individuals who have made such discoveries or had such inspirations lies not in the making or the having, but in how they were able to recognize the importance of what was there and translate it into a meaningful contribution to human scientific knowledge. The discoveries of Øersted, Becquerel, Wilson, and Penzias became important because their discoverers were alert to the implications of what they had stumbled on. In a similar fashion, the fate of an inspiration rests entirely with the person to whom it came. The ultimate triumphs of the theories of Newton and Einstein are due in no small part to those individuals' willingness to ''take the heat'' for their then ''unconventional'' views of the physical world.

All of which raises a haunting question for scientists. For all the accidental discoveries and inspirations pursued in science, how many more have been lost owing to the inattentive and inept investigator or the scientific thinker unwilling to be receptive to inspirations that are seemingly too fantastic to have any possible scientific validity?

PHYSICS FACET

1.2 Science and the "New Age"

One of the more extraordinary aspects of contemporary society is the steadily growing social movement that has been very simply called the "New Age." Despite protestations to the contrary by its adherents, most of the constituent elements of the New Age are in fact not very new at all. Astrology has been around at least since the third millenium B.C. and the present "channelers" are little more than recycled versions of the mediums and spiritualists of yesteryear.

For the most part, scientists would be quite content to simply ignore the New Age adherents were it not for their intrusions into a number of areas that lie within the scientific realm. For example, a number of books on astrology exist in which that ages-old art form is characterized as the "space age" science, the "divine" science, and the like, despite the fact that astrologers do not embrace the scientific method in any meaningful way and that the purported celestial influences and synchronicities are eminently lacking in reproducible supportive evidence.

The New Age "energy" and "vibrations" supposedly emanating from the deep, microscopic recesses of a variety of mineral crystals have thus far eluded both the efforts of the experimental physicist to detect them and the conservation laws by which they should be governed.

UFOs and the beings they presumably transport to earth from the far stretches of one or more "dimensions," "planes," or "universes" routinely violate our scientific perceptions of space and time. These talented "extraterrestrial/extradimensional" craft, for example, are able to execute high-velocity sweeps and incredibly sharp turns through the earth's atmosphere without so much as a singed porthole, squished rivet, or sonic boom.

For the scientist reluctant to appeal to what are in effect supernatural explanations for natural phenomena, such fla-

grant violations of nature's operations can be greeted only with the highest degree of skepticism. The physical world and the human mind that comprehends it are extraordinarily wonderful entities and there is certainly much to be discovered about both. Indeed, amidst the fantasy, fraud, and hucksterism so deeply ingrained in the New Age there may well be profound and important nuggets of knowledge to be had. For the critical thinker, however, New Age appeals to things such as "spectral beings" from the star Alpha Centauri or a 35,000-year-old ex-warrior from Atlantis who rides laser beams must necessarily be explanations of the very last resort.

continue, unresolved, for centuries more. Differences over scientific views of the physical world, meanwhile, can and have been settled by that final arbiter of scientific theories and hypotheses—experimental observation.

Because experimental results play such a central role in the scientific method, however, it is of fundamental importance that those results be reliable. This is most often far easier said than done. Scientific equipment can and often does malfunction, thus yielding results that are in error and/or spurious. It is no accident of history that the infamous and whimsical Murphy's laws,* with which most of us are familiar in one form or other, arose out of the struggles and frustrations experienced in the laboratory by the experimental scientist. The mark of a truly competent experimentalist is one who can repeatedly and unambiguously glean the real phenomena of nature from his or her scientific apparatus.

The scientific method also presupposes a total honesty on the part of the experimentalist. Scientific laws and theories that are based on data or measurements that are false, altered, or manufactured are necessarily flawed and are hence totally useless to science. Empirical integrity of the highest order is therefore a basic prerequisite if the scientific method is to function properly. For this reason, the instructors in laboratory science courses take a very dim view of "dry-labbing" experiments or otherwise faking laboratory results. One sinister side effect of the increasingly important role played by science and technology in our society has come in the form of the pressures that are sometimes exerted on the experimental scientist to obtain "experimental results" that are deemed expedient for political, economic, or other nonscientific considerations. Fortunately modern science has been able to blunt the effects of such pressure by insisting that any and all experimental results be reproducible not only by the scientist in question but also by other scientists working independently.

Despite its great successes in the physical world, there are areas of human endeavor to which the application of scientific method seems to have limited effectiveness or is even inappropriate. Religious matters stand as the best and most familiar example of the latter. For instance, the scientific method cannot generate definitive commentary on the existence or lack thereof of entities such as a Supreme Being, guardian angels, demons and the like, because, by definition these entities are spiritual in nature and therefore have no physical existence into which the scientist can sink his or her empirical teeth. In a similar fashion, the "New Age" has provided us with a plethora of netherworld beings, alien creatures, and other assorted entities, all of which exist in "dimensions" which render them impervious to the best efforts of observational and experimental science to demonstrate their existence. The very term "supernatural" which is often applied to such entities and their world, immediately conveys to us that such a world or "dimension" lies outside the realm of the natural world and hence of experimental science. Whatever we believe about that supernatural world then becomes just that—belief. It is tempting to say, as many do, that anything which cannot be dealt with empirically simply does not exist. However, one must be careful with such views. Barely a century ago, most of the forms of electromagnetic energy such as radio waves, gamma rays, and x rays, which presently provide an important component of our current technology, were totally unknown to the scientists of the day. On the other hand many of the "supernatural" phenomena of past centuries are now recognized as being perfectly natural and explainable within the framework of our modern science. It is for this reason that "supernatural" explanations for observed, but unexplainable, phenomena such as unidentified flying objects—UFOs—are advanced by scientists only as explanations of last resort, if at all.

---

*Some of Murphy's "laws" include: In any field of endeavor, anything that can go wrong, will go wrong; nature always sides with the hidden flaw; and inanimate objects always operate against you.

The success enjoyed by the scientific method in the so-called "exact sciences" such as physics and chemistry has inspired attempts, particularly in the present century, to extend that method to "human" phenomena. The result has been the development of the social sciences in which investigators have endeavored to understand better human personality and human interactions through the application of the scientific method. Problems arise almost immediately, however, due to the fact that most often the factors that determine human personality and interaction such as love, ambition, patience, fear, and intelligence do not readily lend themselves to quantitative measurement. For example, your interactions with friends and fellow students when you are falling deeply in love are far different from those same interactions when you are coming off the breakup of a relationship. From a scientific point of view it would be nice to be able to measure quantitatively the amount of love, etc., in a given person on a given day. The ongoing controversy over the interpretation and validity of the so-called "IQ" tests, which are presumably designed to measure quantitatively the inherent intelligence of a given individual, stands as example of the difficulties involved with such a task.

As a result, most of the laws developed by the social sciences are statistical in nature and can be used only to describe the behavior of large groups of people. Some social scientists, dismayed by this result, feel that the scientific method was developed by physical scientists for the physical sciences and as such is not appropriate for human personality and behavioral studies. What is needed in their view is a "social" scientific method designed by social scientists for the social sciences. Others believe that at some point in the future the problem of quantitatively measuring the qualities of the human psyche will be satisfactorily solved. Hence, the scientific method as it now stands is a totally appropriate tool in gaining an understanding of how human beings operate. These individuals also note that at the atomic level even the exact sciences have been forced to resort to statistical descriptions of the behavior of individual entities (see Chapter 13). It is a debate whose outcome will be fundamental to the future direction and progress of the social sciences, but it is also one that will not be settled here, or any time soon!

From the time of Galileo onward, the scientific method has enabled us to expand our perception and comprehension of the physical world at a rate unparalleled in all of human history. Out of a myriad of scientific experiments, human beings have pieced together a picture of the universe that is as interesting and as fascinating as any that could be conjured up from our most imaginative fantasies. In the chapters that follow we will explore the portions of that portrait or our universe that lie within the province of that most fundamental branch of scientific endeavor that we call physics.

1. What contributions did each of the following make to the development of the philosophy of the scientific method?:
   a. Pythagoras   **b.** William of Occam   **c.** Aristotle
   d. Roger Bacon   **e.** Galileo

2. Compare and contrast the following: principle of universality, principle of simplicity, correspondence principle. Why is each important in the scientific method?

3. What is the difference between a law, hypothesis, and a theory in science?

4. Of what use are predictions in the scientific method?

THOUGHT QUESTIONS

1. What historical factors other than those described in the text do you think could have contributed to the lack of scientific progress during the Dark Ages?

2. Discuss some areas of human endeavor that lie outside of the scope of science.

3. Describe an area of conflict and controversy between science and religion in our present society.

4. Describe a view of ''reality'' that is different from that proposed by the scientific method.

5. How might you go about quantitatively measuring an abstract quality of human psyche, such as love or hate?

6. What harm can be done by reporting false or poor data in a scientific report? How does the scientific method correct such errors and fraud?

7. Compare and contrast the philosophical differences between science and religion.

8. Describe the difficulties that each of the following might encounter in attempting to apply the scientific method to their fields of endeavor: an astronomer, a minister, a social scientist.

9. Discuss why you think the Scientific Revolution occurred in seventeenth century Europe as opposed to another era and/or another location on the earth.

10. Discuss the possibility that the principle of universality might not hold if we were to journey to another solar system or another galaxy beyond our own Milky Way.

11. Give an example of a phenomenon in the physical world that was once thought to be ''supernatural'' but is now explainable scientifically.

12. Using the *Reader's Guide to Periodic Literature* look up and read through a few articles on ''creation science'' or ''creationism,'' and discuss whether or not you think creation science follows the tenets of the scientific method.

13. Describe an instance in which you think pressure might be placed on an experimental scientist to obtain a politically or economically expedient ''scientific'' result.

14. Describe a ''stereotype'' of a scientist that you have come across in the media or in your recreational reading. What characteristics are attributed to this scientist? In your opinion, are such attributes justified? Explain.

15. Suppose someone proposed to you the existence of an ''astral plane'' or unseen dimension to the universe. How, as a physics student, would you respond to such a claim?

16. What aspects of the New Age views do you think lend themselves to scientific testing? How might you conduct such tests?

17. Discuss the idea that science and scientific knowledge are ''dangerous'' for humanity.

18. Do you think the term ''New Age'' is an appropriate one? Explain.

19. Describe the impact on the scientific method if the following were shown to be false:
    a. the physical world can be described by numbers
    b. the principle of universality
    c. the principle of simplicity

20. Where would you place accidental discoveries and inspiration on the diagram shown in Figure 1.10?

21. Do you think that it is possible for scientific progress to occur in a totalitarian state? Explain.

2 NUMBERS AND NATURE

*O*ne need only take a quick glance at the world around us to detect the multitude of mathematical shapes and patterns that exist in the physical universe. ''Natural'' geometry, for example, is manifested in a wealth of structures and phenomena such as the luminous circular disk of the sun, the spirals of snail and nautilus shells, and the hexagons of snowflakes and the honeycomb in a beehive (Figure 2.1). Moreover, such mathematical manifestations occur at all levels of our perceptions of the physical world. They range from the spheroidal viruses and helical-shaped DNA molecules of the microscopic world to the giant elliptical and spiral galaxies existent in the vast and remote universe of the astronomer. It is small wonder, then, that the Pythagoreans promoted the idea that the physical world could be described with mathematical figures, concepts, and principles, a view that continues as one of the basic ideas inherent in our modern scientific method.

2.1 Mathematics as a ''Scientific Paintbrush''

Although much of nature is quite striking in its mathematical patterns, for a significant portion of the physical world, a mathematical description is far more subtle and complex. The process by which the scientist develops complex mathematical descriptions of a phenomenon of nature is not unlike the way an artist

(a)

(b)

(c)

(d)

FIGURE 2.1 ··········

Patterns in Nature   (a) A spiral galaxy; (b) a circular diatom; (c) cubic crystals of the mineral galena and spalerite (lead ore); (d) hexagonal shapes in a honeycomb.

might proceed in creating a drawing of some aspect of nature. One of the very popular and more familiar techniques of learning how to draw has the would-be artist starting a given sketch with a series of simple mathematical figures and curves. If, for example, we wish to make a detailed drawing of the fish shown in Figure 2.2, one way to begin that drawing is to approximate the shape of the fish with two circles and a C-shaped curve. As the drawing develops into a progressively better and more accurate likeness of the fish, the number of mathematical shapes and curves must increase correspondingly. The end result is a drawing that is presumably a fairly accurate rendition of the fish. In a similar fashion, the scientist attempts to "sketch" mathematical "portraits" of natural phenomena using sets of progressively more complex mathematical models and techniques.

In the course of developing such mathematical descriptions, the scientist nearly always makes use of mathematical symbols to represent physical quantities. These symbols are then employed in mathematical statements called *equations,* which are versatile entities capable of being manipulated, transformed, and ultimately solved to yield some piece of desired information, thereby making them the mainstays of any quantitative descriptions of the natural world. In addition, the mathematical symbolism of the scientist allows the laws governing nature to be written in a clear and precise fashion. For example, suppose that we have observed the dating habits of a friend or roommate and have formulated a "law of dates," which can be stated as follows:

The number of dates the student goes out on per week is equal to that student's salary divided by the age of the student's car.

If we let the letter N represent the number of dates per week the student goes out on, S the student's salary, and A the age of the student's car, the above "law of dates" can be very simply written in the form of a mathematical equation as follows

$$N = \frac{S}{A},$$

which is a far simpler and more succinct version than the preceding "word equation."

While most of the mathematical symbolism and techniques employed by physicists in their descriptions of nature lie well beyond the scope of this text, a few fundamental concepts are of considerable importance in any mathematical description of the physical world and will be studied in this chapter.

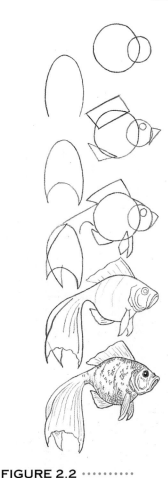

FIGURE 2.2 ·············
Sketching a Fish by Successive Approximations The final fish drawing is accomplished by starting with a set of simple and relatively easily drawn mathematical figures and curves and subsequently improving the representation with additional figures and curves.

1. Give some examples of geometric shapes that occur in nature other than those mentioned in the text.

2. How is mathematics used by physicists?

3. Why do physicists use symbols and equations?

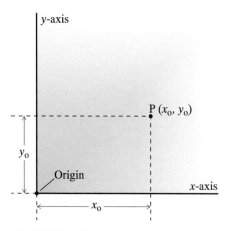

FIGURE 2.3 ···········

Plot of Point P   A plot of a point P with coordinates x_0, y_0. The point is located at a distance x_0 along the x-axis and at a distance y_0 along the y-axis.

2.2 Graphs

In developing a symbolic description of the physical world, it is often very useful to construct a mathematical "picture" of a given relationship between two interdependent quantities in nature. Such a portrait is referred to as a *graph* and represents a powerful tool by which considerable insight can be quickly obtained regarding a given phenomenon.

The most common type of graph begins with two mutually perpendicular lines or axes. The horizontal axis is referred to as the x axis or abscissa and the vertical axis as the y axis or ordinate. A single point on the graph is located by specifying its position along both the x and y axes, as shown in Figure 2.3. A plot of several points, such as those obtained from a set of experiments in which both x and y values were simultaneously determined, would take on an appearance similar to that shown in Figure 2.4(a). A plot of a theoretical mathematical equation involving x and y would result in the solid line shown in Figure 2.4(b).

Graphs can thus be used to compare the behavior of a phenomenon as predicted by one or more theories with a set of points obtained from actual experiments, thereby permitting scientists to distinguish empirically between competing theories for that phenomenon (Figure 2.5). Sets of experimental points plotted onto a graph can also suggest from their pattern or form the type of mathematical model that can be used to account for the empirical results.

Graphs also provide a wealth of information in a concise and convenient form. Consider, for example, the all-too-accurate graph of annual tuition costs versus time at some hypothetical university shown in Figure 2.6. We can use this graph to find the year when tuition costs were at a certain level of expense. Thus if you want to know when you could have gone to this particular school for $2000 per year, you would find from the graph that the year corresponding to a $2000 per year tuition cost would be 1985. On the other hand, when your parents tell you how low their tuition costs were in 1965, a quick check of the graph reveals they paid about $200 in annual tuition that year, about one-tenth what it would be a generation later.

Graphs can also be used to project or extrapolate into the future. Thus if present trends continue, the annual cost of tuition at this particular institution will soar to more than $5000 by 1993. Such projections, of course, always assume that past trends will continue as predicted by the model, and many extrapolations have turned out to be embarrassingly wrong when this assumption did not prove to be valid. Thus, we need to revise and update the model continually with new observations and experiments.

FIGURE 2.4 ···········

Plot of Data Points   A plot of data points obtained by experiment can be approximated by a theoretical curve. In this example the data are best described by a straight line, and the relationship is said to be linear.

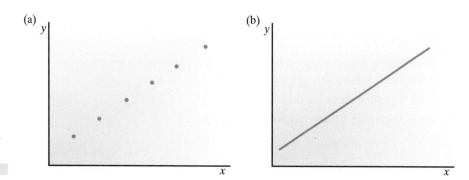

(a)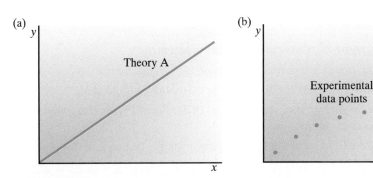

(b)

(c)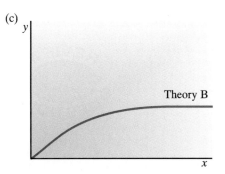

FIGURE 2.5 ············
The Use of Graphs to Discern Competing Theories Theory A predicts that a straight line or linear relationship should exist between the quantities x and y, while theory B predicts that a curved, nonlinear relationship should exist between x and y. The experimental results in this case (center) support theory B.

WORKED EXAMPLE

A physicist obtains the following data for the distance d an object falls in a given time t. Plot these data using d as the y-coordinate and t as the x-coordinate. From your resulting set of points, determine whether or not the graph of t versus d is a straight line.

t	d
0.0	0.0
0.5	1.3
1.0	5.0
1.5	11.3
2.0	20.0

SOLUTION: Plotting the set of points $t = 0.5$, $d = 1.3$; $t = 1.0$, $d = 5.0$; $t = 1.5$, $d = 11.3$; and $t = 2.0$, $d = 20.0$, yields the set of points shown on the graph below. Clearly, these do *not* constitute a straight line.

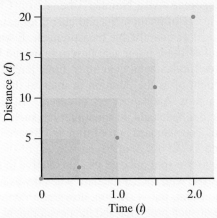

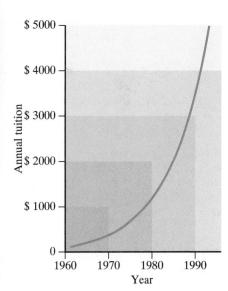

FIGURE 2.6 ············
A Plot of Time Versus Annual Tuition at a Hypothetical University

1. What is a graph? Why are graphs useful to scientists?
2. Explain how points are plotted on a graph.
3. Describe how graphical results can be predicted or extrapolated.
4. Suppose your parents had graduated in 1970. From Figure 2.6, what were their tuition costs in that year?
5. From Figure 2.6, in what year will tuition costs be 20 times what they were in 1965?

1. Plot each of the following sets of x and y values onto a graph:
   a. $x = 0$ **b.** $x = 2$ **c.** $x = 2$ **d.** $x = 0$
      $y = 0$ $y = 0$ $y = 2$ $y = 2$

   What geometric shape, if any, does the set of points seem to form?
2. From the graph of the straight line shown below obtain four sets of x and y values that fall on the line.

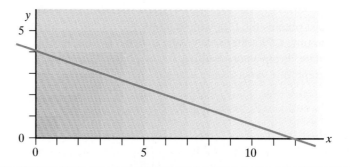

2.3 Basic Quantities in Nature

If we pursue the Pythagorean "All is Number" theme from Chapter 1 a bit further, we find that a wide variety of phenomena and quantities exist in nature that lend themselves to numerical or quantitative measurement. Even more significant is the fact that nearly all of these numerical descriptions can be based directly or indirectly on just three fundamental quantities: time, length, and mass.

Time can be thought of as the duration or interval that exists between two observed events in nature. A quantity or "interval" of time is measured by comparing it to some defined standard time interval, such as the back and forth swing of a pendulum. The standard unit of time in physics is the *second,* which was originally defined as $\frac{1}{60}$ of $\frac{1}{60}$ of $\frac{1}{24}$ (or $\frac{1}{86,400}$) of the average interval of time it takes for the earth to spin once on its axis relative to the sun, or the mean solar day. Problems arose with such a standard when it was found that the earth's rotation rate is not constant, but is slowing down at a rate of about one second per century. Because of such problems and the development of "atomic clocks" (see Physics Facet 13.2), the standard unit of time in physics is now based on

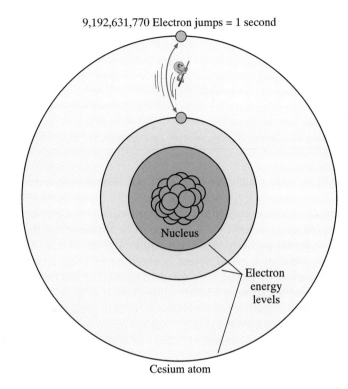

9,192,631,770 Electron jumps = 1 second

Nucleus

Electron energy levels

Cesium atom

FIGURE 2.7 ··········
The Standard of Time   One second is currently defined as the total interval of time for a certain electron in a cesium atom to jump up and down 9,192,631,770 times.

the time interval between successive vibrations of a certain kind of atom. This is a remarkably tiny but precisely measurable segment of time, and as a result, the second is now defined in modern physics as the interval of time required for exactly 9,192,631,770 atomic vibrations of cesium to occur (Figure 2.7).

Length is a quantity used to denote the position or extent of an object in space. A given quantity of length is determined by comparing it to a defined standard of length or reference length. Thus, if we wish to measure the length of an object, we do so by using a ''ruler,'' which is calibrated in terms of standard length units such as inches, feet, or centimeters. Although several different standards of length can be employed in physics, we will use as our reference in this text a standard unit of length called the *meter*. The meter was originally defined during the French Revolution as one ten-millionth of the distance between the earth's north pole and its Equator. Based on that definition, a length equal to the standard meter was delineated by two parallel lines scratched on a platinum-iridium bar kept at the International Bureau of Weights and Measurements at Sevres, near Paris, France. More modern measurements of the earth's dimensions have shown that the standard meter is not exactly one ten-millionth of the pole-to-equator distance on the earth. Moreover, the lines scratched on the standard platinum-iridium bar do not have precisely uniform widths, a fact that causes difficulties if extremely accurate measurements of length are desired. As a result of these problems and the discovery at the end of the last century that the time rate at which light changes its position in space has a value that is the same for any observer in any state of motion (see Chapter 15), physicists now specify the length of the meter in terms of the speed of light. The standard meter is currently defined as the distance traveled by a flash of light in ¹/₂₉₉,₇₉₂,₄₅₈ of a second of time (Figure 2.8). In terms of the more cumbersome system of English units commonly employed in our daily lives, a meter is about 10% longer

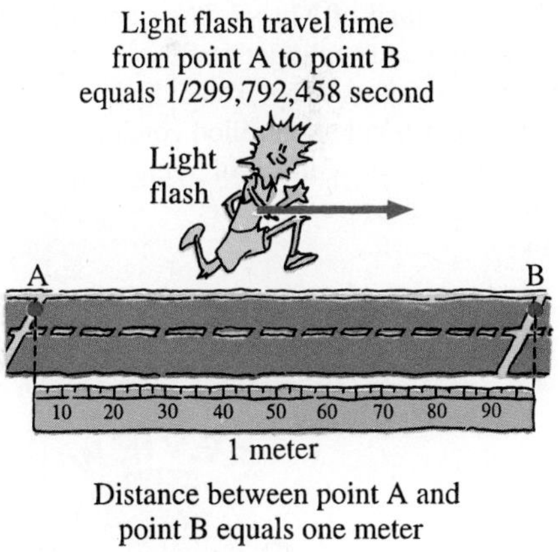

Light flash travel time from point A to point B equals 1/299,792,458 second

Light flash

A   B

10 20 30 40 50 60 70 80 90
1 meter

Distance between point A and point B equals one meter

FIGURE 2.8 ··········
The Standard of Length   One meter is defined as the distance a flash of light can travel in 1/299,792,458 of a second.

TABLE 2.1 Some Equivalents Between the English and Metric Measuring Systems

1 centimeter = 0.394 inches	1 inch = 2.54 centimeters
1 meter = 39.4 inches	1 yard = 91.4 centimeters
1 meter = 1.09 yards	1 yard = 0.914 meters
1 kilometer = 0.621 miles	1 mile = 1.61 kilometers

FIGURE 2.9 ··········

The Standard of Mass The standard kilogram is this cylinder made of platinum-iridium alloy. The cylinder is kept under highly controlled conditions at the International Bureau of Weights and Measures in Sevres, France.

than a yard, and a length of 1000 meters, or one kilometer, is about 0.6 miles (Table 2.1).

The third fundamental measurable quantity used by physicists to describe the physical world is called *mass*. Mass can be thought of as a measure of the amount of "matter" or material particles contained within a given object. As with time and length, the mass of a given object is obtained by comparison with a standard, defined quantity of matter. The standard of mass in the metric system is a cylinder made of a platinum-iridium alloy, which is kept under carefully controlled conditions at the International Bureau of Weights and Measures in France (Figure 2.9). Initially the cylinder was set up so as to have a mass equal to the quantity of mass contained in a *liter* of water or a cube of water 1/10 of a meter on a side. Later, a small discrepancy was detected between the mass of a liter of water and the platinum-iridium cylinder. As a result, the platinum-iridium cylinder is itself now *defined* as having a mass of one kilogram. One *kilogram* is therefore very nearly equal to the mass contained in a one liter volume of water, and one *gram* or 1/1000 of a kilogram is very nearly equal to the mass contained in a cube of water one centimeter or 1/100 of a meter on a side.

Although substances in nature can obviously exhibit a wide variety of physical characteristics considerably different from the platinum-iridium cylinder, the physical property of mass for all substances is independent of any other properties of the substance. This fact permits the physicist to measure the mass of a given substance by comparing the effects of the mass of that substance with those of a standard mass, usually through the use of a scale, or pan balance. Logistically, it would not be convenient to compare a substance's mass each and every time with the standard platinum-iridium kilogram, so physicists have generated a number of "secondary" standard masses. These masses are carefully compared with the standard kilogram in France before being sent off to be used for measurements of mass in various other laboratories around the world.

REVIEW 2.3 QUESTIONS

1. Define time, length, and mass. Why are these regarded as "basic" quantities in nature?
2. What is the definition of the second, meter, and kilogram?
3. Why was the earth's rotation rate abandoned as a fundamental standard of timekeeping?
4. Describe the relationship between the standard meter and the speed of light.

2.4 Units

Suppose one day you walk into a class and your instructor presents you with an exam and tells you that the exam has a time limit of 50. After the initial shock passes of being hit with yet another ''sneak'' quiz, you are now a bit puzzled because even though 50 is a nice, respectable resident of the real number system, a time interval of ''50'' is meaningless in the absence of additional information. If the exam's time limit is 50 seconds, you are already out of time; if it is 50 hours, you probably face perhaps the most murderous take-home exam in your academic career. This example illustrates that whenever a numerical value is assigned to a quantity in nature it must be accompanied by some sort of an indication of what measuring system was employed in expressing that value. In science this is accomplished through the use of units, which are placed after the numerical value of the quantity in question. In the above example, the time interval would have been clearly defined had our hypothetical instructor told the class that the exam time limit was 50 minutes, rather than just 50.

Units are an important part of the numerical description of a given quantity and are subject to the same rules of mathematical manipulations as their numerical counterparts. Thus if we wish to compute the quantity $(3 \text{ meters})^3$, the result is

$$3 \text{ meters} \times 3 \text{ meters} \times 3 \text{ meters} = 27 \text{ (meters)}^3 .$$

Often we will encounter products or ratios of units that differ. Under these circumstances no further unit simplification is possible. Thus 2 kilograms $\times$ 3 meters is equal to 6 kilogram-meters, and 20 meters $\div$ 2 seconds is 20 meters/2 seconds or 10 meters/second. A more familiar version of this concept is the ''fruit theorem'' statement: ''If one divides apples by oranges, the result is apples/oranges.''

Sometimes in the course of our mathematical descriptions, a quantity will emerge in which no units are present. For example, in the calculation (m/sec^2) $\div$ (m/sec)/sec, all of the units cancel out. Such quantities are referred to as *dimensionless* quantities since their mathematical descriptions do not include dimensions.

Very often we need to convert a measurement made in one set of units to a corresponding value in another. To accomplish this, we must first know an equivalent value or *conversion factor* for the two sets of units involved. Thus 1 kilometer = 1000 meters relates distance measured in kilometers to distance measured in meters. One minute = 60 seconds relates time measured in minutes to time measured in seconds. A ratio of the two equivalent quantities such as 1000 meters/1 kilometer or 60 seconds/1 minute is, by definition, equal to one and we may therefore use such ratios to convert the value of a given quantity expressed in one unit to the same value expressed in another. Thus if we wished to convert a distance of 4 kilometers into the equivalent distance expressed in meters we would proceed as follows

$$4 \text{ kilometers} \times \frac{1000 \text{ meters}}{1 \text{ kilometer}} = 4000 \text{ meters} .$$

Notice that kilometers/kilometers = 1 and these units ''cancel out'' of the final result. In like fashion, a 0.5-kg steak would have a mass in grams given by

$$0.5 \text{ kg} \times \frac{1000 \text{ grams}}{1 \text{ kilogram}} = 0.5 \times 1000 \text{ grams} = 500 \text{ grams}$$

Our 50-minute test from above would have a time interval expressed in seconds of

$$50 \text{ minutes} \times \frac{60 \text{ seconds}}{1 \text{ minute}} = 3000 \text{ seconds}$$

and in hours of

$$50 \text{ minutes} \times \frac{1 \text{ hour}}{60 \text{ minutes}} = \frac{50}{60} \text{ hour} = 0.83 \text{ hour} .$$

When we convert units in this way we are not in any way changing the quantity involved; we are merely expressing its value in a different size unit.

WORKED EXAMPLE

The furlong is a distance of ⅛ of a mile. What is the length of the furlong expressed in kilometers?

SOLUTION: The appropriate conversion factors are: 1 furlong = ⅛ mile and 1 mile = 1.61 km. One furlong is thus converted into kilometers as follows:

$$1 \text{ furlong} = 1 \text{ furlong} \times \frac{1/8 \text{ mile}}{\text{furlong}} \times \frac{1.61 \text{ km}}{\text{mile}}$$

$$= 1 \text{ furlong} \times \frac{1/8 \text{ mile}}{\text{furlong}} \times \frac{1.61 \text{ km}}{\text{mile}}$$

$$= 1 \times \frac{1}{8} \times 1.61 \text{ km}$$

$$= 0.21 \text{ km} .$$

1. Explain why units are an important aspect of the numerical description of a given quantity.

2. Describe how units are treated mathematically.

3. What is a conversion factor and why is it important?

1. One league is a distance of about three miles. What is a distance of 20,000 leagues equal to in kilometers?

2. Calculate the number of seconds in a day.

PHYSICS
FACET

2.1 Art, Mathematics, and Physics: Representations of Reality

Beauty, we are told, is in the eye of the beholder. Much the same can be said for that philosophically elusive entity called reality. The nature of reality has most likely been debated in some form or other as long as human beings have existed on the surface of the earth. As we have already seen, reality for the scientist in general and the physicist in particular consists of that which can be observed, measured, and experimented on. But is the reality of the physical world, about which the scientific method has permitted us great insight and understanding, the totality of reality? Or do other frontiers exist beyond the boundaries of scientific empiricism that are as worthy of exploration as those that lie within that realm? The artist, of course, will answer resoundingly in the affirmative, for it is the artist who seeks to capture the abstract emotion of a moment, scene, or person and convert it into a concrete work of art. It is the artist who seeks to translate the abstract qualities of the world of dreams, imagination, and the subconscious into the physical reality of a sketch painting, or sculpture.

For the scientist, the reality of the physical world must necessarily be explored with the greatest possible objectivity and detachment. In dealing with the physical world, our feelings, intuition, and common sense, very often lead us down a wrong path. The "common sense" view of the earth, for example, is that of a body that neither revolves nor rotates, yet the scientific method has unequivocally demonstrated that it does both. Such objective detachment, however, is denounced by the artist who must necessarily become as subjectively and emotionally immersed in his or her work as possible in order to be successful. Such differences can produce very different perceptions of reality. For the insensitive, a starry night is, to the naked eye, little more than a scattering of a few hundred points of light silhouetted against a sea of blackness, but for an artist such as a Vincent Van Gogh, that same starry night can be experienced as a vast, dynamic reality filled with whirling orbs, stellar eruptions, and celestial flows, beneath which human habitations are huddled in awe and wonder.

Astride this gap stands the mathematician who is a curious blend of scientist, artist, and logician, who can be as wildly speculative as any artist, but who can also exhibit thinking as precise as that of the physicist. Mathematics can tell us all about the properties of numbers, abstract spaces having infinite dimensions, and indeed about the topic of infinity itself. It is this reservoir of mathematical knowledge from which the physicist can draw in attempting to describe the physical world. Physicists are not the only ones who make use of mathematical knowledge. In a more indirect fashion, artists routinely employ mathematics in their use of perspective, geometric structuring, vanishing points, and the like.

For all of their differences, the artist, mathematician, and physicist share the crucially important common bond of inspiration. For reasons that are unclear to us, a few humans have been visited with inspiration that has enabled them to take a fundamentally different view of reality, whether it be the scientific reality of a Newton or an Einstein, the mathematical reality of a Pythagoras or Euclid, or the artistic reality of a Monet or Picasso. Perhaps at some future date that same inspiration will enable us to view a far more complete reality of which each of the above is but aspects and components.

Van Gogh, Vincent. The Starry Night. (1889). Oil on canvas, 29 × 36¼" (73.3 × 92.1 cm). Collection, The Museum of Modern Art, New York. Acquired through the Lillie P. Bliss Bequest.

2.5 Scientific Notation

As we begin to assign numbers to our measurements of the physical world, we learn very quickly that enormous ranges of magnitudes exist in numerical values. The diameter of the nucleus of an atom, for example, is about 0.000000000000001 meter, while the distance to the nearest star, Alpha Centauri, is about 40,000,000,000,000,000 meters. Such numbers become particularly awkward if we wish to use them in calculations of one sort or other.

To reduce such difficulties, scientists have devised what is referred to as *scientific notation* in which a given numerical value is expressed as a number between 1 and 10, times ten raised to the appropriate power. Thus the number 350 can be thought of as the product of 3.5 times 100. Since 100 is equal to 10 $\times$ 10 or 10^2, we may write the number 350 as 3.5 (the number between 1 and 10) times 10^2 (ten raised to the appropriate power) or 3.5×10^2. The number above and to the right of the 10 is called the exponent or "power" and tells us how many places one must move the decimal point in the number to express it in conventional form. The sign of the exponent tells us the direction in which to move the decimal point. If the sign is positive, the direction is to the right and if the sign is negative, the direction is to the left. Thus if we have a number expressed in scientific notation, such as 4.5×10^3, the conventional version of that number would be 4.5 with the decimal point moved three places to the right or 4.500 or 4500. Note that if no other digits are present, zeros are added to fill the missing places.

Similarly, the number 3.8×10^{-4} is expressed in conventional form by swinging the decimal point four places to the left or 0003.8 or .00038. Notice that, once more, zeros are used to fill the missing places.

"Conventional" numbers can be expressed in scientific notation by reversing the above process. For example, if we wish to express the number 620 in scientific notation we would first "swing" the decimal point in such a way as to express the number as a number between 1 and 10, in the following manner 620. Since the decimal point was moved two places to the left to get it between the 6 and the 2, to retain the original value of the number, we must multiply 6.2 by 10 raised to the power 2 or 10^2. The number 620 expressed in scientific notation thus becomes 6.2×10^2. In like fashion, the aforementioned diameter of an atomic nucleus expressed in scientific notation would be 1.0×10^{-15} meters and the distance to Alpha Centauri written as 4.0×10^{16} meters.

The advantage of using scientific notation to express numbers is twofold. First, the exponent of 10 effectively absorbs the various "orders of magnitude" or factors of 10 encountered in nature. Thus the diameter of the nucleus of an atom (1.0×10^{-15} meters) can now be expressed in the same convenient format as the distance to the nearest star (4.0×10^{16} meters), despite the fact that the latter number is 31 orders of magnitude or roughly a factor of 10^{31} greater. Moreover, calculations involving such numbers are rather easily performed, since in any calculations the powers of 10 must follow the algebraic laws of exponents which are

$$\textbf{1.} \quad 10^a \times 10^b = 10^{a+b}$$

EXAMPLES: $10^3 \times 10^4 = 10^{3+4} = 10^7$
$10^6 \times 10^9 = 10^{6+9} = 10^{15}$.

2. $10^a \div 10^b = 10^{a-b}$

EXAMPLES: $10^5 \div 10^2 = 10^{5-2} = 10^3$
$10^4 \div 10^9 = 10^{4-9} = 10^{-5}$

3. $(10^a)^b = 10^{a \times b}$

EXAMPLES: $(10^3)^4 = 10^{3 \times 4} = 10^{12}$
$(10^{-2})^5 = 10^{(-2)(5)} = 10^{-10}$.

4. $1/10^a = 10^{-a}$

EXAMPLES: $1/10^5 = 10^{-5}$
$1/10^{-3} = 10^{-(-3)} = 10^3$.

Thus, the value of the speed of light (3×10^8 m/sec) squared would be

$$(3 \times 10^8 \text{ m/sec})^2 = (3 \times 10^8 \text{ m/sec})(3 \times 10^8 \text{ m/sec})$$

$$= 3 \times 3 \times 10^8 \times 10^8 \times (\text{m/sec}) \times (\text{m/sec})$$

$$= 9 \times 10^{16} \text{ m}^2/\text{sec}^2 ,$$

which is a far more convenient way to proceed than by multiplying out $300,000,000 \times 300,000,000$!

Very often scientists will use prefixes to denote the power of 10 by which a particular quantity is to be multiplied. Thus a ''centi''-meter is $1/100$ or 10^{-2} meters, a ''kilo''-gram is 1000 grams, and so on. A list of such prefixes commonly used is given in Table 2.2.

Many more mathematical techniques and approaches, not described here, have flowed from the imaginations of scientists and mathematicians in their quest to describe the physical universe in a quantitative way. So complex and sophisticated have some of these methods become that Pythagoras himself would undoubtedly be very impressed by it all. That venerable old mathematician would almost certainly derive great satisfaction from the spectacular success that his most cherished philosophical principle has come to enjoy in modern times.

TABLE 2.2
Commonly Used Prefixes

Prefix	Power of Ten Denoted
atto-	10^{-18}
femto-	10^{-15}
pico-	10^{-12}
nano-	10^{-9}
micro-	10^{-6}
milli-	10^{-3}
centi-	10^{-2}
deci-	10^{-1}
deka-	10
hecto-	10^2
kilo-	10^3
mega-	10^6
giga-	10^9
tera-	10^{12}

WORKED EXAMPLE

Calculate the value of

$$\frac{(8 \times 10^4) \times (3 \times 10^{-2})}{(6 \times 10^{-3})} .$$

SOLUTION: Rearranging the factors we have

$$\text{value} = \frac{8 \times 3 \times 10^4 \times 10^{-2}}{6 \times 10^{-3}}$$

$$= \frac{24 \times 10^{4+(-2)}}{6 \times 10^{-3}}$$

Continued

WORKED EXAMPLE

Continued

$$= \frac{24 \times 10^{+2}}{6 \times 10^{-3}}$$

$$= \frac{24}{6} \times 10^{+2-(-3)}$$

$$= 4 \times 10^{+2+3}$$

$$= 4 \times 10^{+5}$$

The value of $(8 \times 10^4) \times (3 \times 10^{-2})/(6 \times 10^{-3})$ is thus 4×10^5 or 400,000.

1. What is scientific notation? Why do you suppose it is used by scientists?
2. Describe the steps required to express
    a. a conventional number in scientific notation
    b. a number in scientific notation in conventional form
3. Summarize the laws of exponents and give an example of each.
4. What do each of the following prefixes mean?: centi-, milli-, kilo-

1. Express each of the following as indicated:
    a. the number 0.0084 in scientific notation
    b. the number 4.1×10^4 in conventional notation
    c. the number 150,000,000,000 in scientific notation
    d. the number 4.0×10^{-9} in convention notation
2. Calculate the value of

$$\frac{(3 \times 10^{-3}) \times (2 \times 10^4)^3}{(5 \times 10^2)^2} .$$

Chapter Review Problems

1. Express each of the following numbers in scientific notation:
    a. 450   **b.** 0.00006   **c.** 928,000,000,000
    d. 99   **e.** 0.025   **f.** one-billionth
2. Express each of the following in scientific notation:
    a. three thousand   **b.** seven trillion   **c.** four hundred
    d. six billion   **e.** ninety   **f.** eight million

3. Express each of the following numbers in conventional form:
   a. 1.5×10^8 **b.** 1.7×10^{-6} **c.** 3×10^8
   d. 6.7×10^{-5} **e.** 6.02×10^{13} **f.** 6.6×10^{-15}

4. A total of 9,192,631,770 cesium atom vibrations occur per second. Express this number in scientific notation.

5. Calculate the following:
   a. $3 \times 10^2 \times 4 \times 10^3$ **b.** $5 \times 10^{-1} \times 2 \times 10^4$
   c. $6 \times 10^{23} \times 4 \times 10^{-13}$ **d.** $9 \times 10^{-10} \times 7 \times 10^{20}$
   e. $6 \times 10^1 \times 8 \times 10^{-3}$

6. Calculate the following:
   a. $(4 \times 10^2) \div (3 \times 10^3)$ **b.** $(5 \times 10^4) \div (2 \times 10^{-1})$
   c. $(8 \times 10^{23}) \div (4 \times 10^{-13})$ **d.** $(9 \times 10^{-10}) \div (7 \times 10^{20})$
   e. $(6 \times 10^{-3}) \div (8 \times 10^1)$

7. Calculate the following:
   a. $(2.0 \times 10^{-2})^3$ **b.** $(8.0 \times 10^3)^2$
   c. $(6.0 \times 10^{-1})(2.0 \times 10^3)$ **d.** $(5.0 \times 10^{-3})(4.0 \times 10^{-10})$
   e. $\left(\dfrac{4.0 \times 10^{-2}}{2.0 \times 10^3}\right)^2$

8. Divide 9,000,000,000,000 by 0.000000000003 using (a) standard arithmetic and (b) scientific notation. Comment on the relative difficulty of each method.

9. One angstrom ($1\,\text{Å}$) is a unit of length equal to 10^{-8} cm. How many angstroms are there in a length of one meter?

10. Using the values given in the text for the diameter of an atomic nucleus (1.0×10^{-15} m) and the distance to Alpha Centauri (4.0×10^{16} m), calculate the number of atomic nuclei placed side by side needed to reach all the way to Alpha Centauri.

11. A typical atom is about 10^{-10} m in diameter. How many atoms set side by side would it take before the line of atoms reached a length detectable to the naked eye (0.05 mm)?

12. If the U.S. government spends a trillion dollars a year, how much does it spend each day?

13. If the length of a typical dollar bill is about 15.5 cm, how far would a trillion dollars placed end to end reach out into space? Look up the distances to some of the nearby celestial objects such as the Moon, Venus, and Mars and determine which, if any , of these objects would be reached by such a ''money bridge.''

14. Find the number of millimeters in a gigameter.

15. A box has dimensions of 2 meters by 3 meters by 5 meters. Find the volume of the box. What are the units of your answer?

16. If the distance between two cities is 1000 miles, what is the value of that same distance expressed in kilometers?

17. In the English system of units, the inch, yard, and mile are rough equivalents of the centimeter, meter, and kilometer, respectively. The equivalent of one yard is 36 inches and 1760 yards is the equivalent of one mile. Calculate the number of inches in a mile and the number of centimeters in a kilometer. Comment on why you think scientists prefer the metric system.

18. Which is the longer distance to run, the 1500-meter run in the Olympic Games or the mile run in the Commonwealth Games? Explain.

19. Using the graph shown in Figure 2.6, find the approximate annual tuition a student would have paid in 1980.

20. Using the graph shown in Figure 2.6, in approximately what year will the tuition cost be $4000?

21. Plot each of the following sets of x and y values onto a graph:
    a. $x = 1$ **b.** $x = 2$ **c.** $x = 3$ **d.** $x = 4$ **e.** $x = 5$
    $y = 4$ $y = 5$ $y = 6$ $y = 7$ $y = 8$

    What geometric curve does the resulting set of points seem to form?

THOUGHT QUESTIONS

1. How many grams are contained in a "millikilogram"?

2. Find the number of milliliters in a cubic centimeter.

3. Since the advent of the space age, we have learned from earth satellites that the earth is not a perfect sphere or even a perfect oblate spheroid. Discuss the impact of such a discovery on (a) the original definition of the standard meter and (b) the current definition of the standard meter.

4. Describe one advantage and one disadvantage of defining the standard mass unit in terms of a specified volume of water.

5. Why do you suppose that the one-kilogram standard mass shown in Figure 2.9 is encased in a glass jar?

6. Is volume a fundamental quantity in the physical world? Explain.

7. Explain how you might set up your own standards of time, length, and mass.

8. Why did physicists redefine the "fundamental quantities of nature"? What scientific precautions do you think would have to be taken in such redefinitions?

9. If an academic vice president at the school whose graph is shown in Figure 2.6 told you that the rate of tuition increase is linear, would you believe such a statement? Explain.

10. Explain how graphs can be used for applications other than those described in the text.

11. Why do you think that "kilosecond" as a unit of time has not caught on the way the kilometer and kilogram have?

12. Suppose you have a set of experimental points plotted on a graph, and the plot cannot be explained by any current theory. Describe how you would deal with such a situation within the framework of the scientific method.

13. Can you think of an example in which a graphic projection turned out to be wrong? Explain.

14. If the units of S are in dollars per week and those of A are in years in the "dating law" in section 2.1, are these units consistent with those stated for N? If not, what would you do to remedy the situation?

15. In what way(s) is mathematics a "scientific paintbrush"?

16. Is there such a thing as a secondary standard of length or time? If so, how might they be used in physics?

17. A certain student finds that her grade point average is equal to the time she spends studying divided by the time she spends playing video games at the student union. Write a mathematical equation for this phenomenon.

18. Describe the dangers in projecting or extrapolating graphical results.

19. Theories A, B, and C all predict the graphical behavior between the quantities s and t as shown below. If the lower graph represents a measured set of data points for s and t, which theory provides the best representation of the data? Explain.

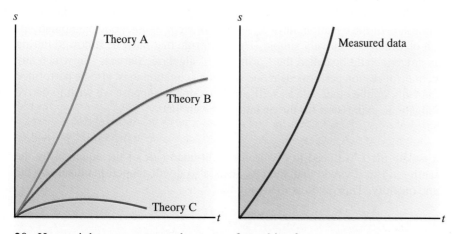

20. How might you measure the mass of an object?

21. The yard was originally defined as the distance from the nose of King Henry VIII of England to the tip of his thumb. In your opinion, is this a good standard of length? Explain.

22. Discuss the possibilities that quantities would exist in nature that are even more fundamental than time, length, and mass.

23. The famed science fiction author Robert Heinlein once wrote:

    Anyone who cannot cope with mathematics is not fully human. At best he is a tolerable subhuman who has learned to wear shoes, bathe, and not make messes in the house.

    Comment on Mr. Heinlein's viewpoint.

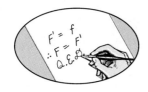

Algebraic Equations

One of the mathematical hurdles for a nonscience student enrolled in a science course comes in the form of the algebraic equation. All too often the symbolism and manipulations associated with algebraic equations generate a degree of confusion, frustration, and dread on the part of the nonscience student to the extent that psychologists have even coined a term for the phenomenon—math anxiety.

In reality, the algebraic equation provides a convenient and precise way of stating the behavior patterns we observe in nature. Energy is equal to the product of the mass and the square of the speed of light. Using the symbols E to denote energy, m to denote mass, and c to denote the speed of light, this result can be very succinctly stated in the form of an algebraic equation as

$$E = mc^2 .$$

Quite often it is desirable to rearrange an algebraic equation so as to solve the equation for a particular quantity. In this process, we isolate the desired quantity on one side of the equation, through a series of steps or algebraic manipulations, each of which is relatively simple. The main rule that must be followed during this process could be called the ''principle of algebraic fairness'' in which any algebraic operation performed on one side of an equation must also be performed identically on the other side as well. As an example, suppose we wish to derive an algebraic expression for the quantity t, given that

$$v = at .$$

To isolate the t, we need to divide the right-hand side of the equation by the quantity a. But if we do that, then we must also divide the left-hand side by the same quantity. This gives us the result that

$$\frac{v}{a} = \frac{at}{a}$$

$$\frac{v}{a} = \left(\frac{a}{a}\right)t$$

$$\frac{v}{a} = (1)t$$

$$t = \frac{v}{a}$$

or

$$t = v/a .$$

Very often, the isolation of a given quantity requires the use of several operations, each of which must be identically performed on both sides of the equation. For example, we are given

$$d = \frac{1}{2} at^2 .$$

The solution of this relationship for the quantity t would proceed by first multiplying both sides of the equation by 2:

$$2 \times d = \frac{1}{2} at^2 \times 2 ,$$

which gives us

$$2d = at^2 .$$

We would then divide both sides of the equation by the quantity a

$$\frac{2d}{a} = \frac{at^2}{a} ,$$

which leaves us with

$$\frac{2d}{a} = t^2 .$$

If we now take the square root of both sides of this equation we obtain

$$\sqrt{\frac{2d}{a}} = \sqrt{t^2} ,$$

which gives us the final result that

$$t = \sqrt{\frac{2d}{a}} .$$

There are many ways to solve similar equations, but as long as one remembers to ''be fair'' to *both* sides of an algebraic equation, then the particular order of steps will not matter and the final result will always be the same.

QUESTIONS

1. Why do you suppose that a given operation must be performed on both sides of a given equation if equality is to be preserved?
2. Can you think of a practical example in which ''it is desirable to rearrange an algebraic equation so as to solve the equation for a given quantity''? Explain.
3. Why do you suppose that physicists make use of algebraic techniques?

PROBLEMS

1. Find a mathematical expression for the quantity a if $F = ma$.
2. Find a mathematical expression for the quantity t if $v = x/t$.
3. Find a mathematical expression for the quantity c if $E = mc^2$.

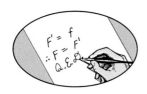

Vector Quantities

Very often physicists encounter quantities in nature for which a numerical value expressed in a well-defined set of units does not provide a complete description of that quantity's behavior and effects. One important class of such quantities is the *vector quantity* or vector. A complete description of a vector quantity requires that both its numerical value, called the magnitude, and its direction be specified. As a result, the effects of both the magnitude and the direction must be taken into account when evaluating even the most basic interactions of vector quantities. Vector quantities can interact with each other as well as with nonvectors in a variety of ways. The investigation of these interactions has given rise to the branch of mathematics called *vector analysis*. Most of the concepts of vector analysis lie outside of our discussion in this text, but knowledge of a few of the basic vector concepts and operations will be helpful in understanding the behavior of our physical world and thus are worthy of our attention.

For our purposes, we will employ a graphical representation of vector quantities in which such quantities are delineated as arrows whose lengths are scaled to their respective magnitudes and whose directions are indicated by the orientation of the arrow relative to some reference direction such as north, the positive *x* axis, etc. An example of this type of graphical representation is illustrated for two vectors A and B in Figure 2.10.

One of the important vector operations is that of vector addition. Graphically, two vectors A and B are added by first placing the tail of vector B at the tip of vector A without changing the orientation of either vector. The vector that results is called a *resultant vector*. It is the vector "arrow" whose tail is located at the tail of vector A and whose tip is located at the tip of vector B (Figure 2.11). The magnitude of the resultant vector is the length between the tip and the tail of the resultant vector measured on the same scale as the graph. The direction of the resultant vector is that of the line from the tail to the tip of the resultant vector.

As an example of this type of vector interaction, consider a person swimming at a rate of 4 miles per hour in a river that itself is flowing at a rate of 3 miles per hours (shown in Figure 2.12). As we shall see in the next chapter, both rates are vector quantities, and the interaction of the swimmer's rate with the flow rate of the river can be represented as an addition of vectors. If a person swims upstream, the resultant vector rate has a magnitude of one mile per hour and is directed upstream. If a person swims downstream, the resultant vector rate has

FIGURE 2.10 ··········

Graphical Representation of Two Vector Quantities   Vector A has a magnitude of 8 and a direction along the positive *x* axis. Vector B has a magnitude of 5 and a direction of 45 degrees above the positive *x* axis.

FIGURE 2.11 ··········

Addition of Vectors A and B from Figure 2.10   The tail of the resultant vector is located at the tail of vector A and its tip at the tip of vector B. The resultant vector has a magnitude of 12 and is directed at an angle of approximately 18 degrees above the positive *x* axis.

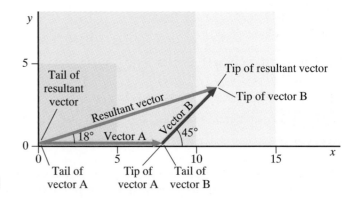

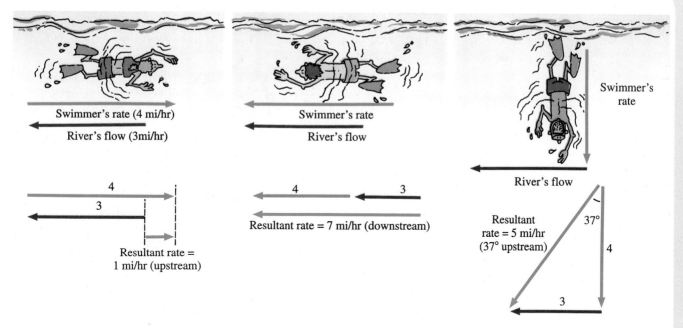

FIGURE 2.12 ·········
**Vector Addition of the Rates of Motion for a River Flow and a
Swimmer**   (a) Swimming upstream; (b) swimming downstream; (c) swimming
perpendicular to the stream.

a magnitude of 7 miles per hour and is directed downstream. Finally, if a person
swims perpendicular to the flow, the resultant vector rate has a magnitude of 5
miles per hour and is along a direction of about 37 degrees downstream.

Notice that in all three cases we have taken into account the magnitudes and
the directions of the quantities involved in order to determine the resultant vector
of the swimmer. Most importantly, however, if we were to have an actual person
swimming in a river under the stated conditions, we would find that the description
of the resultant motions of the swimmer as rendered by our vector addition, would
in fact be a very accurate representation of the actual observed motion of the
swimmer. For the scientist, of course, that is the bottom line in any attempt to
represent a physical phenomenon mathematically. Using vectors or models, physicists are able to account correctly and accurately for most of the motion we
observe in daily life.

QUESTIONS

1. Can you think of any quantities in your everyday life that can be described
   as vector quantities? Explain.
2. Can a vector be three dimensional? Explain.
3. Compare and contrast vector addition with algebraic addition.

PROBLEMS

1. Assume that on a graph the positive y axis denotes the vertical direction and
   the positive x axis denotes the direction to the right of the vertical. Find by

graphical techniques the magnitude and direction of the vector resulting from the interaction of vector A, which points to the left of the vertical and has a magnitude of 4, and vector B, which points toward the vertical and has a magnitude of 3.

2. Assume that on a graph the positive y axis denotes the direction north and the positive x axis denotes the direction east. Find by graphical techniques the magnitude and direction of the vector resulting from each of the following pairs of interacting vectors:

   a. Vector A points due east and has a magnitude of 20; vector B points due west and has a magnitude of 15.

   b. Vector A points due north and has a magnitude of 8; vector B points due north and has a magnitude of 4.

   c. Vector A points due south and has a magnitude of 9; vector B points due east and has a magnitude of 12.

3. The captain of a ship wishes to sail from an island to a port across a strait. If the ship is moving at 12 kilometers/hour along a line directly toward the port and a current in the channel is moving at 5 kilometers/hour perpendicular to that line, at what rate and direction does the ship end up moving through the water?

3 TRANSLATIONAL MOTION

No phenomenon is more intrinsic to the physical world than that of motion. Movement of one sort or another is encountered at every level of our perception of the observable universe from the tiniest subatomic particles racing through "inner space" to the largest clusters of galaxies hurtling toward the farthest reaches of outer space. In spite of this wealth of experience, however, an overall viable description of this important aspect of nature eluded the greatest thinkers in the world for centuries. A profound change came to this state of affairs when in 1687 the Englishman Sir Isaac Newton published his premier work *Mathematical Principles of Natural Philosophy* or, more simply, the *Principia*. The scientific revolutionaries of the generation that preceded Newton—Galileo, Kepler, and Descartes, to name a few—had managed to replace the Aristotelian view of motion with a number of *ad hoc* laws and principles in which a specific phenomenon was described with a specific law. Thus Kepler's laws of planetary motion described precisely that—planetary motion. The brilliance of Newton's *Principia* lay in the fact that it galvanized all of these bits and pieces into a simple, elegant, and , most important, highly successful view of motion that was not limited to any specific motion in particular, but could explain a multitude of motion phenomena using only a few basic concepts and principles. This "Newtonian" or "classical" view of motion was the first truly successful scientific theory of any aspect of nature, and even after more than three centuries of scientific progress, it still offers an excellent accounting of the behavior of moving objects in everyday life.

3.1 Basic Newtonian Kinematics

If you were asked to write an informal definition of "motion" or "movement" you would probably respond with something like "motion is the process of changing places" or "movement is the passage of an object from one location to another." Regardless of how you might state your definition, it would have to involve the idea of an object changing places within the framework of a nonmoving backdrop. Such a backdrop is denoted by physicists as a *frame of reference* or a *reference frame* and provides the base against which any position changes are to be measured (Figure 3.1). Most of the time, common sense will dictate how we define our reference frames. If an instructor strides across the front of a classroom, a reasonable assumption is that the instructor is moving against the background of the classroom as opposed to the reverse—a classroom complete with students and attached building that is moving against the reference frame of a "stationary" instructor. Similarly, the state trooper who just pulled you over for doing 30 miles per hour above the posted speed limit has made the decision that your automobile was moving relative to a stationary road and speed limit sign rather than assuming that the road and sign were whizzing by a stationary automobile. Quite often, however, frames of reference that are defined as being at rest can, in fact, possess considerable motion of their own relative to other reference frames. If a passenger walks up and down the aisles of a jetliner while the aircraft is in flight, we think of the passenger moving relative to the stationary backdrop of the jetliner, despite the fact that the aircraft may be moving several hundred miles per hour relative to the earth's surface. Even "terra firma," our most familiar and widely used stationary reference frame, is whirling at a rate

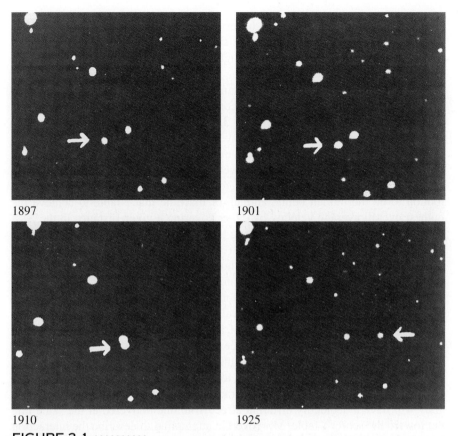

1897   1901

1910   1925

FIGURE 3.1 ··········
The Motion of the Nearby Star *Proxima Centauri* This sequence of photographs
shows the movement of this star against a backdrop of "fixed" stars over a time
interval of 28 years.

of more than 100,000 kilometers per hour about a sun orbiting at more than
700,000 kilometers per hour about the center of our Milky Way galaxy, which
in turn is rushing toward the Virgo "supercluster" of galaxies at millions of
kilometers per hour, and so on.

Thus an "at rest" frame of reference must be decided on and defined. Any
displacement or change in an object's position relative to that rest frame then
constitutes what is known as *translational motion.* Newton's theory provides us
with two basic concepts by which the degree of such motion can be defined and
measured. The first of these is called *velocity* and can be defined as follows:

$$\text{velocity} = \frac{\text{distance object moves in a given direction}}{\text{elapsed time}} \qquad \textbf{3.1}$$

or in algebraic symbols as

$$v = d/t \ .$$

There are several noteworthy characteristics about defining the degree of an
object's motion in this way. First of all, such a definition expresses motion in
terms of measurable quantities, in this instance the fundamental quantities of

length and time, thus allowing an observer to represent numerically the degree of an object's motion. When we make such a measurement, the units in which the velocity is expressed are units of length/time such as miles/hour, centimeters/second, or furlongs/fortnight. In keeping with the fundamental units of the meter and the second, which were described in the previous chapter, the velocities in this text will be expressed in meters/second (m/s). One meter per second is a rate of motion equivalent to approximately 3.5 kilometers/hour or 2.2 miles/hour.

With the above definition of velocity, we can now judge the degree of motion in a far more articulate manner. When we exceed the speed limit in an automobile, for example, we are judged to be covering more distance per unit time than society has deemed safe for the given road. We judge the winner of a race of motion to be the human being, animal, or machine that is able to achieve the highest overall average velocity by either covering a given distance in the least amount of elapsed time or by covering the greatest distance in a given amount of elapsed time. Thus the winner of a 10,000-meter run is the athlete who runs the 10,000 meters in the "fastest" or shortest interval of time, while the driving team that wins the 24-hour LeMans automobile race is the team that drives its machine the farthest distance in the allotted 24 hours of racing time.

A specification of the numerical value of the distance traveled divided by the elapsed time, however, does not of itself constitute a complete description of an object's velocity. An athlete may be able to cover 10,000 meters in a smaller interval of time than the other contestants in the field, but if the runner fails to direct that effort in the proper way, i.e., toward the finish line, the effort becomes meaningless in terms of winning the race. A bowling ball delivered down an alley will have a decidedly different effect from one delivered at the same rate but toward the scorer's table. Moreover, in attempting to describe the interaction of velocities such as a person swimming cross-stream, we find that velocities and their interactions must be described in terms of vector quantities (see Physics Formulation 2.2) in which both a magnitude and a direction are specified for a given velocity. Thus a complete description of an object's velocity requires a value for the magnitude of the velocity, called the *speed* of the object, and a designated direction toward which the object is moving.

A second concept is needed to fully describe motion; specifically, the fact that most objects change their velocities, sometimes speeding up, slowing down, or changing the directions. Athletic coaches, for example, are fond of distinguishing an athlete's "speed" from that athlete's "quickness," where the "quicker" athlete can change whatever he or she is doing over a shorter time interval than one not so gifted in this regard. Obviously such a quality is of great value in virtually every form of athletic activity imaginable, whether it is a volleyball player altering the direction of a spike or a halfback making a crucial cutback on a long run. In a somewhat similar fashion, an automobile may have the ability to attain a high rate of speed, yet will lose out in a "drag" race because it doesn't possess enough "quickness," or ability to achieve that large magnitude of speed in a short enough time interval (Figure 3.2). Newton's theory of motion therefore establishes a second criterion by which the degree of an object's motion can be defined and measured. The concept is, in essence, a formalization of the idea of quickness and is called *acceleration*. The acceleration of an object is defined as follows:

FIGURE 3.2 ⋯⋯⋯⋯

Acceleration Versus Velocity   The Indy-type racing car (top) can move at high velocities but has less acceleration capabilities than the dragster (bottom), which can change its velocity far more rapidly.

$$\text{acceleration} = \frac{\text{change in object's velocity}}{\text{elapsed time}} \qquad \textbf{3.2}$$

or more simply as

$$a = \Delta v/t \ .$$

Like velocity, acceleration is defined in terms of the measurable quantities of length and time. In this case, however, the units of acceleration are those of velocity/time. Since the units of velocity are length/time, one can express acceleration in units of

(length/time)/time

length/[(time) $\times$ (time)]

or simply as

length/(time)^2

For example, an object for which the velocity changes by 10 meters/sec over a time interval of 5 sec experiences an acceleration of 10 (meters/sec)/5 sec or 2 meters/(sec)^2. Acceleration, like velocity, is a vector quantity, and as such, a complete description of acceleration requires that both the magnitude and the direction of the acceleration be specified.

Note that in our definitions of both velocity and acceleration we are in effect dealing with average values for these quantities. If an automobile takes 3 hours to travel 120 kilometers, its average velocity is 120 kilometers/3 hours or 40 kilometers/hour. This value provides us with the overall average velocity, but says nothing about the manner in which the velocity may have changed during the course of the journey. To gain this sort of insight we must examine the average velocity of the auto over shorter intervals of time. In some cases we may need to distinguish between average velocity and the velocity at a particular moment, such as would be indicated on the car's speedometer. As early as the fifth century B.C., the Greek philosopher Zeno of Elea called attention to the fact that any definition of motion that was based on an elapsed time seemed to encounter serious difficulties whenever the value of the elapsed time approached zero (see Physics Facet 3.1). One of the triumphs of Newtonian theory was its success in dealing mathematically with such ''instantaneous'' velocities and accelerations through the use of limits and other methods of the integral and differential calculus developed in the seventeenth century by both Newton and the German mathematician Gottfried Leibniz.

Having decided on how the degree of an object's motion would be defined, Newton then noted that agents existed in nature that were capable of altering the state of an object's motion. These agents are referred to in Newton's theory as *forces*. If, for example, you hold a pencil above the floor, two such agents act on the pencil. The first, called the force of gravity, tends to cause the pencil to alter its state of motion by falling to the floor. Such an alteration of the pencil's state of motion, however, is prevented from occurring by the presence of a second force called a ''frictional'' force, which exists between your fingers and the pencil. If you release the pencil, thereby removing the force of friction, the force of gravity will then cause the pencil to fall to the floor (Figure 3.3).

''Force'' as an undefined and ambiguous concept had been in use for centuries prior to the time of Newton and, for the most part, had little to do with motion phenomena. Such is the case for the ''cosmic'' forces of the astrologer, the ''dark'' and ''light'' forces of witches and warlocks, and a full gamut of all sorts of other ''supernatural'' forces, including ''The Force'' of Star Wars fame.

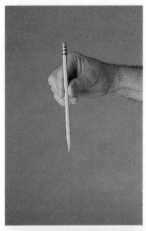

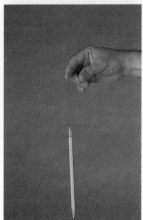

FIGURE 3.3 • • • • • • • • • •
Forces in Newtonian Theory When a pencil is held above the floor, the force of friction between the person's fingers and the pencil counters the force of gravity, and the pencil is prevented from falling to the floor. When the pencil is released and the force of friction is removed, the force of gravity causes the pencil to drop to the floor.

3.1 Achilles and the Tortoise

The Pythagorean proposition that the physical world can be accurately represented and described through the use of mathematics has not been universally accepted by philosophers. One of the earliest challenges to the Pythagoreans came from a group of philosophers who operated out of the Greek city of Elea in southern Italy during the fifth century B.C. The Eleatics, as they were called, claimed that our perceptions of space, time, matter, change, and diversity in nature are false impressions created by our senses, which, through rational thought, can quickly and easily be proven self-contradictory.

Thus, in the middle of the fifth century B.C. the most renowned of the Eleatics, Zeno of Elea, proposed the fol-

lowing dilemma regarding our perception of motion as a change in position. Suppose, Zeno argued, that a race were held between the fleet-footed Greek hero Achilles and a plodding tortoise in which the tortoise was given a 100-cubit* head start. If Achilles were 10 times faster than the tortoise, then in the time that it takes Achilles to run the 100 cubits, the tortoise moving one-tenth as fast can cover 10 cubits. In the time it takes Achilles to run 10 cubits, the tortoise can move 1 cubit, and in the time it takes Achilles to run 1 cubit, the tortoise can move $\frac{1}{10}$ cubit, and so on. Thus, Zeno concludes, Achilles can get infinitely

_____

*An ancient measure of length equal to about 0.5 m.

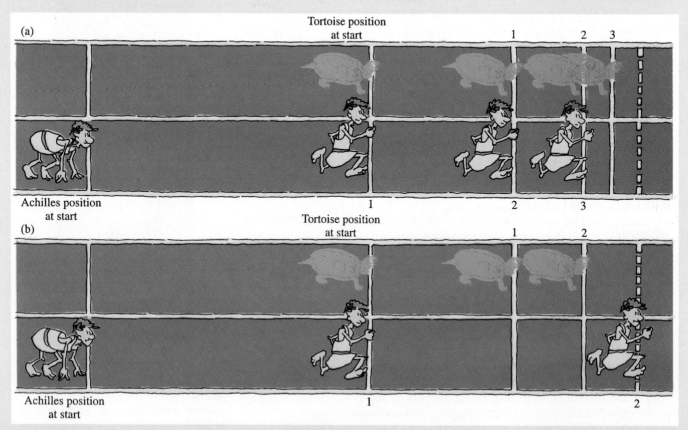

••• **Achilles and the Tortoise** (a) Zeno's version and (b) Newton's version.

Continued

close to the tortoise but can never catch it, let alone pass it. Zeno's arguments notwithstanding, if such a race were to be staged, all of the "smart" betting money would certainly have to be on Achilles. After all, if Achilles is 10 times faster than the tortoise, then from the definitions of velocity, we can conclude that Achilles can either cover 10 times as much distance as the tortoise in the same elapsed time or Achilles can cover the same distance as the tortoise in one-tenth as much elapsed time. Hence, if we divide the motion into 10-second intervals and assume that Achilles can run at a speed of 100 cubits per 10-second elapsed time interval, then at the end of the first such interval we find that Achilles has covered 100 cubits and the tortoise 10 cubits, and Achilles would be 10 cubits behind the tortoise. But over the next 10-second interval, Achilles runs another 100 cubits, while the tortoise moves only another 10 cubits, and sometime during that second elapsed time interval at a distance of a little over 111 cubits, Achilles passes the tortoise, just as a person betting on the race would surmise. If we divide the motion up with a series of one-second elapsed time intervals, the result would be the same, but would occur over a larger number of smaller time intervals.

But suppose, as Zeno argues, that as the interval of elapsed time approaches zero, the motion occurring over that time interval also approaches zero. Since the overall total motion that occurs is the sum of these infinitely small zero-valued increments, the total must itself be zero as well. Indeed, if we photograph someone in motion at various camera shutter speeds, which record the motion over various elapsed time intervals, we find that for the smallest time intervals, the motion appears to be the most "frozen" in its appearance.

This seeming dilemma was not adequately resolved until the time of Newton who, along with the German mathematician Leibniz, was able to prove mathematically that even if the intervals of elapsed time for a given movement were allowed to become infinitely small, a nonzero "instantaneous" change of position associated with an elapsed time of such tiny proportions would still exist. Thus, in the Newtonian view of the race between Achilles and the tortoise, the final result will be the same, namely a victory for Achilles, regardless of whether we consider the race to occur over a series of 10-second elapsed time intervals, a series of one-second elapsed time intervals, or a series of infinitely small elapsed time intervals. The development of the branch of mathematics called *fluxions* by Newton and calculus by Leibniz in the seventeenth century was thus a powerful reaffirmation of the idea that natural phenomena could be mathematically represented in a fashion consistent with our sensory perceptions of those phenomena.

••• **Freezing Action**   A bowler's moving image as seen by a camera over time intervals of (a) 1 second (b) ⅕ second (c) ¹⁄₂₅ second.

The forces proposed by Newton, on the other hand, deal strictly with motion phenomena. Moreover, the Newtonian forces of nature are completely amoral in their influence and efforts. Thus the fall from a high cliff experienced by a great saint due to the action of the earth's gravitational force will be the same as that experienced by the world's worst sinner. Significantly, the forces in Newtonian theory can be mathematically described in terms of measurable quantities. This latter characteristic of Newtonian forces is of crucial importance, since we can, in principle, determine through experimentation what measurable quantities a given type of force may depend on as well as the mathematical nature of that dependence. The force of gravity between two objects, for example, depends inversely on the square of the distance between the objects or $1/(\text{distance})^2$. Thus the forces employed in Newtonian theory, unlike their non-Newtonian counterparts, lie quite within the reach of empirical science, and as a result, physicists have uncovered a number of important interactions in nature that can be described in terms of a mathematical force equation.

The last of the basic concepts required in our discussion of Newtonian theory is that of mass. As we noted earlier, the mass of an object is generally defined in terms of the quantity of matter present in that object. A cubic centimeter of water at a temperature of 4°C has a mass of very nearly 1 gram. A similar cube of water 10 cm on a side, contains 1000 times more water and hence is said to have 1000 times more mass than the 1-cm cube. The mass of an object, however, can also be thought of in terms of the ability of that object to resist a force. This concept of mass is in effect a recognition that certain objects are harder to move than others. Four very large sumo wrestlers, for example, would not have a very hard time pushing a small compact car, but these same four individuals, exerting the same amount of force, would encounter a good deal of trouble if they tried to push a truck (Figure 3.4). Because of this fact, we say that the truck has more mass than the compact car, and thus can resist the force of the wrestlers more effectively than does the car. In a similar vein, the force required to move the more massive cube of water 10 cm on a side is far larger than that needed to move the one cubic centimeter of water.

(a) (b)

FIGURE 3.4 ··········
Mass in Newtonian Theory (a) The four sumo wrestlers can easily lift the compact car but (b) cannot budge the truck with the same force because the truck with its larger mass is better able to resist the force exerted by the wrestlers.

WORKED EXAMPLE

A jetliner can fly at an average velocity of 800 km/hr. How long will it take this plane to fly around the world, assuming that no stops are made for refueling, etc.? Assume that the circumference of the earth is about 40,000 km.

SOLUTION: From the basic definition of velocity we have that:

$$\text{velocity} = \frac{\text{distance traveled}}{\text{elapsed time}} .$$

In this case the plane's velocity is 800 km/hr and the distance traveled is 40,000 km. Thus

$$800 \text{ km/hr} = \frac{40,000 \text{ km}}{\text{elapsed time}} .$$

Solving this equation for the elapsed time, we have

$$\text{elapsed time} = \frac{40,000 \text{ km}}{800 \text{ km/hr}}$$

$$= 50 \text{ hr} .$$

Thus the jetliner would take slightly over two days to circumnavigate the globe.

---

1. What is a frame of reference? Why is the concept important in physics?
2. How do we define and measure motion?
3. Discuss the concept of force as employed in physics.
4. Compare and contrast the definition of mass in Newton's theory of motion with that given in Chapter 2.
5. Explain why each of the following concepts is needed in Newton's theory of motion: velocity, acceleration, force, mass.

---

1. On a cross-country trip four students drove 3000 miles in five days. What was their average velocity?
2. An automobile changes its velocity from 0 to 60 miles/hour in a total elapsed time of 5 seconds. Find the average acceleration of the car.

3.2 Newton's Laws of Motion

Armed with the four basic ideas of velocity, acceleration, force, and mass, Newton then claimed that all motion phenomena could be summarized in three funda-

(a)

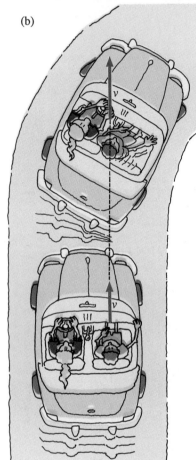

(b)

FIGURE 3.5 ··········
The Law of Inertia The law of inertia for (a) a change in the magnitude of a velocity and (b) a change in the direction of a velocity.

mental statements, called Newton's laws of motion. The first of these is called the law of inertia and is stated by Newton in the *Principia* as follows:

> Every body continues in its state of rest, or of uniform motion in a straight line, unless it is compelled to change that state by forces impressed upon it.

In other words, an object moving at a constant velocity will resist any attempt to change this state of "uniform motion." Numerous everyday examples of the law of inertia exist. When a car starts to move from a standstill, a given passenger will feel a backward "pull" against the seat because the passenger's body, in a state of uniform motion, in this case with a velocity or speed of zero, initially resists the attempt of the car to change that motion. On the other hand, a sudden stop at high velocity finds the passenger's body continuing along in a state of uniform motion at the high velocity until a force is exerted on it by the windshield, dashboard, etc. The use of seat belts or harnesses prevents this from happening by making the passenger in essence an attached part of the vehicle. Because velocity has direction as well as magnitude, a moving object will also resist any attempt to change the direction of the motion. A skier attempting to make a sudden change of direction while coming down the slope finds that such is the stuff of which torn ligaments are made because the body, which is moving in a straight line, is very adamant about not changing that direction of motion, even though the skier's speed has remained nearly constant throughout. If one drives around a mountain curve in such a way that the speedometer reading remains fixed, a "pull" along the seat is still experienced because the inertia of the body's mass tends to keep the passenger moving in a straight line, while the automobile tends to pull the passenger in a curved path (Figure 3.5).

Newton's second law regarding motion can be written as follows:

> Whenever an object is subjected to a new force, it will respond by accelerating in the direction of the force by an amount which is directly proportional to the force and inversely proportional to the object's mass.

This relationship can be written as

$$\text{acceleration} = \frac{\text{force}}{\text{mass}}$$

but is most often expressed in the more familiar form of

$$\text{force} = \text{mass} \times \text{acceleration} \qquad \textbf{3.3}$$

or simply as

$$F = ma \ .$$

In essence Newton's Second Law tells us that an object responds to a given force by changing its velocity. If we exert a force on an object, such as by giving the object a ''push'', its velocity changes. For a given object the harder the push or force, the more pronounced the object's acceleration or change in velocity. On the other hand, the same force applied to a more massive object will result in a smaller change of velocity for that object.

We can see from Newton's second law that the units of force are mass × length/(time)^2. In this context, physicists have defined one newton or 1 N as the amount of force which, when applied to an object having a mass of one kilogram, will cause that object to accelerate at a rate of one meter/sec^2. Thus a force of one N is equal to one kilogram-meter/(second)^2 or kg-m/s^2. As noted earlier, a complete description of any acceleration requires that both the magnitude and the direction of that acceleration be specified. The direction of the acceleration of an object that results from the exertion of a given force depends on the direction in which the force was exerted. Thus, if we are to evaluate the effects of a given force on a given object, we must specify both the magnitude and the direction of that force. In short, Newton's second law requires that forces, like velocities and accelerations, be mathematically described as vector quantities.

Considerable power resides in Newton's second law in that it allows us to test predictions regarding the behavior of objects when they are subjected to one or more forces. For example, suppose that an object having a mass of 5 kg is subjected to a force of 10 N. Newton's second law tells us that this object will respond to such a force by accelerating in the direction of the force by an amount equal to force/mass = 10 N/5 kg = 2 m/s^2. Notice that since 1 N is a kg-m/s^2, N/kg results in units of m/s^2, the units of an acceleration. If more than one force is involved, the magnitude and direction of the resulting acceleration are determined by the magnitude and directions of the *net* force on the object. If all of the forces acting on a given object are exerted colinearly, the net force is obtained by simply adding the forces if they are in the same direction and subtracting the forces if they are in opposing directions. To illustrate, suppose the 10-N force on the object in the preceding example were directed toward the right and an additional force of 30 N were exerted on the object and directed toward the right as shown in Figure 3.6. In this case the *net* force on the object would be 10 N + 30 N or 40 N toward the right. In response to this net force of 40 N to the right, the 5-kg object would accelerate toward the right at a rate of 40 N ÷ 5 kg or 8 m/s^2. If the 30-N force were exerted on the object and directed toward the left, the *net* force on the object would now be 10 N − 30 N = −20 N or 20 N toward the left. In response to this net force of 20 N to the left, the 5-kg object would accelerate toward the left at a rate of 20 N ÷ 5 kg or 4 m/s^2. For the most general cases in which the forces acting on a given object are not colinear, the magnitude and direction of net force must then be calculated by means of vector addition (see Physics Formulation 2.2).

Newton's second law also leads to a distinction between the mass of an object and the gravitational force exerted on that object by the earth or some other body. This latter quantity is referred to as the *weight* of the object. The mass of an object is intrinsically constant in Newtonian theory, but its weight is equal to the product of the object's mass and the gravitational acceleration the object experiences. A 1-kg object placed on the moon will thus have the same 1-kg mass as it does here on the earth, but since the gravitational acceleration on the moon's surface is only one-sixth of that at the earth's surface, the object's weight on the

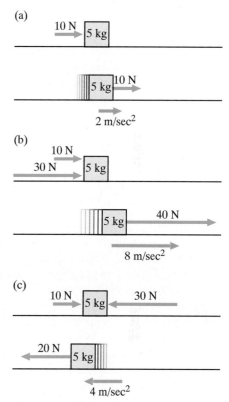

FIGURE 3.6 ⋯⋯⋯⋯

Net Force and $F = ma$ (a) A force of 10 N exerted toward the right on a 5-kg object causes the object to accelerate toward the right at a rate of 10 N ÷ 5 kg = 2 m/sec^2. (b) The 5-kg object responds to the net force of 10 N + 30 N = 40 N to the right by accelerating at a rate of 40 N ÷ 5 kg = 8 m/sec^2 toward the right. (c) The 5-kg object responds to the net force of −20 N or 20 N to the left by accelerating at a rate of 20 N ÷ 5 kg = 4 m/sec^2 toward the left.

3.2 Sir Isaac Newton (1642–1727)

Quite often the greatest figures of human history derive from the humblest of beginnings. So it was with Isaac Newton. Newton barely survived his premature winter birth in a simple farmhouse in the countryside near Woolsthorpe, England. His date of birth on our modern Gregorian calendar was January 4, 1643, but at the time the English, having rejected the "papist" calendar, were still operating on the much older Julian calendric system established in 46 B.C. by Julius Caesar. The delightful result for the romantics among historians, writers, and poets was that by using the calendric system then in use in England, Newton was born on Christmas Day, 1642, a year that also saw the death of Galileo.

Newton's early years offered no portent of the greatness that was to come. After a brief and distinctly unsuccessful stint with agriculture on his mother's farm during his mid-teens, Newton, at the behest of an uncle, enrolled at the University of Cambridge in 1661. Five relatively undistinguished years later, Newton graduated in mathematics. At this point, a plague swept through Cambridge and Newton retreated to his country birthsite at Woolsthorpe. It was in the serenity of the English countryside that Newton found the inspiration that propelled him to the highest levels of scientific immortality. When the plague subsided, Newton returned to Cambridge where he began the task of fleshing out his ideas and insights. Thus did Isaac Newton embark on one of the most profoundly productive periods of achievement enjoyed by any single human being, anywhere, at any time. His work in the field of light and optics was very controversial when first proposed, but is now recognized as being the most complete and advanced of its day. Newton built "with his own hands" the first reflecting telescope, a creation that has become a key instrument for all of modern observational astronomy. He proposed a "corpuscular" or particle theory for the nature of light that predates our modern photons by more than two centuries. His prismatic demonstration that white light is a blend of all of the colors of the rainbow is literally and figuratively one of the classic experiments of scientific history, and Newton's law of cooling was one of the first empirically derived statements concerning the behavior of heat flow in thermodynamics.

To aid in his formulation of a theory of motion, Newton developed the branch of modern mathematics we call calculus and derived the first mathematical description of the law of universal gravitation. But Newton's crowning achievement, indeed the crowning achievement of the Scientific Revolution itself, came when, at the age of 42, he wrote a book he entitled *Principia Mathematica Philosophiae Naturalis (Mathematical Principles of Natural Philosophy)*, which is most often referred to more simply as the *Principia*. Completed in 18 months, the book proposed a view of motion and gravity that was elegantly simple yet eminently successful in explaining these phenomena. Newton did not publish the *Principia* immediately, partly because of his own reluctance and partly because of a lack of funds. Fortunately both obstacles were surmounted when the famed astronomer Edmond Halley implored Newton to

moon's surface will only be one-sixth what it is here on the earth. Since, as we will see, the acceleration of the earth's surface gravity is 9.8 m/s², a mass of 1 kg on the earth will have a weight of 9.8 m/s² × 1 kg or a weight of 9.8 N. This is approximately equal to a weight of 2.2 pounds in the English measuring system (Figure 3.7).

In addition to the law of inertia and $F = ma$, Newton recognized yet a third pattern of behavior in motion phenomena. This behavior, which is referred to as Newton's third law of motion, may be described as follows:

PHYSICS
FACET

Continued

··· **Sir Isaac Newton (1642–1727)**

publish this important work, and in 1687 backed those urgings with a monetary loan to accomplish the task. Newton's reluctance to publish not only the *Principia* but a great deal of his other work as well occasioned a number of bitter feuds, as other scientists and mathematicians such as Robert Hooke and Leibniz independently published scientific results and mathematical solutions that Newton may have actually completed first.

After the publication of the *Principia*, Newton's concerns and attentions took a distinctly nonscientific turn. He suffered a nervous breakdown in 1692 but happily rebounded after several years to serve in English public life, first as a member of Parliament and then as Master of the Mint. He was knighted by Queen Anne in 1705, and at the time of his death in 1727, the Age of Enlightenment in Western Europe was in full swing. Newton's contribution to that era is perhaps most succinctly described by a couplet from the renown English poet Alexander Pope:

> Nature and Nature's laws lay hid in night
> God said, Let Newton be and all was light*

Newton's own response to this sort of adulation can be best summarized by an excerpt from a letter to Robert Hooke, in which the man who had done so much to trigger the onset of an entire intellectual era in Western civilization modestly wrote, "If I have seen further, it is by standing upon the shoulders of Giants."

---

*To which the twentieth century Englishman Sir John Collins Squire whimsically responds:

> It did not last: The Devil howling 'Ho!
> let Einstein be!' restored the status quo.

If an object is subjected to a force it will respond by exerting an equal and opposite force on the agent exerting the force.

This law is often stated simply as "action = reaction." Newton's third law says that when you push an object (action force), you feel a "resistance," which is the reaction force the object exerts on you (Figure 3.8). One of the most familiar examples of Newton's third law is that of a balloon which is inflated and then released. The air escaping out of the nozzle of the balloon pushes off the balloon (action) and the balloon responds by pushing off the air to produce the familiar

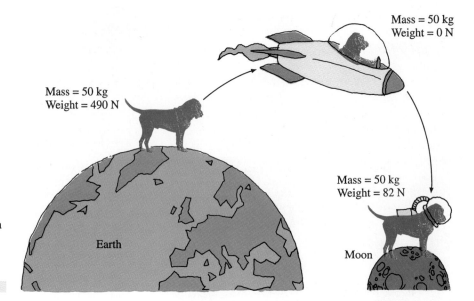

Mass = 50 kg
Weight = 0 N

Mass = 50 kg
Weight = 490 N

Mass = 50 kg
Weight = 82 N

Earth

Moon

FIGURE 3.7 ··········

Mass and Weight in Newtonian Theory   The mass of the space dog is a quantity intrinsic to the space dog while its weight depends on the gravitational force exerted on the space dog.

forward thrust of the balloon (reaction). The escaping air exerts a force on the balloon and the balloon exerts a force on the escaping air. The third law thus tells us that forces in nature must always occur in pairs, action and reaction. Also, in a given interaction, the action force can never act on the same object or entity

FIGURE 3.8 ··········

Action = Reaction   Several examples of systems behaving according to Newton's third law of motion.

as the reaction force. Failure to keep this fact in mind can lead to a certain amount of confusion over how the third law operates. Newton's third law is also sometimes misunderstood due to the fact that the effects of an action or reaction force are not always readily ascertainable. Much of this difficulty can be resolved if we recall that the response of an object to a given force depends on the mass of that object. Thus, even though an individual exerts just as much gravitational force on the earth as the earth exerts on the individual, the mass of the earth is of the order of 10^{22} times as massive as a typical 50- to 100-kg human being. As a result, the earth only accelerates $1/10^{22}$ times as much as the human, or in other words, not very much.

Almost all motion phenomena at the "everyday" level of our perception can be described by these three simple but elegant laws of motion. Interestingly, several aspects of Newton's theory of motion have loose analogues in the non-scientific areas of our lives. For example, administrative types all too often provide a beautiful example of a "bureaucratic law of inertia" for which "habit = psychological inertia" and the state of "uniform motion" is, for all intents and purposes, equal to zero. In a similar fashion, the partners in an emotional relationship can be thought of as affecting each other in a sort of "sociological" action = reaction principle. From a scientific viewpoint, however, the credentials of Newtonian theory have not been gleaned from its simplicity, elegance, or sociological analogues, but rather from the simple fact that it has been eminently successful in describing motion-related phenomena.

One very powerful manifestation of that success has come through the applications of Newton's second law. Suppose we have an object that is subjected to one or more forces. We immediately write from Newton's second law that *F = ma*. We also know that the acceleration is a time rate of change of velocity which, in turn, is a time rate of change of position. Thus, the acceleration can be thought of as a time rate of change of a time rate of change of position, or in mathematical terms, the acceleration can be derived from the position or vice versa. In other words, a very precise mathematical relationship exists between position and acceleration. If we also know the mathematical form of the force(s) involved, we can forge all of this information into an equation of motion. The mathematical solution of this equation of motion yields a description of how the object will move when subjected to the stated forces. This, of course, is an eminently useful thing to be able to do. Much of engineering is concerned with what will happen to a bridge, building, aircraft wing, or a host of other objects if they are subjected to various sets of forces. Newtonian mechanics can, in principle, answer those questions. In practice, however, many such equations of motion are very complicated, and their solutions often require mathematical techniques well beyond the scope of this text. However, a few relatively simple instances of objects in motion are worth considering in more detail.

WORKED EXAMPLE

An object having a mass of 150 kg is subject to a force of 30 N. At what rate does the velocity of the object change?

SOLUTION: From Newton's second law of motion, we know that

Continued

WORKED EXAMPLE

Continued

$$\text{force} = \text{mass} \times \text{acceleration} .$$

For this case, the force is 30 N and the mass is 150 kg. Therefore, the acceleration is given by

$$30 \text{ N} = 150 \text{ kg} \times \text{acceleration}$$

or acceleration = 30 N/150 kg:

$$\text{acceleration} = 0.20 \text{ meters/second}^2 .$$

Thus, the object's velocity changes at a rate of 0.20 meters/second for each second the force is exerted.

1. State Newton's laws of motion.
2. Give an everyday example of each of Newton's laws of motion.
3. Discuss how velocity and acceleration are incorporated into Newton's laws of motion.
4. What role does the concept of force play in Newton's laws of motion?
5. Which of Newton's laws of motion make use of the mass of an object?

1. A certain object is subjected to a force of 20 N and responds by accelerating at a rate of 0.4 m/s^2. What is the mass of the object?
2. A 2000 kg rocket accelerates at a rate of 0.5 m/s^2. What force is exerted on the rocket?

3.3 Motion from a Constant Force

The simplest example of motion resulting from an applied force is that in which an object is subjected to a force that acts in a single direction and does not change its magnitude. The equations that describe the motion of an object under such circumstances are considered in greater detail in Physics Formulation 3.1, but the behavior of such an object can be summarized as follows: If an object is at rest at some zero point reference position and is subject to a constant force, then the object will experience a constant acceleration in the direction of the applied force that will have a magnitude equal to

$$\text{acceleration} = \frac{\text{force}}{\text{mass}}$$

or

$$a = \frac{F}{m} .$$

After a given interval of elapsed time, any object under the influence of this applied force will have acquired a velocity whose direction is the same as that of the applied force and the object's acceleration. The magnitude of this acquired velocity is given by

$$\text{velocity} = \text{acceleration} \times \text{elapsed time}$$

or

$$v = at .$$

The total distance the object will have moved under the influence of the constant acceleration over the given amount elapsed time is

$$\text{distance} = \frac{1}{2} (\text{acceleration}) \times (\text{elapsed time})^2$$

or

$$d = \frac{1}{2} at^2 .$$

The most familiar example of motion under the influence of a constant force, indeed one that we deal with every day of our lives, is that exhibited by free-falling bodies (Figure 3.9). As we have already noted, the natural force that causes objects to fall to the earth's surface is referred to as weight or gravity. Under its influence any object released at or near the earth's surface will fall with a constant gravitational acceleration of $g = 9.8$ m/sec^2 (see Chapter 5). If an object initially at rest begins falling with this constant acceleration, the equations of motion for such a situation tell us that the object will acquire the following velocity:

$$\text{velocity} = g \times \text{elapsed time}$$

or

$$v = gt$$

and will have fallen the following distance:

$$\text{distance} = \frac{1}{2} \times g \times (\text{elapsed time})^2$$

or

$$d = \frac{1}{2} gt^2 .$$

Thus an object dropped from a given height will, after 3 seconds of elapsed time, acquire a velocity of $g \times 3$ seconds, which is

$$\text{velocity} = g \times 3 \text{ sec}$$
$$= 9.8 \text{ m/sec}^2 \times 3 \text{ sec}$$
$$= 29.4 \text{ m/sec} .$$

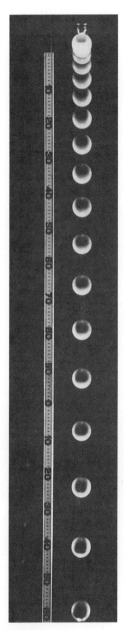

FIGURE 3.9 ··········
A Falling Body   An object released from an at-rest position develops a progressively larger velocity and position displacement as more time elapses, as shown by this falling ball photographed with a flashing strobe light. Each image shows where the ball was when the light was flashed.

The distance the object falls in the 3-second interval is

$$\text{distance} = \frac{1}{2} \times g \times (3 \text{ sec})^2$$

$$= \frac{1}{2} \times (9.8 \text{ m/sec}^2) \times (3 \text{ sec})^2$$

$$= \frac{1}{2} \times 9.8 \text{ m/sec}^2 \times 9 \text{ sec}^2$$

$$= 44 \text{ m} .$$

After 4 seconds of elapsed time the object will have fallen a total distance of 78 m and acquired a velocity of 39 m/sec. Such calculations can be repeated for any value of the elapsed time to yield the speed and/or position for the object at the given time. Thus we can render a complete numerical description of the motion of a free-falling body. We can even deduce the time an object will take to fall to the ground from some specified height by solving the aforementioned distance expression for the elapsed time in the following fashion:

$$\frac{1}{2} \times g \times (\text{time for fall})^2 = \text{distance}$$

$$(\text{time for fall})^2 = \frac{2 \times \text{distance}}{g}$$

$$\text{time for fall} = \sqrt{\frac{2 \times \text{distance}}{g}} .$$

As an example, the time required for an object to fall from the top of a 200-m cliff to the ground below would be

$$\text{time for fall} = \sqrt{(2 \times 200 \text{ m})/g}$$

$$= \sqrt{2 \times 200 \text{ m}/9.8 \text{ (m/sec}^2)}$$

$$= \sqrt{40.8 \text{ sec}^2}$$

$$= 6.3 \text{ seconds} .$$

In all of the calculations above we have assumed that the value of g is constant in its magnitude and direction and that no other forces, such as air resistance, are acting on the falling object in a significant way. In a more advanced course, one can learn how to incorporate these factors into the equations of motion.

A second example of a constant force that can act on a given body is *friction*. Whenever a solid object slides over another solid object, forces arise from interaction between the surfaces of the two objects. These frictional forces depend on the relative smoothness and contact forces between the two objects and act so as to oppose the relative motion between them. Experimental investigations of the behavior of frictional forces indicate that when two solid objects are in contact, a mathematical representation of the resulting force of friction can be expressed as follows:

$$\text{force of friction} = \frac{\text{coefficient}}{\text{of friction}} \times \frac{\text{normal}}{\text{force}} ,$$

where the normal force is the force that is pressing or holding the objects against each other. It is exerted along a line perpendicular to the contact area between the two objects. The coefficient of friction is a dimensionless quantity that usually has a value ranging from 0 to 1, depending on the nature of the surfaces involved. The coefficient of friction between rubber and dry concrete, for example, is very nearly 1.0, but drops off to about 0.3 if the concrete becomes wet and hence more slippery. Under certain conditions, coefficients of friction substantially larger than 1.0 can be produced. For example, if highly purified metallic surfaces are brought into contact in a vacuum, very large frictional forces can be created over the area of contact between the two surfaces. On the other hand, coefficients of friction can be very small. The coefficient of friction between steel and Teflon, for example, is about 0.04, and that between the joints in the higher mammals can be as low as 0.01.

Frictional forces act in such a way as to oppose the motion of the object involved. If one slides a book across the top of a desk, the frictional force between the book and the desk will bring the book to a stop, regardless of the direction in which one slides the book. If the top of the desk is tilted, the presence of frictional forces between the book and the desk can prevent the book from sliding off the edge. The size of the frictional force relative to the other forces acting on the object will control the degree of the object's final motion, if any.

Suppose, for example, we wish to hold a 0.02-kg pen with our fingers as shown in Figure 3.10. The force of gravity on the pen would be

$$\text{force of gravity} = \text{mass of pen} \times g$$
$$= 0.02 \text{ kg} \times 9.8 \text{ m/sec}^2$$
$$= 0.196 \text{ N} .$$

To keep the pen from sliding out of our fingers, the frictional force would thus have to be 0.196 N. Assuming that the coefficient of friction between human skin and a pen surface has a typical value of 0.4, the normal force we would have to exert on the pen with our fingers in order to produce 0.196 N of frictional force would be force of friction/coefficient of friction, or 0.196 N ÷ 0.4 = 0.49 N of normal force. Thus any normal force with a magnitude larger than 0.49 N will serve to hold the pen in place. Because frictional forces appear only as a counter response to motion or possible motion, an excess of contact or normal force over 0.49 N in magnitude will *not* cause the pen to accelerate in an upward

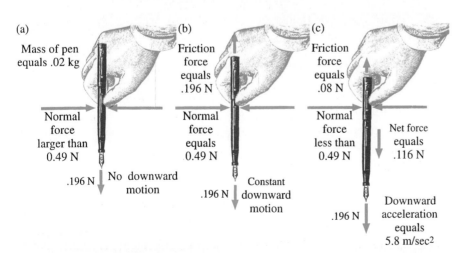

FIGURE 3.10

Friction Versus Gravity (a) The frictional force prevents the pen from moving. (b) The frictional force is just equal to the gravitational force and the pen slides through the person's fingers at a constant velocity. (c) The frictional force is less than the gravitation force and the pen accelerates downward at a rate less than *g*.

direction. If we exert exactly 0.49 N of normal force on the pen, the frictional force will be exactly equal to the gravitational force and the pen will slide through our fingers at a constant velocity, since no net force is causing the pen to accelerate. If we exert less than 0.49 N of normal force on the pen, the frictional force will be less than the gravitational force, and the pen will accelerate downward in response to the net force of weight minus frictional force exerted on the pen. For example, if a normal force of 0.2 N were exerted on the pen, the resulting frictional force between the pen and our fingers would be

$$\text{force of friction} = \text{normal force} \times \text{coefficient of friction}$$
$$= 0.2 \text{ N} \times 0.4$$
$$= 0.08 \text{ N} .$$

Since the force of gravity on the pen is 0.196 N, the net downward force would be $0.196 \text{ N} - 0.080 \text{ N} = 0.116 \text{ N}$, which in turn would result in the pen accelerating downward by an amount equal to force/mass or 0.116 N/0.2 kg or 5.8 m/sec^2. Thus, as a result of the frictional forces, the pen in this case would experience about half of the 9.8 m/sec^2 acceleration that would result if the pen were simply to be dropped.

Because such "motion-inhibiting" frictional forces are always present to some degree whenever two surfaces are in contact, they can become very important in the analysis of moving objects. One familiar example of such a process is the calculation of the initial velocity of an automobile from the length of the skid marks the auto makes when coming to a sudden stop. In such cases the investigator usually assumes that the automobile was moving at some initial velocity that was reduced to zero through the application of a constant frictional force. The length of the skid marks tells the investigator how much the car's position changed while the frictional force was being applied and the magnitude of the frictional force can be obtained from the mass of the car, brand of tires, road conditions, etc. From these data, and using the equations of motion for an object subjected to a constant force, an investigator can deduce the initial velocity at which the car had to be moving in order to leave the observed skid marks while coming to a complete stop.

WORKED EXAMPLE

The force of friction between an eraser and a chalk tray is 1.5 N. If the acceleration of the earth's surface gravity is 9.8 meters/second^2, and the coefficient of friction between the eraser and the tray is 0.8, what is the mass of the eraser?

SOLUTION: The force of friction is given in general by

$$\text{force of friction} = \frac{\text{coefficient}}{\text{of friction}} \times \frac{\text{normal}}{\text{force}} .$$

The coefficient of friction in this case is 0.8 and the force of friction is 1.5 N. Thus, we have

Continued

WORKED EXAMPLE

Continued

$$1.5 \text{ N} = 0.8 \times \frac{\text{normal}}{\text{force}}$$

$$\frac{\text{normal}}{\text{force}} = 1.5 \text{ N}/0.8$$

$$\frac{\text{normal}}{\text{force}} = 1.88 \text{ N} .$$

The normal force is just the force exerted by the earth's gravity on the eraser. From Newton's second law of motion, we have

$$\text{force} = \text{mass} \times \text{acceleration} .$$

The gravitational force on the eraser is 1.88 N and the acceleration of the earth's gravity is 9.8 meters/second². We may, thus, write

$$1.88 \text{ N} = \frac{\text{mass of}}{\text{eraser}} \times 9.8 \text{ meters/second}^2$$

$$\frac{\text{mass of}}{\text{eraser}} = 1.88 \text{ N}/9.8 \text{ meters/second}^2$$

$$\frac{\text{mass of}}{\text{eraser}} = 0.19 \text{ kg} .$$

The mass of this eraser is just under 0.2 kg.

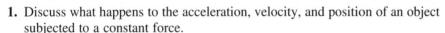

1. Discuss what happens to the acceleration, velocity, and position of an object subjected to a constant force.
2. Give some everyday examples of objects moving under a constant force.
3. Describe the properties of frictional forces.
4. What is a coefficient of friction?
5. Cite several instances where frictional forces can be helpful and several instances where frictional forces are not desirable.

1. An automobile accelerates at a rate of 4 m/sec² from a standing start. Find the following after 6 seconds:
   a. the acceleration of the automobile
   b. the velocity of the automobile
   c. the total distance the automobile travels.
2. A horizontal force of 5 N is required to pull a 1-kg box across a table top. What is the coefficient of friction between the box and the table top?

3.4 Projectile Motion

The simplest example of falling body motion is that for which the motion of the object occurs along a straight line perpendicular to the earth's surface. Suppose, however, that in addition to the vertical motion, the velocity of a falling object were to have a ''sideways'' or horizontal component as well. In such a case, the vertical force of gravity has no effect on the horizontal velocity. If we assume no other horizontal forces are present, the law of inertia tells us that in the absence of such forces, the object's horizontal velocity will not change. The effect of a horizontal velocity then is to cause a change in the horizontal x axis position of the object by an amount (see Physics Formulation 3.1) of

$$x = \frac{\text{horizontal}}{\text{distance}} = v_{ox} t + x_o ,$$

where v_{ox} is the initial horizontal velocity in the x axis direction and x_o is the initial displacement in the x axis direction. The vertical or y axis position meanwhile is given by

$$y = \frac{\text{vertical}}{\text{direction}} = \frac{1}{2} g t^2 + v_{oy} t + y_o ,$$

where v_{oy} is the initial vertical velocity in the y axis direction and y_o is the initial displacement in the y axis direction. The quantity g is the acceleration of gravity at the earth's surface. Using these two relations, and defining the positive y direction as up and the positive x direction as to the right, as we do for a mathematical graph, one can compute for any projectile in free flight a set of x and y values for the object and then plot a two-dimensional path or trajectory for that object's motion.

For example, suppose that an object is released from an airplane flying with a horizontal velocity of 80 m/sec at an altitude of 400 meters. For this case v_{ox} is equal to 80 m/sec, y_o is equal to 400 meters, and x_o and v_{oy} are both equal to zero. Under these conditions, the x and y positions of the object after an elapsed time of 1 second are

$$x = 80 \ (\text{m/sec}) \times 1 \ \text{sec} + 0 = 80 \ \text{m}$$

FIGURE 3.11 · · · · · · · · · ·

The Graph of a Trajectory   In this plot of x versus y, the position of an object released from an airplane flying with a horizontal velocity of 80 m/sec at an altitude of 400 m is shown at one-second intervals. The graph shows that the object strikes the ground at $y = 0$, after about 9 sec of elapsed time of 722 m from the release point.

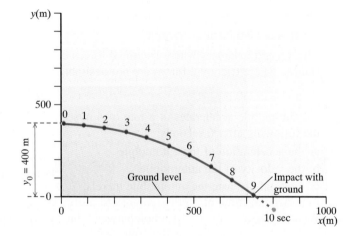

or

$$x = 80 \text{ m}$$

$$y = \frac{1}{2}(-9.8 \text{ m/sec}^2) \times (1 \text{ sec})^2 + 0 + 400 \text{ m}$$

$$y = -4.9 \text{ m} + 400 \text{ m}$$

or

$$y = 395 \text{ m} .$$

Notice that since the direction of the acceleration of the earth's gravity is in a downward or negative y direction, the value of g is equal to -9.8 m/sec^2. For additional elapsed times we obtain the following:

Elapsed Time (seconds)	x (meters)	y (meters)
0	0	400
1	80	395
2	160	380
3	240	356
4	320	322
5	400	277
6	480	224
7	560	160
8	640	86
9	720	3
10	800	-90

If we plot these x values versus their corresponding y values, we obtain the graph shown in Figure 3.11. Such a plot represents the path the object follows from the initial time ($t = 0$) at which it was released from the airplane. From these data we can obtain a number of useful pieces of information about the motion of the object. For example, we find that the object strikes the ground ($y = 0$) just a little over 9 seconds after release at a distance of 722 m in a positive x direction from the release point. In addition to the obvious military applications, one practical use of this sort of analysis is for fighting forest fires, where planes are called on to drop chemicals accurately onto very specific "hot spots" of the fire.

An interesting result from the analysis of this type of trajectory is that the time it takes an object to fall to the ground from a given height is exactly the same time it would take to fall if it possessed no horizontal velocity. Thus if a person were standing on top of a 400-meter cliff and dropped an object off that cliff, the object falling straight down would take the same 9 seconds to fall the 400 meters as the object released from the airplane. The only effect of the horizontal velocity, then, is to displace the location of the impact point in a horizontal direction. If we progressively increase the horizontal velocity, then the horizontal displacement of the object's impact point is also increased but the "fall" time remains the same. If a sufficiently large horizontal velocity is imparted to an object, then the rate at which the object falls back toward the earth's surface is not fast enough to compensate for the rate at which the earth's curvature causes the earth's surface to drop away from the object. Under these circumstances the

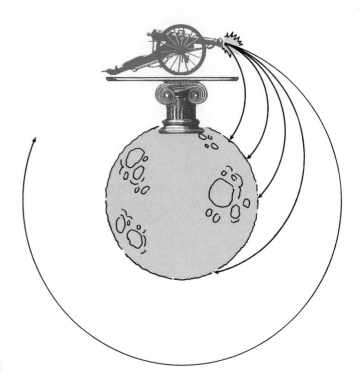

FIGURE 3.12 ···········

''Newton's Cannon'' As the horizontal velocity of a bullet or shell is increased, the horizontal distance the object covers before hitting the planet's surface also increases. For a planet like the earth, if the horizontal velocity exceeds about 8 km/sec, the object will not be able to fall fast enough to reach the planet's curving surface and the object will literally fall into orbit about the planet. As early as 1687 Sir Isaac Newton proposed this possibility in the *Principia* using a diagram similar to this one.

object in effect falls toward an eternally receding surface that it is never able to strike. Such a perpetually falling object is then said to be in orbit about the earth (Figure 3.12). At the earth's surface, the horizontal velocity required to place an object in such an orbital trajectory is approximately 8000 m/sec or a velocity that would get you from New York to Los Angeles in a little over 8 minutes!

WORKED EXAMPLE

A softball pitcher delivers a ball toward home plate at a height of 0.8 m and with a horizontal velocity of 40 meters/second. If home plate is 15 meters distant, at what height does the ball cross home plate?

SOLUTION: Since no forces are acting on the ball along the horizontal direction, the horizontal or x axis position of the ball is given by

$$x = v_{ox}t + x_o .$$

For this case $v_{ox} = 40$ m/sec and $x_o = 0$. Thus

$$x = 40 \text{ (m/s) } t + 0$$
$$= 40 \text{ (m/s) } t .$$

In the case of the vertical or y axis position, an acceleration a_y is present. We therefore write that

$$y = \frac{1}{2} a_y t^2 + v_{ox} t + y_o .$$

Continued

WORKED EXAMPLE

Continued

For this problem we have that $a_y = -g$, $v_{oy} = 0$, and $y_o = 0.8$ m. Thus:

$$y = \frac{1}{2}(-9.8 \text{ m/s}^2)\, t^2 + 0 + 0.8$$

$$= -4.9\ (\text{m/s}^2)\, t^2 + 0.8\ .$$

Using the expressions for *x* and *y*, we calculate the *x* and *y* values for a given set of *t* values. Using the values of $t = 0, 0.1, 0.2, 0.3, 0.4,$ and 0.5 seconds, we can construct the following table:

t	$x = 40$ (m/sec) t	$y = -4.9\ (\text{m/sec}^2)\, t^2 + 0.8$
0 seconds	0 meters	+0.80 meters
0.1	4	+0.75
0.2	8	+0.61
0.3	12	+0.36
0.4	16	+0.02
0.5	20	−0.42

If we plot *y* versus *x* values, we obtain the following graph.

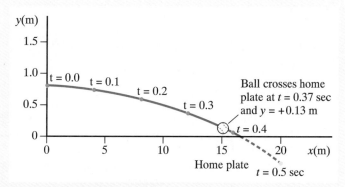

From the graph we can see that the ball crosses the plate at about 0.13 meters above the ground after being in flight for about 0.37 sec.

1. Describe some other applications of the analyses of projectile motion.
2. Sketch a plot of *x* versus *t* for an object that is:
    a. moving at constant velocity in a positive *x* direction, not moving in the *y* direction
    b. moving at constant acceleration in a positive *x* direction, not moving in the *y* direction

REVIEW
3.4
QUESTIONS

c. moving at constant acceleration in a positive *x* direction and at a constant velocity in the positive *y* direction.

3. How might you extend the analyses of projectile motion to three dimensions?

REVIEW 3.4 PROBLEMS

1. An aircraft moving in a horizontal positive *x* direction at a velocity of 100 m/sec releases a sandbag from an altitude of 1500 meters. Calculate the *x* and *y* positions of the sandbag at 2-second intervals from 0 to 20 seconds. Plot your results on a piece of graph paper. From that plot, estimate the time it takes for the sandbag to hit the ground, as well as the location of the impact point. Assume that the release point occurs at *x* = 0, *y* = 1500 m.

2. A bottle rocket is shot straight into the air at an initial velocity of 30 m/sec. Find the *x* and *y* positions of the rocket at 1-second intervals from 0 to 8 seconds. Plot your results on a sheet of graph paper. From your graph, find (a) the maximum height to which the rocket soars, (b) the total amount of time the rocket stays in the air, and (c) the coordinates of the rocket's landing point. Assume that the rocket is launched at *x* = 0, *y* = 0, and *t* = 0.

3.5 Impulse and Momentum

Earlier in this chapter we defined the mass of an object as the ability of that object to resist a force. The four wrestlers mentioned in our discussion of mass could thus lift a compact car far more readily than they could lift a truck, because the truck possesses more mass. In like fashion, if the wrestlers wished to push

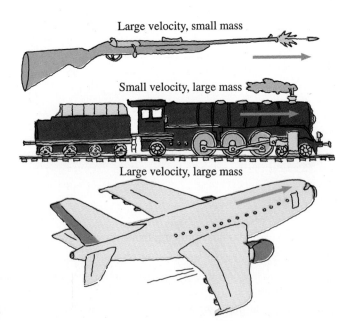

Large velocity, small mass

Small velocity, large mass

Large velocity, large mass

FIGURE 3.13 ··········

Quantities of Motion Large values of linear momentum can arise if the mass and/or the velocity of a given object is large.

both vehicles in a backward direction, they would find that their task would be much easier for the low-mass compact car (assuming, of course, that both the car and the truck were out of gear). But suppose that those same four wrestlers, exerting the same total force, tried to push the car backward when the car itself was rolling forward—this would be a more difficult task. The task would become even harder as the car's forward speed increased. We may deduce from all of this that the ability of an object in motion to resist a force depends not only on its mass but also on its state of motion as well. Newton recognized this effect in his studies of moving objects and in fact makes use of the term ''quantity of motion,'' which he defines as the product of an object's mass and its velocity. This quantity is now called linear or translational momentum and is defined as follows:

$$\text{linear momentum } = \text{ mass} \times \text{velocity} \qquad \textbf{3.4}$$

or

$$p = mv .$$

From this definition we can see that linear momentum, like velocity, is a vector quantity whose description must include the specified direction. Also, a large value for the magnitude of an object's linear momentum can arise from large mass, large velocity, or both. A very large-mass, slow-moving object such as a freight train has a large momentum, but so does a small mass, very rapidly moving object such as a low-mass bullet fired from a high-powered rifle (Figure 3.13). In football, a high-velocity, low-mass linebacker can hit just as hard as a low-velocity, high-mass linebacker. The trick, of course, is to have a team consisting of high-velocity, high-mass linebackers! From the definition of linear momentum, we can also see that this quantity necessarily resides in any moving object.

The use of the concept of momentum can provide us with a great deal of additional insight concerning the behavior of moving objects. Newton's own version of his second law of motion, for example, is framed in terms of momentum as follows:

$$\text{force } = \frac{\text{change in momentum}}{\text{elapsed time}} \qquad \textbf{3.5}$$

or

$$F = \frac{\Delta p}{t} .$$

While the force = mass × acceleration version of Newton's second law stated earlier can only be used to analyze the motion of objects having a constant mass, this more encompassing momentum-based statement permits the analysis of the motion of an object whose mass is changing in time, such as a rocket expending fuel, a dump truck losing sand, etc.

An examination of momentum in light of Newton's third law of motion or the action = reaction principle reveals the most important and fundamental characteristic of all momentum; that is, momentum is conserved. This means that in an interaction between two or more objects, the total amount of momentum associated with the objects in their given state of motion remains unchanged. As an example, suppose we fire a rifle as shown in Figure 3.14. When the rifle is fired, two forces come into play: the ''action'' force exerted by the rifle on the

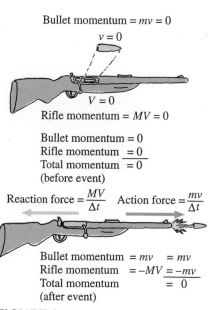

Bullet momentum $= mv = 0$

$v = 0$

$V = 0$

Rifle momentum $= MV = 0$

Bullet momentum $= 0$
Rifle momentum $\;= 0$
Total momentum $\;= 0$
(before event)

Reaction force $= \dfrac{MV}{\Delta t}$ Action force $= \dfrac{mv}{\Delta t}$

Bullet momentum $= mv \;\;= mv$
Rifle momentum $\;= -MV = -mv$
Total momentum $\qquad\;\; = 0$
(after event)

FIGURE 3.14 ··········
Conservation of Linear Momentum Because the sum total of the linear momentum of the rifle and the bullet remains the same before and after the bullet is fired, linear momentum is said to be conserved.

bullet in the "forward" or positive direction, and the "reaction" force exerted by the bullet on the rifle, which produces the "kick" of the rifle in the "backward" or negative direction. We know that these forces are equal and opposite. We can also state that the interval of time over which the bullet is propelled forward is the same as the time interval for the backward "kick" of the gun. Since both the forces and the time intervals over which they act are the same, the change in the magnitude of the momentum experienced by both objects is the same. In one case, however, the momentum change, that of the bullet, is in a positive direction and has a positive value, while the other, that of the rifle, is in a negative direction and has a negative value. If we add the momentum of the rifle and the bullet after the bullet is fired, we find that the total value of the momentum of the rifle-bullet system is equal to zero. But since neither the bullet nor the rifle was moving before the rifle was fired, the velocities and hence the associated momenta of both objects were zero before the gun was discharged. The total momentum of the rifle-bullet system before the rifle was fired was therefore equal to zero, precisely the same as the value for the total momentum of the rifle-bullet system after the rifle was fired. Thus, the total momentum of the system before and after the bullet was fired has the same value—in this case zero—and we say that the total momentum of the system is conserved. As a general principle, then, we can state that for any interaction in which there is no net external force acting,

$$\text{total linear momentum before an event} = \text{total linear momentum after the event} .$$

This principle is called the principle of the conservation of momentum and finds particular usage in the motion analysis of phenomena involving particle interactions such as collisions.

To illustrate how such an analysis would proceed, let us consider Figure 3.15 in which a 1-kg sphere moving at 2 meters/sec to the right collides with a second 1-kg sphere that is at rest. After the collision, the first sphere is now at rest and

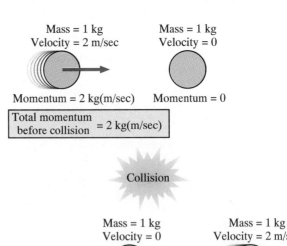

FIGURE 3.15 ···········

Conservation of Linear Momentum in an Elastic Collision—I The sum total of the momentum of the two equal-mass balls before the collision is the same as that after the collision.

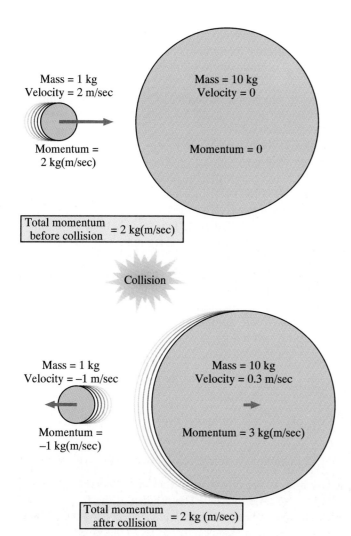

Total momentum
before collision = 2 kg(m/sec)

Collision

Total momentum
after collision = 2 kg (m/sec)

FIGURE 3.16 ....

**Conservation of Linear Momentum in an Elastic
Collision—II** Even though the masses of the two balls are
decidedly different, linear momentum is still conserved in this
collision.

the second sphere is set in motion. What may be said about the resulting motion
of the second sphere? The principle of conservation of momentum tells us that
the total momentum of both spheres is the same before and after the collision
''event.'' Before the collision, the total momentum of the system is that of the
first sphere, which is 1 kg × 2 m/sec or +2 kg-m/sec, plus that of the second
sphere, which is zero because its velocity is zero. After the collision, the momentum of the first sphere is now zero because its velocity is zero. The total
momentum, however, cannot change. Thus, the second sphere must now possess
the total momentum of the system, which is +2 kg-m/sec in a positive direction.
Since the masses of the two spheres are the same, the velocity of the second
sphere after the collision must now be +2 m/sec to the right.

Suppose that the 1-kg sphere now strikes a 10-kg medicine ball from the left
at a velocity of 1 m/sec. Upon colliding with the medicine ball, the sphere recoils
to the left at a velocity of 1 m/sec (Figure 3.16). In this case, the total momentum
before the collision is once more 2 kg-m/sec, and this must again be the total
after the collision. But now the momentum of the sphere after the collision is 1
kg-m/sec in a negative direction or −1 kg-m/sec. The total momentum of the

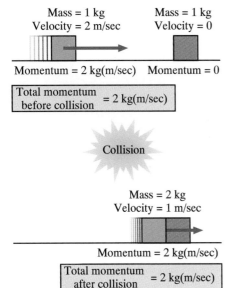

Mass = 1 kg
Velocity = 2 m/sec

Mass = 1 kg
Velocity = 0

Momentum = 2 kg(m/sec)   Momentum = 0

Total momentum before collision	= 2 kg(m/sec)

Collision

Mass = 2 kg
Velocity = 1 m/sec

Momentum = 2 kg(m/sec)

Total momentum after collision	= 2 kg(m/sec)

FIGURE 3.17 ··········
Conservation of Linear Momentum in an Inelastic Collision Two objects that collide and stick together as the result of an inelastic collision still conserve linear momentum.

medicine ball plus the sphere after the collision is $+2$ kg-m/sec, therefore, the momentum of the medicine ball after the collision must now be $+3$ kg-m/sec and its corresponding velocity is its momentum $\div$ mass or 0.3 m/sec in a positive direction.

The preceding examples in which objects bounce off each other are examples of what physicists refer to as *elastic* collisions. A second class of collisions, in which the particles "stick" to each other, are termed *inelastic* collisions. Such collisions can also be analyzed by means of momentum conservation. Let us consider a 1-kg cube moving with no friction to the right at a velocity of 2 m/sec as shown in Figure 3.17. Let us assume that the object now collides and sticks to a second 1-kg cube and the combined pair moves to the right. Once more the initial total momentum is $+2$ kg-m/sec. After the collision, however, there is only a single body with a mass of 2 kg. This body must have a total momentum of $+2$ kg-m/sec and thus the final velocity of the 2-kg mass is 1 m/sec to the right.

Although we have confined our discussion to events along a straight line path, it is possible to analyze more complex particle interactions such as two-dimensional collisions and explosions, by means of vector analysis and making use of the fact that even in these more complicated examples, the Principle of the Conservation of Momentum still holds.

If we multiply both sides of the equation,

$$\text{force} = \frac{\text{change in momentum}}{\text{elapsed time}},$$

by the elapsed time, we obtain the following equation for the change of momentum:

$$\text{change of momentum} = \text{force} \times \text{elapsed time}.$$

Any change in the momentum of an object must therefore be generated by a force acting over an appropriate time interval. The product of the force and the elapsed time over which the force is exerted thus have a special significance in the description of motion phenomena that is referred to as the *impulse*. Given this definition of impulse, the above equation is often stated simply as

$$\text{impulse} = \text{change of momentum}.$$

From these relationships we can now see why, for example, athletes are taught to "follow through" on their tennis, baseball, or golf swings. By following through, the average force exerted on a ball is in effect held in contact with the ball for a longer period of time, thus imparting a larger change of momentum or "carry" to the ball. Similarly, a good follow-through in throwing a frisbee, delivering a bowling ball, or kicking a football reaps far better results than those obtained from a poor follow-through. On the other hand, a karate expert can break a stack of bricks or wood by delivering a considerable change of momentum over a short interval of time. Because the actual time of contact is very small, the force associated with the blow is literally of brick-shattering proportions.

The preceding are all cases in which we seek to impart a change of momentum to an object at rest. In some instances, however, such as our four car-pushing wrestlers, we seek to change the momentum of an object in motion. If the wrestlers are to bring the rolling car to a stop, it is necessary for them to exert their force

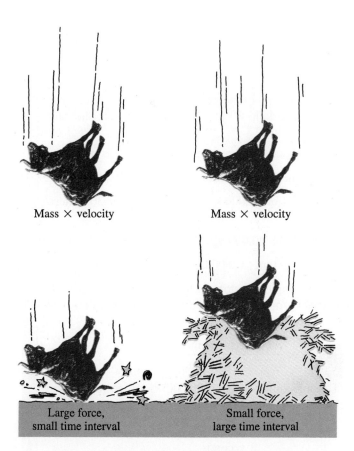

Mass × velocity

Mass × velocity

Large force,
small time interval

Small force,
large time interval

FIGURE 3.18 ··········

Examples of Impulse A change of momentum can be imparted to an object by delivering a large force over a small interval of time or a small force over a long interval of time.

over some specified interval of time. If they wish to stop the car quickly, i.e., over a shorter time interval, they will have to exert a much larger force than if they stop the car more slowly over a longer time interval. The effect of an impact force can be minimized by making use of the concept of impulse. A boxer will "roll" with a punch and a baseball hitter will attempt to fall back and away from a pitcher's "beanball," not only to get out of the way of the punch or pitch but also to *increase* the time of contact, thereby *reducing* the magnitude of the force delivered should the individual get hit. In a similar mode, stunt people falling to the ground will roll their bodies or bend their knees just as they hit the ground so as to increase their contact time. A soft surface that "gives" when a person falls onto it is likely to produce fewer injuries than one that does not give because, once again, the contact time is increased at the expense of the magnitude of the contact force. It is thus far better to fall onto a mattress than onto bare ground, and fewer injuries occur when football is played on a natural turf field that has some "give" to it as opposed to a more rigid field of artificial turf (Figure 3.18).

Up to this point our discussion of motion has dealt with translational motion in which an object moves from one place to another and exhibits an obvious displacement of position. There are, however, instances of motion in which a given object is exhibiting movement, but that movement does not carry the object from one place to another. One example of such motion will be the focus of the next chapter.

WORKED EXAMPLE

Superheroine Physicia has to stop a 20,000-kg runaway train engine moving at 30 meters/second in a time of 3 seconds or less in order to save a school bus full of children. What is the minimum force she must exert to stop the engine in time?

SOLUTION: The linear momentum of a moving object is given by

$$\text{linear momentum} = \text{mass} \times \text{velocity} .$$

For an engine with a mass of 20,000 kg and a velocity of 30 meters/second, we have

$$\text{linear momentum} = 20,000 \text{ kg} \times 30 \text{ meters/second}$$

$$\text{linear momentum} = 600,000 \text{ kg-m/sec} .$$

If the engine is to be completely stopped, the change of momentum would also be 600,000 kg-m/sec. The force required to exact such an impulse is given in general by

$$\text{force} = \frac{\text{change in momentum}}{\text{elapsed time}} .$$

A total change of momentum equal to 600,000 kg-m/sec must occur over an elapsed time of 3 seconds. This, in turn, requires a force given by

$$\text{force} = \frac{600,000 \text{ kg-m/sec}}{3 \text{ seconds}}$$

$$= 200,000 \text{ kg-m/second}^2 .$$

Physicia must exert a force of at least 200,000 N in order to stop the runaway engine in a short enough time to save the school children.

REVIEW 3.5 QUESTIONS

1. What is linear momentum? Describe some of its properties.
2. Give an example of an object that has (a) large momentum and small mass and (b) large momentum and small velocity.
3. What is impulse? How is it related to force and momentum?
4. Describe a practical application of the analysis of colliding bodies.
5. Give an example of an impulse in which (a) the force is large and the elapsed time is small and (b) the force is small and the elapsed time is large.

REVIEW 3.5 PROBLEMS

1. A 2000-kg automobile is moving at a rate of 15 m/sec. What is the momentum of the car? How much force should we exert if the car is to be stopped in 5 seconds?

2. A 0.1-kg ball moving to the left at a velocity of 0.4 m/sec strikes a 0.2-kg ball at rest. If the 0.1-kg ball is at rest after the collision, what is the magnitude and direction of the velocity of the 0.2-kg ball after the collision?

Chapter Review Problems

1. A truck driver drives 1000 km in 15 hours. Find the driver's average velocity over this time interval.

2. If a sprinter can run 100 meters in 10 seconds, how long will it take the sprinter to run one kilometer, moving at this same average velocity?

3. A horse can travel at a rate of 5 km/hr. How far will the horse go in 3 hours, moving at this average velocity?

4. A dragster can go from 0 to 20 m/sec in 5 seconds. What is the average acceleration for this vehicle?

5. If a sprinter can attain an acceleration of 1 m/sec^2, how long will it take the sprinter to attain a velocity of 5 m/sec?

6. If an object weighs 490 N on the earth, what is the mass of the object?

7. A force of 10 N is exerted on a mass of 5 kg. What is the magnitude of the resulting acceleration?

8. An object having a mass of 20 kg is subjected to a force of 4 N. At what rate will the object change its velocity?

9. How much force would be required to accelerate a 2-kg mass 3 m/sec^2?

10. An object initially at rest is accelerated at a rate of 4 m/sec^2. What will the speed of the object be after 3 seconds?

11. How far will an object that is initially at rest be displaced when it undergoes an acceleration of 1 m/sec^2 for 20 seconds?

12. An object initially at rest is subject to a constant acceleration. If the object moves a total of 50 meters in 5 seconds, what is the object's acceleration?

13. Assuming the earth's gravitational acceleration to be 9.8 m/sec^2, how far will an object dropped from a cliff fall in 5 seconds?

14. An object having a mass of 50 kg is subjected to a force of 10 N from the left and 15 N from the right. In which direction does the object accelerate? At what rate?

15. An automobile has a mass of 1500 kg. What is the force of friction between the car and the road if the coefficient of friction between the road and the car's tires is 0.7? Assume that the acceleration of the earth's surface gravity is 9.8 m/sec^2.

16. A 1-kg book rests on top of a table for which the book-table coefficient of friction is 0.3. If the book is subjected to a horizontal force of 10 N, describe the acceleration, if any, the book will exhibit.

17. A freight train engine moving at a speed of 1 m/sec has a mass of 20,000 kg. What is the momentum of the engine? At what speed must a 0.1-kg bullet be fired in order to have the same momentum as the engine?

18. A 5-kg ball is moving at 5 m/sec and is suddenly stopped ''cold.'' What is the change in the momentum of the ball? What impulse did the ball experience? If the ball was stopped in 0.1 sec, what force was exerted to stop the ball?

19. Find the impulse associated with a circus performer with a mass of 75 kg who falls into a net at a velocity of 20 m/sec. If the contact time between the performer and net is 2 seconds, what force did the net exert on the performer? If that same performer hits the ground over a contact time of 0.2 seconds, what average force will be exerted on the performer during the 0.2-second contact time?

20. A 0.1-kg ball moving to the left at 1 m/sec collides with a 0.5-kg ball moving to the right at 0.5 m/sec. After the collision, the 0.5-kg ball is at rest. If the collision is elastic, find the magnitude and direction of the velocity of the 0.1 kg ball after the collision.

21. An object having a mass of 0.1 kg slides to the left at 1 m/sec and collides with a 0.4-kg object sliding to the right at 0.5 m/sec. Upon collision the two objects stick together. At what velocity and in what direction will the stuck-together objects be moving after their inelastic collision?

22. A meteoroid having a mass of 2×10^6 kg is hurtling through space at a rate of 2 m/sec. In what direction and at what speed would you fire a 1000-kg projectile in order to stop the meteoroid completely? Assume that the objects stick together upon collision and assume that no gravitational forces are present.

23. A rifle is fired from the left at a 5-kg stationary block of putty. After the bullet strikes the block, the block and bullet are measured to move to the right at a rate of 0.2 m/sec. If the bullet fired by the rifle has a mass of 0.01 kg, find the muzzle velocity of the rifle.

24. At what speed must a 120-kg linebacker hit an 80-kg halfback moving at 10 m/sec in order to reduce the halfback's speed to zero. Assume that the players ''stick'' to each other after the collision.

1. Is it possible for an object to be subjected to forces and not accelerate? If so, describe the circumstances.

2. If an object is moving at a constant velocity, what, if anything, can be said about the forces acting on the object?

3. In the following graph the velocity of an object is plotted as a function of time. Over what time interval(s), if any, does the object accelerate?

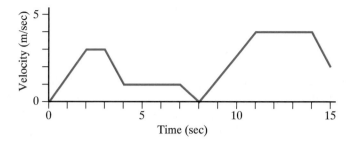

4. Why do two objects having different masses strike the ground at the same time when dropped from the same height?

5. It has been stated that the law of inertia is, in essence, a corollary, or a logical consequence, of Newton's second law of motion. Discuss this idea.

6. Is it possible for a person to "pull oneself up by the bootstraps"? Explain.

7. Suppose your instructor tells you that rockets fly into space by pushing off the earth's surface. How would you respond to such a claim?

8. Discuss Zeno's claim that all motion is an "illusion."

9. Can you think of any example of an object that does *not* obey Newton's laws of motion?

10. Describe some examples of constant forces other than gravity and friction.

11. Describe how you might measure the depth of a deep well or shaft by using a falling object such as a stone.

12. Discuss some practical applications of projectile motion analysis.

13. Discuss some instances in which a high-impulse interaction would be desirable and some in which it would not.

14. Why doesn't an object thrown into the air get left behind by the earth as it rotates on its axis or moves about the sun?

15. Discuss Newton's view that momentum is a "quantity of motion." Might one define such a "quantity" differently? Explain.

16. Describe how you would experimentally determine the value of the coefficient of friction between two surfaces such as rubber and ice.

17. Suppose you discover a substance with an extremely low coefficient of friction. Describe several possible uses for such a substance.

18. Explain in detail why highwire performers at the circus don't get hurt when they fall into a safety net.

19. Describe how you would experimentally measure the acceleration of the earth's surface gravity.

20. Using the series of photographs in Figure 3.1, how would you determine whether or not the star *Proxima Centauri* is accelerating through space?

21. Is is possible for a collision to be "semielastic," i.e., the objects stick together for a time and then separate? If so, would you expect linear momentum to be conserved for such an interaction?

22. An object is tossed into the air three different times and exhibits the trajectories shown below. What, if anything, can be said about the directions of the initial velocities v_{0x} and v_{0y} for each case?

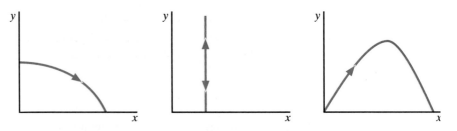

23. Express the units of momentum in terms of meters, kilograms, and seconds.

24. A terrorist constructs a gun made of a low-density composite alloy such that the gun now has a smaller mass than the bullet it fires. Would you want to fire such a weapon? Explain.

25. Watch a cartoon such as Tom and Jerry, the Roadrunner, etc., and describe all of the violations of Newton's laws of motion that occur during the course of the cartoon.

The Equations of Motion for an Object Moving Under a Constant Force

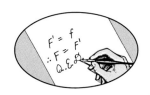

Let us consider an object subjected to a unidirectional or "one-dimensional" force F, which is constant everywhere. From Newton's second law we may immediately write that

$$F = ma,$$

$$a = \frac{F}{m},$$

where m is the mass of the object and a is the acceleration experienced by the object. Since both F and m have constant values, we may conclude that for this system, acceleration a also has a constant value. By definition, acceleration is the change of velocity divided by elapsed time. If the velocity v is attained by the object at some time t and the initial velocity v_0 is the velocity at some starting or initial time t_0 we have

$$a = \frac{v - v_0}{t - t_0}.$$

This equation can be rewritten as follows:

$$a(t - t_0) = v - v_0.$$

The final velocity v is then given by

$$v = a(t - t_0) + v_0.$$

The distance this object moves over a given value of elapsed time $t - t_0$ is simply the average velocity over the elapsed time interval $\times$ the value of that time interval. The average velocity v_{avg} is equal to one-half of the sum of the final velocity and the initial velocity or

$$v_{avg} = \frac{v + v_0}{2}$$

if and only if the acceleration has been constant. The distance D the object moves in the elapsed time $t - t_0$ is then

$$D = \left(\frac{v + v_0}{2}\right)(t - t_0).$$

If we substitute the expression derived earlier for v into this equation, we have

$$D = \left[\frac{a(t - t_0) + v + v_0}{2}\right](t - t_0),$$

which can be rewritten as

$$D = \frac{1}{2}a(t - t_0)^2 + v_0(t - t_0).$$

If we think of the distance the object moves as a "displacement" of the object's position, then for motion along a single direction.

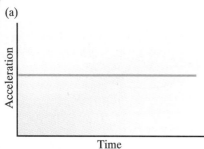

(a)

Acceleration

Time

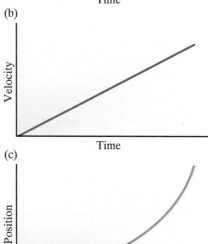

(b)

Velocity

Time

(c)

Position

Time

FIGURE 3.19 ··········
Graphs of (a) acceleration versus time,
(b) velocity versus time, and
(c) displacement versus time for an object
moving under the influence of a constant,
unidirectional force.

$$D = d - d_o,$$

where d is the final position of the object at the end of the elapsed time interval and d_o is its position at the beginning of the elapsed time interval. The final position of an object subjected to a one-dimensional force having a constant value if thus given by

$$d = D + d_o$$

or

$$d = \frac{1}{2} a (t - t_o)^2 + v_o (t - t_o) + d_o.$$

We now have mathematical expressions that enable us to describe the position and velocity at any time of any object moving under the influence of a constant, single-dimensional force (Figure 3.19). All we need to know are the mass of the object, the magnitude and direction of the force exerted on the object, the object's initial position and velocity, and the initial time from which we wish to describe the object's motion. Most of the time these latter three "initial" quantities can be set equal to zero with a suitable choice of frame of reference.

QUESTIONS
··········

1. Which, if any, of Newton's laws of motion have been employed in this discussion?
2. Describe an example of an object moving under a constant force in which the following are *not* equal to zero: (a) d_o, (b) v_o, (c) both d_o and v_o.
3. Are there any applications other than the analysis of falling objects for the equations derived in this discussion? Explain.

PROBLEMS
··········

1. In the long-running TV Christmas classic, "How the Grinch Stole Christmas," Mt. Crumpit is purported to have an altitude of 10,000 feet (about 3050 meters), yet an ornament falls from the top of Mt. Crumpit and is heard to hit the ground 4 seconds later. Is this result consistent with the stated altitude of Mt. Crumpit? If not, how might you explain the discrepancy?
2. How long will it take a rock to fall from the top of Mt. Everest (elevation = 8840 meters) to sea level? Assume that g is constant throughout the fall and there is no air resistance.
3. An explorer wishes to descend into the mouth of a cave on a rope 100 meters long. The explorer drops a pebble into the cave and hears a splash 6 seconds later. Would you descend into the cave with the 100-meter rope? Explain.

4 CIRCULAR MOTION

FIGURE 4.1
Circular Motion in Nature The circular winds within this tornado funnel can reach velocities as high as 600 km/hour.

𝒥n the preceding chapter we were concerned with the properties of an object's translational motion in which the object changed its position from one place to another against a stationary backdrop or reference frame. A skier going down a hillside, for example, exhibits translational motion in moving from the top of the hill to the bottom. Suppose, however, at the beginning of the skier's second run down the same hillside, the skier suddenly slips and tumbles head over heels all the way to the bottom of the hill. The skier's second run still exhibits translational motion from the top of the hill to the bottom but, as the now battered skier would grimly attest, the second run was characterized by an additional "tumbling" motion as well. A rapidly spinning toy top does not experience a translational displacement of its position on the floor or tabletop, but few of us could care to say that the top is devoid of motion. Also, the most salient aspect of a whirlwind is not its overall translational drift across the countryside, but rather the internal motions from which the phenomenon has derived its name (Figure 4.1).

Tumbling skiers, spinning tops, and whirlwinds are all examples of a type of motion that is distinct from translational motion. In its most general form, this different type of movement is referred to as *curvilinear motion* or as *curved motion*. The simplest motion of this type is *uniform circular motion* in which, as the term suggests, objects or parts of objects exhibit a constant rate of circular movement. When one object exhibits circular motion about a second, separate object or point, the motion is referred to as *revolution*. As an example, the earth

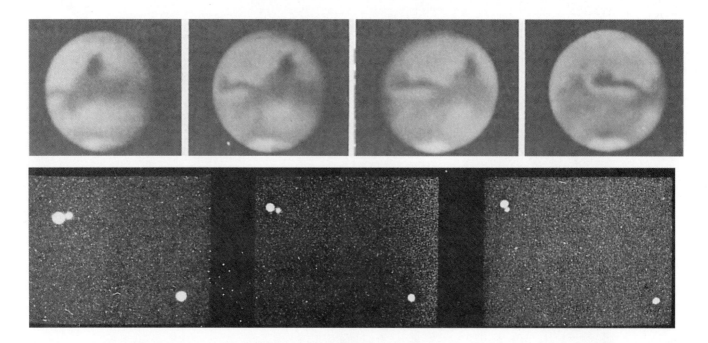

FIGURE 4.2
Rotation and Revolution The upper sequence of photographs of the surface markings of the planet Mars illustrates the rotational motion of Mars over a 10-hour time period. The lower sequence of photographs shows the orbital motion of the star Krüeger 60 B about the star Krüeger 60 A (upper left) over a 12-year interval of time.

and other planets all revolve about the sun in nearly circular orbital paths. When circular motion occurs within an object about an imaginary line thrust through that object, the motion is called *rotation* and the line about which the rotational motion occurs is the object's *axis of rotation* (Figure 4.2). A spinning top thus provides us with one of the more familiar examples of rotational motion. If we take the spinning top onto a merry-go-round, however, the top not only exhibits its spinning rotational motion, but now possesses revolutionary* motion about the center of the merry-go-round as well. The earth, too, exhibits both of these types of circular motion with a regularity sufficiently precise that we are able to define our day as the amount of time it takes the earth to make one complete rotation on its axis. Similarly, our year is defined as the amount of time the earth needs to make a complete orbital revolution about the sun.

4.1 Revolution and Centripetal Force

In the last chapter we found that the law of inertia stated in effect that an object will move in a straight line at a constant speed, "... unless it is compelled to change that state by forces impressed upon it." Thus, if an object changes either or both its speed or the direction of its motion, the law of inertia *requires* that the change be due to the action of a force.

If the object is moving in a curved path at a constant speed, it can be shown (Physics Formulation 4.2) that such movement is the result of a "center-seeking" or *centripetal* force acting on the object. For the simple case of an object moving in a circular path at a constant speed, the direction of this force is toward the center of the circular path and its magnitude is given by

$$\frac{\text{centripetal}}{\text{force}} = \frac{\text{mass}}{\text{of}} \times \frac{(\text{speed of object})^2}{\text{radius of circular path}} \qquad \textbf{4.1}$$

or

$$F = m\left(\frac{v^2}{r}\right).$$

This expression gives the magnitude of the centripetal force, which can be produced by a variety of agents. If an object is being whirled in a circular path at the end of a string, the object is forced into that path by the "center-seeking" tension force in the string. For this case, the centripetal force is the tension. As the planets move about the sun, they are forced into their curved paths by the "center-seeking" gravitation of the sun. The centripetal force for the planetary orbital motions in the solar system is the sun's gravitational force.

Thus, the descriptions of the revolution of one object about another can be handled quite nicely by the principles and techniques employed in dealing with translational motion. Matters for the rotational motion of a solid object, however, are not quite as simple, as we will see in the next section.

---

*The extension of the use of the term "revolution" to its wider, more familiar political context arose in fact out of the seventeenth century scientific, political, and religious controversies stirred by Copernicus's "revolutionary" ideas expressed in *De Revolutionbus Orbium Celestium*.

4.1 "Fictitious" Forces

Perhaps the most bizarre "spin-off" aspect of circular motion lies in its apparent ability to generate "forces" for which there appears to be no natural agent. Because these "forces" do not appear to arise from one of nature's instruments of force such as gravity or friction, such forces are referred to somewhat misleadingly as "fictitious" forces. We will see that these apparent forces can be explained in terms of inertial effects.

To illustrate how such a phenomenon can occur, let us consider a couple inside of an elevator-like compartment, which is at the end of a cable as shown in the figure. If we whirl the elevator in a circular path, the tension in the cable provides the centripetal force by which the elevator is "forced" into a circular path. As occupants of the elevator, the two people inside are also forced into the same circular path. The force that brings about the latter effect, however, does not arise from the tension in the cable, but rather from the floor of the elevator pressing against the couple. The floor of the elevator is in effect exerting a centripetal force on the occupants, causing them to share in the elevator's circular motion. The occupants in turn exert an equal and opposite force on the elevator floor as they are carried along within the elevator. This "away-from-the-center" or "center-fleeing" force is referred to as a *centrifugal force*. The "big picture" view from outside of the elevator tells us that the centrifugal force is simply the result of action equals reaction. The floor pushes up on the couple, and the couple pushes down on the floor with equal and opposite force. The individuals inside of the elevator, however, perceive themselves as being subjected to a "downward" force identical in its effects to a gravitational force. The centrifugal force experienced by the couple in the elevator is thus very real to them and, in the absence of a wider perception of their surroundings that would include the detection of the cable, there would be no obvious source or agent by which this force could be produced.

Extensive use is made of such centrifugal forces in the biological and medical sciences through the use of a device called a *centrifuge*. A centrifuge revolves a given sample in a circular path at very high angular velocity. The rapid rotation produces a large "centrifugal" force within the sample and results in a "layering effect" called *sedimentation,* which can be employed to separate particles of various sizes, fluids of differing densities, as well as other useful

functions. On a larger scale, centrifuges are also used to simulate the high-magnitude accelerations that astronauts and various pieces of equipment might encounter on space flights.

In 1835 the French engineer-mathematician Gustave-Gaspard Coriolis recognized the existence of a second type of fictitious force effect associated with rotating objects and which is referred to in his honor as the *Coriolis effect.* To

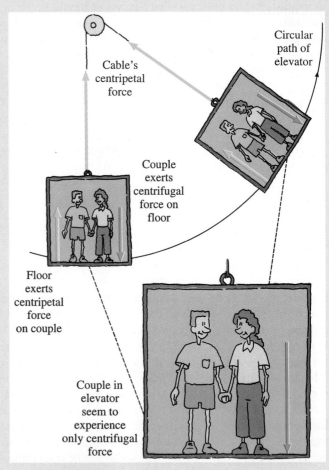

··· Centrifugal Force As the couple in the elevator is whirled around, the elevator exerts a centripetal force on the couple. Since action = reaction, the couple exerts an equal and opposite force on the elevator floor. For the couple in the elevator, the only obvious force is the latter force, which is referred to as a centrifugal force.

Continued

illustrate how the Coriolis effect works and Coriolis "forces" are produced, imagine you are on a city park merry-go-round and wish to play catch with a friend across from you on the merry-go-round. If the merry-go-round is not rotating, the ball can be thrown back and forth along a diameter of the merry-go-round without any unusual effects. Suppose, however, that the merry-go-round is rotating as you try to throw the ball to your friend. If you throw the ball directly at your friend, the merry-go-round will rotate beneath the ball as it moves across toward that person. For an observer looking down on the merry-go-round, the flight of the ball is along a straight line path. To your friend on the other side of the merry-go-round, however, the ball appears to curve away as the merry-go-round rotates, as shown in the figure on the right. The ball thus appears to the two of you on the rotating merry-go-round to change the direction of its velocity. In the Newtonian view of motion, this change of velocity can only arise from the effect of a force being exerted on the ball. As in the case of centrifugal forces, however, there is no natural agent to which we can attribute such a deflective force. Indeed, if we can somehow obtain the more comprehensive view, in this case by looking down on the merry-go-round from above, we find that the "force" seen in the rotating reference frame of the merry-go-round is absent or "fictitious" in the nonrotating view.

Because the earth rotates beneath air masses and ocean currents as both move across the earth's surface, the Coriolis effect is of considerable importance to meteorologists, atmospheric scientists, and oceanographers in their attempts to understand and describe atmospheric and oceanic motion phenomena. The famed trade winds and prevailing easterlies and westerlies that flow in diagonal directions at the earth's middle latitudes are in effect pole-to-equator air flows, which are seemingly deflected into their curved paths by the action of the Coriolis effect.

Thus fictitious forces are not really forces at all, but are the result of an object's inertia, which tends to keep the object moving in a straight line. If the observer is not moving in a straight line as well, then the object will appear to be changing direction. The reality of the situation is that there is a direction change, but it resides with the observer and not the object.

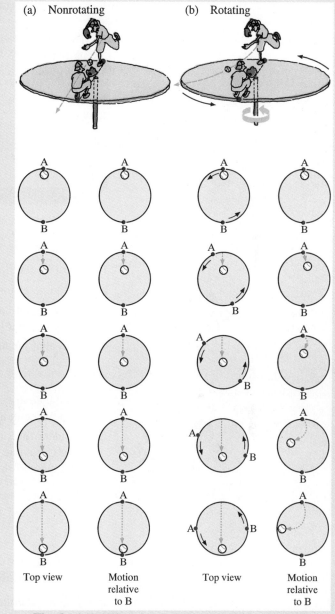

••• **The Coriolis Effect** (a) A ball thrown from position A across the diameter of a nonrotating merry-go-round will arrive at position B without any unusual effect. (b) If the merry-go-round is rotating, a ball thrown in exactly the same way will appear to deflect as the merry-go-round rotates beneath the ball.

> **WORKED EXAMPLE**
>
> A 0.5-kg toy bird is whirling at the end of a string 0.8 m long at a speed of 2 m/sec. Find the magnitude and direction of the centripetal force required to keep the toy bird moving in a circular path.
>
> **SOLUTION:**   The centripetal force is given in general by
>
> $$\text{centripetal force} = \text{mass} \times \frac{\text{speed}^2}{\text{radius of path}}.$$
>
> The toy's speed is 2 m/sec, the radius of its circular path is 0.8 m, and its mass is 0.5 kg. The centripetal force required to keep the toy bird moving in its circular path is thus
>
> $$\text{centripetal force} = 0.5 \text{ kg} \times \frac{(2 \text{ m/sec})^2}{0.8 \text{m}}$$
>
> $$= \frac{0.5 \times 4}{0.8} \text{ N}$$
>
> $$= 2.5 \text{ N} .$$

REVIEW 4.1 QUESTIONS

1. What is a centripetal force?
2. Give some examples of forces that can serve as centripetal forces.
3. Why does the law of inertia require the presence of a centripetal force if circular motion is to occur?
4. Verify that the unit of centripetal force as defined by Eq. 4.1 is the newton.

REVIEW 4.1 PROBLEMS

1. Find the centripetal force needed to keep a 0.1-kg object moving at a speed of 0.5 m/sec in a circular path having a radius of 0.4 m.
2. A student finds that a centripetal force of 2 N is required to keep an object moving at a rate of 1 m/sec in a circular path having a radius of 0.2 m. What is the mass of this object?

4.2 Measurement of Circular Motion

In trying to describe circular motion, one very quickly finds that the techniques used to portray the phenomenon of linear motion require a certain degree of modification. In the case of translational motion, the movement of an object can be defined in terms of the displacement of the object's position with respect to a stationary frame of reference or coordinate system. Circular motion, on the other hand, is far more readily definable in terms of an ''orientation displacement'' or angular displacement relative to a defined stationary reference frame, usually in the form of a reference orientation line (Figure 4.3). If you wish to turn your back relative to someone you dislike, for example, you would do so by changing your angular orientation by one-half of a complete rotation relative to that individual.

Moreover, from geometry we can also obtain a numerical value for the change in angular orientation. In this case, since one complete rotation constitutes a change in angular orientation of 360°, the one-half of a rotation required to turn your back on the target of your scorn would constitute a rotation of ½ of 360° or 180°. Angular displacements can thus be measured in degrees of angle where one complete revolution or rotation corresponds to an angular displacement of 360°, a quarter revolution or rotation to a 90° angular displacement, and so on. In this text, such angular displacements will be most often measured in *radians* (see Physics Formulation 4.1) in which one radian is equal to about 57.3°. Angular displacements are said to be positive if the turn is in a counterclockwise direction as viewed from the top of the object and negative if the turn is in a clockwise direction.

Certain kinds of pivoting, however, have a more dramatic flair than others. If, for example, you sought to show special displeasure toward an individual, you might want to make an especially rapid turn. We can quantify the degree of such circular motion by defining the *angular velocity* of an object as

$$\text{angular velocity} = \frac{\text{displacement in angular orientation}}{\text{elapsed time}},$$

where the displacement in angular orientation is simply the angle through which an object rotates or revolves in the given interval of elapsed time. If the angular displacement is changing in a counterclockwise direction when viewed from above, then the angular velocity is defined as having a positive value. If the angular displacement is changing in a clockwise direction when viewed from above, then the angular velocity is said to be negative. Angular velocity is measured in units of angle/time such as degrees/minute or radians/sec. Often larger angular velocities such as those that occur in automobile engines are expressed in revolutions per minute or RPM where each rotation or revolution, of course, corresponds to an angular displacement of 360° or 2π radians. An angular velocity of 5000 RPM would thus correspond to an angular velocity of

$$5000 \, \frac{\text{rotations}}{\text{min}} \times \frac{360°}{\text{rotation}} = 1{,}800{,}000 \text{ degrees/minute} \, ,$$

which is a much less convenient numerical value to consider. As in the case of translational velocity, the angular velocity is defined in terms of measurable quantities, in this case the difference in angular orientation and an interval of elapsed time.

If an object's angular velocity is observed to change over some interval of time, then the object is said to exhibit an *angular acceleration,* which is formally defined as

$$\text{angular acceleration} = \frac{\text{change in angular velocity}}{\text{elapsed time}} \, .$$

(a)

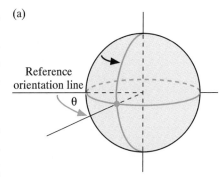

Reference orientation line

θ

(b)

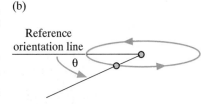

Reference orientation line

θ

FIGURE 4.3 • • • • • • • • • •
Angular Displacement (a) Angular displacement for a rotating object and (b) a revolving object. In each case, the angular displacement is denoted by the Greek letter θ.

Thus if an object changes its angular velocity from 0 to 90 deg/sec over a time interval of 5 seconds, the angular acceleration is then 90 (deg/sec)/5 sec or 16 deg/sec^2. A change of 30 radians/sec in 6 seconds constitutes an angular acceleration of 30 (radians/sec)/6 sec or 5 radians/sec^2.

WORKED EXAMPLE

A gear increases its angular velocity from 0 to 150 radians/sec in 30 seconds. What average angular acceleration did the gear experience?

SOLUTION: In general, the angular acceleration is given by

$$\frac{\text{angular}}{\text{acceleration}} = \frac{\text{change in angular velocity}}{\text{elapsed time}}.$$

The change in the gear's angular velocity is 150 radians/sec. This change occurred over a total elapsed time of 30 seconds. The angular acceleration of the gear is thus

$$\frac{\text{angular}}{\text{acceleration}} = \frac{150 \text{ radians/sec}}{30 \text{ seconds}}$$

$$= 5 \text{ radians/sec}^2 .$$

REVIEW 4.2 QUESTIONS

1. What is uniform circular motion? How does it compare with translational motion? Give some examples of circular motion occurring in nature.

2. Compare and contrast rotation and revolution. Give an example of each kind of motion.

3. Define each of the following terms: angular displacement, angular velocity, and angular acceleration. How is each a measure of an object's rotation or revolution?

4. Give an example of an object that exhibits (a) angular displacement, (b) angular velocity, and (c) angular acceleration.

5. What is a radian? a degree? an RPM? How is each a measure of angular displacement?

REVIEW 4.2 PROBLEMS

1. A rotating object has an angular velocity of 10 radians/sec. After one minute, through what angle has the object rotated as expressed in (a) radians, (b) degrees, (c) revolutions?

2. An object is subject to angular acceleration of 100 deg/sec^2. How much does the object's angular velocity change in a time interval of 3 seconds?

4.3 The Analysis of Rotational Motion

With the concepts of angular displacement, angular velocity, and angular acceleration formally defined, the development of a description of circular motion and its related phenomena proceeds in a way very much like that employed in our characterization of translational motion. In that discussion, we invoked motion-producing entities called *forces*. In a similar fashion, we can argue for the existence of similar agents in nature that can produce or prevent circular motion.

Suppose, for example, you are in an auditorium sitting out the last couple of minutes of one of the most excruciatingly boring large class lectures you have ever endured in your entire life. As the presentation mercifully comes to a close, you make haste for the nearest exit and encounter a large door complete with a horizontal bar handle. On the other side of the door is a patiently waiting friend. As you approach the door, you are prepared to exert a certain amount of force in order to swing the door open. If you exert that force on the part of the door handle nearest the hinged side of the door, there is little if any swinging or rotational response from the door (Figure 4.4). A force exerted halfway between the ends of the handle will produce more rotational response, but the easiest way to push the door open is to exert your force on the horizontal handle as far from the hinged side as possible. On the other hand, the large-sized, large force-producing wrestler or football player seems to be able to push the door open from any place along the handle by exerting a larger magnitude force on that door. We can conclude from all of this that the generation of rotational motion depends on two factors: the magnitude of the force that is applied to the object and the distance between the force and the axis about which the object is to rotate.

The *lever arm* of a force is defined as the shortest distance between the axis of rotation and the line along which that force acts (Figure 4.5). Therefore, the above observations can be described by introducing the concept of a *torque,* which is mathematically expressed as follows:

$$\text{torque} = \text{force} \times \text{lever arm of that force} \qquad \textbf{4.2}$$

or

$$\tau = F \times d .$$

Torques are thus the agents by which changes in rotational motion are produced or prevented, just as forces serve as the agents for producing or preventing translational motion. Larger torques exerted on the same object will produce larger changes in the object's rotational motion. Moreover, the larger torque can be generated by increasing either or both the magnitude of the applied force and the lever arm over which the force operates. The football player with his larger force acting over a given lever arm can thus produce more torque on the door than a person exerting a smaller force over the same lever arm; hence, the football player can open the door more readily. Meanwhile a given force exerted over a larger lever arm, can produce a larger torque than the same force acting over a smaller lever arm, thereby permitting the person exerting the force over the larger lever arm to push the door open more easily. Retaining the convention adopted for angular motion, a torque is said to be positive if it acts to produce a counterclockwise change in the object's rotation and negative if it produces a clockwise

FIGURE 4.4 ··········
Torque Ken is exerting a force farthest from the door's hinges in order to produce a large lever arm and a large torque on the door. By doing so he generates the greatest rotational response from the door making it easier to open with his exerted force.

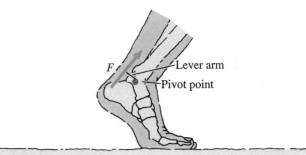

FIGURE 4.5

Lever Arms The lever arm always passes through the pivot point and is always perpendicular to the force F.

change. Since torque is the product of force and length, the units of torque are N-m, a fact that will have considerable significance for us later in the text.

There is yet another factor that controls the degree to which one can produce rotational motion. If the door in our example were made of steel, the rotation of the door generated by a given torque applied to the handle would be decidedly different from that generated by the same torque applied to a door having the same dimensions, but constructed of wood. Obviously the more massive steel door would respond far less to the applied torque. We may thus conclude that the mass of an object plays at least some part in determining the degree to which rotational motion will be produced by a given torque. Suppose, however, we replace the steel door in the above example with one having the same mass but constructed with a window such that most of the mass is located closer to the door's hinges (Figure 4.6). If we tried to push or swing open each of these equal-mass doors, we would find that it is much easier to open the door for which the mass is concentrated toward the hinged side.

Thus the ability of an object to resist the effects of a given torque depends not only on the total mass of the object, but also on how that mass is distributed within the object relative to the axis about which the object is to be rotated. One may infer from all of this that in every object a sort of "rotational mass" exists that is a measure of the object's ability to resist the effects of a given torque.

The rotational mass is essentially similar to the concept of mass in the analysis of translational motion and is referred to as the object's *moment of inertia*. The characteristics of the moment of inertia for a given object can be mathematically summarized by the following equation:

$$\text{moment of inertia} = \text{constant} \times \text{mass} \times (\text{size})^2 \ ,$$

where the "size" is some characteristic dimension of the object such as the width of a door, length of a rod, radius of a disk, ring, or sphere, etc. The constant has a numerical value that depends on the shape of the object and the location of the axis about which the object is to be rotated. Some representative objects and their associated moments of inertia are presented in Figure 4.7. For example, the moment of inertia for a hinged door of uniform mass distribution would be ($\frac{1}{3}$) $\times$ mass of the door $\times$ (width of the door)^2. Even objects having the same size and shape can have considerably different moments of inertia if the mass distribution within the object is significantly different. The moment of inertia for a uniform sphere, for example, is 0.6 of that for a hollow sphere having the same size and mass. A hollow sphere with all of its mass concentrated at its surface would thus be almost twice as hard to rotate as a sphere whose mass is uniformly distributed through its interior, even though their masses and radii are the same.

Angular velocity, angular acceleration, torque, and moment of inertia can be regarded as rotational counterparts to velocity, acceleration, force, and mass, and can likewise be forged into a rotational version of Newton's laws of motion. The "rotational" law of inertia thus becomes

> Every body continues in its state of rest or of uniform rotation unless it is compelled to change that state by torques impressed upon it.

FIGURE 4.6 ··········
Moment of Inertia The door on the left is more readily opened because more of its mass is closer to the axis of rotation than that of the door on the right whose mass is uniformly distributed.

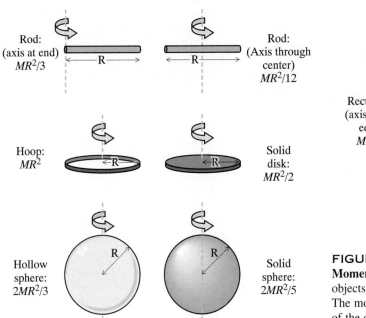

Rod:
(axis at end)
$MR^2/3$

Rod:
(Axis through center)
$MR^2/12$

Hoop:
MR^2

Solid disk:
$MR^2/2$

Hollow sphere:
$2MR^2/3$

Solid sphere:
$2MR^2/5$

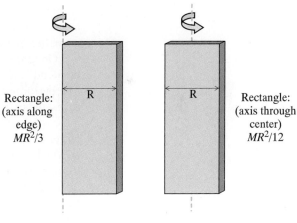

Rectangle:
(axis along edge)
$MR^2/3$

Rectangle:
(axis through center)
$MR^2/12$

FIGURE 4.7 ··········
Moments of Inertia The moments of inertia for various objects having various mass distributions and axes of rotation. The moments of inertia are expressed in terms of the mass M of the object and the pertinent dimension R.

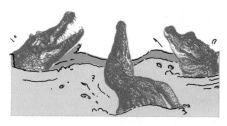

FIGURE 4.8 · · · · · · · · · ·
An Aerial Tightrope Walker   By extending his arms, the performer creates for himself a larger moment of inertia, thereby reducing his tendency to rotate off the tightrope.

In short, an object rotating uniformly, i.e., with a constant angular velocity, will "resist" any attempt to change that state of uniform rotation. The well-known tendency for a rapidly rotating bicycle wheel to damage your hand if you reach out to stop it can be explained by the fact that the bicycle wheel is simply seeking to maintain its state of rapid rotation. The flywheels employed in a variety of engines and machines take advantage of this behavior in that the rotational inertia developed in a flywheel will keep the flywheel rotating smoothly in between the successive "driving" positions of the engine's cycle. The flywheel of a steam engine or steam locomotive, for example, permits the rotational motion of the engine's drive shaft to continue between the successive driving "puffs" of the engine.

The rotational equivalent to Newton's Second Law may be stated as:

If an object is subjected to a net torque, it will respond with an angular acceleration in the direction of the torque by an amount equal to torque divided by moment of inertia.

This law of rotational motion is most often stated simply as

$$\text{torque} = \text{moment of inertia} \times \text{angular acceleration} \qquad \textbf{4.3}$$

or

$$\tau = I\alpha \; .$$

Thus, just as an object subjected to a force will respond by changing its velocity, so too will an object subjected to a torque respond by changing the angular velocity at which it is rotating. If we go through a "revolving" door that is initially at rest, the torque we exert on the door as we pass through will cause the angular velocity of the door to change from zero to some nonzero value. If we are in a hurry as we go through the door and thereby exert more torque, the door will respond by rotating more rapidly. On the other hand, if the door's moment of inertia is large, the door's rotational response to a given torque will be less than that which would occur for the same torque applied to a door having a smaller moment of inertia.

One rather interesting example of the "rotational second law" at work can be found in the performance of aerial tightrope walkers at the circus (Figure 4.8). The performers carry either long poles or keep their arms outstretched. In both cases, the performers are in effect increasing their moments of inertia by placing more mass farther away from a potential axis of rotation about the tightrope. The larger moment of inertia thus reduces the angular acceleration and hence the rate at which an angular velocity develops to the point where if a performer begins to rotate off the tightrope, there is enough time for the performer to readjust his or her balance before rotation pitches the performer onto the sawdust or safety net below.

As with the translational version of Newton's second law, the rotational version can be used to generate equations of rotational motion. These equations, in turn, can be solved to generate a description of the rotational behavior of a given object when subjected to a known set of torques. Once more, such capabilities are of considerable interest to the engineer who may wish to know how well the rotating flywheel of a certain machine, the rotating turbine of a certain generator, or the rotating crankshaft of a certain engine will hold up if made of a given material and subjected to a variety of torques.

The simplest description of the rotational motion exhibited by an object is that for an object having a known moment of inertia subjected to a constant torque. For such a case, the angular acceleration experienced by the object has a constant value that is equal to the torque divided by the moment of inertia or,

$$\text{angular acceleration} = \frac{\text{torque}}{\text{moment of inertia}} .$$

If the object is initially at rest rotationally and has no initial angular displacement, then after a given interval of elapsed time the object will have acquired an angular velocity

$$\text{angular velocity} = (\text{angular acceleration}) \times (\text{elapsed time})$$

and will have turned through an angular displacement of

$$\text{angular displacement} = \frac{1}{2} (\text{angular acceleration}) \times (\text{elapsed time})^2 .$$

These equations allow us to describe the state of the object's rotational motion after any interval of elapsed time. Thus if an object with a moment of inertia of 3 kg-m^2 is subjected to a torque of 12 N-m for 5 seconds, the object will experience an angular acceleration of

$$\text{angular acceleration} = \frac{12 \text{ N-m}}{3 \text{ kg-m}^2}$$
$$= 4 \text{ radians/sec}^2 .$$

The angular velocity after 5 seconds is

$$\text{angular velocity} = (4 \text{ radians/sec}^2) \times 5 \text{ sec}$$
$$= 20 \text{ radians/sec} .$$

The total angle through which the object turns in 5 seconds is

$$\text{angular displacement} = \frac{1}{2} (4 \text{ radians/sec}^2) (5 \text{ sec})^2$$
$$= 50 \text{ radians} ,$$

which corresponds to 2865 degrees or about eight complete revolutions of the object.

Of course, far more complex rotational systems exist than the one we have just considered. Some objects, for example, exhibit a phenomenon called *differential rotation* in which different regions of the object rotate at different angular velocities. The sun is one such object. At its equator, the sun rotates with an angular velocity of about 14.5 degrees/day, while at solar latitudes of 75° north and south of the solar equator, the sun rotates with an angular velocity of about 10 degrees/day. Hence, the sun does not behave as a solid, rigid rotating object. Note that, although the mathematical techniques required to describe this and other similarly complicated rotational phenomena are far more sophisticated than are appropriate for the level of this text, the basic principles employed in such analyses are nonetheless identical to those described here.

WORKED EXAMPLE

A wheel having a moment of inertia of 0.5 kg-m² is subjected to a constant torque of 5 N-m for a period of 10 seconds. After the 5 seconds have elapsed, what is the wheel's angular acceleration, angular velocity, and angular displacement?

SOLUTION: For a rotating object, we know in general that

$$\frac{\text{angular}}{\text{acceleration}} = \frac{\text{torque}}{\text{moment of inertia}} \, .$$

In this case, the torque is 5 N-m and the moment of inertia is 0.5 kg-m². The resulting angular acceleration is thus

$$\frac{\text{angular}}{\text{acceleration}} = \frac{5 \text{ N-m}}{0.5 \text{ kg-m}^2}$$
$$= 10 \text{ radians/sec}^2 \, .$$

The angular velocity is given in general by

$$\frac{\text{angular}}{\text{velocity}} = \frac{\text{angular}}{\text{acceleration}} \times \frac{\text{elapsed}}{\text{time}} \, .$$

During an elapsed time of 5 seconds, the wheel experiences an angular acceleration of 10 radians/sec² and will have acquired an angular velocity at the end of the interval of:

$$\frac{\text{angular}}{\text{velocity}} = 10 \text{ radians/sec}^2 \times 5 \text{ seconds}$$
$$= 50 \text{ radians/sec} \, .$$

The angular displacement is given in general by

$$\frac{\text{angular}}{\text{displacement}} = \frac{1}{2} \left(\frac{\text{angular}}{\text{acceleration}} \right) \times (\text{elapsed time})^2 \, .$$

For the wheel in question, we have

$$\frac{\text{angular}}{\text{displacement}} = \frac{1}{2} \times 10 \text{ radians/sec} \times (5 \text{ seconds})^2$$
$$= 125 \text{ radians} \, .$$

In summary, the wheel's angular acceleration is 10 radians/sec², its angular velocity is 50 radians/sec, and its angular displacement is 125 radians.

REVIEW 4.3 QUESTIONS

1. What is a torque? How are torques and forces similar? How do they differ?
2. Compare and contrast the mass of an object and its moment of inertia.

3. State the rotational law of inertia. Compare this statement with Newton's law of inertia for translational motion.

4. Describe the rotational equivalent to Newton's second law. Of what use is this statement?

5. Give an example of a device that makes use of torques in its operation.

REVIEW 4.3 PROBLEMS

1. A wheel having a moment of inertia of 3 kg-m² is subjected to a constant torque of 9 N-m for a period of 5 seconds. After the 5 seconds have elapsed, what is the wheel's (a) angular acceleration, (b) angular velocity, and (c) angular displacement?

2. A spinning wheel is observed to change its angular velocity by a total of 20 radians/sec in a span of 2 seconds. If the torque exerted on the wheel is 5 N-m, what is the moment of inertia of the wheel?

4.4 Angular Momentum

Nearly all of us at one time or another has tried to make it out of an exit past a closing door, only to get hit as the door arrived at the exit before we did. If something of the sort has happened to you, then you have quite literally run into an example of the most fundamental characteristic associated with a rotating or revolving object.

If we performed a series of "getting hit by closing doors" experiments, we would find that the impact suffered from a door having a larger moment of inertia would be larger than that from a door having a smaller moment of inertia. Also, the impact suffered from a rapidly closing door would be greater than that from a more slowly closing door and so on. As we continued with our experiments, we would begin to see that there exists a "quantity of rotational motion" that is the rotational analogue to linear momentum. Hence, it is called rotational or angular *momentum*. For an object having a given mass that is moving at a given speed in a circular path having a given radius, the magnitude of the angular momentum is defined in general as

$$\text{angular momentum} = \text{mass} \times \text{speed} \times (\text{radius of path}) \qquad \textbf{4.4}$$

or

$$L = mvr .$$

For an object revolving about a second object, the determination of the magnitude of the angular momentum is a relatively straightforward process. As in the example of Figure 4.9, if a knight is able to swing a 7.2-kg metal ball at the end of a 2-meter chain at a rate of 15 m/sec, the knight can generate a total angular momentum of

$$\text{angular momentum} = 7.2 \text{ kg} \times 15 \text{ m/sec} \times 2 \text{ m}$$
$$= 216 \text{ kg-m}^2/\text{sec} .$$

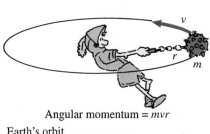

Angular momentum = mvr

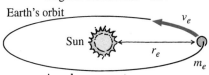

Angular momentum = $m_e v_e r_e$

FIGURE 4.9 ··········
Angular Momentum for Revolving Objects

FIGURE 4.10 ··········
Angular Momentum for a Rotating Object   The rotating object can be thought of as consisting of an array of tiny angular momentum elements, which when mathematically summed yield the result that the total angular momentum of the object is the product of the object's moment of inertia and its angular velocity.

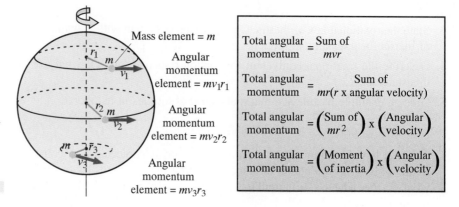

We can compute the angular momentum for an extended object by imagining that the object is composed of a large number of tiny bits or elements of mass. Each of these circles about the object's axis of rotation at some speed and distance characteristic of that mass element. Each of these mass elements has associated with it an element of angular momentum that is equal to the product (mass element) $\times$ (speed) $\times$ (distance from axis). The total angular momentum of the rotating object is then equal to the sum of all of the object's elements of angular momentum (Figure 4.10). If such a summing operation is carried out, it can be shown that the angular momentum for a rotating object is given by

$$\text{angular momentum} = (\text{moment of inertia}) \times (\text{angular velocity}) \quad \textbf{4.5}$$

or

$$L = I\omega .$$

Thus if an object rotates at a rate of 5 radians/second and has a moment of inertia of 7 kg-m², the angular momentum associated with such an object would be

$$\text{angular momentum} = 7 \text{ kg-m}^2 \times 5 \text{ radians/sec}$$
$$= 35 \text{ kg-m}^2/\text{sec} .$$

Note that because of their dimensionless nature, radians do not appear as units in any quantities other than those pertaining directly to angular measure, such as angular displacement, angular velocity, and angular acceleration.

In our discussion of linear momentum we found that in the absence of external forces the total linear momentum of a given system was conserved. So too, we find a similar conservation law for angular momentum. This is called the principle of the conservation of angular momentum and can be stated as follows: If no net torque is acting on a system, then

$$\begin{array}{ccc} \text{total angular momentum} & = & \text{total angular momentum} \\ \text{before an event} & & \text{after the event} \end{array} .$$

For revolving objects this requires that

$$\begin{array}{ccc} \text{mass} \times \text{speed} \times (\text{radius of path}) & = & \text{mass} \times \text{speed} \times (\text{radius of path}) \\ \text{(before event)} & & \text{(after event)} \end{array} .$$

For rotating objects,

$$\begin{pmatrix} \text{moment of} \\ \text{inertia} \end{pmatrix} \times \begin{pmatrix} \text{angular} \\ \text{velocity} \end{pmatrix} = \begin{pmatrix} \text{moment of} \\ \text{inertia} \end{pmatrix} \times \begin{pmatrix} \text{angular} \\ \text{velocity} \end{pmatrix} .$$

(before event)   (after event)

Perhaps the most familiar example of the conservation of angular momentum is that of a spinning ice skater (Figure 4.11). If the skater is spinning with her arms extended and then brings her arms closer to her body, her rate of rotation is markedly increased. This effect can be readily explained by the fact that the skater's angular momentum must be the same regardless of whether her arms are extended or not. When the skater is spinning with her arms extended, the mass in her arms is relatively far away from her axis of rotation and hence she possesses a comparatively large moment of inertia and a small angular velocity. As the skater brings her arms to her side, however, her moment of inertia decreases. Because her angular momentum must be the same before and after the ''event'' of her placing her arms by her side, this decrease in moment of inertia must be compensated for by an increase in the skater's angular velocity.

As a numerical illustration, suppose that the skater's moment of inertia is 10 kg-m^2 when her arms are extended and 5 kg-m^2 when her arms are at her side. If the skater is spinning with an angular velocity of 3 radians/sec when her arms are extended, then her angular momentum before the ''event'' is the product of her moment of inertia and her angular velocity, which is

$$\begin{aligned} \text{angular momentum} &= 10 \text{ kg-m}^2 \times 3 \text{ radians/sec} \\ &= 30 \text{ kg-m}^2/\text{sec} . \end{aligned}$$

After the ''event'' of bringing her arms to her side, the skater's moment of inertia is now 5 kg-m^2, or half of what it was with her arms extended. The skater's angular momentum, however, remains at 30 kg-m^2/sec. Thus the skater's angular velocity must increase to a value of

$$\begin{aligned} \text{angular velocity} &= \frac{30 \text{ kg-m}^2/\text{sec}}{5 \text{ kg-m}^2} \\ &= 6 \text{ radians/sec} \end{aligned}$$

or double what it was before the skater dropped her arms. The rate of the skater's rotation thus speeds up, which is just what we observe.

One of the more interesting problems in astronomy deals with the transfer of angular momentum from the sun to the planets. As the sun coalesced out of the gas and dust clouds of the Milky Way, it evolved from an ever-shrinking amorphous globule into a gas sphere capable of producing its own light through nuclear processes. In the course of this decrease in size, the sun's angular momentum had to have been conserved, thus requiring an increase in its angular velocity in a process much like the skater's increase in angular velocity. An examination of the sun's rate of rotation, however, reveals that it is not spinning nearly as fast as it should be if angular momentum had been conserved while the sun was forming. The difficulty is partially explained by noting that when the sun was formed, a number of planets were formed along with it, and while the masses of the planets are quite small compared to that of the sun, the radii of their orbital paths are very large. Thus more than 90% of the total angular

Angular momentum = Angular momentum

| Small angular velocity | × | Large moment of inertia | Large angular velocity | × | Small moment of inertia |

FIGURE 4.11 ••••••••••
Conservation of Angular Momentum As the skater brings her arms close to her body, the reduction in her moment of inertia is accompanied by an increase in her angular velocity.

momentum associated with the solar system is associated with the revolutionary movements of the planets rather than the rotational motion of the sun. One of the intriguing questions that continues to puzzle astronomers is how the low-mass planets were able to acquire so much of the total angular momentum as the solar system was forming.

If we descend to the tiniest limits on our scale of physical perception, we find that atoms and indeed even the nuclei and the component parts of the nuclei of atoms possess spin. Hence, these objects also have angular momenta of their own. Angular momentum at this atomic level is conserved just as surely as it is in a spinning skater or spiral galaxy, but, as we shall see later in this text, with some most fascinating and extraordinary ramifications.

The rotational version of Newton's second law can be rewritten in terms of angular momentum as follows:

$$\text{change in angular momentum} = \text{torque} \times \text{elapsed time} .$$

This statement tells us in effect that if we wish to change the angular momentum of a rotating object, we must exert a torque on the object. If no outside torque is exerted on the system, both the angular momentum and the object's axis of rotation remain fixed. It is this property of rotational motion that is employed in various gyroscopic toys and direction-measuring devices such as gyrocompasses and rate of turn indicators in aircraft. The preceding equation also tells us that the magnitude of the change in angular momentum produced by an applied torque depends not only on the magnitude of that torque but also on the interval of time over which the torque is applied. The product (torque × elapsed time) thus has the same significance and implications for rotational motion that impulse has for translational motion. Hence, this product is often referred to as *rotational impulse.*

Returning to our closing door example, suppose that door has a moment of inertia of 4 kg-m^2 and is swinging shut at an angular velocity of 1.5 radians/second. The total angular momentum of the closing door would be

$$\text{angular momentum} = 4 \text{ kg-m}^2 \times 1.5 \text{ radians/sec}$$
$$= 6 \text{ kg-m}^2/\text{sec} .$$

The impulse required to stop the door from closing, therefore, is also 6 kg-m^2/sec. The longer we have to stop the door from closing, the less torque required to accomplish the feat. To stop the door in 3 seconds, one would have to exert a torque of

$$\text{torque} = \frac{6 \text{ kg-m}^2/\text{sec}}{3 \text{ sec}}$$
$$= 2 \text{ kg-m}^2/\text{sec}^2 .$$

Recalling that 1 N = 1 kg-m/sec^2, this is a torque of 2 N-m. On the other hand, if the door slams shut, its swing is stopped over a time interval as short as $\frac{1}{50}$ sec, in which case the torque now jumps to 6 kg-m^2/sec ÷ $\frac{1}{50}$ sec = 300 N-m, a value 150 times larger than that required to stop the door in 1 second of elapsed time. You now have numerical justification for the qualitative observation that you don't want to have your fingers caught by a fast-slamming door!

WORKED EXAMPLE

An object has a moment of inertia of 4 kg-m^2 and is spinning at a rate of 60 radians/sec. If the object suddenly reduces its angular velocity to 10 radians/sec, what is the new value of its moment of inertia? Assume angular momentum is conserved.

SOLUTION: Since angular momentum is conserved, we have that

$$\begin{array}{ccc} \text{moment of} \\ \text{inertia} \end{array} \times \begin{array}{c} \text{angular} \\ \text{velocity} \end{array} = \begin{array}{c} \text{moment of} \\ \text{inertia} \end{array} \times \begin{array}{c} \text{angular} \\ \text{velocity} \end{array} .$$

(before event)   (after event)

Before the event, the moment of inertia of the object was 4 kg-m^2 and its angular velocity was 60 radians/sec. After the event, the object's angular velocity was 10 radians/sec. Thus we may write that

$$4 \text{ kg-m}^2 \times 60 \text{ radians/sec} = \begin{array}{c} \text{moment of} \\ \text{inertia} \end{array} \times 10 \text{ radians/sec} .$$

(after event)

The moment of inertia after the event is thus

$$\begin{array}{c} \text{moment of inertia} \\ \text{(after event)} \end{array} = \frac{4 \text{ kg-m} \times 60 \text{ radians/sec}}{10 \text{ radians/sec}}$$

or

$$\begin{array}{c} \text{moment of inertia} \\ \text{(after event)} \end{array} = 24 \text{ kg-m}^2 .$$

Thus the moment of inertia of the object increases by a factor of 6, the exact factor by which the angular velocity decreased.

---

1. What is angular momentum? Why is this concept important?

2. Compare and contrast the equation for the angular momentum of a rotating object with that for a revolving object. Discuss the difference.

3. Give an example of an object or system in which angular momentum is conserved.

4. Describe an object or system that has a large angular momentum and a small rate of rotation.

5. Discuss the relationship, if any, between linear momentum and angular momentum.

REVIEW
4.4
QUESTIONS

REVIEW 4.4 PROBLEMS

1. The earth orbits the sun at a distance of about 150 billion meters and has an orbital velocity of about 30,000 m/sec. What is the earth's orbital angular momentum?

2. Two rotating objects have the same angular momentum and the same radius. Object A is a hollow sphere and is rotating with an angular velocity of 30 radians/sec. If object B is a solid sphere, what is its angular velocity?

4.5 Statics and Stability

Up to this point we have considered objects that were in some state of either translational, revolutionary, or rotational motion. Many situations exist, however, in which an object or structure such as a building or bridge may be subjected to a number of forces and torques and yet exhibit neither translational nor rotational

FIGURE 4.12 ··········

Translational Statics   If an object such as the Magdeburg sphere shown here is not exhibiting translational motion, then the forces acting on the object are balanced in every direction.

motion. This important branch of physics is called *statics,* and deals with the interplay of forces and torques when no motion is present.

For an object that displays no translational motion (Figure 4.12), we may state

The net sum of all forces on the object along *any* direction is zero.

Thus, should a force of 10 N be exerted on a box in the positive *x* direction, if the box is to remain motionless, a force of 10 N must be exerted in the negative *x* direction. Similarly, if the earth's gravity pulls on the box with a force of 15 N downward, the ground must push up on the box with a force of 15 N if no motion is to occur. In many cases, forces are exerted on an object that are neither perpendicular nor parallel to each other. In such cases the forces involved must be broken up into "components" of parallel and perpendicular forces, which can then be used to analyze the given situation.

Rotational statics lends itself to a similar kind of analysis. If an object displays no rotational motion then the net sum of all torques on the object in *any* direction is zero.

For no rotational motion to occur, the clockwise torques on an object must be balanced by the counterclockwise torques on that object. The clockwise torque exerted by gravity acting on a lever arm about the axis of rotation of the castle drawbridge shown in Figure 4.13, for example, is balanced by the counterclockwise torque exerted by the opposite bank of the moat on the end of the drawbridge. As a result, the drawbridge does not rotate. If, however, an unbalanced torque—in this case in a counterclockwise direction—is exerted by the drawbridge chains, then the bridge will, in response, rotate in the same counterclockwise direction and the bridge will be raised.

The principles of rotational statics also give rise to the concepts of center of gravity and stability. Every object has associated with it a *center of gravity,* which is defined as the point about which no net gravitationally induced torque will exist on the object. In essence, if an object is supported from its center of gravity it will not rotate as a result of torques induced by gravitational forces. Moreover, all of the weight of the object can be thought of as being centered on this point. Suppose, for example, we consider the uniform meterstick shown in Figure 4.14. If we attach a support to the meterstick at the 30-cm position, an excess of the meterstick's mass will be to the right of the string and the gravi-

FIGURE 4.13 ··········
Rotational Statics If an object such as this drawbridge is not rotating, then the torques acting on the object are balanced in every direction.

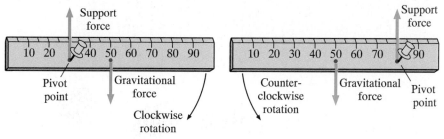

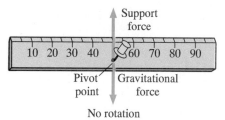

FIGURE 4.14 ··········
The Center of Gravity for a Meterstick Attempts to support the meterstick by pivot points located at 30 cm and 80 cm result in a rotation of the meterstick. If the meterstick is supported from a pivot point located at 50 cm, no rotation occurs and the meterstick is said to be supported from its center of gravity.

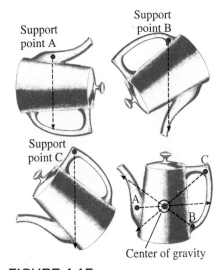

FIGURE 4.15 ···········
The Determination of the Center of Gravity If an object hangs freely from a support point, the action of the earth's gravity will always pass through its center of gravity. By hanging the object from three separate support points, the object's center of gravity can be graphically located as shown.

tational force on that mass will induce a clockwise torque about the support point. On the other hand, if we support the meterstick at the 80-cm position, there is now an excess of mass to the left of the support point and the gravitational force on that mass will induce a counterclockwise torque about the support point. Suppose, however, we support the meterstick at the 50-cm point. Now there will be equal and therefore balanced excess torques on the meterstick and it will not rotate.

The center of gravity of an object is of great importance in physics because in the translational motion of an object it will behave as if all of its mass were concentrated at that center of gravity. The location of an object's center of gravity can be obtained by the process illustrated in Figure 4.15. When the object under consideration is suspended from some arbitrary pivot point, gravitational torque will cause the object to rotate into a position such that the object's center of gravity will lie somewhere along a line that extends directly downward from the pivot point. If the process is repeated for two additional pivot points, the position of the center of gravity of the object will be located at the intersection point of the three downward pointing lines. For some objects, the center of gravity may be situated outside of the physical boundaries of the object (Figure 4.16). The center of gravity of a hollow sphere, for example, is located at the hollowed-out core region of the sphere where no actual mass exists.

If a downward line from an object's center of gravity lies within the support base of that object, the object is said to be ''stable'' or ''balanced'' and will not tip over due to the earth's gravity. On the other hand, if the downward line from an object's center of gravity falls outside of the object's support base, the object

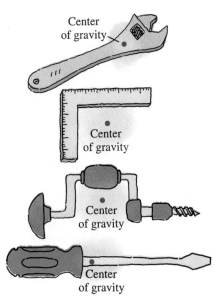

FIGURE 4.16 ···········
The Centers of Gravity for Some Common Tools Notice that it is possible for the center of gravity of an object to be outside of the object's physical boundaries.

FIGURE 4.17 ···········
Stability If a downward line from the center of gravity of an object or animal falls within the support base area, the object or animal is stable and will not tip over. If a downward line from the center of gravity of an object or animal falls outside of the support base area, the object is unstable and will attempt to rotate or fall to a position of stability.

is said to be "unstable," and such an object will tip itself over to reach a stable configuration (Figure 4.17). Objects that are otherwise stable can, of course, be tipped over when subjected to nongravitational torques. Objects with low centers of gravity and large-area support bases are found to be the least susceptible to such tipping. Racing cars are thus characterized by low-silhouette, large-base-area designs so that they can take turns or "corner" at much faster speeds without tipping over.

Although we have now spent two entire chapters describing a variety of motion phenomena, additional examples of movement still exist in nature that cannot be readily classified as either translational or circular motion. The "popping back up" motion of a sponge that has been squished, the "back and forth" motion of a pendulum, and the distinctive "vibratory" motion exhibited when a taut string or wire is plucked are some examples of such motion. Before considering such phenomena in more detail, however, we will take a bit of a "motion time out" to examine a few concepts in physics that will serve us very well when these motion-related topics come up for discussion again later in the text.

WORKED EXAMPLE

An object is subjected to a clockwise torque of 4 N-m and a counterclockwise torque of 12 N-m. Find the magnitude and direction in which a torque must be applied to prevent the object from rotating. If the lever arm of the applied torque is 2 m, what force is needed to prevent the object from rotating?

SOLUTION: The amount of unbalanced torque is (12 N-m) − (4 N-m) = 8 N-m in a counterclockwise direction. From the definition of torque, we have

$$\text{torque} = \text{force} \times \text{lever arm of force} .$$

If the unbalanced torque of 8 N-m is to operate over a lever arm of 2 m, then the force required is given by

$$8 \text{ N-m} = \text{force} \times 2 \text{ m} .$$

Thus,

$$\text{force} = \frac{8 \text{ N-m}}{2 \text{ m}}$$
$$= 4 \text{ N} .$$

A force of 4 N operating over a lever arm of 2 m will prevent the object from rotating.

1. Describe the conditions that must be satisfied for an object to exhibit neither translational nor rotational motion.

2. What is meant by statics? What are some practical applications of this branch of physics?

3. Discuss what is meant by the center of gravity of an object. Why is it an important concept in physics?

4. When is an object said to be stable?

5. If all the objects shown below have a uniform mass distribution, which, if any will topple over? Explain.

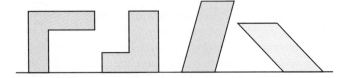

1. A torque of 16 N-m is applied to an object in a clockwise direction. How much force should be applied to that same object over a lever arm of 2 m in order to keep the object from rotating? In what direction should the force be applied?

2. A torque of 8 N-m is applied to an object in a counterclockwise direction. An engineer seeks to prevent an object from rotating with a force of 2 N. Over what lever arm must this force be applied?

Chapter Review Problems

1. If a person makes a quarter-turn, find that person's angular displacement in (a) revolutions, (b) degrees, and (c) radians.

2. A top spins 120 radians in 8 seconds. What is the average angular velocity of this top?

3. A wheel on a piece of machinery rotates a total of 36,000 degrees in 300 seconds. Find the average angular velocity in (a) degrees/second and (b) RPM.

4. A billiard ball turns a total of 80 radians in 0.4 seconds. What is the average angular velocity of the ball in radians/second?

5. A turbine has its angular velocity increased from 0 to 10,000 radians/second in 80 seconds. What average angular acceleration did the turbine experience?

6. The drive shaft of an automobile engine increases its angular velocity from 60 radians/sec to 480 radians/sec in 3 seconds. What is the average angular acceleration for the shaft?

7. Find the torque that would be exerted on an object by a 7-N force operating over a 5-meter lever arm.

8. A wheel having a moment of inertia of 2 kg-m^2 is subjected to a constant torque of 8 N-m for a period of 10 seconds. After the 10 seconds have elapsed, what is the wheel's (a) angular acceleration, (b) angular velocity, and (c) angular displacement?

9. What is the magnitude and direction of the centripetal acceleration necessary to confine the motion of a 0.5-kg mass moving at a speed of 0.1 m/sec to a circular path 0.2 m in radius?

10. If the speed of the 0.5-kg mass in Problem 9 were increased to 0.4 m/sec, what would the radius of the new circular path be?

11. Calculate the moment of inertia for a solid disk having a mass of 0.4 kg and a radius of 0.5 meter. Repeat the calculation for a ring having the same total mass and radius as the solid disk. How do your results compare?

12. A door has a width of 0.6 m and a mass of 10 kg, which is uniformly distributed throughout the door. Calculate the effect on the door's moment of inertia if (a) the mass were doubled and (b) the width were cut in half.

13. A skater with a moment of inertia of 3 kg-m^2 stretches out his arms so that his moment of inertia is now 5 kg-m^2. If the skater was initially spinning at 4 radians/sec, what is his angular velocity now?

14. An object is subjected to a 15-N force directed to the right and a 16-N force directed to the left. Find the magnitude and direction of the force needed to prevent this object from exhibiting any translational motion.

15. A torque of 5 N-m is applied to an automobile wheel with a moment of inertia of 2 kg-m^2. What angular acceleration results?

16. A torque of 6 N-m is applied to an object having an angular momentum of 12 kg-m^2/sec. How long will it be before the object's angular momentum is reduced to zero by this torque?

17. An object is subjected to a clockwise torque of 5 N-m and a counterclockwise torque of 8 N-m. Find the magnitude of and direction in which a torque must be applied if the object is not to rotate.

18. Suppose a door with a width of 0.4 m is subject to a torque of 10 N-m. How much torque would you have to apply to the door to keep it from swinging? How much force would be required if you applied the force (a) 0.4 m from the hinged side and (b) 0.1 m from the hinged side? Which torque application requires less force? Is this what you actually observe?

19. A 40-kg student on the edge of a merry-go-round 10 meters in diameter is moving at a speed of 0.8 m/sec. What is the magnitude and direction of the centripetal force required to keep the student moving in a circular path? What is the magnitude and direction of the centrifugal force exerted by the student on the merry-go-round? What are the agents, if any, for these forces?

20. A 3-kg toy plane is whirling at the end of a string 1 meter long at a speed of 10 m/sec. How long must the string be if we wish to whirl the plane at the speed of sound, about 300 m/sec, and still have the angular momentum of the system conserved?

21. A wheel rotates through 1000 radians before it is slowed to a complete stop. If the stopping process took 2 seconds to accomplish and the applied torque was constant throughout, find (a) the angular deceleration of the wheel and (b) the original angular velocity of the wheel.

22. Two masses of 0.1 kg and 0.2 kg are connected by a string through a hollow tube as shown at the right. At what speed must the 0.1-kg mass be whirled in order to keep the 0.2-kg mass at rest? Assume that the 0.1-kg mass is moving in a circular path with a radius of 0.2 m.

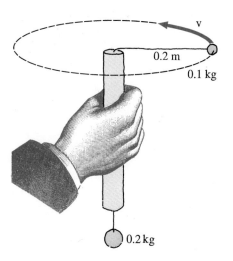

THOUGHT
QUESTIONS

1. At time $t = 0$, the hands on a clock read 12:00 noon. In the next hour, which of the following will the hands on the clock exhibit? (a) angular displacement, (b) angular velocity, (c) angular acceleration. Will these quantities be positive or negative?

2. Suppose you were observing a perfectly smooth ball bearing that had no surface spots or other markings. Could you tell whether or not the ball bearing was rotating? If so, how?

3. A star is observed to move through space along a path that appears as follows:

What, if anything, is unusual about such motion? How would you explain its occurrence?

4. Show that the expression for the centripetal acceleration given in the text has the correct units for an acceleration.

5. An automobile going around a corner exhibits circular motion as it makes its turn. What is the source of the centripetal force in this case?

6. What are the units of moments of inertia?

7. How would you experimentally determine the value of an object's moment of inertia?

8. A solid disk and a ring are released from the top of a hill. If the radius and mass of the solid disk and the ring are the same, which object will roll down the hill faster? Explain.

9. If an object with a moment of inertia I is rotating at an initial angular velocity ω_0, is oriented at an initial angular displacement of θ_0, and is subjected to a constant torque τ, then show that the angular acceleration α after an elapsed time t will be

$$\alpha = \frac{\tau}{I} \; ;$$

the angular velocity ω after an elapsed time t will be

$$\omega = \alpha t + \omega_0 \; ;$$

and the angular displacement θ after an elapsed time t will be

$$\theta = \frac{1}{2} \alpha t^2 \times \omega_0 t + \theta_0 \; .$$

(*Hint:* See Physics Formulation 4.1.)

10. Why do you think that the disks of the planets Jupiter and Saturn appear to be flattened in the photographs of these planets shown below?

 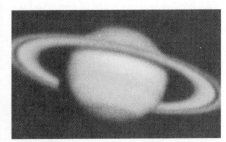

11. An object is observed to rotate at a constant angular velocity. What, if anything, can be said about the torques acting on this object?

12. Find a nearby railroad track and, taking care to watch for trains, try to walk along the top of one of the rails with (a) your arms at your side and (b) with your arms extended. Which is easier to do? Why do you think this is so?

13. How might you state the equivalent of Newton's third law for a rotating object? Is there any evidence for the existence of such a law for rotating objects? Explain.

14. What is a centrifuge? Describe how such a device operates.

15. How might you create "artificial gravity" on board a space station in outer space?

16. If you wished to create a "weightless" condition on board an aircraft, how should you fly the plane? Explain.

17. If you wanted to *increase* your weight on board an aircraft, how should you fly the plane? Explain.

18. Describe as many forces and torques as you can think of that might be acting on the railway crane shown below.

19. If the Leaning Tower of Pisa shown at left has a uniform distribution of mass, show that the Tower will not tip over. How much more can the tower lean before it will indeed tip over?

20. Determine the location of the center of gravity of this textbook. Place a label on the cover to denote the location you obtained.

21. Is it possible for an object to rotate so rapidly that it will break apart? If so, describe the circumstances under which this could occur.

22. Below is a diagram of the wind patterns on the earth's surface. Describe how the Coriolis effect may have played a role in shaping these wind patterns.

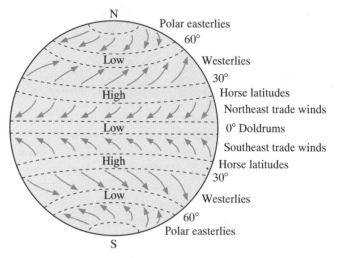
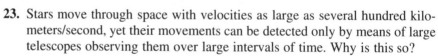
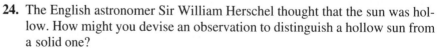

23. Stars move through space with velocities as large as several hundred kilometers/second, yet their movements can be detected only by means of large telescopes observing them over large intervals of time. Why is this so?

24. The English astronomer Sir William Herschel thought that the sun was hollow. How might you devise an observation to distinguish a hollow sun from a solid one?

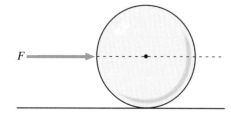

25. A pool player strikes a ball with a force F directly through the ball's center of gravity as shown at left. When so struck, the ball is observed to roll. Explain why this happens.

26. Why doesn't a sphere tip over?

27. Suppose you are whirling a stone on the end of a string and suddenly the string snaps. Describe the path the stone takes after the string snaps, and explain this behavior in terms of one or more of Newton's laws of motion.

28. How might you set up or define a ''standard unit'' of torque?

PHYSICS FORMULATION 4.1

Radian Measure

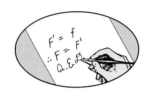

Most of us are familiar with the use of degrees to denote the sizes of angles. The corners of a square, for example, form 90° angles with each other. An equal-sided or equilateral triangle has 60° angles at each of its vertices. In dealing with circular and rotational motion, however, it is useful to measure angles in a somewhat different fashion.

To describe this different way of measuring sizes of angles, let us first consider the circle shown in Figure 4.18 in which two radii have been drawn at an angle θ to each other and extend from the center of the circle to points A and B on the circumference of the circle. The distance along the circumference of the circle between points A and B is referred to as the arc length subtended by the angle θ. It can be shown in geometry that as the value of the angle θ is increased, the arc length s will increase by the same factor. Thus, if θ is doubled, the value of s will also double and so on. This geometric property can be employed to our advantage by defining *one radian* as the angle subtended by an arc length equal to the radius of the circle involved. Any angle θ then can be expressed in radian measure as the total arc length/radius, or

$$\theta \text{ (in radians)} = \frac{s}{r}.$$

For an entire circle, the circumference is $2\pi r$ and θ in radians for such an arc length is $2\pi r/r = 2\pi$ radians. Since there is a total of 360° in a complete circle we have

$$2\pi \text{ radians} = 360°$$

and

$$1 \text{ radian} = 360°/2\pi$$

or

$$1 \text{ radian} = 57.3° .$$

Also

$$1° = 1 \text{ radian}/57.3°$$

or

$$1° = 0.0175 \text{ radians} .$$

Note that since the radian is defined as a length/length, angles measured in radians have no units in terms of the fundamental quantities of length, time, and mass.

One convenient use of radian measure can be made when we wish to determine the distance to an object of known size or the size of an object having a known distance. Suppose, for example, we wish to determine the distance D to a certain building which has a height H of 200 meters and subtends an angle of 0.01 radians (about 0.57°) on the skyline of a large city (Figure 4.19). By assuming that the building can be approximated by a segment of the circumference of a circle whose radius is equal to the distance to the building, we can write that

$$D = \frac{H}{\theta} \text{ (in radians)} = \frac{200}{0.01}$$

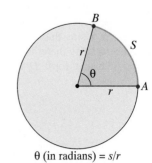

$$\theta \text{ (in radians)} = s/r$$

FIGURE 4.18 ··········
Radian Measure The angle θ is said to have a value of one radian when the arc length s is equal to the radius of the circle. The value of any angle θ in radians is then arc length/radius or s/r.

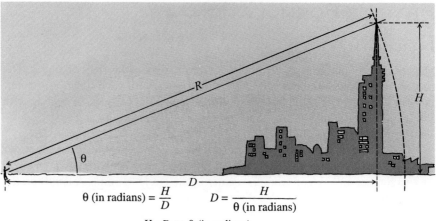

FIGURE 4.19 ··········

Determining Sizes and Distances from Approximate Radian Measures   If the angle θ is less than about 0.1 radian or 5.7°, the value of *H* is approximately the same as the arc length subtended by the angle θ along the circumference of a circle of radius *B*. Using this approximation, we may therefore write the equations indicated for θ, *D,* and *H.*

$$\theta \text{ (in radians)} = \frac{H}{D} \qquad D = \frac{H}{\theta \text{ (in radians)}}$$

$$H = D \times \theta \text{ (in radians)}$$

or

$$D = 20000 \text{ m} = 20 \text{ km} .$$

In like manner we can determine the height of an object of known distance by simply solving the above equation for *H* to obtain

$$H = D \times \theta \text{ (in radians)} .$$

Notice that *H* and *D* must always be in the same units of length if θ is to be expressed in radians.

   If an object is moving along a circular path, then the angle θ traversed in a given elapsed time *t* can be expressed as

$$\frac{\theta}{t} = \frac{s}{r \times t} .$$

The quantity θ/elapsed time, however, is simply the angular velocity and *s/t* is the magnitude *v* of the velocity of the moving object. Thus we have

$$\text{angular velocity} = \frac{v}{r}$$

and

$$v = r \times \text{angular velocity} .$$

From this relationship we can see that for a given angular velocity, the magnitude of the velocity of the associated circular motion increases with the size of the radius, a fact that can be observationally verified by anyone who has ever been at the end of the whip in that chaos-generating children's game "crack the whip."

QUESTIONS
·········

1. Compare and contrast radians and degrees.

2. Why do you suppose that physicists use radian measure?

3. Why must *H* and *D* be in the same units of length in the following relation?

$$H = D \times \theta \text{ (in radians)}$$

PROBLEMS

1. A wheel rotates through an angle of 1000 degrees. How far has the wheel rotated in radians? How many rotations has the wheel made?

2. If the earth spins once on its axis every 24 hours, find the earth's angular velocity in radians/second. If the earth's equatorial radius is 6.4×10^6 meters, at what linear velocity does a point on the earth's equator rotate?

3. An aircraft having a length of 80 meters subtends an angle of 0.02 radians as it passes directly overhead. What is the altitude of the plane?

PHYSICS FORMULATION 4.2

Centripetal Acceleration

When an object moves along a curved path of any kind, it must necessarily be subject to some form of acceleration. This is because, even though the constituent magnitude of the object's velocity may remain constant, the direction of the object's velocity and hence the velocity itself is changing. Such a change, of course, constitutes an acceleration.

The most familiar example of an acceleration that produces a change in an object's direction of motion without a change in the magnitude of the object's velocity occurs when an object moves along a circular path at a constant speed. For such motion, called uniform circular motion, the direction of the object's velocity is always pointed along a line tangent to the circular path at the object's position along the path.

To obtain a mathematical description of the acceleration required to produce such motion, let us assume that an object is moving at a speed v along a circular path having a radius of r as shown in Figure 4.20. As the object moves from position 1 to position 2 in that diagram the direction of the velocity vector v_1 is altered from that of velocity vector v_2 by an angle of θ. Under these circumstances, the velocity vector v_2 can be thought of as being the resultant of a vector sum of v_1 and a vector Δv as shown in Figure 4.20. The vector Δv thus represents the vector change in the object's velocity as it moves from position 1 to position 2, and the associated acceleration is Δv/elapsed time. It can be shown that this acceleration is always directed toward the center of the circular path. Because of its "seeking the center" property, this type of acceleration is referred to as *centripetal acceleration*.

To obtain the magnitude of the centripetal acceleration let us consider the vector triangle shown in Figure 4.20, which has sides v_1, v_2, and Δv and contains the angle θ. If the angle θ is small, Physics Formulation 4.1 tells us that we may write the following equation

$$\theta = \frac{\Delta v}{v}$$

and thus

$$\Delta v = \theta \times v \, ,$$

where Δv is the magnitude of the vector Δv and v is the magnitude of both vector v_1, and vector v_2. The magnitude of the centripetal acceleration a_c is Δv divided by the elapsed time t it takes the object to move from position 1 to position 2, or

$$a_c = \frac{\Delta v}{t} \, .$$

Substitution of the previous expression for Δv into this equation yields

$$a_c = \frac{\theta \times v}{t} \, .$$

From Physics Formulation 4.1, the angular velocity θ/t is given by

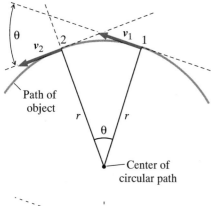

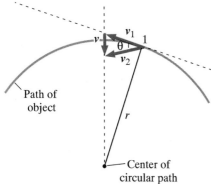

FIGURE 4.20 ··········
The Vector Change in Velocity for Simple Circular Motion   The change in orientation between vector velocity v_1 and vector velocity v_2 is equal to the vector change Δv between the two velocities.

$$\frac{\theta}{t} = \frac{v}{r}.$$

If we substitute v/r for θ/t in the earlier expression for a_c, we obtain

$$a_c = \frac{v^2}{r}.$$

Thus, if an object is moving in a circular path at a constant speed, it must do so as a result of a centripetal acceleration directed toward the center of the circular path and possessed of a magnitude equal to v^2/r. One of the mainstay concepts of the Newtonian view of motion is that if an acceleration is present in an object's motion, there must be some sort of net force being exerted on the object to create that acceleration. For circular motion this net force is called the *centripetal force*. If an object of mass m is moving in a circular path, the centripetal force has a magnitude of $m \times a_c$ or $m\,v^2/r$ and, like the centripetal acceleration, is directed toward the center of the circular path.

Centripetal forces can be exerted in a variety of ways. The moon is forced into its nearly circular orbit about the earth by the action of the mutual gravitational force between the earth and the moon. The stones in a sling are held in a circular path as they are whirled around by the tension forces in the sling, and the various parts of a spinning top are held in place by the microscopic structural forces that enable the top to retain its solid form.

QUESTIONS

1. Explain why an object moving in a curved path must necessarily be accelerating.
2. If an object is moving in a curved path, must it always be subjected to a force? Explain.
3. Is it possible for a centripetal acceleration to exist that is not directed toward the center of the object's curved path? Explain.

PROBLEMS

1. The earth moves about the sun in a roughly circular orbit at a rate of about 30,000 meters/second at a distance of 150 billion meters. What centripetal acceleration is required to hold the earth in such an orbit? What is the source of this acceleration?
2. At what angular velocity must a space station 200 meters in diameter be rotated in order to produce an artificial gravitational acceleration of 9.8 meters/sec^2?
3. Around the turn of the century, one suggestion for the centripetal force agent needed to hold an electron in orbit about an atomic nucleus was gravity. If the electron in a hydrogen atom has a mass of 9×10^{-31} kg and orbits a 1.7×10^{-27} kg proton at a distance of 5×10^{-11} meters and a speed of 2 million meters/second, calculate the centripetal force needed to hold an electron in orbit about the nucleus of a hydrogen atom.

5 GRAVITY

FIGURE 5.1 · · · · · · · · · ·
Jupiter's Galilian Satellites An important indication of the existence of a universal gravitational force came from observations of Jupiter's major satellites. Galileo correctly perceived this system to be dynamically analogous to a Sun-centered planetary system.

*U*ntil now, we have taken the limited view that objects under the influence of the earth's gravity behave as if they were subject to a constant gravitational acceleration of $g = 9.8$ m/sec^2. Certainly Newton's most splendid insight into the workings of nature was his realization that gravity was a far more universal phenomenon than what we observe of it at the earth's surface. Newton recognized that the gravitation causing an apple to fall to the earth is the same gravitation causing the moon to "fall" about the earth in its orbit and the planets to "fall" about the sun in their orbits. Indeed, claimed Newton, anything in the universe that possesses mass also exerts gravitational force (Figure 5.1). Thus, gravitation is responsible for planets orbiting about the sun, stars orbiting other stars, and even entire galaxies of stars orbiting each other. On the most grandiose of scales, gravitation is now believed to be the cause of the observed slowdown in the overall expansion of the entire universe. Gravitation also holds the distinction of being the first of the fundamental interactions of nature to be understood in a comprehensive way.

5.1 The Universal Law of Gravitation

For centuries, it was known that the Moon moves about the Earth in a roughly circular path at a nearly constant speed. Galileo's observations demonstrated a similar kind of motion for the planets Mercury and Venus about the Sun and for the moons of Jupiter about that planet. In the first two decades of the seventeenth century, the German mathematician Johannes Kepler forged his laws of planetary motion in which he demonstrated that not only Mercury and Venus orbited the Sun, but that the Earth and the rest of the planets did too.

Newton surmised that these orbital motions must be the result of the action of some sort of centripetal force. Otherwise, reasoned Newton, the Moon, planets, and the satellites of Jupiter would have long since departed from their primaries in straight line paths at constant speeds as a result of the law of inertia.

If such a "gravitational" force indeed existed, what sorts of characteristics might it possess? Important clues were provided by Kepler. After wading through the myriad of observational data left to him by the Danish astronomer Tycho Brahe, Kepler was able to carve out three laws of planetary motion:

1. All of the planets move about the Sun in elliptical orbits with the Sun located at one of the focal points of the ellipse.

2. The planets move about the Sun in such a way that the area of orbit swept out by the planet per unit time has a constant value (law of equal areas).

3. The ratio of the cube of the mean distance between a planet and the Sun to the square of the planet's sidereal period has the same value for every planet.

In particular, one can express the last of these laws, a relationship called the *harmonic law,* as follows:

$$\frac{(\text{mean distance})^3}{(\text{sidereal period})^2} = \text{constant value} .$$

If one assumes that the planets travel about the Sun in near-circular orbits at nearly constant speed, then the speed of a given planet can be obtained from

$$\text{speed of planet} = \frac{\text{circumference of orbit}}{\text{sidereal period}}.$$

From plane geometry, the circumference of a circle is equal to $2\pi \times$ radius, thus

$$\text{speed of planet} = \frac{2\pi \times \text{radius of orbit}}{\text{sidereal period}}$$

and

$$(\text{speed of planet})^2 = \frac{4\pi^2 \,(\text{radius of orbit})^2}{(\text{sidereal period})^2}.$$

For a circular orbit, the mean distance of a planet and its orbital radius is the same, and Kepler's harmonic law can be written

$$\frac{1}{(\text{sidereal period})^2} = \frac{\text{constant value}}{(\text{radius of orbit})^3}.$$

If we substitute this equation into the previous one, we have

$$(\text{speed of planet})^2 = \frac{4\pi^2 \,(\text{radius of orbit})^2 \times \text{constant value}}{(\text{radius of orbit})^3}$$

or

$$(\text{speed of planet})^2 = \frac{4\pi^2 \times \text{constant value}}{\text{radius of orbit}}.$$

The centripetal acceleration holding the planet in its path is

$$\text{centripetal acceleration} = \frac{(\text{speed})^2}{\text{radius}}.$$

Substituting the expression for $(\text{speed of planet})^2$ that we obtained from Kepler's harmonic law into this equation gives us

$$\text{centripetal acceleration} = \frac{4\pi^2 \times \text{constant value}}{(\text{radius of orbit})^2}.$$

Although we have employed a simple motion in obtaining this equation, it should be noted that through the use of calculus, Newton was able to generalize this result so as to include Kepler's elliptically shaped orbits as well.

Thus Newton was able to demonstrate that his gravitational force exhibited a mathematical behavior, which is called an *inverse square law* (Figure 5.2). For a quantity subject to such a law, the value of that quantity depends on $1/(\text{distance})^2$. In the case of gravitational acceleration, if the distance between the objects doubles, then the gravitational acceleration is only $1/(2)^2$ or ¼ as large. For triple the distance, the gravitational acceleration is $1/(3)^2$ or ⅑ as large, and so on. Thus, Newton was able to state that for two objects A and B separated by a given distance,

$$\begin{matrix}\text{acceleration of gravity} \\ \text{between A \& B}\end{matrix} \quad \text{depends on} \quad 1/(\text{distance})^2.$$

From Newton's second law of motion, we know that an accelerating object is experiencing a net force that is the product of its mass and its acceleration. If

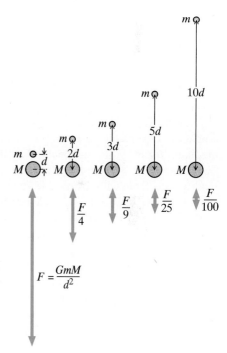

FIGURE 5.2 ··········
**The Inverse Square Law for
Gravitation** Newton was able to
demonstrate that the gravitational
interaction between two masses varies
inversely with the square of the distance
between them, that is $1/(\text{distance})^2$.

5.1 The Discovery of Neptune

The discovery in 1846 of the planet Neptune, the eighth planet in order from the Sun, represents one of the more impressive triumphs of Newton's theories of motion and gravitation.

The story of Neptune's discovery actually began over a half-century earlier when, on the night of March 13, 1781, an English astronomer named William Herschel discovered, quite by accident, the seventh planet from the sun, a world called Uranus. At first Herschel thought he had discovered a comet, and in fact reported it as such to the Royal Astronomical Society. After several months of observations, however, it became clear that the new object was not moving in a highly elongated elliptical orbit, which was typical of a comet, but rather in a nearly circular orbit whose radius was nearly twice that of the orbit of Saturn, the outermost of the then-known planets. Herschel had made the first discovery of a planet since prehistoric times.

Further observations of Uranus, however, indicated that it was not quite moving in the manner predicted for it by Newton's gravitational theory. By the early 1840s the discrepancy between the observed position of Uranus and that predicted for it by gravitational theory amounted to about two arcminutes, an angular difference large enough to be readily detectable with the naked eye. In an era where telescopic observations could be made hundreds of times more accurately than those of the unaided eye, this was an intolerable disagreement. Some scientists and mathematicians of the day questioned whether the universal law of gravity was really universal. It might be, they argued, that at the distance of Uranus from the Sun gravity gets "tired" and does not mathematically behave in the same way as for

distances closer to the Sun. Such a claim was, of course, in direct conflict with the principle of universality and was to be embraced only as a last, desperate resort.

An alternate approach was taken by two mathematicians, one an Englishman named John Adams, and the other a Frenchman named Urbain Leverrier. Adams and Leverrier hypothesized the existence of a "trans-Uranian" planet whose orbit lay outside that of Uranus and whose gravitational effects were the cause of Uranus's misbehavior. Unaware of each other's work, Adams and Leverrier each calculated the celestial position of their trans-Uranian planet and presented their results to observational astronomers beginning in October 1845. Because of the skepticism of the astronomers involved, it was not until September of the following year that Leverrier was able to get an astronomer named Johann Galle at the Berlin Observatory to seek out the proposed planet. Within an hour of the first night of search, September 23, 1846, Galle found that the new planet within a degree of the position predicted for it by both Adams and Leverrier.

The predicted discovery of Neptune gave Newton's theories one of their finest hours. Two individuals using only their wits and their pens had successfully predicted from Newtonian theory, not only the existence of a planet no one knew had existed, but almost exactly where it could be found in the heavens.* Leverrier, in fact, never once laid

---

*There is some evidence to suggest that Galileo may have made an inadvertent observation of Neptune while studying Jupiter's satellites in January 1613. Uncharacteristically, he did not follow up on his suspicions, thereby preserving the mathematical triumph of Adams and Leverrier that was to come two centuries later.

---

object A is experiencing a gravitational acceleration from object B, which depends on 1/(distance)^2, it must also be experiencing a gravitational force characterized as follows:

$$\text{gravitational force of object B or object A} \quad \text{depends on} \quad \frac{\text{mass of object A}}{(\text{distance})^2}$$

PHYSICS FACET

Continued

eyes on the planet whose existence he had so successfully deduced.

Flushed with his success, Leverrier proceeded to tackle the problem of another troublesome inhabitant of the solar system. The planet Mercury is the innermost of the planets and, like Uranus, its motion was behaving in a fashion inconsistent with Newton's gravitational theory. Leverrier attempted to solve the discrepancy in exactly the same way

in which he had been so triumphantly successful in the case of Uranus. He postulated the existence of a gravitationally perturbing ''intra-Mercurial'' planet and published his predictions in 1859. This time, however, no planet was to be found, and a scientifically successful explanation for Mercury's anomalies would have to await the coming of Einstein and his views of gravitation a half-century later.

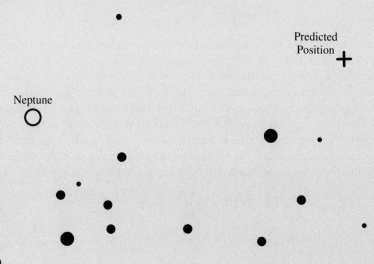

(a)

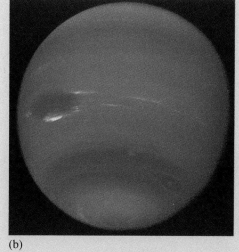

(b)

··· **The Discovery of the Planet Neptune** (a) In the mid-1840s John Adams of England and Urbain Leverrier of France predicted the existence of a new planet and even predicted where in the sky it could be found. In 1846, the German astronomer Johann Galle found the planet within a degree of the position predicted for it by both Adams and Leverrier. (b) Neptune as photographed by *Voyager II* in August 1989.

or

$$F_{BA} \quad \sim \quad \frac{M_A}{d^2} .$$

However, Newton's third law of motion requires that for every action force there is an equal and opposite reaction force. Thus, object A must exert an equal and opposite gravitational force on object B, which is the product of the mass of object B and the gravitational acceleration between the two objects. The gravitational force of object A on object B may hence be characterized as follows:

gravitational force of object A on object B	depends on	$\dfrac{\text{mass of object B}}{(\text{distance})^2}$

or

$$F_{AB} \quad \sim \quad \frac{M_B}{d^2} \ .$$

The mutual gravitational force between objects A and B, therefore, depends on the mass of object A, the mass of object B, and $1/(\text{distance})^2$. Newton mathematically incorporated all of these characteristics into a single equation, which is called the *universal law of gravitation* and can be stated as follows:

$$\begin{array}{c} \text{force of gravity} \\ \text{between two} \\ \text{objects} \end{array} = \text{constant} \ \dfrac{\begin{array}{c}\text{mass of} \quad \text{mass of} \\ \text{first object} \times \text{second object}\end{array}}{\left(\begin{array}{c}\text{distance between} \\ \text{objects}\end{array}\right)^2} \qquad \textbf{5.1}$$

or

$$F_{\text{gravity}} \ = \ G \ \frac{m_A m_B}{d^2} \ .$$

The constant G is called the *universal gravitation constant* and has a value of 6.672×10^{-11} N-m^2/kg^2 (see Physics Formulation 5.1).

The universal law of gravitation thus provides us with the amazing result that there is a mutual gravitational force between any given object and every other object in the universe! It should be quickly noted, however, that even though such gravitational forces exist, most of them are vanishingly small in their magnitudes. Suppose, for example, that a student owns a stereo with a mass of 10 kg and it is located 4 meters across the room from the student's desk. If the student's mass is 50 kg, then the force of gravity between the student and the stereo is

$$\begin{aligned} F_{\text{gravity}} &= 6.672 \times 10^{-11} \ \frac{\text{N-m}^2}{\text{kg}^2} \times \frac{10 \ \text{kg} \times 50 \ \text{kg}}{(4 \ \text{m})^2} \\ &= 6.672 \times 10^{-11} \times \frac{500}{16} \ \text{N} \\ &= 209 \times 10^{-11} \ \text{N} \\ &= 2.1 \times 10^{-9} \ \text{N} \ . \end{aligned}$$

By contrast, the gravitational force exerted on the stereo by the earth (mass = 6×10^{24} kg and radius = 6.4×10^6 m) is given by

$$\begin{aligned} F_{\text{gravity}} &= 6.67 \times 10^{-11} \ \frac{\text{N-m}^2}{\text{kg}^2} \times \frac{6 \times 10^{24} \ \text{kg} \times 10 \ \text{kg}}{(6.4 \times 10^6 \ \text{m})^2} \\ &= \frac{400 \times 10^{13}}{41 \times 10^{12}} \ \text{N} \\ &= 98 \ \text{N} \ , \end{aligned}$$

and that of the earth on the student is

$$F_{\text{gravity}} = 6.67 \times 10^{-11} \ \frac{\text{N-m}^2}{\text{kg}^2} \times \frac{6 \times 10^{24} \ \text{kg} \times 50 \ \text{kg}}{(6.4 \times 10^6 \ \text{m})^2}$$

$$= \frac{2000 \times 10^{13}}{41 \times 10^{12}} \text{ N}$$

$$= 490 \text{ N} .$$

Thus, the gravitational forces between the earth and the student and the earth and the stereo are roughly one hundred billion times more than that between the student and the stereo. Hence, it is the earth-student and earth-stereo gravitational interactions that are most readily observed.

WORKED EXAMPLE

The planet Neptune is about 17 times more massive than the Earth and is 30 times farther away from the Sun. Compare the magnitude of the gravitational force between the Sun and Neptune relative to that between the Sun and the Earth.

SOLUTION: The force of gravity between the earth and the sun is

$$F_{\substack{\text{gravity} \\ \text{Earth-Sun}}} = G \frac{\substack{\text{mass of} \\ \text{Earth}} \times \substack{\text{mass of} \\ \text{Sun}}}{\left(\substack{\text{Earth-Sun} \\ \text{distance}}\right)^2} .$$

The force of gravity between Neptune and the sun is

$$F_{\substack{\text{gravity} \\ \text{Neptune-Sun}}} = G \frac{\substack{\text{mass of} \\ \text{Neptune}} \times \substack{\text{mass of} \\ \text{Sun}}}{\left(\substack{\text{Neptune-Sun} \\ \text{distance}}\right)^2} .$$

Since the distance between Neptune and the sun is 30 times the Earth-Sun distance and Neptune's mass is 17 times that of the Earth, we may write

$$F_{\substack{\text{gravity} \\ \text{Neptune-Sun}}} = G \frac{17 \; \substack{\text{mass of} \\ \text{Earth}} \times \substack{\text{mass of} \\ \text{Sun}}}{\left(30 \times \substack{\text{Neptune-Sun} \\ \text{distance}}\right)^2} .$$

Thus,

$$F_{\substack{\text{gravity} \\ \text{Neptune-Sun}}} = \frac{G \times 17 \; \substack{\text{mass of} \\ \text{Earth}} \times \substack{\text{mass of} \\ \text{Sun}}}{900 \left(\substack{\text{distance between} \\ \text{Earth and Sun}}\right)^2} .$$

Continued

WORKED EXAMPLE

Continued

Dividing this equation by the expression for the Earth-Sun gravitational force, yields

$$\frac{F_{\substack{\text{gravity} \\ \text{Neptune-Sun}}}}{F_{\substack{\text{gravity} \\ \text{Earth-Sun}}}} = \frac{\left[\dfrac{17 \times G \text{ mass of Earth} \times \text{ mass of Sun}}{900 \text{ (Earth-Sun distance)}^2} \right]}{\left[\dfrac{G \times \text{ mass of Earth} \times \text{ mass of Sun}}{\text{(Earth-Sun distance)}^2} \right]}$$

$$= \frac{17}{900} = 0.019 \ .$$

The gravitational force between Neptune and the Sun is about 2% of that between the Earth and the Sun.

REVIEW 5.1 QUESTIONS

1. What is the universal law of gravitation?
2. Describe how each of Newton's laws of motion is used in the development of the universal law of gravitation.
3. What is an inverse square law?
4. How did Newton employ Kepler's harmonic law in the development of his universal law of gravitation?

REVIEW 5.1 PROBLEMS

1. Two spheres having masses of 100 kg and 200 kg are separated by a distance of 0.8 m. What is the magnitude of the gravitational force between these spheres? How does this force compare in magnitude with the weights of the spheres?
2. If the earth were to be moved to a distance that is 20 times farther away from the sun, does the force of gravity between the earth and the sun increase or decrease? By what ratio?

5.2 Surface Gravity

Having formulated his law of universal gravitation, Newton then proceeded to investigate some of its implications and consequences. For example, consider a person who is standing on the surface of the earth as shown in Figure 5.3. Relative to that person, various regions of the earth's interior lie in various directions at various distances. Thus, each mass element within the earth will have its own gravitational interaction with the person on the surface. What, then, is the *net*

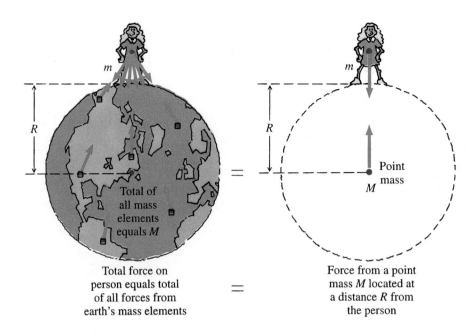

FIGURE 5.3 ··········

The Point Mass and Surface Gravity If an object has a mass distribution that is spherically symmetric, the object gravitationally acts as if all of its mass were concentrated at a point at the center of the object called the center of gravity. Thus, the acceleration of gravity at the surface of the object is GM/R^2 where R is the object's surface radius and M is its total mass.

result of all of these gravitational forces? To answer this question, Newton first assumed the earth could be thought of as being made up of a very large number of tiny gravitating mass elements. Through the use of his ''fluxions'' mathematics or calculus, Newton was mathematically able to obtain the total gravitational effect from this infinitude of mass elements within the earth. The result is amazingly simple. An object having a spherical array of mass gravitationally behaves as if all of its mass were concentrated at a point at its center. This meant that gravitating spheres such as the earth, moon, and sun could be regarded simply as gravitating points in space. Thus, the net gravitational force experienced by a person standing on the earth's surface is exactly equivalent to the gravitational force exerted on that person by a ''mass point'' or *center of gravity* having the same mass as the earth located at a distance equal to the radius of the earth. The force of gravity between the entire earth and a person or object at the earth's surface is then

$$\text{force of gravity} = G \frac{\text{mass of earth} \times \text{mass of object}}{(\text{radius of earth})^2}.$$

From Newton's second law of motion, we also have that force of gravity on the object is the product of the object's mass and the acceleration of gravity at the earth's surface. Thus, we may write that

$$\text{force of gravity} = \text{mass of object} \times \begin{array}{c}\text{acceleration}\\\text{of earth's}\\\text{surface gravity}\end{array}.$$

If we equate these two expressions for the force of gravity on the object, we find

$$\text{mass of object} \times \begin{array}{c}\text{acceleration}\\\text{of earth's}\\\text{surface gravity}\end{array} = G \frac{\text{mass of earth} \times \text{mass of object}}{(\text{radius of earth})^2}.$$

Thus,

$$\begin{matrix}\text{acceleration} \\ \text{of earth's} \\ \text{surface gravity}\end{matrix} = G\,\frac{\text{mass of earth}}{(\text{radius of earth})^2} \qquad\qquad \textbf{5.2}$$

or

$$g = G\,\frac{M}{R^2}\;.$$

Measurements made of the earth's radius and mass (see Physics Formulation 5.2) indicate that the earth has a radius of about 6.37×10^6 m and a mass of 5.98×10^{24} kg. If we substitute these values into the preceding expression for the acceleration of the earth's surface gravities, we obtain

$$\begin{matrix}\text{acceleration} \\ \text{of earth's} \\ \text{surface gravity}\end{matrix} = 6.67 \times 10^{-11}\,\frac{\text{N-m}^2}{\text{kg}^2} \times \frac{5.98 \times 10^{24}\text{ kg}}{(6.37 \times 10^6\text{ m})^2}$$

$$= \frac{39.9 \times 10^{13}}{40.6 \times 10^{12}}\text{ m/sec}^2$$

$$= 9.8\text{ m/sec}^2\;,$$

which is the value we observe.

Even though the value of R changes somewhat due to topography, the rotational flattening of the earth, etc., the resulting variations in the acceleration of the earth's surface gravity or simply its *surface gravity* amount to only a few tenths of a percent. This is the justification for the approximation made in Chapter 3 that the magnitude of g has a constant value of 9.8 m/sec^2.

Notice also that the value of the mass of the object does not appear in the final expression for that object's gravitational acceleration. In other words, the gravitational acceleration experienced by a falling body does not depend on the mass of that body. Any object of any mass will, therefore, fall to the ground at exactly the same rate, in the absence of other forces such as air resistance and the like. From this result, Newton was able to provide the mathematical basis

FIGURE 5.4 · · · · · · · · · · ·

Gravity Inside of a Planet If one burrows deep inside of a spherically symmetric mass distribution, the object will gravitationally act as if all of the mass within the person's distance to the center of the object were concentrated at a point at the center of the object. The net gravitational effect from all of the mass remaining outside of that distance is zero.

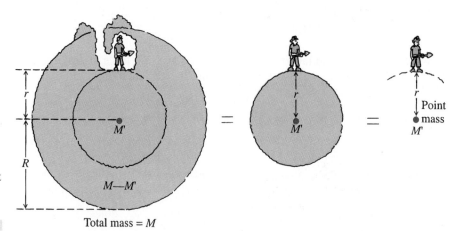

Total mass = M

for Galileo's empirical observation at the turn of the seventeenth century that bodies of different masses dropped from the same height strike the ground at the same time. It should be emphasized that we are dealing here with the *rate* at which a given object falls. The weight of the object, or the gravitational force exerted on the object by the earth, is equal to mass $\times$ *g* and, therefore, *does* depend on the object's mass in accordance with Newton's second law of motion.

For spherical objects in space such as the stars and planets, we find that a wide range of surface gravities exists. The moon has a surface gravity, for example, that is about ⅙ as much as the earth, while the acceleration of gravity in the sun's outer atmospheric layers is nearly 30 times that of the earth. Some celestial objects exist that have very large masses compacted into very small spheres, with the result that their surface gravities are billions of times larger than what we experience here on the earth's surface.

To this point, we have discussed the gravitational acceleration that exists on the surface of a spherical array of mass. Suppose an object is not confined to the earth's surface, but is situated at some distance *d* from the center of the earth, which is significantly different from the surface radius *R*. If the object is located inside of the earth via a tunnel or mine shaft, Newton was able to demonstrate the rather interesting result that any mass further than the object's distance from the center of the earth would have no net gravitational effect on the object. In essence, it would be as if the object were experiencing the surface gravity of a "new" sphere having a radius equal to the object's distance to the center of the earth and a mass equal to the mass contained within that sphere (Figure 5.4). On the other hand, if the object is located at a distance *d* greater than the surface radius, the earth behaves gravitationally as if it were a point mass located at a distance *d* from the object. Under such circumstances, the universal law of gravitation tells us the following about the acceleration of gravity:

$$\frac{\text{acceleration}}{\text{of gravity}} = G \, \frac{\text{mass of earth}}{(\text{distance})^2} \, .$$

Newton seized on this result to test his theory of gravitation on the moon (Figure 5.5).

Having established that the gravitational pull of the earth could be represented by a gravitating point mass at the earth's center, Newton then noted that the moon's mean distance from the earth's center of gravity was some 384,400 km, or about 60 times farther than a person standing on the earth's surface at a distance

Moon:
Distance = 384,000 km
Acceleration = 0.272 m/sec²
60× farther from earth's
center of gravity than apple

v Falls with 1/3600 of apple's acceleration

Apple:
Distance = 6400 km
Acceleration = 9.8 m/sec²
60× closer to earth's
center of gravity
than moon
Falls with 3600× more acceleration than moon

Earth's surface

To earth's center of gravity

FIGURE 5.5 ··········

Newton's Verification of the Universal Law of Gravitation An apple falling to the earth's surface at a distance of roughly 6400 km does so with an acceleration of 9.8 m/sec². The moon, at a distance of 384,000 km is about sixty times further from the earth's center of gravity than the apple. Therefore, Newton argued that the acceleration with which the moon "falls" in its orbit about the earth should be according to the Universal Law of Gravitation, be about $1/(60)^2$ or ⅟₃₆₀₀ as much as the acceleration experienced by the apple. This reasoning predicts a value for the moon's acceleration of (9.8 m/sec²)/3600 or 0.0272 m/sec², the value that is in fact obtained from direct observation.

of 6370 km from the earth's center of gravity. If the universal law of gravitation were correct, then an object 60 times farther from the earth's center of gravity ought to experience a gravitational acceleration of $1/(60)^2$ or $\frac{1}{3600}$ as much as the 9.8 m/sec^2 experienced at the earth's surface. Thus, Newton predicted that the moon should be "falling" about the earth with a gravitational orbital acceleration of

$$\text{moon's orbital acceleration (predicted)} = \frac{9.8}{3600} \text{ m/sec}^2$$

or

$$= 0.00272 \text{ m/sec}^2 .$$

To calculate the observed value of the moon's gravitational orbital acceleration, Newton assumed that the moon's motion could be accurately approximated by simple circular motion. Under such an assumption, the total distance that the moon travels in one orbit is merely the circumference of a circle having a radius equal to the earth-moon distance, or

$$\text{distance moon travels in one orbit} = 2\pi \times \text{radius of orbit}$$

$$= 2\pi \times 3.844 \times 10^8 \text{ m}$$

$$= 24.15 \times 10^8 \text{ m} .$$

The time it takes the moon to traverse one orbit about the earth is equal to 27.32 days, or about 2.36×10^6 seconds. The average speed of the moon in its orbit, therefore, is

$$\text{average speed of moon} = \frac{\text{total distance for one orbit}}{\text{time for one orbit}}$$

$$= \frac{24.15 \times 10^8 \text{ m}}{2.36 \times 10^6 \text{ sec}}$$

$$= 10.23 \times 10^2 \text{ m/sec} .$$

Recalling Eq. 3.1, the centripetal acceleration experienced by the moon is

$$\text{centripetal acceleration} = \frac{(\text{speed of moon})^2}{\text{radius of orbit}} .$$

Substituting the values for the average speed of the moon and its orbital radius into this equation yields

$$\text{centripetal acceleration} = \frac{(10.23 \times 10^2 \text{ m/sec})^2}{3.844 \times 10^8 \text{ m}}$$

$$= \frac{104.7 \times 10^4}{3.844 \times 10^8} \text{ m/sec}^2$$

$$= 27.2 \times 10^{-4} \text{ m/sec}^2$$

$$= 0.00272 \text{ m/sec}^2 .$$

Thus, the observed value of the moon's centripetal acceleration is identical to that predicted by the universal law of gravitation. This magnificent verification was only the first of a long and storied string of triumphs for this simple but elegant law of nature.

WORKED EXAMPLE

The moon is about $3/11$ the size of the earth and has a mass that is $1/81$ that of the earth. How does the moon's surface gravity compare to that of the earth?

SOLUTION: The moon's surface gravity is given by

$$g_{moon} = \frac{G \times \text{mass of moon}}{(\text{radius of moon})^2} \ .$$

The earth's surface gravity is given by

$$g_{earth} = \frac{G \times \text{mass of earth}}{(\text{radius of earth})^2} \ .$$

If we form the ratio g_{moon}/g_{earth}, we obtain

$$\frac{g_{moon}}{g_{earth}} = \frac{\left[\dfrac{G \times \text{mass of moon}}{(\text{radius of moon})^2}\right]}{\left[\dfrac{G \times \text{mass of earth}}{(\text{radius of earth})^2}\right]}$$

$$= \frac{\text{mass of moon}}{\text{mass of earth}} \left(\frac{\text{radius of earth}}{\text{radius of moon}}\right)^2 \ .$$

Since the (mass of moon/mass of earth) $= 1/81$ and (radius of earth/radius of moon) $= 11/3$, we have that

$$\frac{g_{moon}}{g_{earth}} = \frac{1}{81} \times \left(\frac{11}{3}\right)^2$$

$$= \frac{1}{81} \times \frac{121}{9}$$

$$= \frac{121}{729} = 0.17 \ .$$

Thus, the acceleration of the moon's surface gravity is about $1/6$ that of the earth.

1. What is meant by the center of gravity?
2. Suppose you were to journey toward the earth's center from the moon. Describe how the earth's gravitational acceleration would change as you move from the moon to the earth's surface and then burrow toward the earth's center.

REVIEW 5.2 QUESTIONS

3. Explain why objects of unequal mass fall to the surface of the earth at the same rate.

4. Describe how Newton tested his universal law of gravity.

5. Justify the assumption made in Chapter 3 that the value of g is constant at the earth's surface.

REVIEW
5.2
PROBLEMS

1. The earth's orbital radius is about 1.5×10^{11} meters and its orbital period of one year is equal to about 3.2×10^7 seconds. Find the centripetal acceleration of the earth's orbital motion.

2. What is the gravitational acceleration experienced by a satellite in orbit about the earth at a distance one-half that of the moon?

5.3 The Two-Body System

A gravitating system of considerable importance to astronomers is the two-body system, which, as the name implies, is a system in which two objects are moving in each other's gravitational force (Figures 5.6 and 5.7). Since the mathematical form of the universal law of gravitation is known, we can generate an equation of motion for the two objects involved. This equation can then be solved, thereby providing us with the characteristics of the motion that occurs in such a system.

First of all, if a two-body system is moving through space, the overall motion of the system behaves as if the sum of both masses were concentrated at a point mass called the *barycenter,* which is located along a line connecting the two centers of gravity of the bodies. The position of the barycenter along this line is given by

$$\frac{\text{distance to object 1}}{\text{distance to object 2}} = \frac{\text{mass of object 2}}{\text{mass of object 1}}.$$

Thus, if an object is twice as massive as a second object, the barycenter of their mutual revolution will be located along a line joining them at a distance that is twice as far from the smaller mass as from the larger mass (Figure 5.8). Clearly

FIGURE 5.6

The Two-Body System Two spherically shaped objects gravitationally behave as if they were point masses separated by the distance between their respective centers of mass.

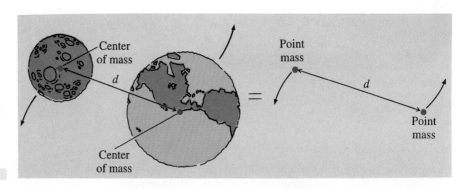

Center of mass

d

Center of mass

Point mass

d

Point mass

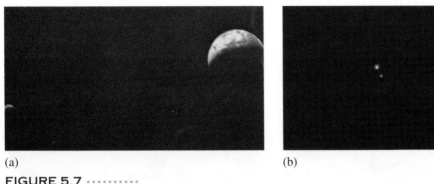

(a)   (b)

(c)

FIGURE 5.7 ··········

Some Examples of Two-Body Systems   (a) The earth-moon system; (b) the binary star system Zeta Ursa Majoris (Mizar); (c) the galaxies NGC 5426 and NGC 5427.

if one object has a very large mass compared to the second object, the barycenter will lie very close to the center of gravity of the massive object. The earth, for example, is about 81 times more massive than the moon. Thus the barycenter of the earth-moon system is 81 times closer to the earth's center of gravity than to that of the moon. Since the earth-moon distance is about 384,000 km, the earth-moon barycenter is located 4680 km from the earth's center of gravity, or about 1690 km below the earth's surface!

One also finds in a two-body system that the motion of one object about the other must occur along one of the following types of mathematical curves: circle, ellipse, parabola, or hyperbola (Figure 5.9). No other path or orbit is possible for a two-body system. Circular and elliptical orbits are sometimes called ''capture'' orbits because of the fact that one body is in effect ''trapped'' by the other's gravity. The moon orbiting the earth, the earth orbiting the sun, and stars orbiting other stars are all examples of objects moving in elliptical and circular orbits.

Kepler's first law of planetary motion can then be demonstrated within the larger framework of Newton's theory of motion and the universal law of gravitation. We also find, however, that two additional types of orbits are possible that are not included in Kepler's first law of planetary motion. These are the parabolic and hyperbolic orbits, which are often referred to as encounter orbits. For these types of orbits, a given object sweeps by the second body in a one-time-only encounter, such as the Voyager space probe flybys of the planets Jupiter, Saturn, Uranus, and Neptune.

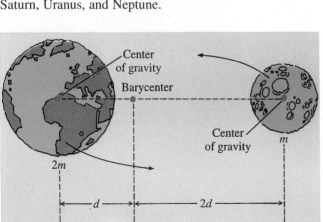

FIGURE 5.8 ··········

The Barycenter   If two bodies are revolving in each other's gravity, they will do so about a point along a line joining the two centers of mass. The distance of the point from the two bodies varies inversely with the masses of the bodies. In this case, since the mass of the larger body is twice that of the smaller, the location of the barycenter is twice as far from the smaller mass as it is from the larger mass.

FIGURE 5.9 • • • • • • • • • • •

Possible Orbits for a Two-Body Gravitating System If two objects move about each other and their mutual gravitation is the sole force present, the only possible orbital paths are the circle, ellipse, parabola, and hyperbola. The closed circular and elliptical orbits are called *capture orbits* while the open parabolic and hyperbolic orbits are called *encounter orbits*.

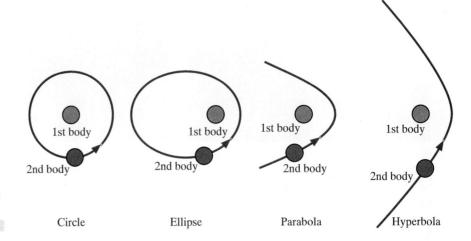

Circle   Ellipse   Parabola   Hyperbola

On a somewhat smaller scale, the paths of projectiles can also be described in terms of the orbits permitted in a two-body system. Any given projectile can be thought of as forming a two-body system with the earth's center of gravity. As such, the projectile must "fall" along one of the two-body orbital paths. In the case of projectile motion, however, the earth's surface "gets in the way" of the projectile as it attempts to fall along its orbit. The result is an "intercepted" orbit, which we perceive as the arc of the object's trajectory (Figure 5.10). For objects moving close to the earth's surface over distances that are small compared to the earth's size, however, the more complicated two-body analysis of the object's motion can, to a high degree of accuracy, be replaced by one based on the simpler Chapter 3 assumptions that the object is moving over level ground and is subject to a constant downward acceleration equal to *g*.

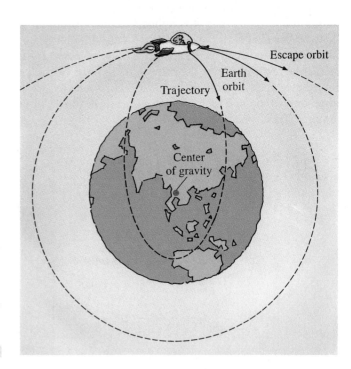

FIGURE 5.10 • • • • • • • • • •

Trajectories as Two-Body Orbital Paths Any object undergoing projectile motion forms a two-body system with the earth's center of gravity. Thus, the trajectory must be one of the orbital paths permitted in a two-body system. If the earth's surface gets in the way of the orbital path, we view the path as a trajectory. If the earth's surface does not get in the way of the orbit, the object will either fall into orbit about the earth or leave the vicinity of the earth altogether.

5.2 Tides

Anyone who has spent time along the seacoast has had the opportunity to observe the movement of the earth's waters, which are referred to as the tides. Twice during a time period of just over a day, the surface of the ocean at a given location flows to a "high-tide" level and then ebbs to a "low-tide" level, with the difference in levels as much as 15 meters or more.

For centuries it had been suspected that the tides were somehow related to the moon, but the justifications for such an influence were buried in the astrological morass of other presupposed stellar and planetary influences. Thus when Kepler expressed his belief that the moon's influence was responsible for the tides, Galileo chided him for having "given his ear and assent to the moon's predominancy over the water, and to occult properties and such like trifles." In the end, however, it was Kepler and not Galileo who

proved to have the correct view on this issue, albeit not for astrological reasons.

Once more it was Isaac Newton who provided the sound theoretical basis for the ocean's tidal movements. Although the gravitational effect of a given spherical mass distribution can be approximated by a mass point, the finite size of the object can result in a gravitationally induced distortion of its shape from a second object. To understand how this can occur, let us consider an object A, shown in the figure on the left, which is orbiting about a second gravitating object B. Because of the inverse square law for gravity, a mass element on the side of A nearest object B will be subject to a higher gravitational force than a mass element on the far side of A from B because the near-side mass element is one diameter of A closer to object B. As a result of the inverse square law for gravity, the near-side mass

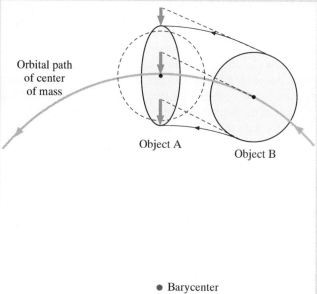

• Barycenter

••• **Tidal Distortion** If an extended object is subject to a gravitational force from a second object, the regions of the extended object closest to the second object "fall" at a greater gravitational acceleration than those farther away. The net result is a stretching or "tidal distortion" of the object into an elongated shape.

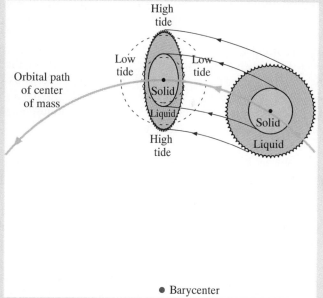

• Barycenter

••• **The Earth's Tides** The water on the earth's surface responds far more to the moons tidal force than does the earth's solid surface. As a result, the water forms into an elongated shape but the earth's surface does not, therefore producing the tidal bulges we perceive as the high-tide regions.

Continued

element will experience a slightly larger gravitational acceleration than the far-side mass element. In effect, the near side of object A "falls" about B more rapidly than does the far side. This difference in gravitational acceleration results in a "stretching effect" or *tidal distortion,* which causes the shape of object A to become elongated as shown in the figure on the left.

If the object is rigid, the actual distortion from such tidal force is minimal. The tidal distortion in the solid surface of the earth from the moon's gravity, for example, is less than 25 cm. On the other hand, if the substance involved has a very low rigidity, such as water, then the tidal distortion will be considerably larger. For a combination of rigid and "easy-flow" material, such as a solid earth covered with a layer of water, the result is an essentially undistorted solid sphere surrounded by an elongated surface as shown in the figure on the right. The regions on the earth's surface located at one or the other of the water's elongations or tidal bulges experience high ocean tides, while those situated at the "shallow" parts of the water's surface profile experience low tides. As the earth spins on its axis relative to the moon, each portion of the earth rotates under the various regions of the water's tidally distorted profile. Thus, a typical location on the earth will experience two high tides and two low tides in a given 24 hour and 50 minute long rotation of the earth relative to the moon.

The sun exerts a similar but smaller set of tidal forces over the earth's surface and when the sun, moon, and earth lie roughly along the same line, the tidal bulges produced by the sun and moon reinforce one another to produce the highest tides, called spring tides. When the sun and moon are at right angles to each other as seen from the earth, their respective tidal bulges are also at right angles, with the result that the water depth is fairly evenly distributed over the earth's surface. These tides are called neap tides and are the lowest tides that can occur.

Of course, many other factors can contribute to the size of ocean tides, including the friction between land and water, the tilt of the earth's axis of rotation, and the topography of the given region. Thus, extraordinarily high tides can occur in bodies of water such as the Bay of Fundy or the Bering Strait, which are somewhat enclosed by dry land. In these regions, tides of more than 15 meters occur on a regular basis.

Because every object has a finite size, tidal distortions, like gravitational forces, exist between any two gravitating objects in the universe. If, however, the distance between the objects is large compared to their linear dimensions, the tidal forces become vanishingly small. Thus the earth's surface radius is about 0.02 that of its distance to the moon, while the earth's surface radius is 1.0×10^{-5} that of the distance of the comparable-mass object Europa, which is a member of the Jupiter satellite system. As a result, Europa's tidal effects on the earth's oceans are virtually nonexistent.

Analysis of the motion in a two-body system also indicates that an orbiting body will move about the second body in such a way that a line joining the two bodies will sweep out a constant area of orbit per unit time regardless of the objects' relative positions. This result is simply Kepler's second law of planetary motion and is often referred to as the *law of equal areas.* The actual size of the orbital area swept out per unit time varies from object to object, but for a given object this rate of swept out orbital area must always remain the same. For example, if a given asteroid sweeps out 2×10^{22} square meters of orbit during a given year in its orbit about the sun, it will sweep out that same orbital area in any other one-year time period. A second asteroid, however, might sweep out 5×10^{22} square meters of orbit in a given year, but it would then have to continue to orbit at that same rate of 5×10^{22} square meters of orbital area per year.

For objects moving in noncircular orbits, the law of equal areas in effect dictates that the object must speed up in its orbit when the objects are close together and slow down when they are far away if a constancy in the orbital area swept out per unit time is to be maintained (see Figure 5.11). Thus, as the famed Halley's comet moves in its highly elongated elliptical orbit about the sun, it only spends a few months of its 76-year orbital period in the vicinity of the sun. As the comet moves back into the deep recess of the solar system, however, its increased distance from the sun permits it to sweep out its constant orbital area per unit time at a lower orbital speed.

Further mathematical analyses of the motions in a two-body system result in the formulation of a more comprehensive version of Kepler's harmonic law. For a two-body system in which the orbits are circular or elliptical, we can show that the mean distance, time for one orbit, and the sum of the masses of the two bodies involved are related as follows:

$$\frac{(\text{mean distance})^3}{(\text{time for one orbit})^2} = \frac{G}{4\pi^2} (\text{mass 1} + \text{mass 2}) .$$

If we consider a planet-sun two-body system and note that the sun's mass is far larger than that of any of the planets, we find that we can, to a high degree of accuracy, neglect the mass of the planet in the right-hand side of the preceding equation. Thus

$$\text{mass of sun} + \text{mass of planet} \approx \text{mass of sun}$$

and

$$\frac{(\text{mean distance})^3}{(\text{time for one orbit})^2} = \frac{G}{4\pi^2} \left(\begin{array}{c} \text{mass of} \\ \text{sun} \end{array} \right) .$$

The quantity on the right is thus very nearly the same for every planet, which is what Kepler found in his harmonic law.

Thus far in our discussion of the two-body system, we have naively assumed that the bodies involved behave as point masses and that no forces are present aside from the mutual gravitation of the two bodies involved. This is not the case in the real universe. Most bodies, for example, do not have perfectly spherical mass distributions. Instead, there exist within these objects small-scale departures or asymmetries from this ideal. As a result, the point mass representation for the gravitational behavior of such objects is one that is not exact. Moreover, according to the universal law of gravitation, every mass in the universe interacts gravitationally with every other mass. In almost every instance, such interactions are vanishingly small. For some situations, however, gravitational interactions with bodies other than the two primary objects can be significant. If such additional gravitational interactions are relatively small compared to the mutual gravitation of the two primary objects, they are referred to as perturbations and manifest themselves as small but observable departures from the motion expected in a ''pure'' two-body system. The moon's orbit about the earth, for example, has a roughly elliptical shape, as predicted by the analysis of the earth-moon system as a two-body system. The moon's motion, however, also exhibits small-scale departures from a strictly elliptical orbit owing to the gravitational effects of the earth's slightly nonspherical mass distributions as well as the gravitational effects of some of the other planets in the solar system.

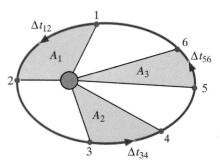

FIGURE 5.11 · · · · · · · · · · ·

The Law of Equal Areas   An object in orbit sweeps out a constant amount of orbital area per unit time. If each of the time intervals Δt is the same, then the area of orbit swept out during each of those time intervals must also be the same. Thus, if $\Delta t_{12} = \Delta t_{34} = \Delta t_{56}$, then $A_1 = A_2 = A_3$.

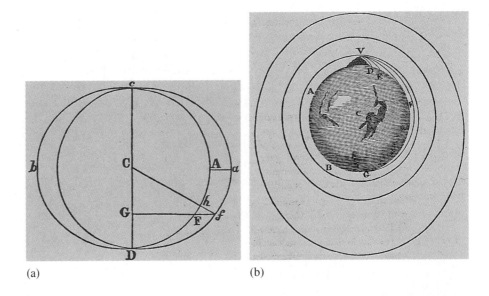

(a)   (b)

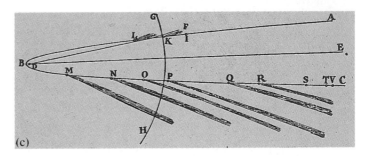

(c)

FIGURE 5.12 ··········
Illustrations from Newton's Works on Gravitation   (a) the ocean's tides; (b) ''Newton's cannon''; (c) the orbit of Halley's comet.

For some systems, such as star clusters and galaxies, the gravitational effects of additional bodies are not small compared to the mutual gravitation of the two primary objects. Such systems are referred to as ''many-body'' systems and any description of the motions of an object in a many-body system must necessarily take into account all of the gravitational influences from all of the bodies present. Unfortunately, there are no simple mathematical techniques for dealing with such systems. As a result, these types of systems must be analyzed by means of computers and complex mathematical techniques.

From its conception, Newton's universal law of gravitation has enjoyed phenomenal success in the analysis of the behavior of gravitating bodies. Within the pages of a section of the *Principia* entitled ''The System of the World'' are to be found a correct explanation for the degree of the earth's flattening, oceanic tides, a valid technique for launching an artificial satellite, and an accurate rendition of the orbit of Halley's comet at a time when comets were widely regarded as supernatural portents of doom, death, and destruction (Figure 5.12). The universal law of gravitation has lead to the discoveries of planets, satellites, and stellar companions. It has permitted the motions of the moon and planets to be predicted with an accuracy such that they form the basis for an entire system of timekeeping called ephemeris time. It even inspired the following ode from Newton's close friend and second Astronomer Royal of England, Sir Edmund Halley:

. . . But now, behold
Admitted to the banquets of the gods,
We contemplate the politics of heaven;
And spelling out the secrets of the earth,
Discuss the changeless order of the world.

Such was the enthusiastic belief in the human capacity to comprehend the workings of the physical universe during the European Enlightenment. Although we have since discovered that the "gods" have yet to display to us their entire menu, our experience with gravitation holds the promise that as more of the banquet is served, we will be equal to the task of "dining" on it.

WORKED EXAMPLE

The planet Jupiter orbits the Sun once every 11.9 years at a distance of 7.8×10^{11} meters. Assuming that Jupiter moves in a circular orbit, answer the following:

a. At what rate does Jupiter sweep out its orbital area?

b. How long will it take Jupiter to sweep out one trillion square meters of orbit?

c. How much area will Jupiter sweep out in one century?

SOLUTION: (a) The rate at which Jupiter sweeps out orbital area is

$$\text{rate} = \frac{\text{area of orbit}}{\text{elapsed time}} .$$

If the orbit has a radius of 7.8×10^{11} m, then the area is that of a circle having

$$\begin{aligned}
\text{area} &= \pi \times (7.8 \times 10^{11})^2 \text{ m}^2 \\
&= \pi \times 60.8 \times 10^{22} \text{ m}^2 \\
&= 191 \times 10^{22} \text{ m}^2 .
\end{aligned}$$

Since the time it takes Jupiter to sweep out the total area of its orbit is just its sidereal period, we have

$$\begin{aligned}
\text{rate} &= \frac{191 \times 10^{22} \text{ m}^2}{11.9 \text{ years}} \\
&= 16.1 \times 10^{22} \text{ m}^2/\text{year} .
\end{aligned}$$

(b) The time required to sweep out a given area is

$$\text{time} = \frac{\text{area to be swept out}}{\text{rate}} .$$

The time needed to sweep out one trillion square meters or 1.0×10^{12} m^2 is then

Continued

WORKED EXAMPLE

Continued

$$\text{time} = \frac{1.0 \times 10^{12} \text{ m}^2}{16.1 \times 10^{22} \text{ m}^2/\text{yr}}$$

$$= 0.062 \times 10^{-10} \text{ yr}$$

$$= 6.2 \times 10^{-12} \text{ yr} \ .$$

Since one year is about 3.2×10^7 sec

$$\text{time} = 6.2 \times 10^{-12} \text{ yr} \times \frac{3.2 \times 10^7 \text{ sec}}{\text{yr}}$$

$$= 19.9 \times 10^{-5} \text{ sec}$$

$$= 0.002 \text{ seconds}$$

(c) In one century or 100 years, Jupiter will sweep out a total area of

$$\text{area} = \text{rate} \times \text{elapsed time}$$

$$= 16.1 \times 10^{22} \frac{\text{m}^2}{\text{yr}} \times 100 \text{ yr}$$

$$= 16.1 \times 10^{24} \text{ m}^2 \ .$$

1. Compare and contrast Kepler's original laws of planetary motion with those obtained from the universal law of gravitation.

2. Give some examples of two-body systems.

3. What is a barycenter? How does it compare with a center of gravity?

4. Explain what is meant by (a) the harmonic law and (b) the law of equal areas.

5. What are the differences, if any, between orbital motion and projectile motion?

1. An orbiting satellite is observed to sweep out 5 million square meters of its orbit in 4 hours.

   a. How long will it take the satellite to sweep out 30 million square meters?
   b. How much area will the satellite sweep out in a day?

2. Look up the mean distance from the sun and sidereal periods for two of the planets in the solar system and calculate the ratio (mean distance)^3/(sidereal period)^2 for each one. Discuss how the results compare.

Chapter Review Problems

1. Two spheres, each having a mass of 100 kg, are separated by a distance of 0.3 m. What is the magnitude of the gravitational force between these spheres?

2. Two identical spheres are to be placed one meter apart. How massive must the spheres be in order to have a mutual gravitational force of 1 N?

3. If the acceleration of the earth's gravitation is 9.81 m/sec^2 at a distance of 6.4×10^6 meters from the center of the earth, at what distance from the earth's center will the acceleration of gravity be 1.09 m/sec^2?

4. Compare the force of gravity between the earth and the sun as it now exists to what it would be if:
   a. the earth's mass were tripled.
   b. the sun's mass were cut in half.
   c. the earth-sun distance quadrupled.
   d. the earth-sun distance were $\frac{1}{5}$ as much as it is now.
   e. conditions (a), (b), and (c) all occurred at once.

5. Calculate the force of gravity between the earth and the sun if the mass of the sun is 2×10^{30} kg, the mass of the earth is 6×10^{24} kg, and the radius of the earth's orbit is 1.5×10^{11} m.

6. The radius of the earth's orbital path is about 1.5×10^{11} m, and the time it takes the earth to complete one orbit is 3.2×10^7 seconds. Find the centripetal acceleration of the earth in its orbit about the sun. How does this value compare with the acceleration of the earth's surface gravity?

7. A physicist working in a laboratory on a different planet finds that the gravitational force between a 1.0×10^{-6} kg sphere and a one kilogram sphere separated by 0.1 m is 6.67×10^{-19} N. What is the value of G for this experiment? How does this compare with the value of G observed here on the earth?

8. The planet Mars has a mass of about 6×10^{23} kg, and when it is at its closest approach to Earth has a distance of about 6×10^{10} meters. At the instant of birth, a 3-kg child is about 0.2 m from the center of gravity of a 50-kg mother. Compare the gravitational force between the mother and the child with that between the child and the planet Mars at closest approach to the Earth. On the basis of your calculations, what would you say to a person who claims that the gravitational force of Mars has an ''influence'' on the child at the time of birth?

9. A softball has a mass of 0.25 kg and a diameter of 0.09 meters. What is the surface gravity of the softball?

10. The moon has a mass of about $\frac{1}{81}$ that of the earth, and the surface gravity about $\frac{1}{6}$ that of the earth. By what factor must the moon's radius be enlarged (or reduced) in order for the moon's surface gravity to be equal to that of the earth?

11. Mount Everest, the highest point on the earth's surface, has an altitude of 8882 meters. The deepest part of the ocean, the Marianas Trench of the Pacific Ocean, has a depth of 11,022 meters. If the mean radius of the earth's sea level is equal to 6.371×10^6 m from the earth's center, how accurate is the assumption that the earth's gravitational acceleration is constant on the earth's surface?

12. An astronaut lands on an unknown planet and finds that an object dropped from a height of 2 meters takes one second to hit the planet's surface. If the planet has a radius of 1.3×10^6 meters, find
   a. the planet's surface gravity.   **b.** the mass of the planet.

13. If the acceleration of the earth's gravitation is 9.8 m/sec² at a distance of 6.4 × 10⁶ m from the center of the earth, what is the mass of the earth in kilograms? How does this value compare with the total mass of six billion human beings, each of whom has an average mass of 50 kg?

14. An astronaut rockets to an asteroid where the acceleration of the surface gravity is ¹⁄₁₀₀ that at the earth's surface. If the asteroid has a radius that is also ¹⁄₁₀₀ that of the earth, how does the mass of the asteroid compare with the mass of the earth?

15. Suppose the earth suddenly shrank to ¹⁄₂ its present radius and retained the same mass. How would the earth's surface gravity be altered, if at all?

16. Two meteoroids are orbiting each other at a distance of 200 meters. If one of the meteoroids has a mass of 400 kg and the other a mass of 1600 kg, find the location of the barycenter of this system.

17. An astronomer notices that the barycenter of a certain binary star is located 50 million kilometers from star Alpha and 500 million kilometers from star Beta. Find the ratio of the masses of these two stars.

18. Verify by calculation the statement in the text that the barycenter for the earth-moon system is about 1690 km below the earth's surface.

19. The planet Jupiter has a mass of about 2 × 10²⁸ kg. The Sun has a mass of about 2 × 10³⁰ kg and a radius of 7 × 10⁸ m. Does the Sun-Jupiter barycenter lie outside of the Sun's outer layers? If not, how much more (or less) mass does Jupiter need for this to be true?

20. NASA wishes to build a deep-space binary satellite, one of whose members is to have a mass of 200,000 kg. If NASA wants the barycenter located 0.5 kilometers from the 200,000-kg satellite component and 1.0 kilometer from the second satellite component, what should the mass of the second satellite component be?

21. An artificial satellite sweeps out 100 million square meters of its orbit per hour. How much area does the satellite sweep out in one day? How long will it take the satellite to sweep out one billion square meters?

22. An asteroid is observed to sweep out 13 × 10¹⁶ m² during the year 1993. What area did the asteroid sweep out in the year 1776? If the total area of the asteroid orbit is 65 × 10¹⁶ m², how long will it take the asteroid to make one complete orbit of the sun?

23. In 1957, the Soviet satellite *Sputnik I* was placed into an orbit about the earth 240 km above the earth's surface. If *Sputnik I* orbited the earth once every 90 minutes and the earth has a radius of about 6.37 × 10⁶ meters, find the mass of the earth in kilograms. You may assume that the mass of *Sputnik I* is negligible compared to the mass of the earth.

THOUGHT QUESTIONS

1. What indications do we have that a universal gravitation exists?

2. While on a date your sweetheart says, ''I am 'gravitationally' attracted to you.'' Is this a large attraction? Explain.

3. Suppose that Kepler had found a harmonic law that stated

$$\frac{(\text{sidereal period})^4}{(\text{mean distance})^4} = \text{constant value} .$$

   What would this law imply concerning the variation of gravitational acceleration with distance?

4. What aspects of the scientific method did Kepler and Newton employ in their work on planetary motions?

5. Can you ever get so far away from an object that its gravitational interaction with you disappears completely? Explain.

6. Show that the units of G are consistent with the law of universal gravitation.

7. Do objects or entities exist in nature that do not exert gravitational forces? If so, describe them.

8. Compare and contrast the view of surface gravity presented in this chapter with that of Chapter 3. Are these two views consistent?

9. Suppose an artificial planet existed in which all of its mass was contained in a spherical shell. Describe what the gravitational acceleration would be for such an object (a) at large distances from the planet; (b) on the surface of the planet; and (c) inside of the surface of the planet.

10. Discuss ways in which the earth's surface gravity might be important to the life forms on our planet.

11. Scientists are sometimes asked the question ''Why is there gravity?'' Is the answer to such a question within the realm of the scientific method? Explain.

12. Compare and contrast the concept of center of gravity as described in Chapter 3 with that discussed in the present chapter.

13. Suppose a one-meter-diameter shaft were bored through the earth's center and out the other side. If a marble were dropped into this shaft, describe the marble's subsequent motion in the shaft.

14. How are tidal forces related to gravitational forces?

15. Of what importance are the ocean's tides to life on the earth?

16. Under what conditions might large tides be raised on celestial bodies? Can you think of any instances where such conditions might be satisfied?

17. Do you think it is possible for tidal forces to be so strong as to cause a body to be torn apart? If so, under what circumstances?

18. Describe how you would launch a rocket that you wished to (a) strike a military target thousands of miles away on the earth's surface; (b) go into an orbit about the earth; and (c) journey to the moon.

19. Is the speeding up and slowing down of orbiting objects according to the law of equal areas consistent with the universal law of gravitation? Explain.

20. Do you think that the law of equal areas applies to all of the permitted orbits in the two-body system?

21. Why doesn't the harmonic law apply to parabolic or hyperbolic orbits?

22. Discuss the relationship between projectile motion and the motion in a two-body system.

23. Suppose you place two basketballs a couple of meters apart on the floor. Is this a two-body system? Explain.

24. Why was the discovery of Neptune so important for the universal law of gravitation?

25. Look up some of the myths and legends regarding comets. With such a background, discuss the importance of Halley's successful analysis of the motion of Halley's comet using Newtonian mechanics.

26. If the radius of the earth were smaller, but its mass stayed the same, would the acceleration of the earth's surface gravity be larger or smaller? Explain.

27. What observational evidence do we have that the universal law of gravitation is indeed universal?

28. Suppose that the earth suddenly increased its mass by a factor of 10. What effect, if any, would this have on the motion of the moon?

29. Is it possible to place a satellite in orbit about the earth in such a way that it will be constantly above the same location on the earth? If so, under what circumstances?

30. Describe how one could use the observed period and mean radius of the orbit of an artificial satellite to determine the mass of the earth.

PHYSICS FORMULATION 5.1

The Gravitational Constant

In developing the universal law of gravitation, Newton left one unsolved problem for his scientific successors. That law tells us that a given object gravitationally attracts another object with a gravitational force F_G given by

$$F_G = G \frac{m_1 m_2}{d^2} ,$$

where m_1 and m_2 are the masses of the two objects and d is the distance between them. The problem that Newton left behind was the determination of the value of the quantity G, the gravitation constant.

In principle such a determination ought to be a relatively simple process. If we first solve the universal law of gravitation for G, we obtain

$$G = \frac{F_G d^2}{m_1 m_2} .$$

To determine the value of G, one needs only to measure the amount of gravitational force exerted between two objects of known masses separated by a known distance. It was quickly discovered, however, that gravitation is a comparatively weak interaction of nature in which very large masses are required to produce significant gravitational forces. For example, even though the earth's mass is impressively large compared to that of a human being, the forces that can be produced by our bodies permit us to overcome briefly that gravity in a variety of ways such as by jumping, climbing stairs, etc. In particular, a force produced between two "laboratory-sized" masses is many orders of magnitude smaller than the gravitational force existent between the earth and an object at the earth's surface. For more than a century after the publication of the *Principia*, physicists struggled with the problem of accurately measuring forces of such tiny magnitude. Finally in 1798, an English recluse named Henry Cavendish put together an ingenious device called a torsion balance, which permitted him to make the measurements necessary to obtain an accurate value of G.

A schematic diagram of the Cavendish apparatus is shown in Figure 5.13. It consists of two small spheres of known masses m which are attached to a thin, almost "massless" rod. This "barbell" arrangement is then suspended in a horizontal position by a very fine fiber to which a flat mirror has been attached. Two large spheres of known mass M are placed at known center-to-center distances from the smaller spheres. The gravitational interaction between the two pairs of spheres produces a change in the orientation of the barbell, which can be accurately measured by noting the change of position of a beam of light reflected off the flat mirror. This amount of deflection depends in a known way on the amount of force exerted on the barbell. Thus a measurement of the amount of the light beam deflection can be used to obtain the magnitude of the force F_G exerted between the two pairs of spheres. Because the values of m, M, and r are already known, the value of G can be readily calculated. Measurements of this type have yielded a value for G equal to 6.672×10^{-11} N-m^2/kg^2.

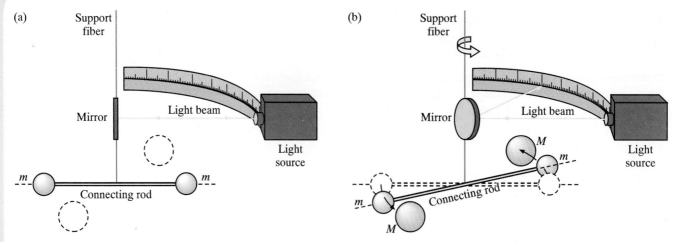

FIGURE 5.13 ··········
The Cavendish Apparatus   (a) Two known masses m connected by a thin rod are supported by a fiber to which a mirror is attached. (b) When the two known masses M are placed at known distances from the smaller masses, the mutual gravitation between m and M causes the support fiber to turn in a measurable way. From the amount of the turn the force between m and M can be calculated and a value for G determined.

QUESTIONS

1. What is the Cavendish apparatus? How does it enable one to determine the value of $G?$

2. Does the earth's gravity have any effect on the Cavendish method for determining the value of $G?$ Explain.

3. Can you think of another way that the value of G could be measured? If so, how?

PROBLEM

1. In a Cavendish experiment, a physicist notes that the force of gravity between two 50-kg balls separated by 0.4 m is 1.2×10^{-6} N. Find the value of G for these data.

"Sizing Up" the Earth

The earth's gravitation has permitted scientists to deduce a number of interesting and important factors concerning the earth and its interior. The approximate size and shape of the earth have been known for centuries. The ancient Greeks deduced the spherical nature of the earth from the curved shape of the earth's shadow at the time of a lunar eclipse. In the third century B.C., the Greek Eratosthenes was able to measure the size of the earth through the use of simple geometric arguments. A schematic of Eratosthenes's method is presented in Figure 5.14. On the first day of summer it was known that the sun was directly overhead at the Egyptian town of Syene, near modern Aswan. On that same date of the year Eratosthenes observed that the sun's rays made a 7.5° angle with an obelisk at the city of Alexandria some 800 km to the north. By assuming that the sun's high rays were parallel as they reached the earth, Eratosthenes was able to calculate the circumference and hence the radius of the earth to an accuracy within a few percent of the modern value of 6.37×10^6 meters.

Once the distance to the center of the earth is known, the mass of the earth can be obtained with an ingenuous appeal to the universal law of gravity first

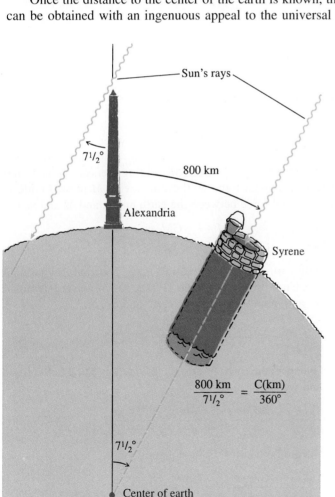

$$\frac{800 \text{ km}}{7\frac{1}{2}°} = \frac{C(km)}{360°}$$

FIGURE 5.14 ··········

The Size of the Earth In the third century B.C., the Greek mathematician Eratosthenes was able to determine the size of the earth from the shadows cast by the sun's rays on various parts of the earth's surface.

FIGURE 5.15 ··········

"Weighing" the Earth   Two identical masses m are placed on both sides of a pan balance. A large mass M is then placed at a known distance beneath one of the masses. A small mass m' is then added to the other side to bring the pans into balance. From the known values of the masses and distances involved the mass of the earth can be determined.

proposed in 1881 by the German physicist Philipp von Jolly. In this method, two identical masses m are placed on the sides of a pan balance as shown in Figure 5.15. A very large mass M is then placed under one of the two masses at a known distance d. The additional gravitational force between m and M causes the balance to tip down toward the side with M present. To balance the effect, a mass m' is added to the side opposite from M to bring the system back into equilibrium. The total gravitational force on the side having M present (side 1) is the sum of the gravitational force between the earth and m and M and m, or

$$\underset{\text{(side 1)}}{F_{\text{grav}}} = \frac{GMm}{d^2} + \frac{GM_E m}{R^2}$$

where R is the radius of the earth and M_E is the earth's mass. The total gravitational force on the side of the balance opposite M (side 2) is the sum of the gravitational force between the earth and m and between the earth and m', or

$$\underset{\text{(side 2)}}{F_{\text{grav}}} = \frac{GM_E m}{R^2} + \frac{GM_E m'}{R^2} \ .$$

Since the system is in balance, the force of gravity on side 1 is equal to the force of gravity on side 2, and we have

$$\frac{GM_E m}{R^2} + \frac{GM_E m'}{R^2} = \frac{GMm}{d^2} + \frac{GM_E m}{R^2} \ .$$

Subtracting $GM_E m/R^2$ from both sides of the equation yields

$$\frac{GM_E m'}{R^2} = \frac{GMm}{d^2}$$

or

$$M_E = \frac{R^2}{d^2} \frac{mM}{m'} \; .$$

Since all of the quantities on the right-hand side of the equation are known or have been measured in the experiment, the value of M_E, the mass of the earth, can be calculated. Using such techniques, the mass of the earth is found to be 5.976×10^{24} kg.

In this discussion, we have assumed that the earth is perfectly spherical in its shape and has a mass distribution that is perfectly spherical in its symmetry. Neither of these assumptions is strictly true. Because of its rotational motion, the earth has taken on the shape of a slightly flattened sphere, or an *oblate spheroid,* with a polar radius of 6.36×10^6 meters and an equatorial radius of 6.38×10^6 meters. This corresponds to a departure from sphericity of about 0.3%. The departures from spherical symmetry in the earth's mass distribution are found to have a similar magnitude.

QUESTIONS

1. How is the universal law of gravitation employed in determining the mass of the earth?
2. In the apparatus shown in Figure 5.15, what effect, if any, would there be on any results owing to the mutual gravitational force between the masses on sides 1 and 2 of the balance?
3. Can you think of any other way that the earth's mass could be measured? If so, explain your method.

PROBLEM

1. An astronaut journeys to an asteroid having a radius of 10,000 m and places two identical 5-kg masses on each side of a pan balance located on the asteroid's surface. A 1000-kg mass is placed 30 cm below one of the pans. To rebalance the pans, the astronaut must add a total mass of 1.5×10^{-5} kg to the pan opposite that with the 1000-kg mass beneath it. Find the mass of the asteroid.

6

WORK, ENERGY AND POWER

"All nature," wrote the English poet Samuel Coleridge, "seems at work." Certainly there is much activity in and about our lives that we refer to as work. In the spring and summer we can watch ants and worker bees carry the "work ethic" to its ultimate. Most of us work diligently at hobbies, sports, or just keeping physically fit. Presumably all of you are working hard in this course. Obviously the term "work" can be applied to various activities that are otherwise distinguished by the fact that they are "purposeful." Unfortunately one person's purposeful activity, such as a college professor's research or writing, may well seem devoid of meaning to the professor's bored students, or even to that professor's colleagues. Despite such subjectivity, work is nonetheless the agent by which "things get done" in our society and indeed in nature itself. It is this link between work and the natural world that beckons us to lend a certain measure of precision to the concept of work. As we proceed with this endeavor, we will discover that work is but one manifestation of an even more encompassing concept, one that is perhaps the most fundamental in all of scientific thought. At this point we will have had our first encounter with the entity we call *energy*.

6.1 Work

In the "everyday" world we think of work as "purposeful activity" or "effort expended to do something." In physics we specify exactly what we mean by "expended effort" and the "something" we want accomplished by that effort. To this end, then, the work done on an object is formally defined in physics as the product of a force acting on the object and the distance through which the object moves along the direction of that force (Figure 6.1). More simply

$$\text{work} = \text{force} \times \text{distance object moves} \qquad \textbf{6.1}$$

or

$$W = F \times d \, .$$

The "expended effort" is thus any force applied to an object and the "something" we wish to have the force do is to move the object. From this formal definition of work, we see that in general this quantity has the units of force $\times$ distance, and in particular can be expressed in newton-meters or N-m. The name for this particular unit for work is called the joule. One *joule* (J) is defined as the amount of work done when one newton of force acts through a distance of one meter. Thus $1 \text{ J} = 1 \text{ N-m}$ and is roughly the amount of work done if you were to lift this book 0.1 m or 10 cm above a desk top or hoist a 12-oz. can of soda 0.3 meter, about the distance from your chest to your lips.

The most common case in which work is performed is in lifting an object against the gravitational force of the earth. If we carry a box of books up a flight of stairs to a new apartment, for example, we are performing work because the books are being moved along the distance through which the lifting force acts. If the weight of the books is 300 N and we are carrying this weight up a 4-meter-high flight of stairs, the distance the box of books moves against the earth's 300-N gravitational force is 4 meters. The work expended in moving the books up the flight of stairs is force $\times$ distance moved, which in this case is 300 N $\times$ 4 m or 1200 joules. We also know from experience that we would have to do more work carrying the books up several flights of stairs as opposed to one. Five

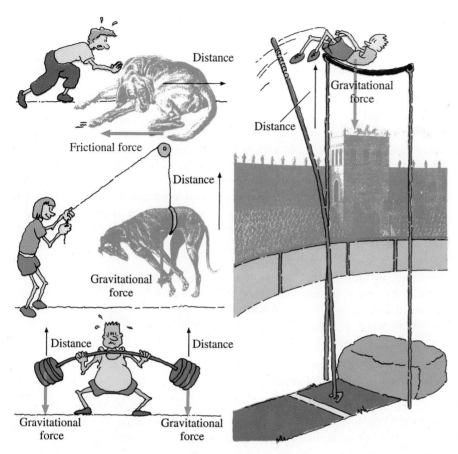

FIGURE 6.1 · · · · · · · · · ·
Work in Physics In each of these examples a force is being exerted through a distance, and work is thus being expended or produced.

4-meter flights of stairs, for example, would have a total upward distance of 20 meters, and the corresponding work required to move the 300-N load of books up such a set of stairs would be 300 N × 20 m or 6000 joules, which is indeed a larger amount of work compared to that expended in moving the books up one 4-meter flight of stairs.

Suppose now we lift the load of books off the floor and then hold it for several minutes while our new roommate tries to decide where to put them. In lifting the books off the floor, we are, of course, performing work. Once the books stop moving, however, no further work is done in the technical sense, despite the fact that our muscles may be agonizing in their effort to hold the books off the floor while our roommate struggles over the location of their final resting place. For work to be done, some sort of movement or change has to result.

A second form of work commonly encountered is the work required to overcome frictional forces. Frictional forces, as we have seen, act to prevent motion from occurring. If a box is placed on a ramp, for example, it will not slide down the ramp if the frictional forces between the box and the ramp are large enough to prevent it from doing so. If a load of books is placed on our new apartment floor and pushed along the floor from one place to another, one must exert a force on the books continuously because frictional forces between the floor and the box are continuously resisting the motion. We must therefore exert a force on the load of books all along the path between the two locations. In short, we have to do work on the books to move them from one place to the other.

To illustrate numerically, suppose we wish to push our 300-N load of books 3 meters across the room. If the coefficient of friction between the floor and the box is 0.6, then the frictional force is 300 N × 0.6 or 180 N. If we push the box 3 meters against this force, we will have to expend a total of 180 N × 3 m or 540 J of work to do so. That same amount of work would serve to lift the same 300-N load to a height of 540 J/300 N or 1.8 m. It can be seen from this type of calculation that frictional forces often consume a considerable amount of work in our attempts to overcome them, and as such are sometimes a bane to working humans and machines alike. As a result, the reduction of frictional forces in work-producing machinery is of paramount importance to engineers and technicians.

Recall that, we have met the N-m before. It is the unit by which we specify torques. This equivalence in the units by which work and torque are specified suggests that the two quantities might somehow be related, but in fact they are not. The force that produces torque must be directed at a right angle to the lever arm involved, while the force that performs work must act along the direction the object moves.

Although they are not the same as work, torques can nonetheless be of considerable use in the transmission of work through a device called a lever. The various classes of levers are considered more fully in Physics Formulation 6.1, but for any lever in general the applied force required to lift a given weight and the weight to be lifted are related as follows

$$\frac{\text{applied}}{\text{force}} \times \frac{\text{applied force}}{\text{distance}} = \text{weight} \times \frac{\text{distance weight}}{\text{is lifted}},$$

where the applied force distance and the distance the weight is lifted are proportional to the respective lever arms of the applied force and the weight. Thus for a given lever an applied force of 15 N acting along an applied force distance of 4 m will be able to transmit a total of 15 N × 4 m or 60 J of work to the weight. If the weight is 30 N this amount of work will permit the weight to be lifted a total of 60 J/30 N or 2 m (Figure 6.2). The lever is not giving us ''something for nothing.'' Even though the 15-N force can be used to lift the 30-N weight, the 15-N force must act through twice the distance the 30-N weight is lifted. The total work put into the lever, therefore, is the same as the work we get out of the lever. The lever arms of the two torques involved merely permit us to exchange a larger applied force for a larger distance. Levers thus demonstrate

FIGURE 6.2 ··········

A Simple Lever　The 30-N gorilla can be lifted 2 m by the 15-N applied force, but only if the latter force acts through a distance of 4 meters. Thus the work done by the 15-N applied force (60 J) is the same as that required to lift the 30-N gorilla to a height of 2 m (60 J).

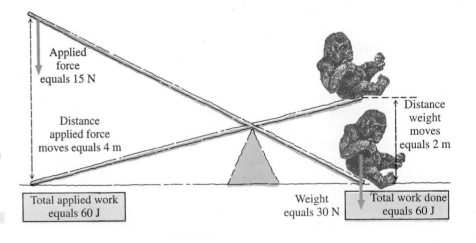

Applied force equals 15 N

Distance applied force moves equals 4 m

Distance weight moves equals 2 m

Total applied work equals 60 J

Weight equals 30 N

Total work done equals 60 J

that work is a "transferable" quantity in that it can be delivered from one place to another. The person pushing down and exerting work at one end of a lever is delivering that work to the other end of the lever to perform work on the load or weight involved. If work can be transferred from one place or object to another through the use of torques in a lever, one might ask if it is possible for work to be transferred and delivered in other ways as well. It was one of the key questions of the Scientific Revolution and the answer is a resounding "Yes!"

WORKED EXAMPLE

A force of 50 N is exerted on a given object.

a. How much work is done if the object moves 15 meters along the direction of the force?

b. How far can the object move if 5000 J of work are available to move it?

SOLUTION:

a. The work done is given in general by

$$\text{work} = \text{force} \times \text{distance moved} .$$

In this case the force is 50 N and the distance moved is 15 meters. The work done is thus

$$\text{work} = 50 \text{ N} \times 15 \text{ meters}$$
$$= 750 \text{ N-m} = 750 \text{ J} .$$

b. If 5000 J of work are available then we have

$$5000 \text{ J} = 50 \text{ N} \times \text{distance moved}$$

thus

$$\text{distance moved} = 5000 \text{ J/50 N}$$
$$= 5000 \text{ N-m/50 N}$$
$$= 100 \text{ m} .$$

1. What is the "physics" definition of work. How does this definition compare with our "everyday" usage of the term?

2. Cite an instance where a force is exerted but no work is performed.

3. What is a joule?

4. Torque and work have the same units, i.e., N-m. Are these two quantities the same? If not, discuss their differences.

5. What is a lever? What does a lever do for the person using it?

REVIEW 6.1 QUESTIONS

1. A 25-kg suitcase is to be lifted up a flight of stairs. How far above the ground floor can the suitcase be lifted before 10,000 J of work are expended?
2. A total of 3000 J of work is required to move an object 25 meters. What force was acting on the object?

6.2 Energy

We now come to what scientists generally regard as the most fundamental and unifying quantity in nature, a quantity we call *energy*. For simple systems, energy is a measure of the ability of an object, person, or system to deliver or perform work. When one pushes down on the end of a lever, one can be thought of as expending energy to do the work of lifting the weight at the other end of the lever. Although energy is the agent by which work is performed, it is a far more comprehensive quantity than work. Like force, momentum, and torque, among others, energy is an abstract quantity that has enjoyed considerable success in explaining and representing the way nature is observed to operate.

We can observe and measure the effects of the presence of energy over a wide range of phenomena, one of which is the performance of detectable work, but we cannot point to a given entity and say "this is it." Yet, as we shall see, energy is to be found at all levels of the physical world, sometimes in rather subtle, "disguised" forms. Moreover, countless scientific experiments the world over have led to the result that although energy can be transferred, transformed, and even dissipated to infinitesimally small quantities, it cannot be created or destroyed by any process in nature of which we are aware. In other words if energy is gained by a system it has to come from somewhere and if it is lost by a system it has to go somewhere. If we require the "somewhere" to be in the observable universe, then the total reservoir of energy in the observable universe must have a value that remains constant in time. This profound property of energy is expressed on a less grandiose level in the form of the *principle of conservation of energy*, which can be stated as follows:

> The total amount of energy in a closed, isolated system remains the same, even though the energy in the system may be converted from one form to another.

The closed, isolated provision ensures that the system cannot import or export any energy to any nearby system. Under these circumstances, the principle of the conservation of energy tells us that the total energy in a given system that is closed off and isolated from the rest of the universe will not change before, during, or after an event or interaction of any kind.

At first glance, such a claim may seem somewhat puzzling. For example, if we lift our box of books, we have done work on the box and, by definition, have also expended energy on the box as well. If energy is conserved, then, where did the expended energy go in the course of our lifting? To answer this question, remember that energy can take on many different forms and thus can transform itself from one form of energy to another. In its lifted position the box of books has the "potential" to do work on other objects. Were we to drop the box of books onto the applied end of a lever, for example, it could cause the weight in the load end to be raised, thereby performing work on the weight. We can, therefore, say that the work done in lifting the box of books is stored in the books as *potential energy*. Potential energy is thus an energy of position or of configuration. A grand piano teetering at the top of a flight of stairs, a stretched spring, or water at the top of a dam are all examples of systems that have stored potential energy as a result of their position or configuration.

The most familiar form of potential energy is that resulting from elevating an object in the earth's gravitational field. If we lift an object to a given height, the work done in general is

$$\text{work} = \text{force} \times \text{distance object moves} .$$

In particular,

$$\frac{\text{work done lifting an object}}{\text{to a given height}} = \text{force} \times \text{height} .$$

The force involved in this case is that due to the earth's gravity and is equal to the weight of the object or mass $\times$ g. Thus we have

$$\frac{\text{work done lifting an object}}{\text{to a given height}} = \text{mass} \times g \times \text{height} .$$

The work energy expended in lifting the object has been transferred to the object in the form of gravitational potential energy, which allows us to state

$$\begin{array}{c}\text{gravitational}\\\text{potential energy}\\\text{acquired by object}\end{array} = \text{mass} \times g \times \begin{array}{c}\text{height to which}\\\text{object is lifted}\end{array} \qquad \textbf{6.2}$$

or

$$\text{PE} = mgh .$$

The gravitational potential energy gained or lost by a given object is always measured by the final change in height that the object experienced, regardless of the path taken by the object to gain or lose that change in height or what the actual height of the object was at the start. The gravitational potential energy acquired by a 400-kg piano that is moved up to an apartment 25 meters above the street level, for example, is (Figure 6.3):

$$\begin{array}{c}\text{gravitational}\\\text{potential energy}\end{array} = 400 \text{ kg} \times 9.8 \text{ m/sec}^2 \times 25 \text{ m}$$

$$= 9.8 \times 10^4 \text{ J} .$$

If our intrepid movers wished to move the piano another 25 meters above the street level they would have to expend another 9.8×10^4 J of energy relative to the 25-meter level for a total of 1.96×10^5 J of energy required to lift the piano 50 meters above the street. These changes in potential energy are independent of the path taken by the piano. The piano will have acquired the same 1.96×10^5 J of gravitational potential energy regardless of whether the piano was lifted the 50 meters directly up to the apartment by an outside crane or lifted 50 meters more circuitously and tediously via the apartment building stairwell. It should be noted that we have again assumed a constant value for g, the acceleration of the earth's gravity. As long as we are at or near the earth's surface,

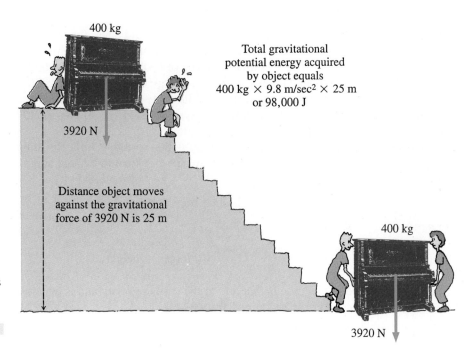

FIGURE 6.3

Gravitational Potential Energy   The work done to push the piano up the stairs is acquired by the piano as gravitational potential energy.

such an assumption is quite accurate and can be used to simplify calculations of gravitational potential energy.

If we wish to lift objects such as rockets and their payloads to heights well above the earth's surface, then it becomes necessary to use a more generalized expression for gravitational potential energy. Unfortunately, the gravitational force between two objects depends on the distance between the two objects in accordance with the universal law of gravitation. As a result, as one object moves relative to another object, not only is the distance between them changing but also the gravitational force between them as well. Thus if we were to push a mass such as a rocket from the earth's surface to a distance twice the earth's surface radius, we would find that over the course of our pushing, the gravitational force would have decreased to $\frac{1}{4}$ of its original value.

To calculate the total amount of work done in such cases, one must calculate the small increments of work done by pushing the object small increments of distance and then adding up the increments with a total amount of work according to the mathematical principles and techniques of calculus. From such analyses, we find in general that to change the distance between two gravitating objects having masses m and M from a distance d_1 to distance d_2 it is necessary to expend a total amount of work equal to

$$\text{work} = GmM \left(\frac{1}{d_1} - \frac{1}{d_2} \right) .$$

Thus the gravitational potential energy acquired by an object of mass m when it is moved from a distance d_1 to d_2 away from a second mass M is

$$\begin{array}{l} \text{gravitational} \\ \text{potential energy} \\ \text{acquired by object} \end{array} = GmM \left(\frac{1}{d_1} - \frac{1}{d_2} \right) ,$$

where d_2 is larger than d_1. For the case in which d_1 and d_2 are very nearly equal to the earth's radius R, this expression reduces back to Eq. (6.2).

For centuries it has been recognized that there is something intrinsically different about getting hit by a falling safe as opposed to a falling tennis ball. Aristotle's oft-pilloried claim that heavier objects fall to the ground faster than lighter ones was an early, albeit unsuccessful attempt to explain this effect. Nearly two millennia later, the German mathematician Gottfried Leibniz claimed that moving objects possessed a *vis viva* or "living force," which was a measure of the ability of an object to do damage when it collided with a second object. Newton was able to describe only partially the phenomenon with his "quantity of motion" or momentum concept. The elusive quarry sought by these and others is the quantity we now refer to as the object's energy of motion or *kinetic energy*.

The concept of kinetic energy can be formally established by noting that it takes work to get an at-rest object to begin moving at some nonzero velocity. From the basic definition of work we have

$$\text{work} = \text{force} \times \text{distance object moves} .$$

If an object initially at rest is subjected to a constant force, the distance the object travels in a given elapsed time is

$$\text{distance object moves} = (^1/2)(\text{acceleration})(\text{elapsed time})^2$$

and the force = mass × acceleration. If we substitute these latter two expressions into the basic definition for work, we have

$$\text{work} = (\text{mass} \times \text{acceleration}) \times [(^1/2)(\text{acceleration})(\text{elapsed time})^2]$$

or

$$\text{work} = (^1/2)(\text{mass}) \times (\text{acceleration} \times \text{elapsed time})^2 .$$

But acceleration × time is equal to the velocity the object will have acquired from the constant force after the given elapsed time. We may therefore write

$$\text{work} = (^1/2)(\text{mass})(\text{velocity})^2 .$$

The work performed in setting the object in motion is now present in the object as an energy of motion or kinetic energy. Hence, we have that

$$\text{kinetic energy} = (^1/2)(\text{mass})(\text{velocity})^2 \qquad\qquad \textbf{6.3}$$

or

$$KE = (^1/2)mv^2 .$$

The kinetic energy associated with a moving object is thus somewhat similar to the object's linear momentum in that both quantities depend on the object's mass and velocity, but the kinetic energy of an object is more strongly dependent on its velocity. A 75-kg object moving at a velocity of 10 m/sec, for example, would have a kinetic energy of

$$\text{kinetic energy} = \frac{1}{2} \times 75 \text{ kg} \times (10 \text{ m/sec})^2$$
$$= 3750 \text{ J} ,$$

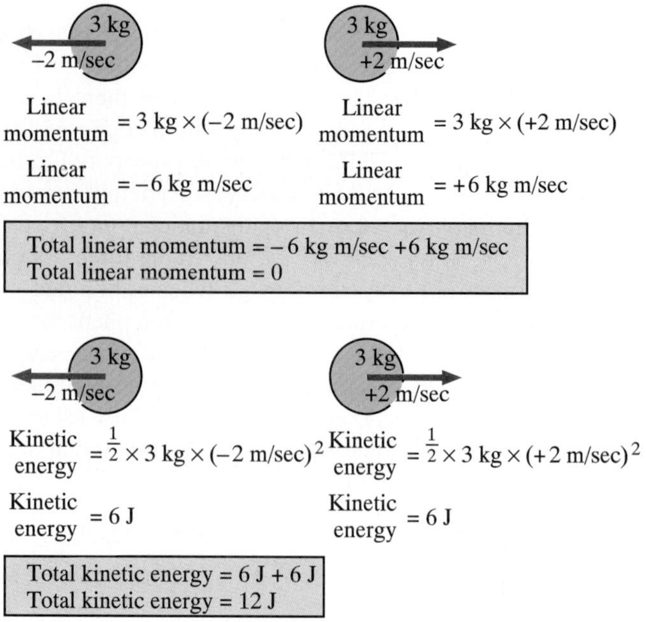

$$\frac{Linear}{momentum} = 3 \text{ kg} \times (-2 \text{ m/sec})$$

$$\frac{Linear}{momentum} = -6 \text{ kg m/sec}$$

$$\frac{Linear}{momentum} = 3 \text{ kg} \times (+2 \text{ m/sec})$$

$$\frac{Linear}{momentum} = +6 \text{ kg m/sec}$$

Total linear momentum $= -6$ kg m/sec $+6$ kg m/sec
Total linear momentum $= 0$

FIGURE 6.4 ··········

Momentum and Kinetic Energy Although linear momentum and kinetic energy both depend on the mass and the velocity of an object, the total linear momentum for this system is equal to zero, while the total kinetic energy has a positive value.

$$\frac{Kinetic}{energy} = \frac{1}{2} \times 3 \text{ kg} \times (-2 \text{ m/sec})^2$$

$$\frac{Kinetic}{energy} = 6 \text{ J}$$

$$\frac{Kinetic}{energy} = \frac{1}{2} \times 3 \text{ kg} \times (+2 \text{ m/sec})^2$$

$$\frac{Kinetic}{energy} = 6 \text{ J}$$

Total kinetic energy $= 6$ J $+ 6$ J
Total kinetic energy $= 12$ J

and a linear momentum of

$$\text{linear momentum} = 75 \text{ kg} \times 10 \text{ m/sec}$$
$$= 750 \text{ kg-m/sec} .$$

If we were to increase the velocity of the object by a factor of 4 to 40 m/sec, the linear momentum would then have a value of

$$\text{linear momentum} = 75 \text{ kg} \times 40 \text{ m/sec}$$
$$= 3000 \text{ kg-m/sec} ,$$

which is an increase of four times over its previous value. On the other hand, if the velocity of the object were to increase to 40 m/sec, the kinetic energy would be

$$\text{kinetic energy} = \frac{1}{2} \times 75 \text{ kg} \times (40 \text{ m/sec})^2$$
$$= 60,000 \text{ J}$$

or an increase of 16 times more than its previous value. Moreover, to describe the momentum of an object completely, both the magnitude and the direction of the object's momentum must be specified. Kinetic energy is not a vector quantity as is momentum and as such it can be completely described by specifying only its magnitude. Thus a system consisting of two 3-kg objects, each of which is moving at 2 m/sec, but in exactly opposite directions from each other as shown in Figure 6.4 will have a total linear momentum of

$$\text{linear momentum} = 3 \text{ kg} \times 2 \text{ m/sec} + 3 \text{ kg} \times (-2 \text{ m/sec})$$
$$= 0 .$$

On the other hand, the total kinetic energy of this system is

$$\text{kinetic energy} = \frac{1}{2} \times (3 \text{ kg})(-2 \text{ m/sec})^2 + \frac{1}{2} \times (3 \text{ kg})(-2 \text{ m/sec})^2$$

$$= 6 \text{ J} + 6 \text{ J}$$
$$= 12 \text{ J} .$$

Note that regardless of the sign of the object's velocity, kinetic energy will, unlike momentum, always have a positive value.

One of the more spectacular instances of kinetic energy in action can be seen in the performance of circus acrobats who vault one another up to impressive heights through the transfer of kinetic energy. The instrument of the transfer is a simple lever, which is called a catapult. As one performer jumps down onto one end of the catapult, the performer's kinetic energy increases with the increasing velocity of the fall. As the performer lands on her end of the catapult, her kinetic energy is transferred to the performer on the opposite end of the catapult, and he leaves his end of the catapult in an upward direction with the kinetic energy that has been transferred to him. Very often the performer on the "receiving end" will supplement his kinetic energy by timing his own jump off the catapult at the exact instant he receives the kinetic energy from the opposite side of the catapult, thereby enhancing his upward vault.

If objects exhibiting translational motion can have an energy associated with that motion, it is reasonable to assume that rotating objects might also possess some sort of "rotational kinetic energy" or *rotational energy*. To investigate this possibility, let us consider one bit or element of mass in a solid rotating object as shown in Figure 6.5. As the object rotates, the mass element will have a kinetic energy equal to $(\frac{1}{2}) \times (\text{mass}) \times (\text{velocity})^2$. For circular motion, the magnitude of the velocity = (angular velocity) × (radius of circular path). Hence,

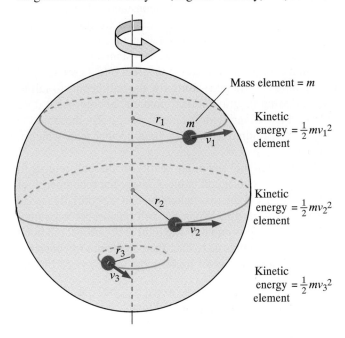

Mass element = m

Kinetic energy element $= \frac{1}{2}mv_1^2$

Kinetic energy element $= \frac{1}{2}mv_2^2$

Kinetic energy element $= \frac{1}{2}mv_3^2$

Total rotational energy = sum of $\frac{1}{2}mv^2$

Total rotational energy = sum of $\frac{1}{2}m(r \times \text{angular velocity})^2$

Total rotational energy = $\frac{1}{2}(\text{sum of } mr^2) \times (\text{angular velocity})^2$

Total rotational energy = $\frac{1}{2}(\text{moment of inertia}) \times (\text{angular velocity})^2$

FIGURE 6.5 ············

Rotational Energy   A solid rotating object can be thought of as being composed of tiny bits of mass, each of which has associated with it a small bit or element of kinetic energy. When all of these elements of kinetic energy are added up for the object, the total is equal to ½(moment of inertia)(angular velocity)^2, which is the rotational kinetic energy or rotational energy of the object.

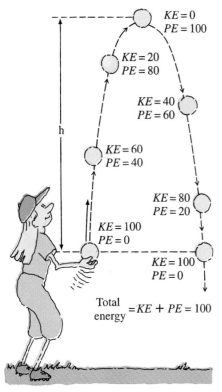

$KE = 0$
$PE = 100$

$KE = 20$
$PE = 80$

$KE = 40$
$PE = 60$

$KE = 60$
$PE = 40$

$KE = 80$
$PE = 20$

$KE = 100$
$PE = 0$

$KE = 100$
$PE = 0$

h

Total
energy $= KE + PE = 100$

FIGURE 6.6 ···········
Conservation of Energy When the ball
first leaves the student's hand from an
upward toss, the total energy of the ball
is all in the form of kinetic energy. At the
top of its trajectory, all of the kinetic
energy is now in the form of gravitational
potential energy. As the ball falls back
down, the potential energy of the object
is transformed back into kinetic energy.
Notice that at any time in the ball's flight
the sum of the ball's gravitational
potential energy and its kinetic energy
has the same value.

each element of mass has associated with it a bit of kinetic energy given by

$$\text{kinetic energy of mass element} = (1/2)(\text{mass})\left(\text{angular velocity} \times \text{radius of circular path}\right)^2 .$$

If we add up the kinetic energies of all the mass elements in the rotating object, we obtain the total amount of energy associated with the object's rotational motion. When such a mathematical summing process is carried out, the result can be stated as follows

$$\text{rotational energy} = (1/2)(\text{moment of inertia})(\text{angular velocity})^2 .$$

Thus a spinning object has associated with it an energy of rotation that depends on the object's moment of inertia and the angular velocity of the rotation in a fashion analogous to the relationship between an object's kinetic energy and its mass and velocity.

Having described three types of energy, we are now in a position to illustrate how one form of energy can be transformed into another. Suppose a student imparts a certain amount of kinetic energy to a ball by throwing it straight up into the air at some initial upward velocity (Figure 6.6). At the instant the ball leaves the student's hand, it has a kinetic energy of $(1/2)(\text{mass})(\text{velocity})^2$. As it rises, however, it is in essence being lifted against the earth's gravity just as surely as if we were actually pushing it ourselves. Thus, work is being expended in order for the ball to rise. But where does the energy for this work come from? If we watch the ball's upward flight, we see that it is constantly slowing down. As its velocity decreases, there is a corresponding decrease in its kinetic energy. Finally, the ball comes to the point where the available kinetic energy is all used up and the upward velocity has decreased to zero. At this instant all of the kinetic energy that was imparted to the ball has now been converted into a gravitational potential energy of $(\text{mass}) \times (g) \times (\text{height})$. But now the ball begins to fall to a lower height, thereby decreasing its gravitational potential energy. As it does so, it begins to pick up speed and along with it kinetic energy. By the time the ball returns to our hand, it has acquired a velocity equal to that with which it was initially tossed upward.

In other words, as the gravitational potential energy of the ball returns to zero, the kinetic energy returns to its original value. Clearly the total energy of the ball is equal to its kinetic energy at the time when the ball was initially tossed into the air and at the time it was caught. When the ball reaches its maximum height, the total energy of the system is equal to the total gravitational potential energy of the ball. This latter fact enables us to calculate the height to which the ball can rise if it is given a certain amount of initial kinetic energy. At the top of the ball's flight

$$\frac{\text{gravitational potential}}{\text{energy of ball}} = \frac{\text{initial kinetic}}{\text{energy of ball}}$$

or

$$\text{mass} \times g \times \text{height} = \frac{1}{2} \times \text{mass} \times (\text{initial velocity})^2 .$$

The ball therefore rises to a height given by

$$\text{height} = (\text{initial velocity})^2/2g \ .$$

Interestingly, this equation tells us that the height to which an object rises does not depend on its mass but depends fairly strongly on the magnitude of the initial upward velocity of the object. This equation also raises the intriguing possibility of hurling an object upward so fast that the height attained will be large enough to allow the object to escape the earth's gravity altogether. The idea of such an "escape velocity" is discussed in more detail in Physics Formulation 6.2.

If we were to make very careful measurements of the tossed ball's height and velocity throughout its flight, and from these data calculated the ball's gravitational potential energy as well as its kinetic energy, we would find that the sum total of both energies at *any time* in the ball's flight remains at exactly the same value. Kinetic energy has been transformed into gravitational potential energy and back again, but the total energy of the system has stayed the same.

Another form of energy transformation can be illustrated by the arrangement shown in Figure 6.7 in which a person has to go up a flight of stairs to get to the top of a large wheel. If a mass (the rock) is lifted up the flight of stairs to a given height, it will acquire an increase in gravitational potential energy equal to mass $\times$ g $\times$ height. If the mass is now hung on the wheel in such a way that it in effect becomes a part of the wheel, the earth's gravity will pull the mass to the base of the wheel, and in doing so will cause the wheel to rotate. As the mass decreases its height, the rate at which the wheel rotates will increase in such a way that the total of the gravitational potential energy of the mass and the rotational energy of the wheel remains constant. When the mass reaches the base of the wheel, all of its gravitational potential energy will have been transformed into the rotational energy of the wheel, and the wheel will therefore have acquired a total rotational energy given by

rotational energy of wheel = initial potential energy of mass

or

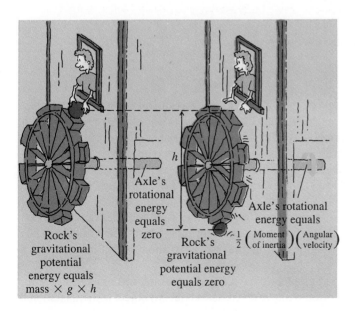

Rock's gravitational potential energy equals mass $\times$ g $\times$ h

Axle's rotational energy equals zero

Rock's gravitational potential energy equals zero

Axle's rotational energy equals $\frac{1}{2}\left(\begin{matrix}\text{Moment}\\\text{of inertia}\end{matrix}\right)\left(\begin{matrix}\text{Angular}\\\text{velocity}\end{matrix}\right)$

FIGURE 6.7 · · · · · · · · · ·

Transformation of Gravitational Potential Energy into Rotational Energy   As the rock falls through the height h, it converts its gravitational potential energy into the rotational energy of the wheel and axle.

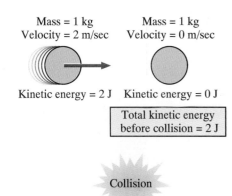

Mass = 1 kg
Velocity = 2 m/sec

Mass = 1 kg
Velocity = 0 m/sec

Kinetic energy = 2 J   Kinetic energy = 0 J

Total kinetic energy
before collision = 2 J

Collision

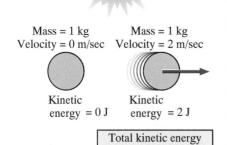

Mass = 1 kg
Velocity = 0 m/sec

Mass = 1 kg
Velocity = 2 m/sec

Kinetic
energy = 0 J

Kinetic
energy = 2 J

Total kinetic energy
before collision = 1 J

FIGURE 6.8 ···········

Kinetic Energy in an Elastic Collision In an elastic collision, the total kinetic energy of the system is the same before and after the collision, and the total kinetic energy is conserved.

$$\left(\frac{1}{2}\right) \left(\begin{array}{c}\text{moment of}\\ \text{inertia}\end{array}\right) (\text{angular velocity})^2 = (\text{mass}) \ (g) \left(\begin{array}{c}\text{initial height}\\ \text{of mass}\end{array}\right) .$$

If the mass somehow falls off the wheel as it swings past the base of the wheel, in the absence of other forces or torques, the wheel will continue to rotate at an angular velocity whose magnitude is governed by the above energy conservation condition. Devices such as this which transform gravitational potential energy into rotational energy have played an important role in human technology for centuries. Even now they find considerable usage in contemporary society in the form of water wheels and hydroelectric turbines.

As one final example of energy transfer let us return to the examples of the colliding objects used in Chapter 3 to illustrate the conservation of linear momentum and reanalyze those collisions in terms of the kinetic energies involved. In our first collision example, a 1-kg billiard ball moving at 2 m/sec strikes a 1-kg billiard ball that is initially at rest (Figure 6.8). The total kinetic energy of the system before the collision is the kinetic energy of the first billiard ball or $\frac{1}{2} \times (1 \text{ kg}) \times (2 \text{ m/sec})^2 = 2$ J, since the kinetic energy of the second billiard ball before impact is $\frac{1}{2} \times (1 \text{ kg}) \times (0) = 0$ J. The principle of the conservation of linear momentum tells us that after the collision the velocity of the first ball is zero and that of the second ball is 2 m/sec (see Chapter 3). The total kinetic energy after the collision is then

$$\text{kinetic energy} = 0 + \frac{1}{2}(1 \text{ kg})(2 \text{ m/sec})^2$$
$$= 2 \text{ J} .$$

Thus not only is linear momentum conserved in this case, but the system's total amount of kinetic energy as well. If, however, we calculate the total kinetic energy of the Chapter 3 inelastic collision in which the two 1-kg participants stick to one another in the collision, we find that the total kinetic energy of the system before the collision is still 2 J (Figure 6.9), but after the collision the resulting 2-kg object has a velocity of 1 m/sec. The total kinetic energy of the system after the collision is thus just equal to the kinetic energy of the "stuck-together" masses, which is

$$\text{kinetic energy} = \frac{1}{2} \times (2 \text{ kg}) \times (1 \text{ m/sec})^2$$
$$= 1 \text{ J} .$$

The total kinetic energy of the system after the collision is thus less than that before the collision, and the total kinetic energy is not conserved. This result provides us with a formal way of distinguishing elastic collisions from inelastic ones. For an elastic collision, kinetic energy is conserved. For an inelastic collision, the total kinetic energy of the system is not conserved. The principle of the conservation of energy, however, requires that the total energy of the system be conserved. We are thus left with the proposition that the missing 1 J of energy has somehow been energetically absorbed by the stuck-together masses, an assumption that implies the existence of a different energy form that is not obviously kinetic, potential, or rotational in nature, but is in some way "internal" to a given mass.

In the preceding paragraphs we have found that some sort of energy accompanies every type of motion we have discussed so far. An object with translational motion also has kinetic energy. A spinning object has rotational energy, and even an object that is merely lifted becomes infused with gravitational potential energy. We even found a case where objects can take on a form of "internal" energy seemingly unrelated to any energy form we have yet encountered. If you are beginning to get the idea from all of this that there may not be a single phenomenon in the entire observable universe that is not somehow energy related, you would be right.

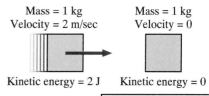

Total kinetic energy before collision = 2 J

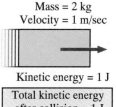

Total kinetic energy after collision = 1 J

FIGURE 6.9
Kinetic Energy in an Inelastic Collision In an elastic collision, the total kinetic energy of the system is not the same before and after the collision and the total kinetic energy is not conserved. The missing energy is transformed into a different form of energy.

WORKED EXAMPLE

How fast must a 200-kg automobile travel in order to have a kinetic energy of 25,000 joules?

SOLUTION: The kinetic energy of any object is given by

$$\text{kinetic energy} = \frac{1}{2} \times \text{mass} \times (\text{velocity})^2 .$$

If we wish that an object having a mass of 2000 kg be moving at a velocity sufficient that it possess a kinetic energy of 25,000 joules, then

$$25,000 \text{ J} = \frac{1}{2} \times 2000 \text{ kg} \times (\text{velocity})^2 .$$

Solving this equation for the velocity, we obtain:

$$\frac{1}{2} \times (2000 \text{ kg}) \times (\text{velocity})^2 = 25,000 \text{ J}$$

$$1000 \text{ kg} \times (\text{velocity})^2 = 25,000 \text{ J}$$

$$(\text{velocity})^2 = 25,000 \text{ J}/1000 \text{ kg}$$

$$(\text{velocity})^2 = 25 \text{ meters}^2/\text{second}^2$$

$$\text{velocity} = 5 \text{ meters/second} .$$

In order for the 2000-kg automobile to have a kinetic energy of 25,000 joules, it must be moving with a velocity of 5 meters/second.

1. Compare and contrast work, kinetic energy, gravitational potential energy, and rotational energy.

2. Give everyday examples of objects that have kinetic energy, gravitational potential energy, and rotational energy.

3. If height is expressed in meters, mass in kilograms, and the acceleration of the earth's gravity in m/sec²m, show that the unit of potential energy is the joule.

REVIEW 6.2 QUESTIONS

1. How much kinetic energy would be required to lift a 15-kg object to a height of 20 meters above the earth's surface?
2. An 8-kg object with a moment of inertia of 10 kg-m² is rolling along with an angular velocity of 0.5 radians/sec and a linear velocity of 0.25 m/sec. What is the total energy of this object?

6.3 The Work-Energy Theorem

In the preceding section we noted that the quantitative potential energy gained or lost by a given object depends only on the final change of height experienced by that object, regardless of the path taken to move the object from its beginning

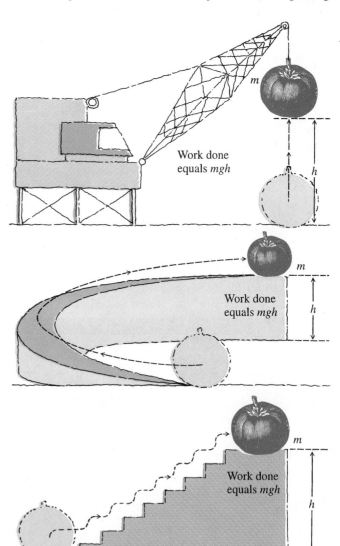

FIGURE 6.10

Conservative Forces For a conservative force such as the earth's gravity, the total amount of work done to lift the object to the height h is the same, regardless of the path taken in that endeavor.

position to its end position. When the work done against a force such as gravity does not depend on the path over which the object moves, that force is said to be a *conservative force*. For example, if we were to lift a large mass against the force of gravity to some height, the work done would be equal to mass × g × height, regardless of whether the object were lifted vertically to that height using a crane, pushed up to that height along a gradually sloping frictionless ramp, or raised bit by bit to that height along a flight of stairs (Figure 6.10).

In contrast, forces exist, most notably friction, in which the work done against the force *does* depend on the path taken to move the object from one position to another. Suppose, for example, we wish to move a box from corner A of a square room to corner D as shown in Figure 6.11. Experience tells us that it is far easier to move the box directly from corner A to corner D, rather than through corners B and C to get to D. If we calculate the work done in moving the box directly from corner A to corner D, we have

$$\text{work} = \text{force} \times \text{distance object moves} .$$

If the force we are working against is friction, we can write that

$$\text{force of friction} = \frac{\text{coefficient}}{\text{of friction}} \times \text{normal force} .$$

The normal force in this case is just the weight of the box, mass × g, and thus

$$\text{force of friction} = \frac{\text{coefficient}}{\text{of friction}} \times \text{mass} \times g .$$

If the room has a side length of *S*, then the work performed in sliding the box directly from corner A to corner D is

$$\frac{\text{work}}{\text{(A to D)}} = \frac{\text{coefficient}}{\text{of friction}} \times \text{mass} \times g \times S .$$

If, on the other hand, we wish to slide the box from corner A to corner B, then from B to C, and finally from C to D, the total path length will be 3 × S and the total work done along this path is

$$\frac{\text{work}}{\text{(A-B-C-D)}} = \frac{\text{coefficient}}{\text{of friction}} \times \text{mass} \times g \times 3 \times S$$

or

$$\frac{\text{work}}{\text{(A-B-C-D)}} = 3 \times \frac{\text{coefficient}}{\text{of friction}} \times \text{mass} \times g \times S .$$

Thus, in accordance with what we observe, it takes three times as much work to slide the box to corner D by moving it through corners B and C than it does by sliding the box directly from corner A to corner D. When the work done against a given force in order to move an object from one position to another depends on the path taken, as in the above example, such a force is said to be *nonconservative*.

The principle of conservation of energy requires that all of the work done on an object against either a conservative or nonconservative force be transformed into an equal amount of energy. If the work is done against a conservative force, then it can be shown that the work done on the object must be completely transformed into a change in the object's total mechanical energy. In general, this change of energy is split between a change in the objcct's kinetic energy

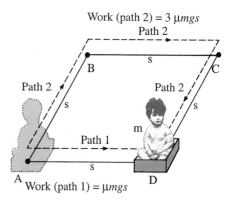

Work (path 2) = 3 μmgs

Work (path 1) = μmgs

FIGURE 6.11
Nonconservative Forces
Nonconservative forces have the property that the work done on an object against the given forces depends on the path taken in moving the object. In this case it takes three times as much work to push the box against the force of friction from A to D via B and C as it does to push the box directly from A to D.

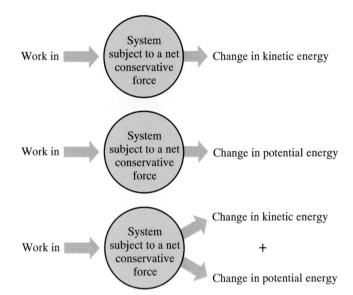

FIGURE 6.12 ··········
The Work-Energy Theorem   If work is done on an object against a conservative force, the object will experience a change in either or both its kinetic energy and its potential energy.

and a change in its potential energy. This result is called the *work-energy theorem* (Figure 6.12) and can be summarized as follows

> The work done on an object by conservative forces is equal to the change in its kinetic energy plus the change in its potential energy.

Mathematically, we have

$$W = \Delta\text{KE} + \Delta\text{PE} .$$ **6.4**

If the change in the object's potential energy is zero, then all of the work done on the object appears as a change in the object's kinetic energy. Thus if we wish to calculate the amount of work required to hurl a 0.2-kg baseball at a rate of 40 m/sec, then from the work-energy theorem we have

$$\text{work} = \Delta\text{KE} + \Delta\text{PE} .$$

Assuming that there is no change in the potential energy of the baseball, we have

$$\text{work} = \frac{1}{2} \times \text{ mass } \times \text{ (velocity)}^2 + 0$$

$$= \frac{1}{2} \times 0.2 \text{ kg } \times (40 \text{ m/sec})^2$$

$$= 160 \text{ J} .$$

On the other hand, if there is no change in the kinetic energy of the object, then the work done on the object appears as a change in the object's potential energy. Suppose, for example, we were to raise a 2-kg mass very slowly to a height of 10 meters. During this process the change in the kinetic energy of the object is virtually zero and we now have from the work-energy theorem that

$$\text{work} = 0 + \Delta\text{PE}$$

$$= 0 + \text{mass} \times g \times \text{change in height}$$

$$= 0 + 2 \text{ kg } \times 9.8 \text{ m/sec}^2 \times 10 \text{ m}$$

$$= 196 \text{ J} .$$

It should be emphasized that the work-energy theorem applies *only* to work done against conservative forces. If the force is nonconservative, then the work done on the object can be transformed into forms of energy other than kinetic energy and potential energy.

WORKED EXAMPLE

How much work is required to get a 50,000-kg jet aircraft up to a speed of 300 m/sec if the aircraft is initially at rest?

SOLUTION: Assume that all of the work expended on the jet aircraft is transformed into kinetic energy. The total kinetic energy of the plane when it reaches a speed of 300 m/sec is

$$\text{kinetic energy} = \frac{1}{2} \times \text{mass} \times (\text{velocity})^2$$

$$= \frac{1}{2} \times 50,000 \text{ kg} \times (300 \text{ m/sec})^2$$

$$= 25,000 \times 90,000 \text{ J}$$

$$= 2.5 \times 10^4 \times 9 \times 10^4 \text{ J}$$

$$= 22.5 \times 10^8 \text{ J} .$$

Thus the amount of work required to get the jet aircraft up to a speed of 300 m/sec will be over 2 billion joules.

1. What is a conservative force? What is a nonconservative force? Give an example of each type of force.

2. What is the work-energy theorem? How is it related to the principle of conservation of energy?

3. Given an example in which the work done on an object is converted into (a) a change in kinetic energy only, (b) a change in potential energy only, and (c) a change in both kinetic and potential energy.

1. A baseball player expends 100 J of work while throwing a 0.4-kg softball. If all of the work is transformed into the kinetic energy of the softball, how fast will the ball travel?

2. A 500-kg piano is slowly lifted up a 30-meter flight of stairs. How much work is done?

6.4 Power

If a student wishes to carry a box of books from the street up a flight of stairs into his or her new apartment, the amount of work done in moving the books to their new location is the same whether the student walks up the stairs, runs up the stairs, or stops several times to rest while going up the stairs. In each case, the final distance the books have been lifted against the earth's gravitational force is the same. Yet there is no question that there would be some very distinct differences in the way we would feel after moving the books in each of these ways. Similarly, when we do exercises or aerobics, we get a far better ''workout'' if we perform them more rapidly than if we perform them less rapidly.

In forging a complete description of the expenditure of work and energy, therefore, it is necessary to take into account the time interval over which a given amount of work is performed or energy is expended. To this end, we define *power* as the time rate at which work is performed or energy is released:

$$\text{power} = \frac{\text{work done}}{\text{elapsed time}} = \frac{\text{energy transformed}}{\text{elapsed time}}.$$

Power thus has the units of work/time, which for this text would be expressed in N-m/sec or simply J/sec. A power output or expenditure of 1 joule/sec is defined as one *watt* of power. If a 12-oz. can of soda were raised 0.3 meter in one second, the power expended would be about one watt or 1 W. A commonly used unit of power in our society is the *horsepower,* which was originated in the 18th century by the Englishman James Watt. Watt found that the ''typical'' horse could perform work over a ''reasonable'' length of time at a rate equivalent to lifting a 550-pound weight at a rate of one foot per second. In the metric system, one horsepower corresponds to a power output of about 746 watts or 0.7463 kilowatts.

To illustrate the relationship between power and energy in numerical fashion, let us recall from an earlier section that the amount of work expended in lifting a 300-N load of books up a 4-meter flight of stairs was 300 N × 4 m or 1200 J of work. If the task was accomplished in 60 seconds by walking up the stairs with the load, the average power expended would be 1200 J/60 sec or 20 watts. If the task was accomplished in 5 seconds by running up the stairs with the load, the average power increases to 1200 J/5 sec or 240 watts (Figure 6.13). The same task performed over 5 minutes or 300 seconds of elapsed time by someone who makes frequent stops on the way up the stairs would require an average power of only 1200 J/300 sec or 4 W. We can thus see why our bodies would detect a noticeable difference among these three ways of moving books, even though the total work expended in each instance was 1200 J.

In their billings, power companies most often make use of a unit called the *kilowatt-hour,* which is defined as the expenditure of 1000 watts of power over a one-hour interval of time. Since the watt is measured in energy/time, the kilowatt-hour is in reality a measure of (energy/time) × time or simply energy. The actual expenditure of energy associated with the kilowatt-hour can be calculated as follows:

$$\text{energy expended} = 1 \text{ kilowatt-hour}$$
$$= 1000 \text{ watts} \times \text{hour}$$

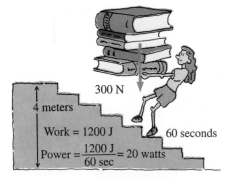

FIGURE 6.13 ···········

Work and Power Although the total amount of work done in each case is 300 N × 4 m = 1200 J, the power expended by the student on the bottom is 12 times higher because the work was done by that student in $\frac{1}{12}$ of the time it took the student on the top.

$$= 1000 \ \frac{\text{joules}}{\text{sec}} \times \text{hour}$$

$$= 1000 \ \frac{\text{joules}}{\text{sec}} \times 1 \ \text{hour} \times \frac{3600 \ \text{sec}}{\text{hr}}$$

$$= 3.6 \times 10^6 \ \text{joules} .$$

Thus one kilowatt-hour corresponds to an energy expenditure of about 3.6 million joules.

Our concern in contemporary society with the technological concept of power stems from the fact that machines which are said to be more "powerful" develop larger amounts of power and thus can get work done faster than machines which are less "powerful." The technology of the past two centuries has led to the creation of a wide variety of engines and machines having a considerable amount of power. We can presently move millions of kilograms of earth, transport millions of people millions of miles, and otherwise perform a myriad of tasks with speeds unprecedented in the history of human beings. But if machines and engines are to perform work at an ever-increasing rate, then the energy required to run these machines and engines must also be furnished at an ever-increasing rate. If the power or rate at which energy can be supplied to our society falls behind the power requirements of that society, then a so-called "energy crisis" or "energy crunch" results. The various options of power production available to an ever-burgeoning human population will be further discussed, but only after we have described a number of other ways in which the energy we seek to employ appears in the physical world.

WORKED EXAMPLE

Water flows over a 50-meter-high waterfall at a rate of 2000 kg every 4 seconds. What is the potential power output of the waterfall?

SOLUTION: The gravitational potential energy of 1000 kg of water at a height of 50 meters is

$$\text{potential energy} = \text{mass} \times \text{height} \times g$$
$$= 2000 \ \text{kg} \times 50 \ \text{m} \times 9.8 \ \text{m/sec}^2$$
$$= 980{,}000 \ \text{J} .$$

Since this amount of potential energy is available every 4 seconds, we have

$$\text{power} = \frac{\text{potential energy}}{\text{elapsed time}}$$
$$= \frac{980{,}000 \ \text{joules}}{4 \ \text{seconds}}$$
$$= 245{,}000 \ \text{watts} .$$

The waterfall can thus deliver a total of 245,000 watts of power.

6.1 Machines

A machine may be simply defined as a device that can convert or direct energy from a given source or supply into "useful" work or energy. A machine can thus be thought of as a sort of "connecting link" between the energy source or "prime mover" and the work we wish to have done. Moreover, we would like our machines to be able to somehow deliver an enhanced amount of force to whatever task is at hand. To these ends, human beings have devised five "simple machines" (shown in the figure) that in one way or another provide the basis for every moving part of every mechanical device thus far produced by human technology.

The lever is perhaps the most familiar of the simple machines. A worker who wishes to lift a heavy rock can do so by exerting a force on one end of a lever, thus causing the rock to lift off the ground. The lever is somewhat restricted, however, in that the presence of obstructions such as the ground limits the distance through which the applied force can be exerted, thereby limiting the amount of work that the lever can perform.

This limitation can be surmounted by using the second of the simple machines, an arrangement referred to as a wheel and axle. The wheel and axle system is in effect a lever that can apply a force over whatever distance we choose by simply exerting the force in a circular path along the rim of a wheel as shown in part b of the figure. Such an arrangement gives rise to a remarkably versatile device that cannot only be used to lift objects, but can be employed in a variety of other ways as well.

Another variation of the levering process is to be found in a device called a pulley. A pulley consists of a freely rotatable wheel mounted on some sort of frame or support as shown in part c of the figure. The torques are exerted by means of a flexible rope, wire cable, etc., passing over the rim of the pulley. With a suitable arrangement of pulleys, the force required to lift a given weight can be considerably reduced as is the case for the block-and-tackle arrangements used to pull engines out of automobile frames and other similar tasks.

A fourth technique employed to perform a given amount of work by exerting a comparatively small amount of applied force involves the use of a "machine" called an inclined plane. This device and its blade-like sibling, the wedge, deliver a lifting force (or splitting force in the case of the wedge) by exerting a reduced force along a path slanted to the direction of the lift or split.

If we construct a "circular" inclined plane or wedge the result is the screw, the last of the simple machines. Screws perform their lifting or splitting by means of a force directed in a circular path and are thus related to inclined planes in a fashion very much analogous to the relationship that exists between the lever and wheel/axle. Augers, vices, jackscrews, and propellers are but a few of the devices based on these "wrap-around" inclined planes. More complicated machines can, of course, be constructed from variations and combinations of these five fundamental devices.

A given machine's usefulness is most often measured in terms of two quantities, efficiency and mechanical advantage. The *efficiency* of a machine is defined as the ratio of the useful work produced by the machine to the total energy put into running the machine, or

$$\text{efficiency} = \frac{\text{useful work produced}}{\text{total energy input}}.$$

Obviously the principle of the conservation of energy prohibits any machine from having an efficiency greater than one, because that would imply that the given machine could create energy. Unfortunately, as we shall see later on, we cannot even construct a machine that is 100% efficient because of the fact that there will always be some losses due

PHYSICS FACET

Continued

to friction, etc., which will drop the efficiency to a value less than unity.

A second ratable characteristic of machines arises from the fact that they can deliver a *mechanical advantage* in which the machine's load force is larger than its applied force. Thus a force exerted at the end of a long lever arm can counter a much larger force or weight at the end of a much shorter lever arm. The mechanical advantage of a given machine is formally defined as the ratio of the machine's load force to its applied force or:

$$\text{mechanical advantage} = \frac{\text{load force}}{\text{applied force}}.$$

Unlike the efficiency, the mechanical advantage of a machine can exceed one because forces need not be conserved.

A worker who wishes to lift a large weight with a small applied force can do so, but only if that force is exerted through a much larger distance since

$$\begin{matrix}\text{small applied} \\ \text{force}\end{matrix} \times \begin{matrix}\text{large applied} \\ \text{force} \\ \text{distance}\end{matrix} =$$

$$\begin{matrix}\text{large load} \\ \text{force}\end{matrix} \times \begin{matrix}\text{small load} \\ \text{force} \\ \text{distance}\end{matrix}.$$

Such machines, in effect, trade off the distance through which the applied force must be exerted for the privilege of being able to use the smaller applied force.

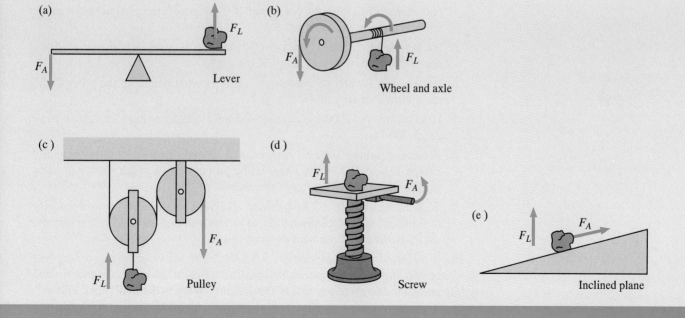

(a) Lever

(b) Wheel and axle

(c) Pulley

(d) Screw

(e) Inclined plane

REVIEW 6.4 QUESTIONS

1. Describe what is meant by power. How is power related to work and energy?
2. What is a watt? A kilowatt-hour? How are these two units related?
3. Describe some examples of power production in your everyday life.

REVIEW 6.4 PROBLEMS

1. A certain machine can deliver 5000 joules of work in 5 seconds. What power does this machine have?
2. A certain machine is rated at 8000 watts of power. How long will it take this machine to perform 100,000 joules of work?

Chapter Review Problems

1. A student takes a case of soda weighing 27 N up an elevator from the ground floor to the roof of a physics building 60 meters high. How much work was done in this process?

2. A student expends 2000 J of work carrying a bicycle 16 meters upstairs. What is the weight of the bicycle?

3. A machine can produce 10,000 J of work before running out of fuel. How high can the machine lift a 120-N load before it runs out of fuel?

4. An automobile weighing 10^4 N is at one end of a lift. If a mechanic can exert a total of 250 N on his end of the lift, how far down must he push on his end of the lift to raise the car a total of 6 meters off the ground? How much work is done in the process?

5. An automobile having a mass of 10^3 kg moves along a highway at 20 m/sec. What is the kinetic energy of the car?

6. If the car in Problem 5 encounters a hill, to what height up the hill will the car's kinetic energy allow it to go?

7. How fast must a 2000-kg automobile travel in order to have a kinetic energy of 25,000 J?

8. A piano weighing 2000 N is moved up a 30-m flight of stairs. What potential energy is acquired by the piano in the process? How much work was done getting the piano to the top of the stairs?

9. To what height will a 2-kg ball rise if it is given an initial upward velocity of 3 m/sec? Calculate the height to which a 20-kg ball will rise under the same conditions. Comment on your result.

10. If its initial upward velocity is 0.5 m/sec, at what velocity will a 3-kg ball be moving when it reaches its maximum height? What is the kinetic and potential energy of the ball at this instant? What is the ball's total energy?

11. A baseball pitcher can hurl a baseball at a velocity of 40m/sec. If the pitcher throws the baseball straight up in the air, to what height can the baseball go before falling back to the earth?

12. A machine delivers 1000 joules of work but uses up 2500 joules of energy in doing so. What is the efficiency of this machine? How much energy should be used if we sought to have the machine's efficiency equal 90%?

13. A box with a weight of 20 N is pushed across a floor a total of 6 meters. If the coefficient of friction between the box and the floor is 0.6, how much work is done moving the box?

14. How much rotational energy is associated with an object having a moment of inertia of 7 kg-m^2 and rotating at a rate of 12 radians/second?

15. The flywheel of a certain machine rotates at a rate of 200 radians/second. If the amount of rotational energy possessed by the flywheel is 5000 joules, what is the moment of inertia of the flywheel?

16. The flywheel of a 0.4-kg toy ''friction'' car is rotated at a rate of 100 radians/second. If the moment of inertia of the flywheel is 4×10^{-5} kg-m^2, how much rotational energy does the rotating flywheel possess?

17. If the car in Problem 16 is set down on the floor and all of the flywheel's rotational motion is transformed into kinetic energy how fast will the toy car move when it is set down?

18. On earth, an astronaut tosses a ball straight up with an initial upward velocity of 4 m/sec. If she performs that same act on the moon where the acceleration of gravity is ⅙ that of the earth, compare the initial kinetic energies of the two tosses.

19. Suppose that the student in Problem 1 makes it to the roof of the physics building in 2 minutes. How much power was expended?

20. Look at the last bill you received from your local power company. How many joules of energy did you use during the billing period? How did you calculate this value?

21. A 100-watt light bulb radiates 100 watts of power. How long can such a bulb run on a kilowatt-hour of energy?

22. A 3-kg ball is tossed into the air with an initial velocity of 6 m/sec. If the power delivered to the ball was 100 watts, over what time interval did the power delivery occur?

23. An astronaut tosses a 4-kg ball upward from the surface of a distant planet. The ball rises to a height of 5 meters. If the initial velocity of the toss was 3 m/sec, find the acceleration of the planet's surface gravity.

24. A 2-kg ball moving at 3 m/sec strikes a second ball at rest having a mass of 8 kg. If kinetic energy is conserved, find the velocity of the second ball after the collision if the velocity of the first ball is reduced to zero.

THOUGHT QUESTIONS

1. A student picks up a load of books, walks across the room with the books at the same height, and sets them down. When was work performed in this sequence of actions?

2. Much of the ''New Age'' philosophy deals with ''new'' forms of energy. Is it possible for as-yet undiscovered forms of energy to exist? How might you test for their existence?

3. If velocity is expressed in m/sec and mass in kg, show that the unit of both kinetic energy and rotational energy is the joule.

4. Describe what you think would happen to the wheel in Figure 6.7 if the mass remained attached to the wheel after it passed the wheel's base.

5. Is it possible for an object to possess momentum but not kinetic energy? Kinetic energy but not momentum? Would your answers change if a system of two or more objects were involved? Explain.

6. A car starts rolling down a hill from point A. How far along the hilly terrain can the car roll without power? Assume no forces other than gravity are present.

7. Give an example of an everyday device which employs one or more of the simple machines shown in the figure in the Physics Facet in this chapter.

8. Can you think of any additional types of simple machines other than those shown in the figure in the Physics Facet?

9. If a worker carries a load from one end of a building to another without changing the height of the load off the ground, does the worker perform any work? Explain.

10. Since energy cannot be ''seen,'' is the concept of the existence of such an entity in the universe really consistent with the principles of the scientific method? Explain.

11. In the text it was stated that the height to which a given object will rise when tossed upward depends only on the upward velocity imported to the object. Is this consistent with the everyday observation that it's harder to toss a safe 5 meters into the air than it is to toss a softball that same 5 meters? Explain.

12. What relationship, if any, exists between force and potential energy?

13. Describe a machine or other device that makes use of the following energy transformations:
    a. gravitational potential energy into rotational energy
    b. kinetic energy into gravitational potential energy
    c. rotational energy into kinetic energy.

14. Why can the mechanical advantage of a machine be greater than 1, while the efficiency cannot?

15. A wheel starts rolling down from the top of a hill. Halfway down, which of the following does the wheel have?:
    a. gravitational potential energy

b. kinetic energy

c. rotational energy

d. power.

16. Does the height to which a ball will rise depend on the power delivered to the ball? Explain.

17. Do you think work, energy, or power are vector quantities? Explain.

18. James Watt, an English inventor at the turn of the nineteenth century and for whom the watt is named, defined the ''horsepower'' of a machine in terms of the rate at which an average horse could perform work. Do you think this is a very good unit to use for power measurement? If not, what would you do to make it better?

19. Would you expect power to be conserved in nature? Explain.

20. Do you think it is necessary or desirable for a civilization to have ''powerful'' machines and other devices? Explain.

21. Describe in general two possible ways to increase the power production of a given machine.

22. Can an object that is decelerating deliver either work or power? Explain.

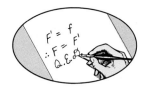

Levers

Levers are among the most ancient of the ''labor-saving'' devices developed by human cultures around the planet. For centuries these simple mechanisms have served in a most imaginative inventory of tasks, ranging from lifting heavy loads to providing the basis for the ''oar power'' employed to drive ships across vast windless stretches of sea. The most renowned mathematician of antiquity, Archimedes of Syracuse, made extensive use of levers as he almost single-handedly held off the mighty Roman army for three years during the siege of Syracuse in the third century B.C.

To illustrate how these enduring devices operate, let us consider the ancient Egyptian worker shown in Figure 6.14 who is employing a lever-like device known as a shaduf. If the worker wishes to lift a rock, quantity of water, or other load with the shaduf, he must exert some sort of downward force on his end. If the load has a given mass, then the force of gravity on the load or load force F_L is equal to $g \times$ load mass and the load torque exerted by the load at a distance d_L from the pivot point or *fulcrum* of the lever is $F_L \times d_L$ in a clockwise direction. As long as the load is on the ground, no clockwise rotation will occur, but if the worker can exert a torque larger than that of the load, he can produce a counterclockwise rotation of the system and thereby lift the load weight. The condition for which this lifting effect will just begin to occur is

$$\text{worker's applied torque} = \text{load torque} .$$

If the worker can exert an applied force of F_A over an applied lever arm distance of d_A, then we have

$$\text{applied force} \times \text{applied lever arm} = \text{load force} \times \text{load lever arm}$$

or

$$F_A \times d_A = F_L \times d_L .$$

This relationship shows that a given weight can be lifted by a smaller magnitude force acting over a longer applied lever arm. If, for example, the lever arm of a given weight is one meter and the worker wishes to lift the weight with a force ¼ that of the weight, he must do so by using a lever arm four times longer than that of the weight, which for this case would be an applied lever arm four meters long.

The placement of the fulcrum point relative to the applied and load forces gives rise to three basic classes of levers. Class I or first class levers have the fulcrum located between the applied and load forces, such as the case for the shaduf. Class II or second class levers have the load force located between the applied force and the fulcrum, while class III, third class levers, have the applied force located between the load force and the fulcrum (Figure 6.15).

The work done by pushing down on one end of a lever with an applied force F_A is $F_A \times x_A$ where x_A is the distance over which the applied force is exerted. In response to the applied force the load will move upward by an amount x_L (Figure 6.16). Since the angle θ through which the lever rotates is the same on either side of the fulcrum, we may write (see Physics Formulation 4.1) that

$$\theta = \frac{x_L}{d_L} = \frac{x_A}{d_A} .$$

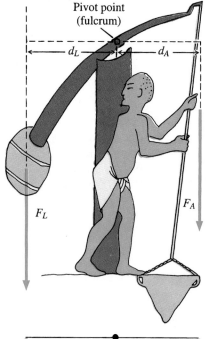

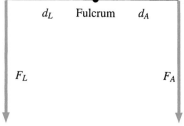

FIGURE 6.14 ··········

A Simple Lever This lever device from ancient Egypt is called a shaduf. To lift the weight, the worker must exert an applied torque $F_A \times d_A$ that is greater than the load torque $F_L \times d_L$.

Thus

$$d_A = d_L \frac{x_A}{x_L} .$$

Substituting this expression into the previous condition for lifting we have

$$F_A \times \left(d_L \times \frac{x_A}{x_L}\right) = F_L \times d_L ,$$

which can be written as

$$F_A \times x_A = F_L \times x_L$$

or

$$\frac{\text{applied}}{\text{force}} \times \frac{\text{applied force}}{\text{distance}} = \frac{\text{load}}{\text{force}} \times \frac{\text{load force}}{\text{distance}} .$$

This very important result tells us that for any kind of lever the total work put into the lever is theoretically the same as the total work that comes out. All the lever does is transfer the work from one place to another and allow the worker to trade the magnitude of the applied force for a larger applied force distance, or vice versa.

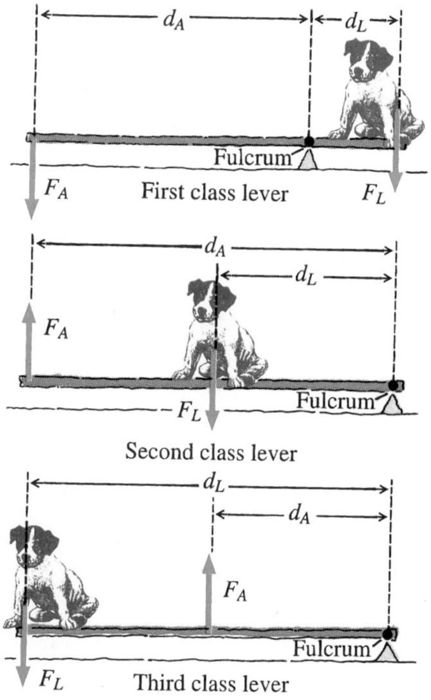

QUESTIONS

1. Explain the role of torques in the operation of levers.
2. Give some examples of the practical applications of levers.
3. Can additional classes of levers exist besides the Class I, II, and III levers? Explain.

PROBLEMS

1. Freddie has a mass of 50 kg and sits on a teeter-totter 2 meters from the

FIGURE 6.15 ....
The Various Classes of Levers The types of levers are defined in terms of the relative position of the fulcrum, load force, and applied force along the lever.

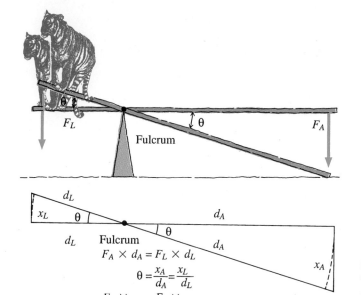

FIGURE 6.16 ....
The Basic Equations Governing Levers

fulcrum. If Sam, with a mass of only 35 kg, wishes to lift Freddie off the ground, how far from the teeter-totter fulcrum should Sam sit?

2. A worker wishes to lift a 200-kg stone with a lever having a load lever arm of 3 meters and an applied lever arm of 15 meters. If the worker pushes down a total distance of 0.5 meters with a force of 200 N, how high will the stone be lifted off the ground?

3. Archimedes once said, "Give me a lever long enough and a place to stand, and I can lift the world." If the "weight" of the earth is about 6×10^{25} N and the fulcrum for such a lever were to be located 7×10^6 meters from the earth's center, how far would Archimedes have to stand from the fulcrum point in order to lift the earth with a force of 200 N? How does this distance compare with the distances to the planets, stars, and galaxies?

Escape Velocity

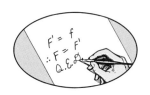

Suppose we wish to hurl an object such as a rocket up into the air in such a way that it would be able to escape the earth's gravity. One way of obtaining such a result would be to impart to the rocket enough initial velocity so that earth's gravitational force cannot slow it down enough to prevent the rocket from escaping from the earth. The minimum velocity required for such an ''escape'' is called the *escape velocity* and can be calculated from the principle of energy conservation (Figure 6.17).

When an object of mass m_1 is moved from a distance r_1 to a distance r_2 from a second mass m_2, the change in gravitational potential energy ΔPE is given by

$$\Delta PE = Gm_1m_2 \left(\frac{1}{r_1} - \frac{1}{r_2} \right) ,$$

where G is the universal gravitation constant. If the rocket's mass is m and the earth's mass and radius are M and R, respectively, then the gain in the rocket's gravitational potential energy when it is launched from a distance R to an infinitely large distance will be

$$\Delta PE = GMm \left(\frac{1}{R} - 0 \right)$$

since at large values of r_2, $1/r_2$ is for all intents and purposes equal to zero. This gain in the rocket's gravitational potential energy must, according to the principle of the conservation of energy, come from somewhere. That ''somewhere'' is the kinetic energy KE that the rocket possesses as a result of being accelerated up to its escape velocity, v_e. That kinetic energy is

$$KE = \frac{1}{2} mv_e^2 .$$

If the rocket has just enough kinetic energy to escape the earth's gravitation, then the rocket's initial kinetic energy is equal to the final gain in the rocket's gravitational potential energy and we have

$$\frac{1}{2} mv_e^2 = \frac{GMm}{R} .$$

If we solve this equation for v_e, we obtain

$$v_e = \sqrt{\frac{2GM}{R}} .$$

Notice that the velocity of escape from an object such as the earth depends only on the mass and radius of the object to be escaped from and not on the mass of the escaping object.

QUESTIONS

1. Why doesn't the escape velocity depend on the mass of the escaping object?
2. Explain how the principle of the conservation of energy is used to obtain the equation for the escape velocity.

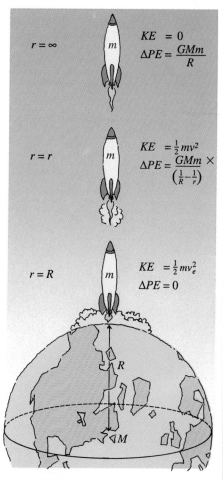

FIGURE 6.17 ···········

Escape Velocity The rocket leaving the surface of the planet with mass M and radius R will be able to escape to an infinitely large distance from the planet if its initial velocity is equal to or larger than the quantity $\sqrt{2GM/R}$, the planet's escape velocity.

3. If the escape velocity is not dependent on the mass of an object, why is it that larger mass payloads require larger rockets to escape the earth's gravity?

PROBLEMS

1. Find the escape velocity of a 10-kg ball having a radius of 0.2 meter.
2. The earth has a mass of 6×10^{24} kg and a radius of about 6.4×10^6 meters. Find the escape velocity of the earth.
3. Show that the expression for the escape velocity can be written as

$$\text{escape velocity} = \sqrt{2 \times \frac{\text{surface}}{\text{gravity}} \times \frac{\text{radius of}}{\text{object}}}.$$

7 PROPERTIES OF MATTER

Up to this point we have taken a somewhat limited view of matter in the universe in that we have considered its properties primarily in terms of motion and motion-related phenomena. In doing so, we have found that matter possesses an intrinsic and universal property that we call mass. We have used the concept of mass to describe a wide variety of phenomena including gravitational interactions, momentum conservation, energy transfer, and the like. When viewed in this context, a one-kilogram snowball is pretty much equivalent to a one-kilogram ball of iron. Yet our observations tell us that the ball of iron and the ball of snow are distinctly different in many obvious ways. What is it then that makes them different? In this chapter we will begin to look at this question.

7.1 The Atomic Theory

One of the central questions relating to matter is the question of what happens if one takes a substance, such as a page from this book, and cuts it into ever smaller pieces. As we have already seen, the Greek thinker Democritus claimed in the fifth century B.C. that if an object were chopped up into even smaller pieces, one would eventually come to the "ultimate" indivisible particle of matter—the atom. For over two millennia this view of matter took a distant back seat to the Aristotelian continuous "four-element" hypothesis. By the turn of the seventeenth century, however, the idea of four elements and continuous, forever divisible matter was beginning to crumble before the onslaught of the Scientific Revolution.

In the latter part of the seventeenth century, the English scientist Sir Robert Boyle, for example, found that air could be readily compressed and expanded. If matter were continuous as the Aristotelian view suggested, it would be most difficult to explain such behavior. If, however, the air were composed of atoms, then the compression and expansion could be accounted for by assuming that one is not expanding and compressing the air itself, but rather the space in between the atoms that comprise the air. Moreover, chemists of the seventeenth and eighteenth centuries had found that the Aristotelian elements of air, water, and earth were quite capable of being broken down into simpler substances and were thus not "elements." The last blow came when it was shown that the fourth Aristotelian element, fire, wasn't even a chemical substance at all, but was instead a form of released energy.

The most striking piece of empirical evidence in favor of Democritus's view of matter, however, came over a century after Boyle when the French chemist Joseph Louis Proust announced his discovery in 1797 of the so-called "law of fixed proportions" or law of definite proportions. Proust found that regardless of how a given chemical compound was prepared, the component elements in that compound were always present in exactly the same fixed proportions. In water, for example, for every weight unit of hydrogen there were eight of the same weight units of oxygen. If a certain quantity of water were found to have a total of 1 kg of hydrogen, it would also have 8 kg of oxygen. If 4 kg of oxygen were present, then 0.5 kg of hydrogen would be present, and so on. Other substances exhibited other fixed proportions. This phenomenon did not occur in so-called "mixtures," such as the earth's atmosphere, in which two or more substances coexist without reacting chemically. Proust also found that the proportions he observed were always small integral ratios, such as 4.0 parts to 3.0

instead of say 4.2 to 2.9. In short, Proust had found that whenever the basic elements reacted chemically, they fit themselves together in certain definite and unique integral proportions.

Less than a decade later, the English chemist John Dalton advanced a modern version of Democritus's atomic theory. Dalton's theory similar to that of the Greek thinker in that it retained Democritus's idea of indivisible atoms as the building blocks of matter, but incorporated some important differences as well. Dalton proposed the following:

1. Each chemical element is made up of individual billiard ball-like atoms, each of which shares the properties of that element.

2. Different elements are characterized by atoms having different masses. For example, an oxygen atom is 16 times more massive than a hydrogen atom.

3. Chemical compounds form when atoms of different elements combine into clumps of atoms called molecules. A molecule of water is thus formed when two hydrogen atoms combine with an oxygen atom.

Just as it was necessary for Dalton to amend the atomic theory of Democritus in light of the empirical evidence available in 1803, so the discoveries of nineteenth and twentieth century physics, which are more fully pursued in later chapters, have brought about even further and more pronounced changes in our perceptions of the atom since Dalton's time.

We now know, for example, that atoms are almost unimaginably small (see Physics Facet 7.1.) Approximately one million atoms side by side would cover a distance roughly equal to one of the printed periods in this text! Because they are so small, atoms are able to evade even the tiny wavelengths of light by which our most powerful optical microscopes are able to detect and resolve objects in the microbial world. Nevertheless, improved instrumentation such as the so-called "electron" microscope (see Chapter 14) has allowed us to "see" many of these atoms and molecules (Figure 7.1).

We have also learned that Dalton's billiard ball atoms possess a structure of their own whose complexity is considerable. As a first approximation to this structure, one can imagine that atoms are composed of three basic types of "building block" particles: protons, electrons, and neutrons. Protons and neutrons occupy the center or nucleus of an atom and are simultaneously small in size and relatively large in mass. Electrons, meanwhile, are comparatively low-mass particles and, depending on their number, occupy one or more concentric shells or "energy levels," which are centered on the nucleus (Figure 7.2).

Protons and neutrons are each about 1800 times more massive than an electron. Thus the bulk of the mass contained in an atom is highly concentrated over a relatively small volume of space at the center of the atom. While protons and neutrons are relatively large mass particles atomically speaking, on an "everyday" scale they have masses of only 1.7×10^{-27} kg. In other words, it would take about 10^{27} protons or neutrons to form into a total mass of one kilogram. To place this number into perspective, the number of grains of sand on all the world's beaches has about this same value of 10^{27}!

Each chemical element in nature is characterized by its *atomic number* or number of protons in its nucleus. The element hydrogen, for example, has one proton in its nucleus, the element helium has two, and oxygen has eight. The element uranium has a 92-proton nucleus. The *mass number* of an atom is defined as the total number of protons and neutrons contained in the nucleus. Thus an

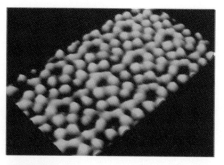

FIGURE 7.1 ··········

Silicon Atoms   Each sphere represents the image of a single silicon atom as "seen" by an electron microscope (see Chapter 14). The silicon atoms shown in this photograph have diameters of about 4×10^{-10} meters.

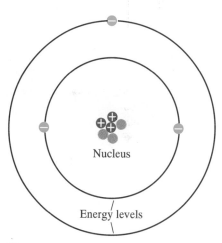

⊕ Proton
● Neutron
⊖ Electron

FIGURE 7.2 ··········

A Schematic Diagram of a Lithium Atom   Stable atoms of the element lithium consist of a three-neutron, three-proton nucleus, which is surrounded by these electrons. Similar structures exist for the atoms of other elements.

7.1 Sizes of Atoms and Molecules

The determination of the actual sizes of objects as tiny as individual atoms and molecules presents a most interesting challenge to the experimental physicist and chemist.

In one of the more ingenious methods, a single drop of oil having a known chemical composition is dropped onto a clean water surface. Upon hitting the water, the oil drop quickly spreads out into a thin, flat layer, which is roughly circular in shape. If the experiment if repeated, it is found that drops having the same volume will produce circular films having the same diameters, and any attempt to stretch a given oil layer out to a larger area produces ruptures in that layer. It is therefore reasonably certain that such a layer is one oil molecule diameter thick.

Because the layer is in the shape of a thin, flat cylinder, the volume of the layer is given by

$$\text{volume} = \pi \times (\text{radius})^2 \times (\text{thickness}) .$$

If we solve this equation for the thickness of the layer, we obtain

$$\text{thickness} = \frac{\text{volume}}{\pi \times (\text{radius})^2} .$$

All of the quantities on the right-hand side of this equation are known, and the thickness of the layer and hence the diameter of the molecules that comprise the layer can be readily calculated.

For example, suppose that a drop of solution in which the volume of stearic acid is equal to $1.0 \times 10^{-11} \text{m}^3$ is dropped onto a water surface and spreads out into a circular film having a diameter of 0.8 cm or 0.08 m. The radius of the film is thus 0.04 m and the thickness is

$$
\begin{aligned}
\text{thickness} &= \frac{1.0 \times 10^{-11} \text{ m}^3}{\pi \times (0.04 \text{ m})^2} \\
&= \frac{1.0 \times 10^{-11} \text{ m}^3}{\pi \times .0016 \text{ m}^2} \\
&= \frac{1.0 \times 10^{-11} \text{ m}^3}{5.0 \times 10^{-3} \text{ m}^2} \\
&= 0.2 \times 10^{-8} \text{ m} \\
&= 2.0 \times 10^{-9} \text{ m} .
\end{aligned}
$$

Thus, the diameter of the stearic acid molecules in the layer is about 2×10^{-9} m. Because stearic acid is a somewhat large and complex organic compound, we would expect the diameters of this molecule's constituent carbon, oxygen, and hydrogen atoms to be even smaller. More sophisticated measuring techniques indicate that the typical diameter of an individual atom such as carbon, oxygen, or hydrogen is about 10^{-10} m.

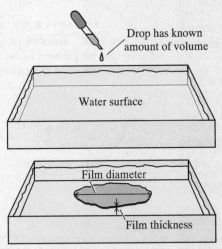

Drop has known amount of volume

Water surface

Film diameter

Film thickness

··· Diameters of Molecules When a drop of oil having a known volume is placed on a water surface, it will spread out to a roughly circular film layer, which is one molecular diameter thick. By measuring the diameter of the film layer, the thickness of the layer and hence the size of the molecules in the layer can be calculated.

oxygen atom with 8 protons and 8 neutrons has a *mass number* of 16 and a uranium atom with 92 protons and 146 neutrons has a *mass number* of 238. Because any given proton, neutron, or electron is identical to any other proton, neutron, or electron, it is possible to transform one element into another by "nuclear" reactions of such particles with atomic nuclei (see Chapter 15).

Atoms of a given element may also exist in various forms of *isotopes* in which atoms have the same number of protons in their nuclei but a different number of neutrons (Figure 7.3). An example of an isotope is the substance called deuterium. Normally a hydrogen atom has only a single proton in its nucleus, but the much rarer deuterium has one proton and one neutron in its nucleus. Thus deuterium still qualifies as a hydrogen atom because it has a one-proton nucleus, but it also has a mass number twice as large as the more common variety of hydrogen in which the neutron is absent. Many isotopes are important because they are "radioactive" and, as such, can be both quite useful and also pose potential dangers to our environment (Chapter 17).

Atoms of a given element can interact with other atoms of either the same element or different elements to form combinations of atoms called *molecules*. In the latter part of the nineteenth century, the Russian chemist Dmitry Mendeleyev noticed that certain groups or "families" of chemical elements existed that exhibited similar types of chemical behavior. For example, the elements magnesium (Mg) and calcium (Ca) react with oxygen (O) in similar ways to form MgO and CaO, respectively. Hence, magnesium and calcium are said to be in the same chemical family. At the same time, it is possible for the element sulfur (S) to react with hydrogen (H) to form H_2S just as oxygen reacts with hydrogen to form H_2O or water. Thus sulfur and oxygen are said to be in the same chemical family.

Using such information, Mendeleyev proceeded to arrange the chemical elements into a "periodic table" (see Appendix C) in which these chemical families are set up in columns in order of increasing atomic number (oxygen, sulfur, selenium, etc.) and in rows according to the basic type of chemical behavior exhibited by each chemical family (alkali metals, halogens, etc.). One column or family of elements, called the "noble" gases, includes elements such as helium, neon, and argon, which are characterized by the fact that they are not likely to react chemically at all!

Many of the properties and characteristics of the periodic table can be accounted for in terms of the valence electrons and electron spaces, which occupy the outermost regions of a given atom and are hence the most likely participants in chemical reactions with other atoms. Sulfur and oxygen atoms, for example, have two valence electron spaces in their outermost regions. Meanwhile, a hydrogen atom has one valence electron in its outermost region. As a result, two hydrogen atoms and their valence electrons react to fill the electron vacancies in a sulfur or oxygen atom, thus forming H_2S or H_2O. The noble gases have no valence electrons nor valence vacancies in their outer regions, explaining why these substances are chemically nonreactive.

Protons and electrons also possess a fundamental property called an *electric charge*. As we shall see in a subsequent chapter, two types of electric charge exist in nature that have been designated positive (+) and negative (−). Electric charges with like signs exert repelling or repulsive forces on one another. Electric charges with unlike signs exert an attractive force on each other. Protons possess a positive electric charge and electrons a negative electric charge. Neutrons are

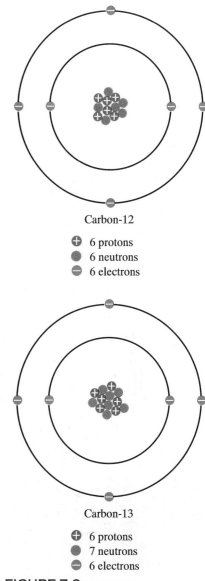

Carbon-12

⨁ 6 protons
● 6 neutrons
➖ 6 electrons

Carbon-13

⨁ 6 protons
● 7 neutrons
➖ 6 electrons

FIGURE 7.3 ·········

Isotopes Isotopes of a given element have the same number of protons, but different numbers of neutrons, resulting in different atomic weights. In this case the isotopes shown for the element carbon have atomic weights of 12 for carbon-12 (six protons, six neutrons) and 13 for carbon-13 (six protons, seven neutrons). Notice that both isotopes have six protons, the atomic number for the element carbon.

in effect composed of a proton and an electron that have been ''squished'' together and thus possess no net electrical charge.

Recent decades of physics research has provided an overwhelming array of empirical evidence in favor of the idea that protons, neutrons, and electrons are themselves composed of even smaller, more elementary particles or ''subatomic'' particles (Chapter 13), which leads one to wonder whether Democritus's ''ultimate billiard ball'' even exists at all, thus giving the last laugh in such matters, at least in part, back to Aristotle.

WORKED EXAMPLE

Hydrogen gas (H_2) consists of two protons. What is the atomic weight of a hydrogen gas molecule? What is its mass? How many hydrogen gas molecules are there in a 1-kg mass of hydrogen gas?

SOLUTION: Since each hydrogen atom has an atomic weight of 1, the total atomic weight for the H_2 molecule would be

$$mass = 2 \times 1.7 \times 10^{-27} \text{ kg}$$
$$= 3.4 \times 10^{-27} \text{ kg} .$$

The number of hydrogen gas molecules in a 1-kg mass sample will be

$$\frac{\text{number of}}{\text{molecules}} = \frac{\text{total mass}}{\text{mass/molecule}} .$$

In this case

$$\frac{\text{number of}}{\text{molecules}} = \frac{1 \text{ kg}}{3.4 \times 10^{-27} \text{ kg/molecule}}$$
$$= \frac{2}{3.4} \times 10^{27} \text{ molecules}$$
$$= 0.29 \times 10^{27} \text{ molecules}$$
$$= 2.9 \times 10^{26} \text{ molecules} .$$

REVIEW 7.1 QUESTIONS

1. What is the atomic theory of matter? In what ways is modern atomic theory different from that of Democritus's theory?

2. Discuss the extent to which the claims made by Dalton in his original atomic theory are still regarded as being correct by scientists today.

3. What are valence electrons? Valence electron spaces? Why are they important in chemical reactions?

1. How much oxygen is contained in a 2-kg sample of water (H_2O)? How much hydrogen?

2. The molecule carbon dioxide (CO_2) consists of one carbon atom and two oxygen atoms. If a sample of CO_2 is found to contain 10 kg of carbon, what is the mass of the oxygen present in this sample?

7.2 Solids

One of the historical difficulties with the atomic theory, indeed one that Democritus himself was not able to address effectively, was the problem of how the billiard ball atoms were able to stick and hold themselves together as solid substances, i.e., substances that retain both their shape and volume. As we have seen, the valence electrons and electron spaces in the outer regions of atoms are extremely important in terms of how atoms bind themselves into molecules and, on a larger scale, into solids.

The simplest type of chemical bonding occurs when an electron is transferred from the outer shell of one atom to the outer shell of another. As a result, both atoms now possess a net electrical charge. The atom that lost the electron now possesses a net positive charge because there is one excess proton. The atom that picked up the electron now has a net negative charge because it has an excess electron. Such atoms with a net electrical charge are called *ions,* and because unlike electric charges in nature attract each other, the positive ion is attracted to the negative ion and the resultant chemical bond is called an *ionic bond.* An example of a substance formed by such ionic bonding is ordinary table salt or sodium chloride (Figure 7.4).

Other atoms ''share'' their valence electrons with each other in a type of chemical bond called a *covalent bond.* The diamond is a familiar example of a solid that is held together by covalent bonding. In some substances such as copper and silver, there is a sort of ''communal'' covalent bonding or *metallic bonding* in which the valence electric charges are free to move about the assembly of atoms easily (Figure 7.5). As a result, heat energy and electric currents are able to flow readily through such metallic solids, which are simply referred to as metals.

A fourth mechanism for bonding arises out of the fact that many molecules may exhibit an asymmetric shape and charge distribution. Such molecules are said to be *polar molecules* and behave as if one end of the molecule were positively charged and the other end negatively charged. As a result, when the polar molecules are close to one another, a weak attractive force exists between the oppositely charged ends of the molecule, which weakly bonds the molecules together and is referred to as the van der Waals's force. Chemical bonding of this type is thus called *van der Waals's bonding* (Figure 7.6).

Most solids exist in the form of arrays or crystals in which their component atoms are arranged in a regular and repeating three-dimensional pattern. Solids

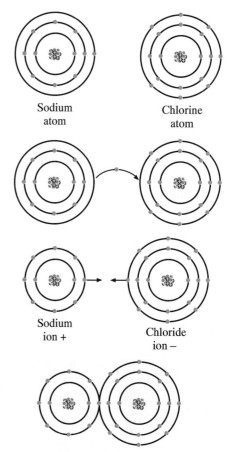

Sodium atom

Chlorine atom

Sodium ion +

Chloride ion −

Sodium chloride

FIGURE 7.4 ··········
Ionic Chemical Bonding   When an electron is transferred from one atom to another, both atoms become ions with unlike electric charges. The ions are then attracted to each other and form an ionic bond such as this bond involving the elements sodium and chlorine.

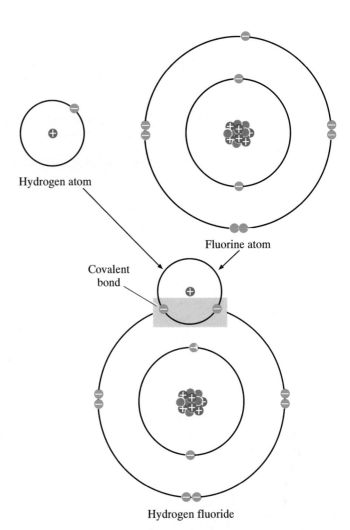

Hydrogen atom

Fluorine atom

Covalent bond

Hydrogen fluoride

FIGURE 7.5 ··········
Covalent Chemical Bonding In a covalent bond, the valence electrons of the elements involved are shared by both constituents. If the valence electric charges are relatively free to move throughout the assembly of the covalently bonded atoms, the result is called metallic bonding. In this case the hydrogen and fluorine atoms share each other's valence electrons to form a hydrogen fluoride molecule.

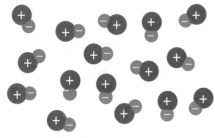

FIGURE 7.6 ··········
Polar of van der Waals's Bonding If a given molecule has an asymmetric shape and electric charge distribution, as do these "generic" molecules, it is possible for weak electrostatic forces to exist between such molecules that serve to hold them together loosely in a polar or van der Waals's bond.

that exhibit such atomic arrays are said to be *crystalline* (Figure 7.7). Gemstones are perhaps the most beautiful and familiar examples of crystalline substances. Other solids, such as glass and plastics, exhibit no definite repeating patterns in the arrangements of their atoms. These types of substances are referred to as *amorphous solids.*

Solids are quite often characterized by the degree to which their atoms are compacted, or the *density* of the solid (Figure 7.8). The density of a solid or, for that matter, *any* substance is formally defined as

$$\text{density} = \frac{\text{mass}}{\text{volume}} \qquad\qquad 7.1$$

or

$$\rho = m/V \ .$$

A "light" solid such as balsa wood contains a great deal of space within its structure and will have a lower density than a solid such as iron whose atoms are more closely packed. Densities are usually expressed in grams/cubic centi-

(a)

(b)

(c)

(d)

FIGURE 7.7

Crystals Crystalline structure for
(a) fluorite, (b) selenite, (c) stibnite, and
(d) emerald.

meter or kilograms/cubic meter. Water, for example, has a density of 1000 kg/m³ or 1 gram/cubic centimeter. Some representative densities of other substances are listed in Table 7.1.

Whenever a force, such as a compression force, is exerted on a solid, that solid will experience changes in its size and shape, the magnitude of which will depend on the nature of the solid. The measure of an object's ability to restore itself to its original shape after a deforming force is removed is called its elasticity and is further pursued in Physics Formulation 7.1.

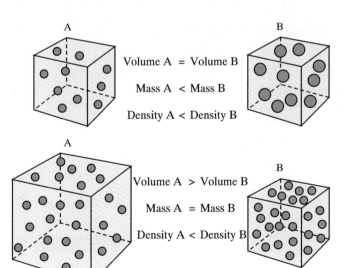

A

B

Volume A = Volume B

Mass A < Mass B

Density A < Density B

A

Volume A > Volume B

B

Mass A = Mass B

Density A < Density B

FIGURE 7.8

Density If two substances have the same volume, the higher density material will have the higher mass. If two substances have the same mass, the higher density substance will have a smaller volume.

TABLE 7.1 **Densities of Various Solids and Liquids**

Material	Densities	
	grams/cm³	*kg/m³*
Gold	19.3	19,300
Mercury (liquid)	13.6	13,600
Lead	11.3	11,300
Silver	10.5	10,500
Iron	7.2	7,200
Diamond	3.3	3,300
Aluminum	2.7	2,700
Concrete	2.3	2,300
Ebony wood	1.2	1,200
Water (liquid)	1.00	1,000
Ethyl alcohol (liquid)	0.79	790
Oak wood	0.72	720
Gasoline (liquid)	0.68	680
Balsa wood	0.13	130

WORKED EXAMPLE

A certain block of material 3 cm × 2 cm × 5 cm in size has a mass of 80 grams. Find the density of the block. From the data in Table 7.1, of what material do you think the block is made?

SOLUTION: The density of any substance is given by

$$\text{density} = \text{mass/volume} .$$

The volume of the block in cubic meters is

$$\text{volume} = 0.03 \times 0.02 \times 0.05 \text{ meters}^3$$
$$= 3 \times 10^{-2} \times 2 \times 10^{-2} \times 5 \times 10^{-2} \text{ meters}^3$$
$$= 30 \times 10^{-6} \text{ meters}^3 .$$

Since the mass is 80 grams or 0.08 kg, we have

$$\text{density} = 0.08/30 \times 10^{-6} \text{ kg/m}^3$$
$$= 8 \times 10^{-2}/3 \times 10^{-5} \text{ kg/m}^3$$
$$= 2.7 \times 10^3 \text{ kg/m}^3 .$$

The block has a density of 2700 kg/m³ and, from Table 7.1, is most likely made of aluminum.

1. Describe the various ways that atoms can combine with each other to form molecules.

2. Discuss the characteristics of a solid. Give some examples of solids in your everyday life.

3. What is the density of a substance? Give some examples of high-density substances and low-density substances.

REVIEW 7.2 QUESTIONS

1. A block of uranium having a volume of 25 cubic centimeters is found to have a mass of 475 grams. Find the density of uranium in grams/cm^3 and kg/m^3.

2. If the world market price of gold is about \$15,000/kg how much mass would one million dollars worth of gold have? How much volume would this gold occupy? Assume that the density of gold is 19,300 kg/m^3.

REVIEW 7.2 PROBLEMS

7.3 Liquids

As we have seen above, solids are characterized by a definite shape and volume because their component atoms and molecules are confined to what are very nearly fixed positions. If enough energy is added to the solid, such as by raising its temperature, then the bonding that gives the solid its rigidity breaks down and is replaced by a much looser type of cohesiveness that is strong enough to hold the substance to a definite volume but not strong enough for it to retain a definite shape. A substance in such a state of affairs is said to be in a *liquid* state.

The cohesive forces within the liquid tend to pull a given atom or molecule away from the surface of the liquid, giving rise to a tendency for the surface of the liquid to contract itself into a surface area that has the smallest possible value under a given set of circumstances. Such a contractive force within a given liquid is referred to as the *surface tension* of that liquid and is responsible for a number of interesting properties exhibited by the liquid state. The spherical shapes of raindrops, for example, arise out of the fact that the surface tension of the water within the drop acts in such a way as to minimize the surface area of the drop. For a given volume, the shape having the smallest possible surface area is a sphere, and drops of liquid therefore take on a spherical shape. Many small insects, most notably the water strider, are able to scurry across the surfaces of ponds and lakes because their weights are not sufficiently large to "break through" the surface tension of the water molecules (Figure 7.9).

In addition to the cohesive forces that give rise to surface tension, liquids also exhibit "adhesive" forces between themselves and the surfaces of their containers. If the liquid is confined to a relatively small tube, then the surface of the liquid will take on a curved shape or "meniscus." The meniscus is concave upward if the adhesive forces between the liquid and the container wall are larger than the liquid's self-cohesive forces, as in the case of water in a glass tube. The

FIGURE 7.9 ·········

The Water Strider Because of the surface tension of the water, this water strider can walk along the surface of the pond without sinking even though its density is larger than that of the water.

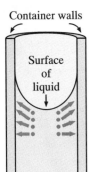 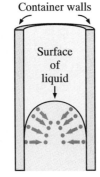

Container walls   Container walls

Surface of liquid   Surface of liquid

FIGURE 7.10

Adhesion and Cohesion The self-cohesive forces of the liquid on the left are less than the adhesive forces between the liquid and the wall; therefore, the surface of the liquid is concave upward. The self-cohesive forces of the liquid on the right are greater than the adhesive forces between the liquid and the wall; therefore, the surface of the liquid is concave downward.

meniscus is concave downward if the adhesive forces are less than the cohesive forces, as in the case of mercury in a glass tube (Figure 7.10).

If a small-diameter tube is placed into a large reservoir of liquid, as shown in Figure 7.11, the level of the liquid inside of the tube can rise above or be pushed below the level of the liquid in the reservoir by a process called *capillary action* or *capillarity*. Capillary action is essentially an interplay of adhesive forces and surface tension. In the case of a narrow glass tube in water, for example, the adhesion attraction of the water molecules for the sides of the tube produces the aforementioned concave upward-shaped water surface inside of the tube. The surface tension of the water, however, seeks to ''flatten out'' the meniscus and in doing so pulls the level of water up to the highest part of the meniscus, provided the cross-sectional area of the tube and hence the amount of water to be pulled up are not too large. The adhesive forces now respond by, in effect, creating another meniscus, which the water's surface tension tries to once more flatten out, thereby drawing the water in the tube to a yet higher level, and so on. The process will continue until the weight of the column of water drawn up into the tube is equal to the total adhesive force between the water and glass.

In a similar manner, the interplay between surface tension and self-adhesive forces can actually push the level of a liquid in a tube to a level below that of the surrounding liquid reservoir. Perhaps the best example of this type of capillary action occurs when a narrow glass tube is placed in a reservoir of mercury. There are also many everyday manifestations of the phenomenon of capillary action. If the corner of a paper towel is placed in contact with a puddle of water, the water will be ''sucked up'' by the towel as a result of this process (Figure 7.12). Capillary action also plays an important role in such diverse phenomena of nature as the transport of aqueous solutions from plant roots to plant foliage, the dynamics of moisture in soils, and blood flow in the capillaries of animals and humans.

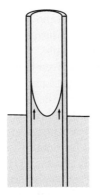 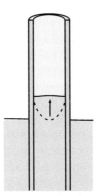 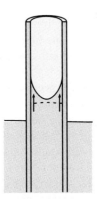 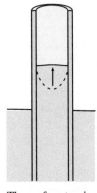

Adhesive forces pull the liquid up the sides of the tube   The surface tension of the liquid ''smooths out'' the liquid's surface   Adhesive forces pull the liquid up the sides of the tube   The surface tension of the liquid ''smooths out'' the liquid's surface

FIGURE 7.11

Capillarity If a liquid is contained in a relatively small-diameter tube, the interplay of adhesive forces and the liquid's surface tension produce an effect called capillarity or capillary action in which the level of the liquid in the tube can rise above or fall below the level of liquid in the surrounding reservoir.

FIGURE 7.12
Capillarity in Action   In this sequence of photographs the small fibers in the paper towel are ''sucking up'' the ink by means of capillary action.

1. Discuss the characteristics of a liquid. Give some examples of liquids in your everyday life.
2. What is surface tension? Give an example of surface tension.
3. Describe the phenomenon of capillary action. Give an example of capillary action that occurs in nature.

REVIEW 7.3 QUESTIONS

7.4 Gases

From the preceding discussion, a liquid can be thought of as a solid whose bonds have been partially broken down because of the addition of some form of energy such as thermal energy. If we continue to add even more energy to the substance, it is possible to break down the cohesive forces in a liquid completely so that the molecules in the substance behave as separate individual entities. In such a state, called the *gaseous* state, molecules now collide with the walls of the container and with each other like billions of tiny billiard balls rattling around on top of a billiard table.

Gases have neither a definite form nor do they occupy a definite volume. If a small quantity of gas is released in a room, it will quickly expand to fill the dimensions of the room, regardless of the room's size and shape. Liquids and solids have definite volumes, which are largely incompressible. The volume of a gas, however, can be readily changed by changing its temperature, forcing the container walls to take on a smaller enclosed volume. It is thus necessary to specify the *state* of a gas by means of a set of *state variables,* which include parameters such as temperature, pressure, and density (see Physics Formulation 8.1).

If a gas is heated to a high enough temperature, the frequency and violence of the collisions between the atoms in the gas cause electrons to be knocked off their parent atoms, resulting in a mixture of free electrons and ionized atoms that

(a)

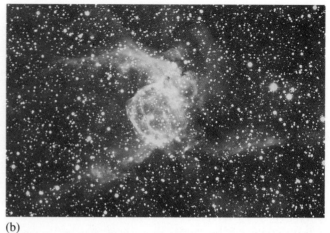

(b)

(c)

FIGURE 7.13 ···········
Plasmas in Nature   (a) The aurora borealis in the earth's upper atmosphere; (b) a glowing interstellar plasma; (c) the glowing plasma of a solar flare.

is called a plasma. Although the overall electric charge on the plasma is still zero, the free electrons and charged ions within the plasma are, like liquid solutions and metallic solids, able to interact with electric and magnetic fields in a variety of interesting and useful ways. For example, plasmas play an important role in producing the light emitted from any kind of fluorescent lamp. Plasmas are also intimately involved in a number of geophysical and astrophysical phenomena such as auroral displays, glowing interstellar gas clouds, and solar flares (Figure 7.13).

REVIEW 7.4 QUESTIONS

1. Discuss the characteristics of a gas.
2. What is a plasma? How is a plasma different from a gas? How is it the same?
3. Describe how gases are important to us in our daily lives.

7.5 Fluids

Because both liquids and gases possess the ability to flow and take on the shape of their containers, it is often useful to describe the behavior of these two states of matter in terms of this shared property of *fluidity*. To this end, scientists and engineers have defined a fluid as any substance that lacks the rigidity of the solid state. Thus any entity that can flow or ''run'' can be classified as a fluid.

Some fluids flow more readily than do others. ''Molasses in January'' does not take on the shape of its container as quickly as the gaseous aroma emanating from a good meal in preparation. The measure of the ability of a substance to

PHYSICS FACET

7.2 Living Matter

One of the more impressive manifestations of the principle of universality is the observation that all matter in the universe, from moon rocks and meteorites to galaxies and quasars, is composed of precisely the same chemical elements that make up all matter here on the earth. Although all matter shares this commonality of composition, a type of matter exists that possesses an additional quality, which is most extraordinary and remarkable. It is the quality we call "life."

Matter that is said to be "living" is distinguished from matter that is "nonliving" or "inanimate" in terms of five so-called "life activities." Any substance that can move, grow in size, take in food or other sustenance, respond to stimuli, and reproduce itself is defined as "living" matter. To be regarded as living, a given object must exhibit *all* of these life activities. An automobile, for example, can move, "eat," and respond to the "stimulus" of one's foot on the break or accelerator, but is not regarded as a living thing, because it doesn't grow in size or reproduce itself.

The life activities thus serve to provide us with criteria by which one can distinguish living matter from nonliving matter, but they do not really address the most interesting questions regarding living material. The instant before a living organism dies, for example, is fundamentally different from the instant after that organism dies, yet the array of atoms and molecules making up the organism does not seem to have changed in a substantial way over this infinitesimally small interval of time. What are the qualities that distinguish the lifeful instant from the lifeless one? If life operates on the basis of some as-yet and undiscovered physical laws of nature, this question may one day be answered satisfactorily. If, however, life is an entity which through its abstract or "spiritual" nature defies measurement and quantitative description, then the issues become ones of philosophy, metaphysics, and religion rather than of empirical science.

In taking the former tack, scientists have sought to explain the origins of living matter on the earth in terms of physical processes that occurred early in the earth's geological history. Such a suggestion has provided considerable fuel for science fiction writers, owing to the fact that if life on the earth received such a start, then the same life-producing processes may well have occurred elsewhere in the universe, thereby resulting in the generation of "extraterrestrial" creatures of one sort or other. Unfortunately, the discussion of the possible existence of life forms elsewhere in the universe has been restricted to products of probabilities in which a probability of occurrence is assigned for each factor believed to be necessary for life forms to have developed in a given location.

 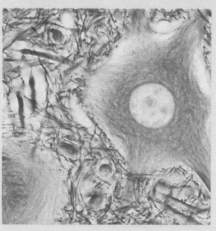

⋯ Living and Nonliving Matter Although there are some similarities between the structural patterns in the rutilated quartz sample (left) and the brain cells (right), the quartz does not exhibit any of the characteristics exhibited by life and is thus regarded as nonliving matter.

PHYSICS FACET

Continued

We have reasonably firm evidence that the total number of stars in our Milky Way Galaxy is of the order of 100 billion and that almost all of these suns are being orbited by one or more planets. At this point matters become highly speculative in terms of what factors are believed necessary for the formation and evolution of living material and what probabilities should be assigned to them. If, for example, we attempt to determine the fraction of the total of all the planets and satellites in the galaxy that exhibit conditions suitable for the development of life forms, we quickly run aground over the question of what those conditions might be. The underwater ocean environment that is eminently suitable for a mako shark is hardly the environment a human being or a dandelion would care to call home. Once established, however, there is considerable evidence that life forms can alter and adapt themselves to changing conditions. Thus, the divergent life-styles of the dandelion and the mako shark are believed to be the result of a process called evolution, in which evolutionary "forces" such as climate, level of radiation, food supply, etc., are thought to have played a crucial role in the creation of the diversity of life forms we now observe on the earth. The exact nature and overall impact of these "forces" continues to be a topic of considerable debate.

Because of the large degree of uncertainties and unknowns involved, attempts to answer the question of the extent to which life exists in the universe and the ways in which it might have evolved must necessarily be highly speculative in nature. As a result, the scientific views on the matter range from the "Star Trek" picture of life existing just about anywhere one cares to look to the opposite "$N = 1$" extreme, which holds that the life forms here on the earth constitute the "only game in town."

The physical evidence is tantalizingly mixed. On one hand, materials have struck the earth from outer space that have been found to contain chemical substances called amino acids, which are complex molecules of hydrogen, carbon, oxygen, and other elements known to be the "last stage" building blocks of living material. At the very least, the presence of amino acids in extraterrestrial meteorites tells us that natural processes are occurring in deep space, which gives rise to hydrocarbon molecules whose complexity is comparable to that of living matter. On the other hand, in our explorations of space, particularly in those of the solar system, we have yet to come across any sign of any type of life form, whether it be microbes in the desert sands of Mars, "balloon spider"-type life forms in the cloud banks of Venus, or any of the myriad of similar speculations that have been made over the past few centuries. Those speculations notwithstanding, the earth's 15-km-thick biosphere remains to date the only location in the observable universe where we know for sure that living matter exists.

resist such fluid flow is called its *viscosity*. Viscosity is a sort of internal friction caused by the interactions among the molecules of the fluid. For substances in which those interactions are relatively weak, such as in a gas, the viscosity is said to be low, and the substance flows easily. For substances having high viscosity, the molecular interactions are much stronger, and the substance does not flow as readily. As a result, liquids tend to have higher viscosities than gases. Air at room temperature, for example, has a viscosity roughly 100 times less than water and 10^{11} times less than that of molasses. Air is thus 100 times more prone to flow as is water and 10^{11} times more "fluid" than molasses.

In describing the properties of fluids, it is most useful to introduce the concept of *pressure* (Figure 7.14), which is formally defined as

$$\text{pressure} = \frac{\text{force exerted by a fluid}}{\text{area over which force is exerted}}$$

or

$$P = F/A \ . \qquad\qquad 7.2$$

Pressure is in essence a force per unit area. It is measured in units of newtons/square meter or pascals, where 1 pascal (Pa) is defined as the pressure exerted by a total force of one newton being exerted over a surface area of one square meter. A given force can thus exert a wide range of pressures, depending on the area over which it is exerted. A woman in stiletto high-heel shoes, for example, exerts dozens of times more pressure on a floor for the force of her weight than she would if she wore tennis shoes having a larger contact area with the floor. In like fashion, a given person wearing ordinary boots will sink deeply into a large snow drift, while that same person wearing wide area snowshoes can walk atop the snowdrift with relative ease because the force of the person's weight is spread out over a larger area. In other words, the pressure exerted on the snowdrift by the individual in snowshoes is far less than that exerted by the same individual wearing boots.

With this definition of pressure in mind, let us consider a liquid in a simple container whose bottom has a given surface area and whose sides are perpendicular to the ground. The volume of the liquid in the container is then equal to the area of the container bottom × the depth of the liquid. The mass of the liquid in the container is equal to the volume of the liquid in the container × the density of the liquid, and the force exerted by the earth's gravity on this mass of liquid is simply mass × acceleration of the earth's gravity. By definition, the pressure exerted by the liquid on the bottom of the container is given by

$$\text{liquid pressure} = \frac{\text{force}}{\text{area}} \ .$$

The total downward force of gravity on the mass is mass × g, hence we have

$$\text{liquid pressure} = \frac{\text{mass} \times g}{\text{area}} \ .$$

The total mass of the liquid is the product density × volume, which allows us to write

$$\text{liquid pressure} = \frac{\text{density} \times \text{volume} \times g}{\text{area}} \ .$$

The total volume of the liquid is the product of depth × area, and so

$$\text{liquid pressure} = \frac{\text{density} \times \text{depth} \times \text{area} \times g}{\text{area}} \ ,$$

which reduces to

$$\text{liquid pressure} = g \times \text{depth} \times \text{density} \ .$$

This relationship tells us that the pressure exerted by a liquid at a given depth can be thought of as being equal to the pressure exerted by a unit area column of that liquid having a vertical dimension equal to the depth of the point. Thus the water pressure at the bottom of a swimming pool 3 meters deep will be

$$\begin{aligned}
\text{pressure} &= 9.8 \text{ m/sec}^2 \times 3 \text{ m} \times 1000 \text{ kg/m}^3 \\
&= 29{,}400 \text{ N/m}^2 \\
&= 29{,}400 \text{ pascals} \ .
\end{aligned}$$

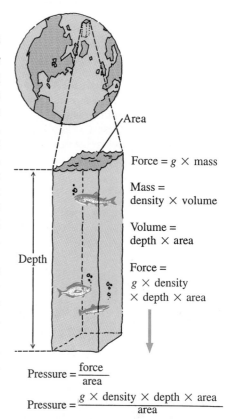

Area

Force = g × mass

Mass = density × volume

Volume = depth × area

Force = g × density × depth × area

Depth

$$\text{Pressure} = \frac{\text{force}}{\text{area}}$$

$$\text{Pressure} = \frac{g \times \text{density} \times \text{depth} \times \text{area}}{\text{area}}$$

Pressure = g × density × depth

FIGURE 7.14 ············
Pressure Pressure is a force per unit area. For a liquid, the pressure at any given depth is equal to the product g × depth × density.

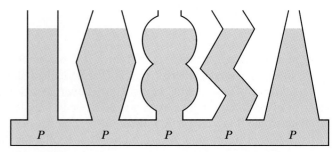

FIGURE 7.15 ⸳⸳⸳⸳⸳⸳⸳⸳⸳⸳

Pascal's Principle and Pascal's Vases   Because the pressure P on an enclosed fluid is transmitted undiminished to all parts of the fluid, the level of the fluid in all of the variously shaped and interconnected "Pascal's vases" will be the same regardless of the shapes of the vases.

P = pressure

It can also be shown that the pressure exerted by a given liquid at a given depth is the same, regardless of the shape of the container. Thus our container could have had any arbitrary shape, and we could have derived the same result, albeit with a bit more complicated mathematics. This leads to the interesting result that if we fill a container consisting of a series of irregularly shaped vessels as shown in Figure 7.15, the liquid will "seek its own level" in each of the vessels, provided that the effects of capillary action are negligible for each vessel. Observations of this effect led the seventeenth century French mathematician Blaise Pascal to formulate Pascal's principle, which states:

> Any change of pressure on an enclosed fluid is transmitted undiminished to all parts of that fluid.

To illustrate one of the consequences of this statement, consider the U-shaped container shown in Figure 7.16. If a force is placed on the small area, then the resulting pressure, according to Pascal's principle, will be transmitted undiminished to the larger area portion of the container. The pressure across the weight resting on the larger area is thus the same. The *total force* on the weight, however, has now increased by an amount equal to the ratio of the large area to that of the small area. It is this property of fluids in general, combined with the high degree of incompressibility of liquids, that underlies the workings of hydraulic lifts, one version of which can be seen in motion the next time your car is put "up on the rack" at your local service station. A more detailed discussion of these devices is to be found in Physics Formulation 7.2.

FIGURE 7.16 ⸳⸳⸳⸳⸳⸳⸳⸳⸳⸳

Pascal's Principle and the Hydraulic Lift   From Pascal's principle, a force exerted over a small area is transmitted as an undiminished pressure to all parts of a larger area, thereby "multiplying" the initial applied force. This is the concept underlying hydraulic lifts and presses.

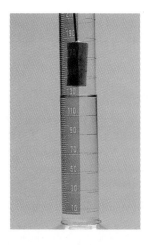 

FIGURE 7.17 ··········
Buoyant Force If an object is submerged in a liquid such as water, the volume of liquid displaced by the object will try to regain its last volume by exerting a buoyant force on the submerged object.

Although our discussion of pressure has thus far centered on liquids, gases, too, are capable of exerting pressures of their own. Perhaps the most familiar example of gas pressure comes to us courtesy of the layer of gas surrounding the earth's surface, which we call the earth's atmosphere. Most of the gas in the earth's atmosphere lies in a layer about 11 km deep and exerts a pressure of about 100,000 newtons per square meter at sea level, a pressure that is defined as *one atmosphere* of pressure.

Recall from Chapter 3 that for every action force there is an equal and opposite reaction force. Fluids obey this law of mechanics in a most interesting fashion. If, for example, you try to push a cork down into a glass of water so that it is submerged, two effects can be observed. First of all, in response to the "action force" of pushing the cork down into the water, there is a "reaction force," called a *buoyant force,* which trys to push the cork back to the surface of the water (Figure 7.17). In addition, while the cork is submerged beneath the water, the level of the water in the glass increases its height in the glass. These two effects, as you might imagine, are not independent of one another. As you push the cork down into the water, some of the water is pushed out of the way or is "displaced" by the cork. The total volume of water displaced by the submerged cork is simply the volume of the cork itself. The water tends to regain its lost volume by exerting a "reaction" buoyant force on the cork, which is directed upward toward the surface of the water. The formal relationship between the buoyant force exerted by a fluid was discovered more than two thousand years ago by the Greek Archimedes who while taking a bath, we are told, hit upon the idea that:

> A fluid will exert a buoyant force on any body floating on or immersed in it which is equal to the weight of the fluid displaced by the floating or immersed body.

*Archimedes' principle,** as it is called, gives us a handle on why some objects float and other objects sink. If we had placed an iron "cork" into the glass that had exactly the same volume as the wooden cork, it would have sunk straight to the bottom, because the weight of the iron "cork" would be more than the water's buoyant force. The wooden cork, however, is a much lighter object and

---

*Upon being struck by this "bathtub" inspiration, scientific legend has it that Archimedes immediately went running naked through the streets of Syracuse shouting "Eureka!" (I have found it!)

(a)

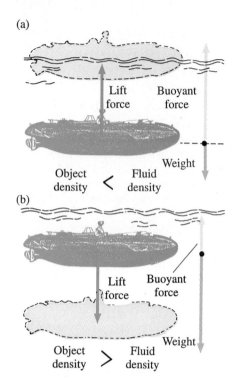

(b)

FIGURE 7.18 ···········
The Mechanics of Floating and Sinking (a) If the buoyant force on an object is larger than its weight, then the lifting force is positive and the object floats. (b) If the buoyant force on the object is smaller than its weight, then the lifting force is negative and the object sinks.

its weight can easily be overcome by the buoyant force of the water. These results can be mathematically summarized by introducing the concept of a *lift force*, which is defined as

$$\text{lift force} = \frac{\text{buoyant}}{\text{force}} - \frac{\text{weight of}}{\text{object}}.$$

If the lift force is positive, then the weight of the object cannot overcome the buoyant force and the object floats on the surface of the fluid. On the other hand, if the weight of the object exceeds the buoyant force, then the lift force is negative and the object sinks (Figure 7.18). One can pursue this line of reasoning a bit further by noting that from Archimedes's principle

$$\frac{\text{buoyant}}{\text{force}} = \frac{\text{weight of displaced}}{\text{fluid}}$$

or

$$\frac{\text{buoyant}}{\text{force}} = g \times \left(\frac{\text{mass of displaced}}{\text{fluid}}\right).$$

From the definition of density, we have

$$\frac{\text{mass of displaced}}{\text{fluid}} = \frac{\text{density}}{\text{(fluid)}} \times \frac{\text{volume}}{\text{(fluid)}},$$

and hence

$$\frac{\text{buoyant}}{\text{force}} = g \times \frac{\text{density}}{\text{(fluid)}} \times \frac{\text{volume}}{\text{(fluid)}}.$$

The weight of the object meanwhile is

$$\frac{\text{weight of}}{\text{object}} = g \times \left(\frac{\text{mass of}}{\text{object}}\right).$$

Again, appealing to the definition of density, we have

$$\frac{\text{mass of}}{\text{object}} = \frac{\text{density}}{\text{(object)}} \times \frac{\text{volume}}{\text{(object)}},$$

and thus

$$\frac{\text{weight of}}{\text{object}} = g \times \frac{\text{density}}{\text{(object)}} \times \frac{\text{volume}}{\text{(object)}}.$$

Since the volume of the object and the volume of the displaced fluid are equal, we can express the lift force as follows

$$\text{lift force} = g \times \frac{\text{density}}{\text{(fluid)}} \times \frac{\text{displaced}}{\text{volume}} - g \times \frac{\text{density}}{\text{(object)}} \times \frac{\text{displaced}}{\text{volume}}$$

or

$$\text{lift force} = g \times \frac{\text{displaced}}{\text{volume}} \times \left[\frac{\text{density}}{\text{(fluid)}} - \frac{\text{density}}{\text{(object)}}\right].$$

Thus, the relative densities of the object and the fluid in which it finds itself will determine the sign of the lift force and hence whether or not the object will float. If the object density is larger than the fluid density, as in the case of our iron "cork" (density = 7900 kg/m^3) in water (density = 1000 kg/m^3), the lift force is negative and the iron "cork" sinks. On the other hand, for the "real" cork (density = 150 kg/m^3) in water, the object density is less than the fluid density. The lift force is now positive and the cork floats. If the object density and the fluid density are the same, then the forces arising from buoyancy and the weight of the object are balanced out and the object will neither float nor sink. Such is the case with marine creatures whose densities are such that they do not have to contend with a net lifting or sinking force as they go about their existence.

Because the lift force depends on the relative densities of the object and the fluid in which it is immersed, objects can be made to float or sink by changing their weight and/or their volume and hence their overall density. An ocean liner or aircraft carrier made of metals that would ordinarily sink are able to float because they are designed with air space, which keeps the overall density of the ship less than that of water. Submarines and certain sea creatures can actually regulate their densities by means of "ballast tanks," which permit them to move upward, downward, or stay at the same depth while in the water (Figure 7.19).

When displaced by an immersed object, gases, like liquids, also exhibit a relative buoyant force on the immersed object, the magnitude of which is also given by Archimedes's principle. Thus, blimps, dirigibles, and hot air balloons in effect are buoyed up by the earth's atmosphere because the overall densities of these airships are less than that of the earth's atmosphere and hence the lift forces on such "lighter than air" craft are positive. The density of the earth's atmosphere, however, decreases with increasing attitude, so that there comes a point in altitude when a "lighter than air" airship is no longer lighter than air. At this point, the weight of the airship is balanced by the buoyant force from the atmosphere and the ship can no longer gain altitude.

Note also that in addition to the difference in densities between the object and the fluid, the magnitude of the lifting force depends on the total displaced volume as well. As a result, the "load capacity" of a ship is often expressed in terms of the ship's "displacement," or volume of water the ship displaces while at sea. If a given ship has a large displacement, then the lifting force experienced by the ship will also be large, thus increasing the ability of the ship to carry larger cargos.

Our discussion thus far has centered primarily on fluids that are motionless or "static." For such fluids the pressure depends only on the depth and the density of the given fluid. If we set a fluid in motion, however, then a principle of fluid physics called *Bernoulli's principle* (Physics Formulation 7.3) tells us that the pressure in a fluid will now depend on the velocity of that fluid. In particular, in those regions where the fluid is moving at relatively high velocity, the fluid pressure will be relatively low. In those regions where the fluid is moving at relatively low velocities, the fluid pressure will be relatively high. Thus the wind that blows past a chimney constitutes a region of relatively high velocity fluid flow in the vicinity of the chimney opening, which in turn results in a reduced fluid pressure at the chimney opening. As a result, the higher pressure air from the fireplace below is pushed up toward the lower pressure air at the chimney opening, thereby preventing the smoke from the fireplace from filling up one's house.

Fluid flow also permits a sailing ship to perform the seemingly impossible task of sailing against the wind. If the sail of the ship is curved and the ship "tacks" against the wind as shown in Figure 7.20, then the wind velocity across the lead surface of the sail is higher than that across the trailing surface of the

FIGURE 7.19
A Nautilus Shell This cross section of the shell of the nautilus of the South Pacific shows how the shell volume is sectioned off in chambers. By pumping liquid in and out of these chambers, the nautilus can control its overall density and is thus able to descend or ascend to whatever depth it chooses.

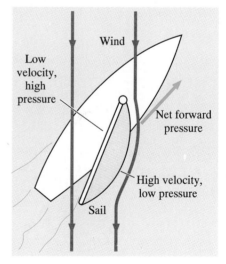

FIGURE 7.20
Sailing with Bernoulli's Principle As the wind flows across the curved portion of the sail, a reduced pressure is created on the leeward side of the sail, thus allowing the ship to sail against the wind.

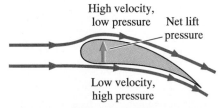

High velocity,
low velocity   Net lift
pressure

Low velocity,
high pressure

FIGURE 7.21 ....
Flying with Bernoulli's Principle The shape of an airplane wing is designed so that the velocity of the air flow above the wing is larger than below. From Bernoulli's Principle the pressure on the top of such a wing is less than that on the bottom of the wing, thus creating a net upward pressure or "lift" to the wing.

sail. As a result, the air pressure behind the sail is higher than that ahead of the sail and air pushes the sail toward the lower pressure.

Another important application of this sort of fluid flow can be found for the wings of aircraft. The typical airplane wing is designed so that the air flowing over the top of the wing has a farther distance per unit time to travel and hence a higher velocity than the air flowing beneath the wing (Figure 7.21). As a result of this fluid velocity difference, the pressure exerted by the air moving above the wing is less than that exerted by the air moving below the wing and there is a net upward "lift" imparted to the wing as the air sweeps past it. If this lifting force exceeds the weight of the airplane, then the airplane is able to fly.

In this chapter we have thus explored a few of the general characteristics of matter. Our understanding of matter and its behavior has been enhanced in no small degree by an appeal to a revised version of Democritus's centuries-old "atomic" concept of matter made up of tiny discrete particles. In the next chapter, we will combine this atomic theory with the concept of thermal or heat energy to extend our understanding of matter and matter-related phenomena even further.

WORKED EXAMPLE

The reservoir behind a dam has a depth of 200 meters. What is the pressure at the base of the dam?

SOLUTION: The pressure at any given depth in a fluid is

$$\text{pressure} = g \times \text{depth} \times \text{density} .$$

If the reservoir consists of water (density $= 1000 \text{ kg/m}^3$), then at a depth of 200 meters, the pressure is

$$\text{pressure} = 9.8 \text{ m/sec}^2 \times 200 \text{ meters} \times 1000 \text{ kg/m}^3$$
$$= 1.96 \times 10^6 \text{ N/m}^2 .$$

The pressure at the base of the dam is just under 2 million newtons/meter^2.

REVIEW **7.5** QUESTIONS

1. Compare and contrast a gas, a liquid, and a fluid.
2. What is pressure? How is pressure related to force?
3. State Archimedes's principle. Discuss an example of Archimedes's principle in action.

REVIEW **7.5** PROBLEMS

1. A swimming pool 10 meters long and 5 meters wide is filled with water to a depth of 1.5 meters. What is the pressure at the bottom of the pool? What is the total force exerted on the bottom of the pool?
2. A circular glass cylinder has an area of 0.4 m². To what height must water be poured into the cylinder in order to obtain a pressure of 100 N/m² at the

base of the cylinder? To what height must water be poured in order to create a total force of 100 N on the bottom of the cylinder?

Chapter Review Problems

1. A certain substance is found to have the following chemical formula: $C_6H_{12}O_6$. What is the atomic weight of this substance?

2. A given sample of ammonia (NH_3) has a mass of 1 kg. Find:
   a. the atomic weight of ammonia
   b. the composition of ammonia by the percentage of the nitrogen and hydrogen atoms present
   c. the composition of ammonia by the percentage of the masses of the nitrogen and hydrogen atom present.

3. A gas is analyzed and found to have an atomic weight of 16. Four-fifths of the atoms composing a molecule of this gas are found to be hydrogen atoms. What is the other element in this molecule?

4. If the typical water molecule has a volume of about 10^{-30} m^3, approximately how many water molecules will fit into
   a. a cubic centimeter
   b. a cubic meter?

5. Show that a density of 1000 kg/m^3 corresponds to a density of 1.0 gram/cm^3.

6. A certain metal sample has a mass of 15 kg and a density of 8000 kg/m^3. What volume does the sample occupy?

7. A total of 7 kg of a certain liquid is poured into a cube-shaped container having sides 0.2 m in length. If the cube is filled to the top, what is the density of the liquid?

8. A gasoline tank has a volume of 5 cubic meters. If the density of gasoline is about 680 kg/m^3, what is the total mass of gasoline this tank could hold?

9. Calculate the pressure of a filled swimming pool having dimensions of 3 m × 4 m and a depth of 2 m. Compare your result with the pressure at the bottom of a lake having an area of 3 km^2 and a depth of 2 meters.

10. A body of water on the planet Zantor has a depth of 20 meters and exerts a pressure of 260,000 N/m^2. What is the acceleration of the surface gravity on Zantor?

11. The atmospheric pressure at sea level in the earth is about 10^5 N/m^2. To what depth must one go in a body of water to experience this same pressure?

12. A 50-kg woman is wearing stiletto high-heel shoes whose heels have a total surface area of about 1.6×10^{-4} m^2. Assuming that the woman's weight is exerted entirely through the heels of her shoes, how much pressure does she exert on the floor?

13. A wooden block with dimensions 0.1 m × 0.1 m × 0.05 m is placed in a bathtub of water. What volume of water is displaced by the block? What is the buoyant force exerted on the block by the water?

14. If the total mass of the block in Problem 13 is 2×10^{-4} kg, will the block float? Explain.

15. A chunk of material is submerged in water and is found to have a weight of 20 N. If its weight out of the water is 30N, find:

    a. the buoyant force of the water

    b. the weight of the water displaced by the object.

16. What was the mass of the water displaced in Problem 15? What is the density of the chunk?

17. A mixture of ethyl alcohol and water is created by adding 2 kg of water to 1 kg of alcohol. Assuming that the total volume of the mixture is equal to the sum of the constituent volumes, find the mean density of the mixture.

18. The bottom of a water tower tank can withstand a total force of 200,000 N. If the bottom of the tank has a total area of 50 m², how much pressure can the tank bottom withstand?

19. If the liquid to be stored in the tank in Problem 18 is water, to what depth can the tank be filled? To what depth if the liquid were mercury?

20. A liquid exerts a total pressure of 100 N/m² at the bottom of a container. If the depth of the liquid is 4 meters, find the density of the liquid.

21. A large supply of water is taken to the moon and poured into a container to a depth of 2 meters. If the moon's surface gravity is ⅙ that of the earth, what is the water pressure at the bottom of the container?

22. A cube of lead 2 cm on a side is immersed in water. If the density of lead is 11,300 kg/m³, find:

    a. the weight of the lead cube

    b. the buoyant force of the water on the lead cube

    c. the magnitude and direction of the lifting force on the lead cube.

    From your result, does the cube float or sink?

23. A 10-kg sample of copper is to be made into a floating tank. If the density of copper is 8890 kg/m³, what is the minimum volume that the copper tank must have in order to be able to float on seawater (density = 1030 kg/m³)?

24. A piston has a surface area of 0.2 m². To what depth must the piston cylinder be filled with lubricating oil having a density of 900 kg/m³ in order to produce a total force of 300 N on the piston?

25. An ore sample is submerged in water and is found to have a weight of 30 N. If its weight out of the water is 35 N, find:

    a. the buoyant force of the water on the sample

    b. the weight of the water displaced by the sample

    c. the volume of the water displaced by the sample

    d. the density of the object.

26. When placed on a water surface, an oleic acid drop forms a circular film layer having a diameter of 12 cm. If the volume of oleic acid in the original drop was 5×10^{-12} m³, find the diameter of an oleic acid molecule.

THOUGHT QUESTIONS

1. How would you prove to a skeptic that matter is made up of atoms that are too tiny to be seen with the unaided eye?

2. Suppose a new element were discovered. Where do you suppose such an element might appear in the periodic table?

3. Can an ion be an isotope or vice versa? Explain.

4. Would iron react with chlorine in the same way that it does with oxygen? Explain.

5. Would you expect the surface tension of a liquid to be more or less at a higher temperature? Why?

6. Why don't solids and gases exhibit surface tensions?

7. Falling raindrops quite often exhibit elongated shapes. Why do you suppose this is so?

8. Why does capillary action tend to occur more readily in smaller diameter tubes?

9. If trees could somehow be grown on the moon, would they grow taller or shorter than they do on earth? Explain.

10. Why do you suppose that the top of a frisbee has a convex-shaped surface?

11. Describe in terms of atomic phenomena why a substance such as table salt dissolves in a liquid such as water.

12. Discuss possible reasons for the fact that liquids exhibit adhesive forces?

13. Suppose you were asked to design an experiment to measure the viscosity of a given fluid. How might you proceed?

14. Quite often just before an important exam or the due date for a term paper, we think of ourselves as being under ''pressure'' or ''stressed out.'' How does this usage compare with the way these terms are defined in the text?

15. It has been suggested that a plasma is a fourth state of matter that is different from the solid, liquid, and gaseous states. Comment on this idea.

16. Show that work energy is conserved in a hydraulic lift or press.

17. Describe some of the factors that might be important in terms of determining whether or not life forms can exist on a given planet or satellite.

18. Would a liquid having a high viscosity be a good choice for use in a hydraulic lift or press? Explain.

19. Describe three practical applications of Bernoulli's principle.

20. Discuss the physics of what happens when a ship sinks.

21. Why do you suppose that the great airships of the 1930s such as the *Hindenberg* and the *Graff Zepplin* were so enormous in size?

22. When Archimedes discovered his buoyant force principle, he was trying to help the king of Syracuse determine the percentages of gold and silver contained in a certain crown without destroying or damaging the crown. How can Archimedes's principle be used to solve this problem?

23. Discuss how iron can be made to float.

24. Describe how high speed, high altitude jet aircraft are able to fly without having very much Bernoulli lift.

25. Describe how it is possible for a boat to sail against the wind.

Elasticity

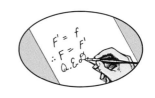

In Chapter 4 we briefly alluded to mechanical systems that exhibited the property of returning to their original shape when deformed in some fashion. The forces that serve to restore the original shape of a deformed object are referred to as *restoring forces*. Restoring forces have the property that the magnitude of the restoring force is directly proportional to the amount of the displacement involved—provided the displacement is not too large. Thus if a spring that normally has a length of 1 meter is stretched to 4 meters, it will exert a restoring force that is twice as large as it would exert if its length was stretched out only half as far, or in this case, to a length of 2 meters.

Interestingly, the chemical bonds that hold the atoms, molecules, and crystals in the fixed configurations characteristic of a solid also display such restoring forces in a great many solids, which are called *elastic solids*. It is almost as if the individual atoms and molecules within an elastic solid were connected by tiny springs, which define an equilibrium configuration for the atoms and molecules of the solid. Any attempt to deform a solid by expanding or compressing these intermolecular springs will cause the solid to exert a restoring force to regain its equilibrium shape. Some materials such as clay and dough have virtually no elastic properties whatsoever and are referred to as *inelastic solids*.

Whenever a solid is deformed, the deformations manifest themselves in one or more of three basic ways. The deformation can occur as a change in length or *length deformation* as in the case of a stretched wire, a change in volume or *bulk deformation* as in the case of a solid that is "squished" from all sides, and a twisting change in the relative positioning of atoms and molecules, or *shear deformation,* as in the case of the shape of a large book being deformed without a change in its volume (Figure 7.22). In each instance the degree to which a given elastic solid responds to a given type of deformation is directly proportional to the magnitude of the deforming force, where the constant of proportionality depends on the material involved.

In the comparatively simple case of linear deformations of elastic solids such as long wires whose lengths are large compared to their diameters, it is useful to express the deforming force as a *stress* or force per unit cross-sectional area of the solid. If the response of the solid is expressed in terms of a fractional change of length or *strain,* then we have

$$\text{stress} = \text{constant} \times \text{strain} .$$

The constant in this instance is called the *elastic modulus* or *Young's modulus,* and its value is a characteristic of the solid involved. Some representative values of the Young's modulus constant are listed in Table 7.2. Since the strain is (change of length)/length we can write for length deformations that

$$\text{stress} = \left(\frac{\text{elastic}}{\text{modulus}}\right) (\text{change of length})/\text{length} .$$

A similar equation can be developed for the relationship between bulk deformation, which is referred to as volume stress, and the (change of volume)/volume ratio or volume strain that a given substance exhibits in response to such a stress. This relationship can be written as

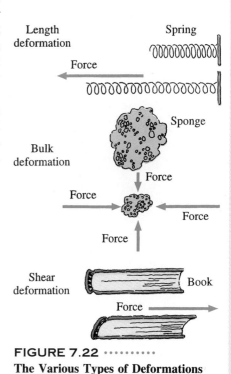

Length deformation   Spring

Force

Bulk deformation   Sponge

Force

Force   Force

Force

Shear deformation   Book

Force

FIGURE 7.22 ··········
The Various Types of Deformations

TABLE 7.2 Various Moduli for Various Solid Materials

Material	Elastic Modulus (N/m²)	Bulk Modulus (N/m²)	Shear Modulus (N/m²)
Diamond	83×10^{10}	55×10^{10}	34×10^{10}
Steel	20×10^{10}	14×10^{10}	8.0×10^{10}
Copper	11×10^{10}	14×10^{10}	4.2×10^{10}
Iron	10×10^{10}	9.0×10^{10}	4.0×10^{10}
Aluminum	7.0×10^{10}	7.0×10^{10}	2.5×10^{10}
Glass	6.0×10^{10}	3.7×10^{10}	2.3×10^{10}
Granite	4.5×10^{10}	4.5×10^{10}	2.0×10^{10}

$$\text{volume stress} = \text{constant} \times \text{volume strain} .$$

In this case the constant is called the *bulk modulus* and, like the elastic modulus, has different values for different substances as seen in Table 7.2. This equation can hence be rewritten as

$$\text{volume stress} = \left(\begin{array}{c} \text{bulk} \\ \text{modulus} \end{array} \right) (\text{change of volume})/\text{volume} .$$

Shear deformations are caused by shear stresses and are expressed in terms of the shear strain according to

$$\text{shear stress} = \text{constant} \times \text{shear strain} .$$

where the constant is now called the *shear modulus* of the substance. Shear strains are defined in terms of a displacement per unit height. If the top of an object having a thickness or height h is displaced a distance Δx from the vertical by a shear stress then the shear strain is $\Delta x/h$. We may thus express the above equation for shear stress as follows

$$\text{shear stress} = \left(\begin{array}{c} \text{shear} \\ \text{modulus} \end{array} \right) (\Delta x/h) .$$

When a deformation causes the structure of a solid to be permanently altered, such as an overstretched spring to the point where it cannot ''spring back,'' the *elastic limit* of the solid is said to be exceeded, and the deformation is no longer proportional to the deforming force. If a given solid is deformed even further, it will ultimately reach its ''breaking point'' and come apart or ''fracture.'' Obviously at this point, the description of the object as an elastic solid has long since ceased to be applicable.

QUESTIONS

1. Give examples of stress, volume stress, and shear stress.

2. Why are the units of elastic modulus, bulk modulus, and shear modulus as listed in Table 7.2 all the same?

3. Discuss how you might experimentally obtain the various moduli listed in Table 7.2.

4. From time to time, the world's great cities have been pummelled in science fiction movies by various creatures of skyscraper size. Discuss whether such creatures could exist in terms of stress strains, elastic limits, fracture points, etc.

PROBLEMS

1. A stress of 10^{10} N/m^2 produces a change of 0.002 m in the length of a 2-meter wire. What is the elastic modulus for the wire?

2. Water has a bulk modulus of 2×10^9 N/m^2. What stress would be required to compress a given volume of water to half its original size?

3. One of the editions of *Webster's Unabridged Dictionary* has a thickness of 0.2 m. If a stress of 10 N/m^2 is placed on the top of the book and produces a displacement of the top cover of 0.02 m relative to the bottom cover, what is the shear modulus of the book?

Hydraulics

One of the more useful applications of Pascal's principles is to be found in a device called a hydraulic lift. The hydraulic lift in effect uses fluid pressure to gain mechanical advantage or "multiply" a given force. The device consists of two pistons having surface areas A_1 and A_2. Each of these surfaces is in contact with a fluid in a closed container having two vertical cylinders, which house the pistons and are interconnected so that the fluid can flow back and forth between the pistons as shown in Figure 7.23. If an applied force F_A is placed on the fluid in piston 1, the pressure delivered to the bottom of the piston is F_A/A_1. Since the fluid is in contact with both pistons, Pascal's principle tells us that this same pressure will exist at every point over the area of the bottoms of both piston 1 and piston 2. The total force F_L delivered to a load placed on the top of piston 2 is then given by

$$F_L = \frac{\text{pressure on}}{\text{piston 2}} \times A_2 \ ,$$

but the pressure on piston 2 is simply F_A/A_1 and we thus have that

$$F_L = \frac{A_2}{A_1} \times F_A \ ,$$

and hence we can multiply the magnitude of an applied force by an amount equal to the ratio of the areas of the two pistons involved. As we force the fluid from the small-area piston over to the large-area piston, the volume of fluid transferred for an applied force distance x_A is

$$\text{volume transferred} = x_A \times A_1 \ .$$

When this volume of fluid reaches piston 2, it will occupy the same volume, but the distance the fluid level moves the load on piston 2 upward, x_L, is equal to volume transferred/A_2 or

$$x_L = \text{volume transferred}/A_2 \ ,$$

FIGURE 7.23 ···········

The Hydraulic Lift If a force F_A is applied to a piston with an area A_1, the pressure produced, F_A/A_1, is then transmitted according to Pascal's principle over the entire area A_2. The total force F_L on A_2 is the product $A_2 \times$ pressure or $A_2 \times F_A/A_1$. The product of the total distance x_A through which the applied force F_A moves or $F_A \times x_A$ is equal to the product of the load force F_L and the distance the load force moves x_L or $F_L \times x_L$.

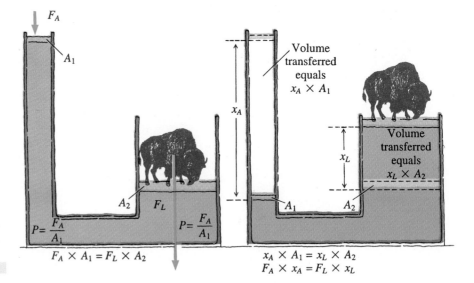

but the volume of fluid transferred is $x_A \times A_1$, and thus the distance the load on piston 2 moves upward, x_L is given by

$$x_L = \frac{A_1}{A_2} x_A \; .$$

The ratio of the areas A_2/A_1 can then be expressed as

$$\frac{A_2}{A_1} = \frac{x_A}{x_L} \; .$$

If we substitute this expression into the relationship between F_2 and F_1 we have

$$F_L = \frac{x_A}{x_L} \times F_A$$

or

$$F_L \times x_L = F_A \times x_A \; ,$$

which tells us that a hydraulic lift (or press) is simply another "trade-off" device in which an increase in applied force distance is exchanged for an increase in the force delivered to a given load.

QUESTIONS

1. Discuss the relationship, if any, that exists between the hydraulic lift and the lever.
2. Using a hydraulic lift, a construction worker can lift a 1000-N weight using a force of only 100 N. Does this lift violate the principle of the conservation of energy? Explain.
3. Can gases be used in a hydraulic lift? If so, under what circumstances?

PROBLEMS

1. What load can be raised by a hydraulic lift having an applied piston surface area of 0.08 m^2 and a load piston surface area of 2 m^2 if the applied force is to be 140 N?
2. An auto mechanic wishes to raise a 1000-kg car off the ground by using a hydraulic lift. If the area of the piston on which the car is situated is 100 times that which the mechanic pushes down on, how much force should the mechanic exert on the second piston?
3. An engineer wishes to design a hydraulic lift with a mechanical advantage of 15 and a load piston surface area of 1.2 m^2. What surface area should the applied piston have in order to satisfy these requirements?

Bernoulli's Principle

The principle of the conservation of energy has enabled us to understand a great many phenomena in nature. Such is the case for the behavior of fluids that are at rest (static fluids) and fluids that are in motion (dynamic fluids).

Consider a fluid moving through a flow tube having a nonuniform cross section as shown in Figure 7.24. At position 1, the fluid has a height h_1, a density ρ_1, and is moving at a velocity v_1 through a cross-sectional area A_1. Similarly, the fluid at position 2 has a height h_2, a density ρ_2, and is moving at a velocity v_2 through a cross-sectional area A_2.

If the fluid is flowing "smoothly," i.e., without turbulent eddies, and has a relatively low viscosity, such as the case for water or air, then from the work-energy theorem, we can state that the work W_{12} done in moving the fluid from position 1 to position 2 is equal to the sum of the changes in the fluid's potential energy and its kinetic energy in moving from position 1 to position 2, or

$$W_{12} = \Delta PE_{12} + \Delta KE_{12} .$$ **7.3**

Imagine now a piston at position 1, which displaces the fluid a distance d_1. This fluid displacement is transmitted to a second imaginary piston at position 2, which causes that piston to move a distance d_2. The force F_1 required to push the piston at position 1 is equal to the product of the pressure P_1 at position 1 and the area A_1, or

$$F_1 = P_1 \times A_1 .$$

In like fashion, the force F_2 exerted on the piston at position 2 is

$$F_2 = P_2 \times A_2$$

where P_2 is the pressure at position 2. The net work done in moving the fluid from position 1 to position 2 is thus

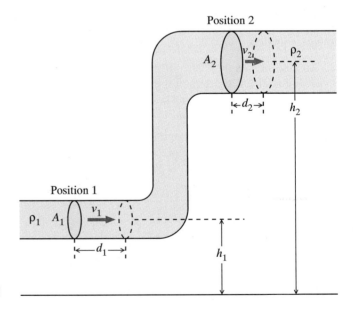

FIGURE 7.24 ·············

Fluid Flow and Bernoulli's Principle The work required to produce a fluid flow is equal to the sum of the change in the fluid's potential energy and the change in its kinetic energy. From this result, one can obtain Bernoulli's principle.

$$W_{12} = P_1 d_1 - P_2 d_2 \; .$$

Substitution of the equivalent expressions for P_1 and P_2 yields

$$W_{12} = P_1 A_1 d_1 - P_2 A_2 d_2 \; .$$

The potential energy of the fluid displaced at position 1 is

$$PE_1 = m_1 g h_1 \; ,$$

where m_1 is the total mass of the fluid displaced at position 1. The fluid displaced at position 2 has a potential energy of

$$PE_2 = m_2 g h_2 \; ,$$

where m_2 is the total mass of the fluid displaced at position 2. The net change in the potential energy ΔPE_{12} of the fluid as it moves from position 1 to position 2 is thus

$$\Delta PE_{12} = m_2 g h_2 - m_1 g h_1 \; .$$

The kinetic energy of the fluid displaced at position 1 is

$$KE_1 = \frac{1}{2} m_1 v_1^2 \; ,$$

where v_1 is the velocity of the fluid flow at position 1. In like fashion, the kinetic energy of the fluid displaced at position 2 is

$$KE_2 = \frac{1}{2} m_2 v_2^2 \; ,$$

where v_2 is the velocity of the fluid flow at position 2. The change ΔKE_{12} in the fluid's kinetic energy in moving from position 1 to position 2 is thus

$$\Delta KE_{12} = \frac{1}{2} m_2 v_2^2 - \frac{1}{2} m_1 v_1^2 \; .$$

Substitution of the expressions obtained for W_{12}, ΔPE_{12} and ΔKE_{12} into the work-energy theorem (7.3) followed by algebraic manipulation yields the following equation

$$P_1 + \left(\frac{1}{2}\right) \rho_1 v_1^2 + \rho_1 g h_1 = P_2 + \left(\frac{1}{2}\right) \rho_2 v_2^2 + \rho_2 g h_2 \; .$$

$$\left(\begin{matrix} \text{first position} \\ \text{in fluid} \end{matrix}\right) \qquad \left(\begin{matrix} \text{second position} \\ \text{in fluid} \end{matrix}\right)$$

This result can be generalized into the following statement:

> The quantity $P + (\frac{1}{2}) \rho v^2 + \rho g h$ has the same value for a given fluid at any position within the fluid.

This principle governing fluid flow is called *Bernoulli's principle* and finds considerable usage in fluid mechanics.

Consider, for example, a static fluid system in which the velocity of the fluid flow is zero. Bernoulli's principle tells us that for such a system the quantity $P + \rho g h$ has a constant value for a given fluid at every point within the fluid, and

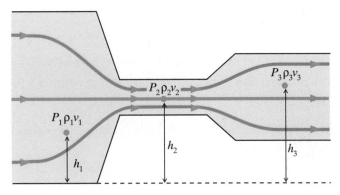

FIGURE 7.25 ⋯⋯⋯⋯

Bernoulli's Principle For any low-viscosity fluid flowing without turbulent eddies, the quantity $P + \frac{1}{2}\rho v^2 + \rho gh$ has the same value at any point in the fluid.

$$P_1 + \tfrac{1}{2}\rho_1 v_1^2 + \rho_1 gh_1 \; = \; P_2 + \tfrac{1}{2}\rho_2 v_2^2 + \rho_2 gh_2 \; = \; P_3 + \tfrac{1}{2}\rho_3 v_3^2 + \rho_3 gh_3$$

$$P_1 + \rho_1 gh_1 \; = \; P_2 + \rho_2 gh_2 \quad .$$

$$\left(\begin{array}{c}\text{first position}\\ \text{in fluid}\end{array}\right) \quad \left(\begin{array}{c}\text{second position}\\ \text{in fluid}\end{array}\right)$$

We can thus see why in the system of Pascal's vases for which $P_1 = P_2$ and $\rho_1 = \rho_2$, that water "seeks its own level." In this case the level "sought" is the height h (Figure 7.25).

For systems in which a fluid is moving along a *level* fluid flow tube, it is the quantity $P + (\frac{1}{2})\rho v^2$ that now has a constant value for a given fluid at every point within the fluid, and we may write that

$$P_1 + \left(\frac{1}{2}\right)\rho_1 v_1^2 \; = \; P_2 + \left(\frac{1}{2}\right)\rho_2 v_2^2 \; ,$$

$$\left(\begin{array}{c}\text{first position}\\ \text{in fluid}\end{array}\right) \quad \left(\begin{array}{c}\text{second position}\\ \text{in fluid}\end{array}\right)$$

which provides a mathematical description of the observed fact in nature that high fluid flow velocities result in reduced pressures for points in the vicinity of the flow such as at the mouth of a chimney, etc.

QUESTIONS

1. How is the principle of the conservation of energy employed to obtain Bernoulli's principle?
2. Describe three observational manifestations of Bernoulli's principle in action.
3. How might Bernoulli's principle change, if at all, for nonlevel fluid flow?

PROBLEMS

1. A fluid flowing in a level tube is observed to have a pressure of 2×10^4 N/m^2 and a velocity of 2 m/sec at a given position in the tube. At a second position, the pressure is found to be 1×10^4 N/m^2. If the fluid density throughout the tube is 1500 kg/m^3, find the velocity of the fluid at the second position.

2. If the velocity of the fluid in Figure 7.24 is 10 m/sec at position 2 and 2 m/sec at position 1, find the pressure at position 1 if the pressure at position 2 is 3×10^4 N/m^2. Assume the density of the fluid is 700 kg/m^3 and that h_1 = 0.5 m and h_2 = 2 m.

3. The velocity of the fluid in Figure 7.24 is 5 m/sec at position 2 and 4 m/sec at position 1. If the pressure at positions 1 and 2 are 40,000 N/m^2 and 25,000 N/m^2, respectively, find the density of the fluid flowing through the tube. Assume that the density of the fluid remains constant throughout the tube and that h_1 = 2 m and h_2 = 5 m.

8 HEAT AND THERMODYNAMICS

nyone who has ever "fried" on a summer afternoon in the desert southwest or suffered through a midwestern winter night has a deep and abiding appreciation of the sensations of "hotness" or "coldness." Hot showers, warm fireplaces, and cold dinners are but a few examples of how hot and cold enter into our daily lives.

If we observe "hotness" and "coldness" more carefully, we begin to uncover a number of rather interesting patterns. If, for example, we open a door joining a room that is hot with one that is cold, we find that the cold room becomes hotter and the hot room colder. It is as if the hotness "flowed" in the direction of the coldness. More of a clue concerning the nature of hotness and coldness can be obtained by stopping a car by slamming on the brakes. When the car stopped, its kinetic energy disappeared. Where did it go? If we check the tires and the brakes of the car, we find that they have become hotter. Similarly, the total kinetic energy of two objects before they are involved in an inelastic collision is larger than the total kinetic energy present after the collision. Where did the kinetic energy difference go? Again we find that the stuck together particles are hotter than they were before they collided. These and many other observations strongly suggest that "hotness" and "coldness" are somehow related to yet another form of energy, which is "internal" to a given object and which we will call "heat" or thermal energy.

8.1 Measurement of Thermal Energy

The temperature of an object can be loosely thought of as a measure of an object's hotness or coldness or as a measure of the ability of an object to transfer thermal energy. A red-hot piece of metal, when plunged into a vat of cold water, will transfer its heat or thermal energy much more readily than would that same piece of metal if it were only slightly warmer than the water. We thus say that the metal had a higher temperature when it was red-hot. Later we will more formally define the temperature of an object or substance in terms of the average kinetic energy of the atoms and molecules present in that object or substance.

The temperature of an object is measured with respect to a substance that is at some defined standard of hotness or coldness. Boiling water is hot. Ice is cold. Different standards give rise to different temperature scales (Physics Formulation 8.1). The most commonly used temperature scale in science is the centigrade or Celsius scale in which 0°C is defined as the temperature of an ice-water mixture at one atmosphere of pressure, and 100°C is defined as the temperature of a water-steam mixture at one atmosphere of pressure. Another temperature scale that enjoys considerable usage in science is the Kelvin scale in which the zero point is that of *absolute zero* or −273.15°C (Physics Facet 8.3) and each degree, called a *kelvin,* is exactly the same size as a Celsius degree. Thus every Kelvin temperature has a positive value, and a temperature of 0°C would be about 273 kelvins on the Kelvin scale.

Temperatures can be measured in a variety of ways (Physics Facet 8.1) but in each case we are in effect measuring some sort of response by a substance to a variation in its temperature. The most familiar response is perhaps the expansion or contraction of a volume of liquid such as water, alcohol, or mercury to an upward or downward change in temperature. One can also employ more exotic sorts of responses such as the variation in the electrical resistance of a platinum wire (Physics Facet 8.1) or the flow of electrical current in a "thermocouple" device.

Once we have established a temperature scale, we are then in a position to define heat quantitatively in terms of that scale. The "British thermal unit" or BTU, for example, is defined as the amount of input thermal energy required to raise the temperature of one pound of water by one degree Fahrenheit. A more convenient unit is the *calorie,* which is defined similarly, but uses metric quantities. The calorie is defined as the amount of input heat necessary to raise the temperature of one gram of water by one degree Celsius. Unfortunately, a second kind of calorie, spelled with a capital C, is used in food and dieting discussions. One Calorie, with a capital C, is equal to 1000 calories, with a small c, or what amounts to a kilocalorie. One useful spinoff of using the Calorie to rate the ability of the human body to store a given food or drink as body fat is that a 150,000 calorie dessert now becomes a mere 150 Calorie dessert if we employ the dietary units!

Notice that in defining the calorie we not only specified the temperature rise but the mass and type of substance as well. A larger rise in temperature requires a larger input of thermal energy. Thus more thermal energy is required to produce a 10°C rise in the temperature of a given object than that required for a 5°C rise. The mass of the object also has to be specified. Two grams of water, for example, would require twice as much thermal energy to produce a temperature change of 1°C as would one gram of water. The substance used must be specified simply because some substances can be heated to higher temperatures much more readily than others. Water can absorb a great deal of thermal energy without displaying a very large rise in temperature, while metals such as iron or copper, on the other hand, will exhibit a very large temperature change in response to a comparatively low input of thermal energy.

We thus find it useful to define the *specific heat* of a substance as the amount of thermal energy required to raise the temperature of one gram of that substance one degree Celsius. As a general rule metals tend to have low specific heats, while liquids tend to have high specific heats (Table 8.1). The amount of thermal energy required to raise the temperature of one gram of ethyl alcohol one degree Celsius, for example, is about 20 times that needed to increase the temperature of one gram of gold by that same one degree Celsius.

The honor of having the highest specific heat of substances commonly found on the earth belongs to water. Because of this fact, it is more difficult for bodies of water to heat up and cool off than the substances that make up their shorelines. This in turn has a great impact on the world's climatic phenomena, since 4/5 of the earth's surface area is covered by this comparatively slow-to-warm, slow-to-cool liquid, while the other 1/5 is occupied by rapid-to-warm, rapid-to-cool solid land masses. Coastal regions of the world, for example, tend to have fewer extremes in their annual temperature variation than do points inland. Thus the mean temperature variation experienced by the coastal city of Seattle is about 13°C over the course of a given year, while that of Denver, which has no large bodies of water close by, is closer to a 40°C variation, or over three times higher. Such differentials in heating and cooling rates for dry land around bodies of water also constitute an important driving mechanism for both the large- and small-scale movements of air masses in the atmosphere of the earth.

The total amount of thermal energy taken on or given off by a given object in a given situation thus depends on the temperature change experienced by the object, the total mass of the object, and the specific heat of the material composing

TABLE 8.1   Specific Heats of Various Substances

Substance	Specific Heat	
	cal/gram-C°	*joules/kg-C°*
Water (liquid)	1.00	4180
Ethyl alcohol (liquid)	0.58	2420
Gasoline (liquid)	0.50	2090
Wood	0.40	1670
Aluminum	0.21	880
Silicon	0.17	710
Concrete	0.16	670
Glass	0.15	630
Diamond	0.12	500
Iron	0.11	460
Copper	0.092	385
Silver	0.056	235
Lead	0.038	160
Mercury (liquid)	0.033	140
Gold	0.031	130

8.1 The Measurement of Temperature

All temperature measurements in science are based on the fact that a wide variety of the observable characteristics of a substance are temperature dependent. For example, the volume of a solid, liquid, or gas varies in a predictable way with a change in temperature. To employ such temperature-dependent characteristics, we must "calibrate" the temperature-measuring entity or "thermometer" by exposing it to two temperatures that are known or have been defined.

Suppose we have a glass tube filled with mercury, as shown in the figure, that we wish to use as a thermometer. To make use of this thermometer, we would first immerse it in an ice-water mixture and carefully note the level of the mercury within the tube. The process would then be repeated for a steam-water mixture and would generate a second level for the mercury. The first level corresponds to a temperature of 0°C and the second level to a temperature of 100°C. The distance between the two marks thus represents a difference of 100°C. If that difference is carefully divided up into 100 equal divisions, then each division corresponds to 1°C. We thus have a thermometric device for which the response of the mercury in the tube has been noted at 0°C and at 100°C.

If we now expose our mercury thermometer to an unknown temperature, the unknown temperature will cause the mercury in the tube to expand to a level in the tube that

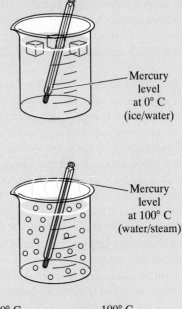

Mercury level at 0° C (ice/water)

Mercury level at 100° C (water/steam)

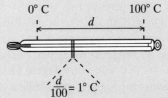

$$\frac{d}{100} = 1° C$$

the object. This total change in thermal energy can be stated as

$$\text{total thermal energy gained or lost} = \text{mass} \times \text{specific heat} \times \text{change in temperature}.$$

The product mass × specific heat is referred to as the *heat capacity* of a given object and is the amount of thermal energy required to raise the temperature of the object 1°C. Thus, if we wish to raise the temperature of an object having a mass of 30 g and a specific heat of 0.1 by 20°C, we must supply a total of 30 g × 0.1 cal/g-°C × 20°C or 60 calories of thermal energy to the object. Conversely, if we supply 45 calories of thermal energy to the same object, it will experience a change in temperature equal to 45 cal ÷ (30 g × 0.1 cal/g-°C) or 15°C.

Continued

corresponds to the volume expansion for mercury at that temperature. Since the level of mercury in the tube has been calibrated for both the zero point and size of one degree for the given thermometer, the unknown temperature can be read directly from the level of the mercury, assuming of course that the expansion of the volume of the mercury in the tube is linear with temperature.

Other characteristics may be employed in a similar fashion. Perhaps the most precise type of thermometer is the platinum resistance thermometer, which is based on the fact that the electrical resistance of platinum wire varies linearly with the temperature of the wire over a very large range of temperatures. Such thermometers are calibrated by noting the electrical resistance of the given wire at two well-defined temperatures. If the platinum wire is now exposed to an unknown temperature, the resistance of the platinum wire

will take on a value corresponding to the temperature in which the wire finds itself. This value of the resistance can be readily converted to a temperature, on the basis of the earlier calibration.

Many times it is not convenient to obtain temperatures from the response of a platinum wire or a tube of mercury. For example, we may wish to measure the temperature of an object such as the sun, whose temperature is so high that even if we had access to it in a laboratory, it would make very short work of any wire or glass tube. Temperatures of these sorts of objects are obtained by examining the characteristics of the radiant energy coming from the object. The characteristics are then compared with similar radiative characteristics of objects having known temperatures. We discuss this procedure in more detail in a later chapter.

••• **Constructing a Thermometer (Facing page)**   A glass tube filled with mercury is placed in an ice-water mixture and the level is marked on the tube. The process is repeated for a steam-water mixture. One one-hundredth of the distance between the marks is equal to 1°C. The thermometer can be marked off accordingly, with 0°C being the level of the mercury when it was in the ice-water mixture and 100°C being the level of the mercury when it was in the steam-water mixture.

If this thermal energy is truly a form of energy, then it should be possible to show that there is an equivalence between thermal energy and other forms of energy. In other words, we should be able to transform thermal energy into other forms of energy and vice versa. It was the English physicist James Joule who conclusively demonstrated this equivalence in a classic series of experiments performed during the 1840s. In honor of Joule's work, the newton-meter unit of energy has been named the joule.

A simple version of Joule's apparatus is shown in Figure 8.1. Joule first submerged a paddle wheel into a water bath. The submerged paddle wheel was then connected to a weight in such a way that when the weight fell, the paddle wheel rotated and performed mechanical work on the water. Joule found that

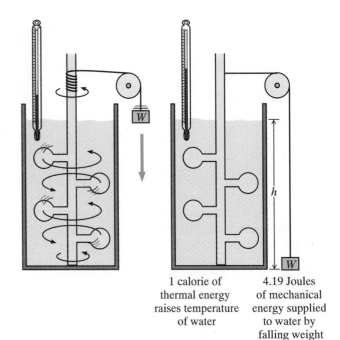

FIGURE 8.1 ··········

The Joule Apparatus   As the weight W falls through the height h, it causes the paddled shaft to rotate. As the shaft turns, the paddles perform work on the water, which is manifested by an increase in the temperature of the water. From this experiment, James Joule found that 1 calorie of thermal energy is the equivalent of 4.19 joules of mechanical energy.

1 calorie of thermal energy raises temperature of water

4.19 Joules of mechanical energy supplied to water by falling weight

after the weight had fallen, the temperature of the water had increased. Knowing the mass of the water, its specific heat, and the observed temperature rise, Joule could calculate the increase in thermal energy of the water bath. But he also knew how much mechanical energy was expended on the paddle when the earth's gravitational force pulled a known mass through a known distance. In this fashion Joule was able to demonstrate the equivalence of mechanical and thermal energy and established the fact that one calorie of thermal energy is the equivalent of about 4.19 N-m or 4.19 joules of mechanical energy.

WORKED EXAMPLE

How much thermal energy will be required to raise the temperature of 2 kg of water from 10°C to 35°C? Assume that the specific heat of water is 4180 J/kg-°C.

SOLUTION: The total thermal energy gained or lost in a heat transfer process is given by

$$\text{change in thermal energy} = \text{mass} \times \text{specific heat} \times \text{change in temperature}.$$

The stated specific heat of water is 4180 J/kg-C°, the mass of the water involved is 2 kg, and the desired temperature change is 35°C − 10°C = 25°C. Therefore,

$$\text{change in thermal energy} = 2 \text{ kg} \times 4180 \text{ J/kg-°C} \times 25°C$$

$$= 209{,}000 \text{ J} \ .$$

Continued

WORKED EXAMPLE

Continued
A total input of 209,000 joules of thermal energy is required to raise the temperature of the 2 kg of water from 10°C to 35°C.

1. What is meant by "temperature"?
2. Describe the temperature scales commonly used in science.
3. Compare and contrast calorie, joule, and Calorie.
4. Discuss the difference between the specific heat of an object and its heat capacity.
5. How did James Joule demonstrate that heat was indeed a form of energy?

1. A 3-kg block of iron initially at a temperature of 30°C is heated with a total of 40,000 J of thermal energy. Find (a) the total change in temperature of the block and (b) the final temperature of the block.
2. A physicist notes that the temperature of a 2-kg block of metal changes by 5°C when heated by a total thermal energy of 2370 J. What is the specific heat of this substance? What is its heat capacity?

8.2 Interactions Between Thermal Energy and Matter

In our initial discussion of work and energy, we defined these quantities in terms of forces and changes of position. From this standpoint, thermal energy at first glance seems to present us with an enigma. As we have just seen, thermal energy is indeed a quantity equivalent to mechanical energy. But if that is the case, we should be able to detect either increased motion in the water or a change in the positioning of the water's mass when a pan of water is heated—we see neither. Where then does the energy go? To answer this question, we must descend to the microscopic level and examine the effects of thermal energy on the individual masses of atoms and molecules.

Suppose we have a solid of some sort that is at a relatively low temperature. If we add thermal energy to that solid, what is the result? As we have already seen, solids can be thought of as structures of billiard balls interconnected by tiny springs. The molecules vibrate about their equilibrium positions, much as the weight on the end of a spring vibrates about its equilibrium position. Any thermal energy added to the system is absorbed at an atomic or microscopic level by the atoms and molecules in the solid. On picking up the energy, they begin

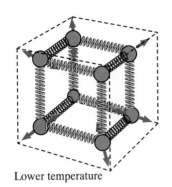

Lower temperature

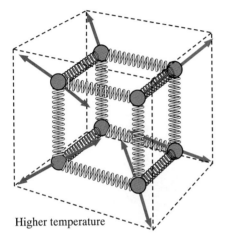

Higher temperature

FIGURE 8.2 ⸱⸱⸱⸱⸱⸱⸱⸱⸱⸱⸱
Thermal Expansion Any thermal energy that is added to a solid causes the vibrational displacements of the individual atoms to increase, which in turn causes the solid to appear to increase its linear dimensions.

to vibrate more rapidly and with a larger displacement from their equilibrium positions. We cannot directly detect this change in the individual vibrational motions of the molecules, but what we can and do detect is a phenomenon called *thermal expansion* (Figure 8.2). By increasing the displacements in the molecular vibrations, the space between the molecules is thus increased, and the overall result is an observable expansion in the size of the solid. The effect is not particularly dramatic but can be detected with relative ease. A meterstick made of aluminum, for example, will increase its length about 1 millimeter if its temperature is increased by 50°C.

We can also see the consequences of thermal expansion phenomena in our everyday life. If you walk down a sidewalk or along a railroad track, you will notice that at certain intervals there are spaces between the concrete slabs or between the ends of the individual rails (Figure 8.3). Sometimes, as in the case of sidewalks, soft material is placed in the spaces to provide a so-called expansion joint. If such spacing were not provided, the segments of sidewalk or rails would expand against one another under the heat of a fiery summer sun and eventually buckle in an undesirable way. The spaces, in effect, provide room for the rails or sidewalk to undergo such thermal expansion safely. Different substances exhibit different amounts of thermal expansion when subjected to the same amounts of thermal energy. Thus, a tight metal lid or cap on a glass container can be loosened when placed in hot water because the metal lid expands more than the glass container it covers in response to the increase in temperature.

The expansion of a solid with increasing amounts of thermal energy does not continue indefinitely. Eventually, the bonds begin to break apart and, although they no longer have equilibrium positions, enough attractive forces still exist between the molecules and atoms to hold the substance together in the form of a liquid. This transformation from a solid to a liquid state is called melting (Figure 8.4).

Associated with this change of state is a quantity called the heat of fusion. The *heat of fusion* of a substance is defined as the amount of energy required to convert one gram of that substance from a solid into a liquid at one atmosphere of pressure. One gram of water requires 80 calories of thermal energy to transform it from a solid to a liquid, while the performance of the same task for lead only requires 6 calories of thermal energy per gram. For a substance having a temperature equal to its melting point, any thermal energy that is added to the

FIGURE 8.3 ⸱⸱⸱⸱⸱⸱⸱⸱⸱⸱
Thermal Expansion in Construction Design The spaces or "expansion joints" between the sidewalk and rail segments permit thermal expansion to occur without causing the rail or sidewalk to buckle.

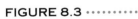

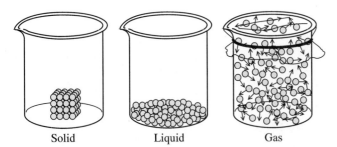

Solid   Liquid   Gas

FIGURE 8.4 ···········

Phase Transitions   As the temperature of a substance is increased, the chemical bonding of the solid state breaks down to cause the substance to melt into the liquid state and then boil off as individual unbonded atoms and molecules in the gaseous state.

substance goes into changing the substance from a solid into a liquid. Until that phase change is completely accomplished, no additional temperature rise will occur. Interestingly, when a substance is cooled and "fuses," i.e., is converted back from a liquid to a solid, it gives up the same amount of energy that was required for the melting process.

For most substances the transformation from a solid into a liquid results in a larger volume per unit mass or in effect a lowered density. Because of the nature of their molecular structure in the liquid state and their crystalline structure in the solid state, some substances, most notably water, actually occupy more volume per unit mass and hence have a lower density in their solid state. As a result, the solid phase of such substances will float on top of the liquid phase. The fact that water exhibits this property is of crucial importance for life on earth. If water behaved as most solids, ice would sink to the bottom of the earth's bodies of water at the onset of freezing weather, and ponds, lakes, and oceans would freeze from the bottom up, with devastating results to the earth's marine life. As it is, any ice that is formed on the surface layers of a body of water remain on the surface. Moreover, the density of water passes through a maximum value at 4°C. As a result, when the surface layers of a body of water are cooled they increase their density until this 4°C temperature is reached. At this temperature, the surface layers of water begin to sink into the warmer, less dense layers below, pushing them to the surface. The process continues until all of the water in the body has been cooled to 4°C. Only after this happens can ice form, and even then only from the top of the body of water downward. Thus creatures in the deeper regions of a lake, river, or ocean can still conduct their affairs in a liquid environment.

Once a given solid is completely melted, it then takes on all of the previously discussed properties of a liquid. The "loose cohesiveness" exhibited by liquids can be accounted for an atomic theory in terms of loose cohesiveness at the microscopic level, i.e., in their chemical bonding. The bonds in a liquid are strong enough for the liquid to retain a definite volume, but not strong enough to retain a definite shape. Like solids, liquids exhibit the phenomenon of thermal expansion, and usually to a much greater degree than do their solid counterparts. It is this property of liquids that makes them so very useful in instruments such as thermometers. For example, as the liquid mercury in the glass tube of a mercury thermometer is heated, it responds to the heat by expanding much more rapidly than does its glass container. In fact, if the thermometer is overheated, the resulting expansion of the liquid mercury can actually cause the tube to burst.

Continued heating of the liquid progressively loosens the molecular cohesiveness within the liquid until finally the *boiling point* of the liquid is reached. At the boiling point, the bonds in the liquid are completely broken down and the atoms and molecules are driven off the surface of the liquid as separate billiard

PHYSICS FACET

8.2 Chemoreception

Unlike their brethren in the solid and liquid states, the atoms and molecules in a gas can travel freely in virtually any direction at speeds of dozens of meters/second. Because of this freedom of motion, atoms and molecules in the gaseous state can serve as a very efficient means of transmitting and receiving information. The overall process by which these gaseous chemical "messengers" are transmitted and received is referred to in biology as *chemoreception* and is one of the more amazing skills possessed by life on this planet.

Chemoreception in some form or other is a phenomenon that is exhibited by nearly all of the higher order animals and by a great many plants and protozoans as well. The process is most often employed by creatures to locate food, to avoid being someone else's food, or to find the love of the creature's life. To accomplish these tasks, the plants and animals involved are endowed with ingenious arrays of chemoreceptors that serve to detect, collect, and analyze a wide variety of atoms and molecules that are often present only at minuscule levels. Perhaps the most familiar example of chemoreception is to be found in our own sense of smell. The odor of burning leaves in the autumn or that of steak sizzling on a summer time barbecue are but two examples of gaseous molecules striking the olfactory receptors in the human nose, thereby triggering neurological responses to the brain for processing and evaluation.

Although the process of chemoreception in mammals is similar, albeit usually more sensitive than that of human beings, other creatures have their detectors set up in other motifs. The tongue of a snake, for example, is a chemical receptor which samples the atoms and molecules in the surrounding air each and every time the tongue flicks out of the snake's mouth. The flatworm has its chemoreceptors located in the form of two projections or "ears" on either side of the head which are sensitive to chemical stimuli such as the molecules from its favorite foods.

Chemoreception processes are of extreme importance in the mating process as well. In what has to be very nearly the ultimate in "chemoeroticism," the female silk moth releases a complex molecule called bombykol into the air during the mating season. When these molecules strike the chemoreceptive antennae of the male silk moth they serve to excite the male, who reorients his flight pattern in the direction from which the bombykol is coming. Measurements have indicated that the antennae of the male silk moth can respond to bombykol impacts as infrequent as one molecule/second out of a total of hundreds of billions of impacts/second on the moth's antennae from other gaseous substances in the earth's atmosphere! Given these kinds of statistics it is little wonder that the phenomenon of chemoreception is among the most fascinating in all the life sciences.

(a)

(b)

(c)

(d)

••• **Molecule Detectors in Nature**   (a) The nose of a cheetah; (b) the antennae of the male silk moth; (c) the "smelling nodes" of the flatworm; (d) the tongue of a sidewinder rattlesnake.

ball-like entities. As in the conversion from a solid to a liquid, a certain amount of energy must go into converting a liquid into a gas. The amount of energy required to convert one gram of a liquid into a gas at one atmosphere of pressure is called the *heat of vaporization.* Heats of vaporization, like heats of fusion, can vary considerably from substance to substance. The heat of vaporization for water is about 540 calories per gram; that of ethyl alcohol is about 204 calories per gram. Helium has a heat of vaporization of only 6 calories per gram. As in the solid-liquid phase change, no temperature change will result from additional thermal energy until the phase change from a liquid to a gas is completely effected. Condensation back to a liquid from a gas will also bring about a release of all of the energy that initially went into producing the liquid-to-gas phase change.

In the gaseous state, there is essentially no bonding of the individual atoms and molecules. Instead they move through space as separate billiard balls having a random distribution of velocities ranging from zero to several kilometers per second. Because of this distribution, a certain fraction of molecules and atoms in a gas, or for that matter even in a liquid, have relatively high velocities. As a result if a ''fast'' liquid molecule happens to be near the surface of the liquid and moving in the proper direction, it can break the surface tension of the liquid and fly off the liquid or evaporate as a gas. Similarly, a certain fraction of the atoms in the earth's atmosphere can escape the earth's gravity and ''evaporate'' into the interplanetary medium. Most of the molecular velocities, however, are centered about a comparatively low mean value that depends on the temperature of the gas. A higher temperature results in a higher mean molecular velocity for the gas. Thus, we see that the addition of thermal energy to a gas in effect results in an increase in the total kinetic energy of the molecules in that gas. For molecular gases such as H_2 and CO_2 not only is the total kinetic energy of the gas molecules increased, but these molecules can also absorb thermal energy and transform it into vibrational and rotational energy as well.

Much of the behavior of gases can be interpreted in terms of this kinetic theory model in which the atoms in a gas are viewed as tiny billiard balls that collide with each other and with the walls of their container. In a gaseous state, these high-speed billiard balls are moving far too rapidly to cling to a definite volume as do the liquids. Instead they relentlessly seek out the walls of their container. Once there, the atoms or molecules collide with a given container wall and experience a net change of momentum as they recoil off that wall. The resulting impulse experienced per unit area of the container wall from large numbers of such impacts is observable to us as the pressure that the gas exerts on the wall.

We can also employ kinetic theory to develop a beginning mathematical description for the behavior of gases. Suppose we wish to increase the pressure on a given mass of gas while maintaining a constant temperature. As the outside pressure is increased, the number of collisions per unit area of wall per unit time from the gas cannot balance this excess pressure and the volume decreases. As the volume decreases, the rate of the wall collisions goes up because the molecules do not have as far to go before hitting a wall. Eventually the collisions reach a high enough value that the outside pressure is balanced. At this point, the volume of the gas is reduced and its density is increased. Conversely, if the pressure is reduced at constant temperature, the volume of the gas increases so as to reduce the rate of the wall collisions. This relationship between the pressure and the volume of a gas at constant temperature is called Boyle's law (Figure 8.5), and can be stated as follows: For a gas at constant temperature

pressure $\times$ volume = constant value .

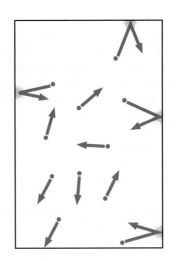
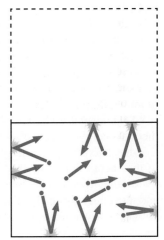

FIGURE 8.5 ··········

Kinetic Theory and Boyle's Law   When the volume of a given quantity of gas is reduced at constant temperature, the number of collisions per unit time between the atoms of the gas and the walls of the container increases. The gas thus exerts a larger pressure on the container walls in such a way that the product pressure × volume remains constant.

In particular, if a gas is initially at a certain pressure and volume and has its pressure and volume changed to new values while the temperature stays the same, then the final pressure and volume must satisfy the following statement:

$$\begin{array}{c} \text{pressure} \times \text{volume} \\ \text{(initial state)} \end{array} = \begin{array}{c} \text{pressure} \times \text{volume} \\ \text{(final state)} \end{array}.$$

Thus, if the pressure on a given gas is doubled from the initial state to the final state, Boyle's law tells us that the volume of the final state must be half of what it was in the initial state, and so on.

Suppose now we wish to increase the pressure of a gas but we will not allow the gas to reduce its volume. How can we raise the rate of wall collisions so as to balance the increase in pressure? We may accomplish this feat by simply raising the temperature of the gas. By doing so we increase the mean velocity

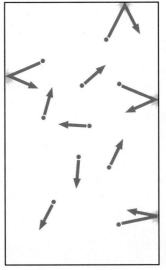
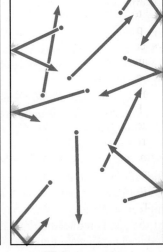

FIGURE 8.6 ··········

Kinetic Theory and Charles's Law   When the temperature of a given quantity of gas is increased at constant volume, the velocities of the atoms of the gas also increase. As a result, the number of collisions per unit time between the atoms of the gas and the walls of the container increases and the gas thus exerts a larger pressure on the container walls in such a way that the quantity volume/temperature remains constant.

Lower temperature   Higher temperature

of the molecules in the gas, thereby decreasing the average time between collision. The rate of the wall collisions thus goes up, and if the temperature is increased by the proper amount, the rate of wall collisions will balance the increase of outside pressure. At constant pressure, the volume of a gas will increase with an increase in temperature and will decrease with a decrease in temperature because now the gas is adjusting for the temperature-induced changes in the rate of wall collisions by altering its volume. This relationship between the Kelvin temperature and the volume of a gas is called Charles's law (Figure 8.6), and can be stated as follows: At constant pressure,

$$\frac{\text{volume}}{\text{temperature}} = \text{constant value} ,$$

and if a gas has its temperature and volume changed at constant pressure, then the final temperature and volume must satisfy the following:

$$\frac{\text{volume}}{\substack{\text{temperature} \\ \text{(initial state)}}} = \frac{\text{volume}}{\substack{\text{temperature} \\ \text{(final state)}}} .$$

Thus, if the temperature of a given gas is tripled at constant pressure, the volume of the final state must be tripled as well. As shown in Physics Formulation 8.1, Boyle's law and Charles's law can be combined into an equation of state for a gas that is written as

$$\frac{\text{pressure} \times \text{volume}}{\text{temperature}} = \text{constant value} ,$$

which tells us that

$$\frac{\text{pressure} \times \text{volume}}{\substack{\text{temperature} \\ \text{(initial state)}}} = \frac{\text{pressure} \times \text{volume}}{\substack{\text{temperature} \\ \text{(final state)}}} .$$

A specification of the pressure, volume, and temperature of a given gas thus constitutes a complete description of the "state" of that gas. Pressure, volume, and temperature are measured in pascals, m^3, and kelvins. However, sometimes gas pressures are expressed in atmospheres. One atmosphere is the pressure of the earth's atmosphere at sea level, a pressure equivalent to about 10^5 N/m^2. Gas volumes can be expressed in terms of the liter, which is the volume contained in a cube 0.1 m on a side. Thus, one liter is 10^{-3} m^3, or a volume slightly larger than a liquid quart in the English system.

From the preceding paragraphs we have seen how a wide range of heat-related phenomena can be explained by viewing these phenomenon from the microscopic level of the atom and molecule. Within this context, one can now formally define the temperature of an object or substance as a measure of the average kinetic energy of the molecules making up that object or substance. Thus the molecules in a high-temperature gas have, on the average, much larger kinetic energies than those bound up in a low- temperature solid.

For additional insights into heat-related phenomena, in the next section we depart the world of atoms and molecules for the time being and return to the "everyday" level of the physical world.

8.3 Absolute Zero

Charles's law for ideal gases tells us that the change in the volume of such an ideal gas at constant pressure varies in direct proportion to the change in the temperature of that gas. In theory, if the temperature of a given gas at constant pressure is allowed to increase indefinitely, then the volume of that gas would also increase correspondingly. On the other hand, if we seek to reduce the temperature of a given constant pressure gas indefinitely, we would find that a theoretical limiting temperature exists, called *absolute zero,* at which the volume of the gas would be zero. Every gas will liquefy or even solidify before its volume will reduce to zero, but the backward extrapolation of Charles's law to zero volume for each and every gas yields the same theoretical value for the zero volume point, $-273.15°C$. It was this fact that led nineteenth century physicists to conclude that "absolute zero" was a temperature possessed of considerable significance.

Since then we have come to interpret absolute zero as the temperature of a given object at which no more thermal energy can be extracted from that object. In short, absolute zero is the lowest temperature that any object in the universe can ever possibly have, and as such, provides a useful zero point for the Kelvin or "absolute" temperature scale in which all temperatures then have positive values.

Experimental studies done in recent years indicate that while it is possible to attain temperatures very close to absolute zero, of the order of 10^{-5} K, it is apparently not possible for an object to actually reach a temperature of "absolute zero." This fact is sometimes referred to as the "third law" of thermodynamics.

Objects and substances that are cooled to temperatures within a few degrees of absolute zero can take on a number of unusual characteristics. At such temperatures, for example, it is possible for liquid helium to exhibit so-called "superfluid" properties in which the liquid helium has zero viscosity. Perhaps the most famous of low-temperature characteristics is the phenomenon of superconductivity in which the electrical resistance of a given substance, usually

a metal, in effect totally vanishes when the substance is at temperatures near absolute zero. In more recent years substances have been developed that exhibit this phenomenon at much higher temperatures, and as such will have considerable impact on our technology.

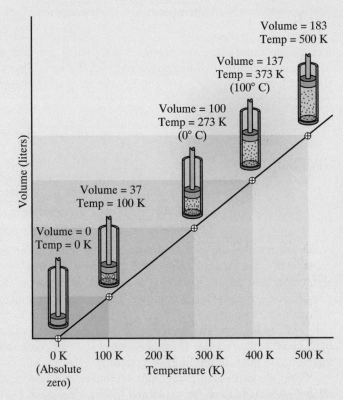

··· **Absolute Zero**   If the temperature of a gas at constant pressure could be reduced indefinitely, a temperature exists at which the volume of the gas would finally reduce to zero were the gas not to liquefy or solidify. This limiting degree of an object's coldness in the universe is called absolute zero.

WORKED EXAMPLE

If the heat of vaporization of water is 540 calories per gram and that of liquid helium is 6 calories per gram, how much liquid helium could be vaporized with the thermal energy needed to vaporize 3 grams of water? Assume both liquids are at their respective boiling points.

SOLUTION: The thermal energy required to vaporize 3 grams of water is given by

$$\text{thermal energy } = \text{ mass } \times \text{ heat of vaporization } .$$

For 3 grams of water having a heat of vaporization of 540 calories/gram, we have

$$\text{thermal energy } = 3 \text{ grams } \times 540 \text{ calories/gram}$$
$$= 1620 \text{ calories } .$$

The amount of liquid helium that can be vaporized with this same amount of thermal energy is

$$1620 \text{ calories } = \text{ mass } \times 6 \text{ calories/gram } ,$$

assuming that the heat of vaporization of liquid helium is 6 calories/gram. Solving for the mass yields

$$\text{mass } = \frac{1620 \text{ calories}}{6 \text{ calories/gram}}$$
$$= 270 \text{ grams } .$$

1. Describe how the atoms and molecules in a substance behave when progressively more thermal energy is added.

2. Compare and contrast the heat of fusion of a substance with its heat of vaporization. Which occurs at the higher temperature? Why?

3. What is the kinetic theory of gases? What assumptions do we make about gases in the kinetic theory?

4. Discuss how the kinetic theory explains the following: (a) boiling, (b) evaporation, and (c) gas pressure.

5. State Boyle's law and Charles's law. How does the kinetic theory account for these laws?

1. How much thermal energy would be required to change the temperature of 5 grams of water from $-2°C$ to $+2°C$? Assume that the heat of fusion of water is 80 calories/gram.

2. A gas sample has a volume of 12 liters. What volume will the gas have if:
   a. its temperature were doubled?

b. the pressure were tripled?

c. both (a) and (b) occurred simultaneously?

8.3 Energy Transport

In the introduction to this chapter, we described an "experiment" in which the door between a hot room and a cold room was opened and, as a result, the hot room cooled off and the cold room warmed up. The conclusion that one can draw from such phenomena is that thermal energy can "flow" from a high-temperature object or place to a low-temperature one. The question we now wish to address, then, is how such an energy flow is accomplished.

Suppose you get up in the morning before classes and pour yourself a very hot cup of coffee. In doing so, let us suppose you absentmindedly leave your spoon in the cup. If you touch the spoon before the coffee has cooled off, you find that the spoon handle has become as hot as the coffee despite the fact that the spoon handle is sticking well out of the coffee. Thermal energy from the high-temperature cup of coffee has been transported somehow into the handle of the spoon. This is an example of energy transport by the process of *conduction.* When the spoon was immersed in the coffee, the thermal energy present in the higher temperature coffee caused the molecules in the lower temperature spoon to increase their degree of agitation. These molecules in turn, by bumping into neighboring molecules, cause the latter to increase their degree of agitation. Eventually this degree of agitation or temperature rise is propagated throughout the spoon.

Whenever thermal energy is transported in this fashion, we refer to the process as *heat conduction* (Figure 8.7). The process stops at the spoon handle/atmosphere boundary because the average spacing of the molecules in the atmosphere is too large to transport the thermal energy effectively any further. The ability of a substance to transport thermal energy in this way depends entirely on the molecular structure of that substance. Solids, such as our coffee spoon, that possess metallic bonding are usually excellent conductors of heat. On the other hand, solids that have comparatively rigid electronic bonding or have a lot of spacing in their crystalline structures are poor thermal conductors. Such substances are often referred to as *insulators.* Examples of solid insulators include wood, brick, and snow. Most liquids and gases are also poor conductors of thermal energy because of the spacing of the atoms and molecules within these substances. Because of such structural differences, a metal such as copper or aluminum transports thermal energy roughly 2000 times more efficiently by conduction than does the earth's atmosphere. Thus, thermal energy transport by conduction is almost exclusively the domain of the metallic bonded solids, and hence the heat conduction process in our stainless steel coffee spoon effectively stops at the spoon handle/atmosphere boundary.

In the preceding paragraph we noted that fluids are almost always poor conductors of thermal energy due to the inherent nature of their atomic structures. The very spacing and loose bonding that render fluids highly inefficient media

FIGURE 8.7 ··········
Conductive Energy Transport The end of the metal tongs feels hot to Larry because thermal energy has been transmitted along an array of bound atoms in the tongs from the hot coals to Larry's hand. Energy transferred in this fashion is said to be accomplished by conductive transport or thermal conduction.

for thermal energy transport by conductive processes, however, provide an ideal setting for a second type of thermal energy transport.

Suppose a layer of fluid comes in contact with a higher temperature surface at the bottom of the layer. The fluid in contact with that surface will begin to heat up. As it does so, the bottom layer expands its volume and becomes less dense. Because this layer is now less dense than the layers of fluid above it, the bottom layer begins to "float" toward the higher and cooler regions of the fluid. The layer that floats upward is then replaced by the cooler and denser layers, which sink down to the hot surface. They too are then heated and float upward. Meanwhile the first layer has now been cooled off by its low-temperature surroundings in the higher fluid layers and now falls back toward the heated surface. In this fashion a cyclic flow of fluid is produced by the unequal heating in the liquid or gas.

As the fluid's temperature increases and decreases during the course of this process, thermal energy is taken on as the fluid heats up at the lower, hotter fluid layers, is transported upward as the fluid rises, and then is deposited into the surrounding medium as the fluid cools off at the higher, cooler fluid layers. Thermal energy transported in this way is said to be transported by *convection*. Uneven heating of the earth by the sun produces a wide range of convective types of meteorological phenomena. The great thunderheads, which are a familiar part of summer time rain storms, are a very dramatic example of convective energy transport (Figure 8.8). The hot summer time sun locally heats up the ground, which in turn heats the overlaying air. The warmer, lower density air rapidly rises to heights of as much as 50 to 60 thousand feet above the earth's surface. In the cooling process, precipitation of the rising air in the form of rain or hail can occur.

From time to time an "anticonvective" situation can develop close to the earth's surface. This situation is called a *temperature inversion*. When such an inversion sets in at a given location, high-density cooler air is close to the ground while warm low-density air is above it at a higher altitude. As a result, both air masses are in effect "where they want to be," and no convective circulation occurs. It is during such temperature inversions over large cities that the city's hot gas pollution such as car exhaust cannot rise past the warm air inversion layer and disperse. As a result, these waste gases become trapped in the inversion layer, often causing the pollution levels in the city's air to reach dangerously high levels.

On a much larger scale, virtually all of the overall circulation patterns in the earth's atmosphere are the result of convective motions in which the air heated up in the higher temperature regions of the earth flows toward the lower temperature regions of the earth's surface. The earth's rotation in turn breaks up these motions into belts of wind, such as the trade winds and prevailing easterlies and westerlies, and belts of calm such as the doldrums and horse latitudes.

Convective effects can also be observed when a container of liquid is heated from the bottom. As the heating progresses, warm liquid rises from the high-temperature bottom of the container, while the lower temperature liquid descends to replace it. In fact, when a cup of coffee or a pan of soup is being heated in such a fashion, we very often try to speed up the convective energy transport process by stirring the liquid in a convective-like "vertical loop" fashion.

Let us now return to our hot spoon in a hot cup of coffee. If we had stopped our hand short of actually touching the hot spoon handle, we would very likely have been able to "feel" the heat coming off the spoon handle, and thereby have

FIGURE 8.8 ··········
Convective Energy Transport The temperature difference between the ground and higher altitudes produces a convective energy flow, which manifests itself in this gigantic "thunderhead" or cumulus cloud.

FIGURE 8.9 ∙∙∙∙∙∙∙∙∙∙
Radiative Energy Transport   There are many instances in which energy is transported by means that are clearly not conductive or convective in nature. Energy transfer in such cases is said to be accomplished by radiative transport. The transport of the sun's energy to the earth is an example of such a radiative process.

been warned of the fate that awaited had we actually tried to touch the handle. How was that sensation of heat transmitted to us? Certainly not by conduction. As we have already noted, ample evidence exists to indicate that air is an extremely poor conductor of thermal energy. Convective currents may exist in the air about the spoon handle, but these would tend to transport the thermal energy upward. By simply moving one's hand about the handle, one quickly finds that the sensation of the spoon handle's heat can be felt from *any* direction off the handle.

We are forced to conclude that yet a third means of energy transport occurs in nature, which we call *radiative transport.* Radiative transport is different from convection and conduction in that it requires no medium through which to propagate. Were we to hold our hand close to the spoon handle while in a vacuum, we would still be able to feel the heat by virtue of this radiative transport. In a more extreme example, to get to the earth's surface, the sun's rays must traverse 150 million kilometers of virtually empty interplanetary space as well as two dozen kilometers of a most effective layer of thermal insulator in the form of the earth's atmosphere. The sun accomplishes this feat through radiative transport (Figure 8.9).

Because radiative transport does not require a medium, it is really not the transport of thermal energy but rather the transport of a different type of energy called radiant energy, which we will describe in more detail in Chapter 14. Interestingly, given the three possible choices for transporting itself—conduction, convection, or radiation—the most efficient method of energy transport will dominate under the given set of circumstances. Thus the thermal energy from the hot coffee was transported to the spoon handle via conductive transport because that was the most efficient mechanism under the circumstances. On the other hand, radiative transport became the most favored mechanism in transporting the thermal energy from the spoon handle to our hand when the mechanisms of conduction and convection were now less efficient under the given conditions.

1. Describe the various ways that thermal energy can be transported.

2. Give everyday examples of each of the processes you described in Question 1.

3. What is an insulator? How do insulators work?

4. Which type of energy transport do you think would be most likely to occur in solids? liquids? gases? Explain.

5. Why do you suppose that "smog alerts" in big cities most often occur when a temperature inversion is present?

8.4 The Laws of Thermodynamics

We have already seen that it is possible to transform mechanical energy into thermal energy by means of friction and similar processes. Indeed, as we shall soon learn, we cannot avoid doing so. But is the reverse true? Can we design some sort of device or "engine" that will convert thermal energy into mechanical energy? This is the central topic of the branch of physics that is called *thermodynamics.*

The basic idea underlying all heat engines is the fact that thermal energy will flow from a high-temperature "reservoir" to one at a lower temperature. As this flow of thermal energy takes place it can, in theory, be converted into mechanical work. For example, suppose we ignite a fuel inside a closed cylinder at one end of which is placed a movable piston (Figure 8.10). Upon ignition of the fuel, gas in the cylinder will be raised to a very high temperature. This hot gas "reservoir" will seek immediately to cool itself off by expanding against

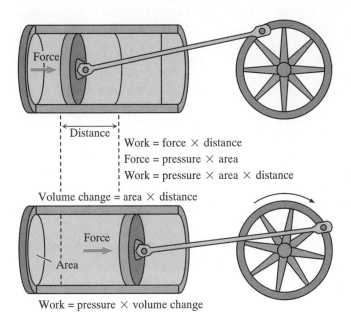

Work = force × distance
Force = pressure × area
Work = pressure × area × distance

Volume change = area × distance

Work = pressure × volume change

FIGURE 8.10 ···········

Work in a Heat Engine   As the volume in the piston cylinder expends at constant pressure, the piston performs an amount of work equal to pressure × change of cylinder volume.

the walls of the cylinder. During the course of this volume expansion, two things will happen. The expanding gas will first of all perform mechanical work. To see how this occurs, suppose that the piston has a certain surface area against which the ignited gas pushes. The total force exerted on the piston by the gas is given by

$$\text{total force } = \text{ pressure } \times \text{ area } .$$

If the piston moves a given distance, the force acts through this distance and performs an amount of work that can be expressed as

$$\text{work } = \text{ force } \times \text{ distance } .$$

The product of the distance and the surface area of the piston, however, is equal to the increase in the volume of the gas in the cylinder and we have

$$\text{volume change } = \text{ distance } \times \text{ area}$$

or

$$\text{distance } = \text{ volume change/area } .$$

Thus,

$$\text{work } = \text{ force } \times \text{ volume change/area}$$

or

$$\text{work } = \text{ (force/area) } \times \text{ volume change } .$$

The force/area, however, is just the pressure of the gas in the cylinder and we finally obtain

$$\text{work } = \text{ pressure } \times \text{ volume change } .$$

From Charles's law we can also see that as the volume of the gas in the cylinder expands at constant pressure, it must necessarily cool back down into a low-temperature reservoir. Cyclic repetition of this process gives rise to a continuing generation of mechanical work. In this case the piston was driven by the pressure produced by a gas explosion and, as a result, this type of heat engine is often referred to as an "internal combustion" engine. The most familiar examples of internal combustion engines are the diesel engines that power trucks and trains and the gasoline engines that run automobiles. The ideal heat engine is one envisioned by the French physicist N. L. Sadi Carnot early in the nineteenth century. In Carnot's idealized cycle, all of the input thermal energy originates from a hot reservoir at a single "hot" temperature and all of the engine's unused or "rejected" thermal energy emerges at a cold reservoir, which is also at a single temperature.

We may also conceive of a type of "heat engine in reverse" or "refrigerator" in which mechanical energy is used to "pump" thermal energy from a low-temperature reservoir to a high-temperature reservoir. When thermal energy is removed from the lower temperature reservoir in this fashion, the temperature of that reservoir will drop to a lower value while that of the higher temperature reservoir will increase to a higher value. There are many instances in which such refrigeration is desirable. Many foods that would quickly spoil at room temperature can be preserved for long periods of time at the cooler temperatures provided and maintained by refrigeration processes. Also, many summer days would be

If no net heat flow between A and T

AND
If no net heat flow between B and T

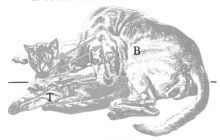

THEN
No net heat flow between A and B

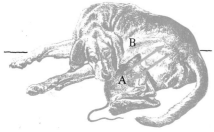

FIGURE 8.11 ··········
The Zeroth Law of Thermodynamics   If body T is in thermal equilibrium with body A and if body T is in thermal equilibrium with body B, then bodies A and B are in thermal equilibrium with each other.

totally unbearable were it not for an air conditioning unit dutifully pumping thermal energy out of our low-temperature reservoir car, classroom, or apartment and into the high-temperature "hot, cruel world" outside, thereby maintaining the inside temperature at a comfortable level.

Over two centuries of human experience in the study of thermal energy has led to the formulation of three basic statements, called the *laws of thermodynamics,* which govern the behavior of heat and heat flow. First of all, if two objects are brought into contact with each other and no net flow of thermal energy occurs between them, then these objects are said to be in *thermal equilibrium* with one another. Given this definition, we can state what is called the *zeroth* law of thermodynamics* as follows:

> If two bodies are each in thermal equilibrium with a third body, then they must be in thermal equilibrium with each other.

This seemingly commonsense statement is a validation of the entire concept of temperature in the sciences. Suppose, for example, we have an object A that is in thermal equilibrium with a second object T as shown in Figure 8.11. If we now find that object T is also in thermal equilibrium with a second object B, then the zeroth law tells us that if objects A and B were to be placed in contact, there would be no net flow of thermal energy between them. In short, all three objects would have the same temperature.

Pursuing this idea a bit further, suppose we allow our object T to come to thermal equilibrium with a "standard" object or system, such as a water-ice mixture, and note the physical characteristics that T takes on while in that particular state of thermal equilibrium. The zeroth law now guarantees that whenever our object T takes on that same set of physical characteristics as a result of being in thermal equilibrium with another object, the temperature of that object is the same as the "standard" object, thereby establishing temperature as a reproducibly measurable quantity.

The second statement one can make concerning the balance of thermal energy is, in essence, a thermal energy version of the principle of the conservation of energy and the work-energy theorem. It is called the *first law of thermodynamics* and states that:

> If a certain amount of energy is added to a given system, the total amount of energy of all forms emerging from that system cannot exceed the amount of energy initially added to the system.

This first law of thermodynamics tells us that in the realm of thermal energy, like any other energy form, we simply cannot construct an energy-creating device of any kind. In particular, we cannot create a heat engine that will perform for us an amount of work that is greater than the amount of energy we initially supply to that engine (Figure 8.12).

But it is worse than that. Extensive observations of thermodynamic systems have led to a third conclusion regarding the behavior of thermal energy, which can be summarized as follows:

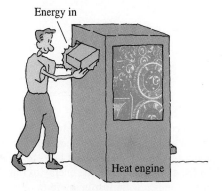

Energy in

Heat engine

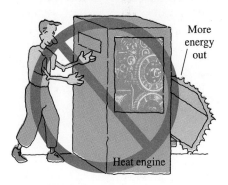

More energy out

Heat engine

FIGURE 8.12 ··········

The First Law of Thermodynamics
No heat engine can be constructed that produces more energy than it takes to run the engine.

---

*The zeroth law was proposed in the 1930s, well after the first and second laws of thermodynamics were proposed and numbered. Because the zeroth law is basic to the first and second laws, scientists have deemed that it should have the lowest number of the three laws of thermodynamics. Rather than changing the numbers on the other two laws, their numbers have been preempted by having a zeroth law.

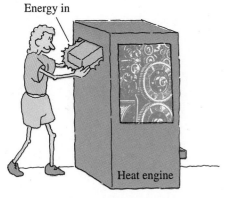

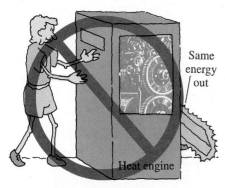

FIGURE 8.13 ············
The Second Law of Thermodynamics
No heat engine can be constructed that produces as much energy as it takes to run the engine. Some of the energy to run the engine is *always* lost in the form of thermal pollution.

If a certain amount of energy is added to a given system, the total amount of work obtained from the system is always less than the amount of energy initially added to the system.

This statement is called the *second law of thermodynamics.* It tells us, among other things, that we cannot design a heat engine that will completely convert all of the input thermal energy into mechanical work (Figure 8.13). Some of the thermal energy will always be left behind. In other words, the second law of thermodynamics tells us we cannot even recover our initial energy investment. We can see the second law at work in automobile engines. When such an engine burns fuel, only about 30% of the energy is converted into the mechanical work that runs the car. The rest of the energy is lost in heating up the engine, friction, and in the exhaust gases emanating from the engine.

The very best we can possibly do in terms of the efficiency of a heat engine or the ratio of work out/heat input is that obtainable from the idealized Carnot engine for which

$$\text{efficiency} = \frac{\text{work out}}{\text{heat in}} = 1 - \frac{\text{temperature of cold reservoir}}{\text{temperature of hot reservoir}}.$$

The rejected or waste thermal energy emanating from our various heat engines serves to heat up the overall environment as ''thermal'' pollution. As our technology has become more extensive, thermal pollution from our heat engines has become an ever-increasing environmental problem.

At the atomic or microscopic level, the second law of thermodynamics can be described in terms of probabilities. It has long been recognized that the number of possible ''worthless'' configurations that can come up on a slot machine or at the poker table far exceeds the number of configurations that constitute a payoff or a good hand. Thus we see far more of the worthless configurations and hands simply because those are the odds. In a like fashion, consider a gas-filled room, into which has been placed an open jar. If we compile a list of all of the possible configurations of the atoms and molecules of a gas in the room, we would find that an overwhelming number of the particle distributions would be those in which particles were found throughout the room. Only a vanishingly small number of possible configurations would occur in which all of the particles in the room were confined to the jar. Thus, the odds are overwhelmingly in favor of finding particles moving ''helter-skelter'' throughout the room, which is precisely what we observe.

The state in which the atoms and molecules would all be located in the same small volume of the jar is an extreme example of what physicists refer to as an *ordered state,* while the ''helter-skelter'' atoms and molecules are an example of a random or *disordered state.* What becomes very obvious very quickly is the existence of a ''flow of orderliness'' from states of higher order and less randomness to states of lower order and more randomness. Thus, if the atoms and molecules in the room are initially confined to the volume of the jar via a lid, they will take on the largest state of disorder possible within the jar. If the lid on the jar is removed, however, then for the atoms and molecules to remain within the confines of the jar no longer represents the configuration of lowest order. The atoms and molecules, therefore, move to take on the more disorderly or random configuration in which they now move ''helter-skelter'' throughout the confines of the entire room. Moreover, if we wish to reverse this flow back to a more ordered state, we must expend energy to do so. Thus, if we wish to reconfine the atoms and molecules back into the volume of the jar, it must be done at a cost of the energy needed to run a pump.

This "tendency toward disorder" is observable at every level of the physical world. Weathering processes reduce relativity ordered mountain ranges into relatively unordered layers of soil and sand. The order of a well-tended garden or flower bed rapidly decays into the disorder of a weed patch if efforts to maintain the original order are not ongoing. The process in which the morning toast is burned to a crisp converts the well-ordered piece of bread into a much less ordered kitchen-sized volume of smoke. Scientists have thus formally defined a quantity called *entropy*, which is in essence a measure of the state of disorder in a given system, or, alternatively, the "probability rating" of the given system. The more disorder present in a system, the larger its entropy, and the more probable its existence. Accordingly, the second law of thermodynamics can be alternatively stated as follows:

> The entropy of a given system tends to increase in time.

Thus, left to natural processes, states of higher order or less entropy will decay into states of lower order and higher entropy. If, for example, an "ordered" form of energy is to be converted into an alternate form of "ordered" energy such as work by some process, a certain fraction of the input energy must decay into less ordered "waste" energy in that process. Therefore, the total entropy present after the process is greater than the total entropy present prior to the process. In a similar vein, it is only by a steady infusion of energy via food and drink that life forms are able to maintain their highly organized and ordered states. Such an infusion, however, can only be accomplished by raising the level of disorder or entropy of the surrounding physical environment. We will return to this brutal thermodynamical fact of life in a later chapter.

WORKED EXAMPLE

A piston having an area of 1.6 m² puts out 150 joules of work while being subjected to a gas pressure of 75 N/m². How far did the piston-move?

SOLUTION: The work done by the piston is given by

$$\text{work} = \text{pressure} \times \text{volume change} .$$

Since the work done at the pressure of 75 Nm² was 150 joules, we have

$$150 \text{ J} = 75 \text{ N/m}^2 \times \text{volume change}$$

or

$$\text{volume change} = \frac{150 \text{ J}}{75 \text{ N/m}^2}$$
$$= 2 \text{ m}^3 .$$

The volume change is also given by

Continued

WORKED EXAMPLE

Continued

$$\text{volume change} = \frac{\text{distance piston moves}}{} \times \text{area of piston}.$$

If the total volume change is 2^3 and the area of the piston is 1.6 m², then we have

$$2 \text{ m}^3 = \frac{\text{distance piston moves}}{} \times 1.6 \text{ m}^2.$$

Solving for the distance the piston moves yields

$$\frac{\text{distance piston moves}}{} = \frac{2 \text{ m}^3}{1.6 \text{ m}^2}$$

$$= 1.25 \text{ m}.$$

REVIEW 8.4 QUESTIONS

1. Explain how thermal energy can be converted into mechanical work. Give an example of such a process.

2. State the laws of thermodynamics. How are they related to the principle of conservation of energy?

3. What is the efficiency of a heat engine? Can the efficiency of a heat engine ever be (a) greater than one? (b) equal to one?

4. Describe thermal pollution. Can such pollution be eliminated from a heat engine? Explain.

5. Discuss the concept of entropy. Give some examples in your everyday life in which entropy is increasing through natural processes.

REVIEW 8.4 PROBLEMS

1. A piston having an area of 0.4 m² is subjected to a pressure of 50 N/m². If the piston moves a total of 0.2 m, how much work did the piston perform?

2. A Carnot engine operates between a hot reservoir of 1200 K and a cold reservoir at 300 K. What is the efficiency of this engine? To what temperature must the hot reservoir be raised (or lowered) in order for the engine to have an efficiency of 90%?

Chapter Review Problems

1. The level of mercury in a certain glass tube is at a height of 5 cm at 0°C and 50 cm at 100°C. What will the level of the mercury be at a temperature of 50°C?

2. If the level of mercury in the glass tube of Problem 1 is 40 cm, what is the temperature?

3. If a gram of water is heated from 10°C to 25°C, how much thermal energy was used in the process?

4. How much thermal energy is taken on by 20 kg of ice if its temperature is raised from −5°C to −2°C? Assume the specific heat of ice is 0.48 that of water.

5. A 5-kg mass in a Joule experiment falls a total distance of 0.4 m. How much mechanical energy is transferred to the water bath. How does this energy manifest itself in an observable way in the water bath?

6. If the water bath in Problem 5 has a mass of 10 kg and was at an initial temperature of 5°C, what is the final temperature of the water bath?

7. How many newton-meters of energy are there in
   a. one joule of thermal energy?
   b. in one calorie of thermal energy?

8. How much energy is required to convert 20 grams of ice at 0°C into water at 0°C?

9. How much energy is required to convert 30 grams of water at 100°C into steam at 100°C?

10. A total of 15 grams of water at 20°C is mixed with 15 grams of water at 30°C. If the final temperatures of the mixture is 25°C, how much thermal energy was gained and lost by each water sample? Is this result consistent with the principle of the conservation of energy?

11. A gas at one atmosphere pressure has a volume of 2 liters. If the pressure is increased to four atmospheres and the temperature remains the same, what is the new volume of the gas?

12. A gas at two atmospheres of pressure has a volume of 3 liters. What pressure must this gas experience in order for its volume to be 12 liters assuming a constant temperature?

13. If the temperature of 5 liters of gas is tripled at constant pressure, what is the volume of the gas after the temperature change?

14. A gas is observed to change its volume from 6 liters to 4 liters at constant pressure. If the initial temperature of the gas was 450 K, what is the final temperature of the gas?

15. Find the pressure in N/m^2 exerted by a gas at 10 atmospheres of pressure.

16. A .02-kg mass of gas at 300 K and one atmosphere of pressure has a volume of 10 liters. What is the density of the gas?

17. If the temperature of the gas in Problem 16 were doubled, and the pressure remained constant, what would the gas density then be?

18. A piston having an area of 2 m^2 is subjected to a pressure of 100 N/m^2. If the piston moves a distance of 0.4 m, find:
   a. the total force on the piston
   b. the total volume change of the gas in the piston cylinder
   c. the work produced by the piston.

19. If a piston having an area of 1.5 m^2 moves 0.2 meters while putting out 30 J of work, at what pressure was the gas in the piston cylinder operating?

20. A total of 1000 J of energy is transferred into a heat engine having a volume of 2 m^3 of gas at a pressure of 400 N/m^2. If the gas volume of the engine doubles in size at constant pressure during this process, how much work does the engine perform? What is the efficiency of this engine?

21. A Carnot engine operates between a hot reservoir at 500 K and a cold reservoir at 200 K. What is the efficiency of this engine?

22. If the engine in Problem 21 is fed 5000 J of thermal energy, how much work will the engine produce? How much thermal pollution energy will the engine produce?

23. The temperature and pressure of a 5-liter sample of gas are both doubled. What is the volume of the sample after this occurs?

THOUGHT QUESTIONS

1. Devise a temperature scale of your own in which you specify a well-defined zero point and the size of each of your temperature degrees.

2. Do you think that the Fahrenheit scale has an absolute zero point? Explain.

3. Look up an extensive table of specific heats in a reference work such as the *Handbook of Chemistry and Physics* and find as many substances as you can that have specific heats larger than water. Are any of them common substances?

4. Discuss a way not mentioned in the text by which you could measure an object's temperature.

5. Why are the specific heats of liquids generally higher than those of solids? Where do you think the values of the specific heats of gases might enter into the picture?

6. What are the differences, if any, between heat and temperature?

7. Can the specific heat and the heat capacity of a given object ever be the same? Explain.

8. The element helium is observed only in a liquid or gaseous phase, even at temperatures very close to absolute zero. What, if anything, does this tell you about helium atoms?

9. Can you think of an aspect of the way gases behave that cannot be explained by kinetic theory? Explain.

10. Why do you suppose that thermal energy flows from a region of high temperature to one of lower temperature?

11. Discuss the phenomenon of frictional heat loss in terms of what is happening at the microscopic or atomic level.

12. In meteorology a "high-pressure system" or region of high atmospheric pressure in the earth's atmosphere is usually associated with cold temperatures, while a "low-pressure system" or region of low atmospheric pressure is associated with warmer temperatures. Why do you think this is so?

13. Can you think of a practical use for substances with
    a. large specific heats?
    b. with small specific heats?

14. Explain in terms of the kinetic theory why we can detect a woman's perfume or a man's after-shave lotion from across a classroom?

15. When a substance is placed into a liquid to form a solution, such as salt into water, etc., the temperature at which the solution freezes is lower than the temperature at which the pure liquid freezes. Why do you think this is so?

16. What is a ''perpetual motion'' machine? Is it possible to create such a machine? Explain

17. Why do you suppose that at night cold-blooded reptiles prefer to lie on solid structures such as roads and patios as opposed to the desert sand?

18. What do you think the most efficient means of thermal energy transport would be for
    a. a solid?   **b.** a liquid?   **c.** a gas?   **d.** a plasma?

19. The laws of thermodynamics have been whimsically summed up as follows:

    You can't win.
    You can't break even.
    You can't get out of the game.

    Discuss this version of the laws of thermodynamics in terms of the presentation made in your text.

20. Show that the equation of state of a gas can be written as

    $$\text{pressure} = \text{constant} \times \text{density} \times \text{temperature} \ .$$

21. Is it possible to have a heat engine that performs work without a volume change anywhere in its cycle? Explain.

22. Suppose in Joule's experiment, he found that no temperature change occurred in the water. What would his conclusion then have been?

23. Do you think it is possible for energy to be transported in nature by a means other than conduction, convection, or radiation? If so, speculate on how such transport might occur.

24. Explain the phenomenon of chemoreception in terms of the kinetic theory of gases.

25. Pick a creature of your choice and describe how its chemical receptors operate.

Temperature Scales

The ability of an object or system to transfer thermal energy is specified in science by a parameter referred to as the *temperature* of that object or system. To measure such a temperature parameter, we must set up and define a temperature scale. This can be accomplished in a variety of ways, but in each instance, the temperature scale must have a specified zero point and a specified increment or "degree" of temperature (Figure 8.14).

The temperature scale most commonly employed in science is the centigrade or Celsius scale developed by the eighteenth century Swedish astronomer-physicist Anders Celsius. The zero point of this scale, or 0°C, is defined as the temperature of an ice-water mixture at one atmosphere of pressure. The size of the centigrade degree of temperature is specified by defining 100°C as the temperature of a water-steam mixture at one atmosphere of pressure. Thus a change in temperature of 1°C represents a temperature change that is ¹/₁₀₀ of the temperature difference between the freezing and boiling points of water.

Interestingly, there is an ultimate limit to the degree of "coldness" that a body can possess. This limiting degree of coldness occurs at − 273.15°C and is referred to by scientists as "absolute" zero (see Physics Facet 8.3). The nineteenth century English scientist Sir William Thompson, who is better known simply as Lord Kelvin, pointed out that a convenient temperature scale was to be had by setting the zero point at absolute zero, thereby guaranteeing a temperature scale whose temperatures would all have positive values. The resulting absolute or

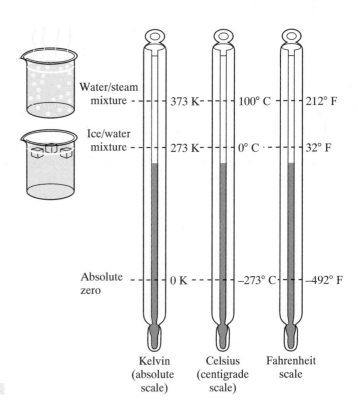

FIGURE 8.14 ···········
Temperature Scales The freezing and boiling points of water as well as absolute zero are shown on the Kelvin, Celsius, and Fahrenheit temperature scales.

Kelvin temperature scale developed by Kelvin thus has its zero point at absolute zero, but retains the degree size of the Celsius scale. In modern technology, however, Kelvin degrees are simply referred to as *kelvins* (K). Thus $0 \text{ K} = -273°C$, $273 \text{ K} = 0°C$, $373 \text{ K} = 100°C$, and in general

$$T_K = T_C + 273 \ ,$$

where T_K is the Kelvin temperature and T_C the corresponding Celsius temperature.

Another temperature scale commonly employed in the nonmetric world is the Fahrenheit scale, which was developed in the eighteenth century by the German physician Gabriel Fahrenheit. Fahrenheit defined the zero point of his scale as the lowest freezing point that he could obtain by mixing salt and water, and 96°F as the temperature of the human body. Like the denizens of the English system of units, temperatures on the Fahrenheit scale are not particularly convenient to use or convert. The freezing point of water, for example, is 32°F and its boiling point on this scale is 212°F. Even one of Fahrenheit's own standard reference points, the temperature of the human body, turns out to be more like 98.6°F rather than the 96°F Fahrenheit envisioned.

Nevertheless, the Fahrenheit scale still enjoys a good deal of current usage, particularly in realms such as cooking and weather and it is thus useful to be able to convert back and forth from Fahrenheit temperatures to Celsius temperatures. If we let T_F represent the Fahrenheit temperature, then a change of 212°F − 32°F = 180°F corresponds to the difference in Fahrenheit degrees between the freezing and boiling points of water on the Fahrenheit scale. This same temperature difference between the freezing and boiling point on the Celsius scale however is, by definition, 100 Celsius degrees. Thus we have that

$$100 \text{ Celsius degrees} = 180 \text{ Fahrenheit degrees}$$

or

$$\frac{5 \text{ Celsius degrees}}{9 \text{ Fahrenheit degrees}} = 1 \ .$$

The location of the freezing point of water on the Fahrenheit scale is at 32°F, while on the Celsius scale it is defined to be 0°C. Therefore to convert an arbitrary Fahrenheit temperature T_F to its corresponding value on the Celsius scale, one must first determine the difference between the given temperature T_F and the freezing point of water on the Fahrenheit scale. This difference is simply $T_F - 32$ in Fahrenheit degrees. To convert this difference to Celsius degrees we have

$$(T_F - 32) \text{ Fahrenheit degrees} \times \frac{5 \text{ Celsius degrees}}{9 \text{ Fahrenheit degrees}} \ .$$

Since the freezing point of water on the Celsius scale is 0, this difference now represents the Celsius temperature T_C and, hence,

$$T_C = (T_F - 32) \times \frac{5}{9} \ .$$

This equation can be rewritten in the following ways

$$T_C = \frac{5}{9} (T_F - 32)$$

and

$$T_F = \frac{9}{5} T_C + 32 \, ,$$

which permit a Celsius temperature to be converted into a Fahrenheit temperature and vice versa.

QUESTIONS

1. Under what circumstances might you use:
   a. the Celsius temperature scale?
   b. the Fahrenheit scale?
   c. the Kelvin scale?

2. Set up and define your own temperature scale. You may not use any of the characteristics of the Fahrenheit, Celsius, or Kelvin scales. Derive an equation relating your temperature to the corresponding temperature on one of these scales.

3. Which temperature is indicative of the hottest object; 20°F, 20°C, or 20 K? Explain.

PROBLEMS

1. At what temperature will the reading on a Fahrenheit thermometer be the same as that on a Celsius thermometer?

2. Find the value of absolute zero on the Fahrenheit temperature scale.

3. Derive an equation relating the Fahrenheit temperature T_F of an object to its Kelvin temperature T_K.

The Equation of State of a Gas

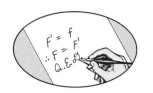

In the course of his experiments on gaseous materials, the English physicist Robert Boyle discovered that if pressure is exerted on a gas at constant temperature, the gas responds by reducing its volume by exactly the same factor that the pressure was increased. Thus if the pressure on a gas at constant temperatures is doubled, its volume will be cut in half. If the pressure is reduced to one-third of its original value, then the volume of the gas is increased by a factor of three, and so on. This relationship between pressure P and volume V is known as Boyle's law and can be written as

$$PV = \text{constant value}$$

for a given gas. If a gas changed from one pressure and volume to a second pressure and volume with the temperature held constant then the pressure P_1 and volume V_1 of the gas in state 1 are related to the pressure P_2 and the volume V_2 of that same gas in state 2 as follows

$$P_1 V_1 = P_2 V_2$$

if $T_1 = T_2$.

If the pressure on a gas remains constant, then the volume of the gas will increase in direct proportion to the absolute or Kelvin temperature T of the gas. If the Kelvin temperature doubles, then so does the volume of gas. If the temperature is cut in half, then the volume of gas suffers the same fate. This behavior pattern was discovered by the French physicist Jacques Charles in 1787 and is known as Charles's law. This law can be written as

$$\frac{V}{T} = \text{constant value} ,$$

which tells us that if a gas is changed from a given volume V_1 and temperature T_1 to a second volume V_2 and temperature T_2 at constant pressure, the two volumes and temperatures must obey the following equation

$$\frac{V_1}{T_1} = \frac{V_2}{T_2}$$

if $P_1 = P_2$.

The only way that a given gas can simultaneously satisfy both Charles's law and Boyle's law is for

$$\frac{P_1 V_1}{T_1} = \frac{P_2 V_2}{T_2}$$

or more generally

$$\frac{PV}{T} = \text{constant value} .$$

These equations permit us to describe mathematically the transition of a given gas from one specified set of pressure, temperature, and volume values to a second set of pressure, temperature, and volume values. Because they describe the

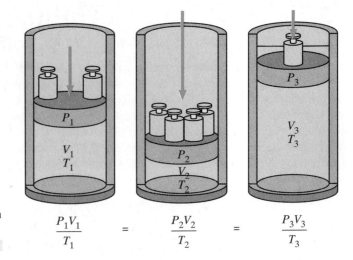

FIGURE 8.15 ··········

The Equation of State of a Gas   If a certain quantity of gas at a given temperature, pressure, and volume has one or more of these ''state'' conditions changed, the changes must occur in such a way that the quantity *PV/T* remains at a constant value.

$$\frac{P_1 V_1}{T_1} = \frac{P_2 V_2}{T_2} = \frac{P_3 V_3}{T_3}$$

''state'' of a given gas, these equations are referred to as the equations of state for the gas (Figure 8.15).

QUESTIONS
·········

1. Suppose that the density of a gas decreases. What will happen to the temperature of the gas if its pressure remains constant?

2. Verify this statement: ''The only way that a given gas can simultaneously satisfy both Charles's law and Boyle's law is for

$$\frac{P_1 V_1}{T_1} = \frac{P_2 V_2}{T_2} .$$

3. Is the value of *PV/T* the same for every gas system? Explain.

PROBLEMS
·········

1. If the pressure on 3 liters of gas at 300 K is tripled, to what value must the temperature of the gas be adjusted in order for the volume of the gas to remain at a value of 3 liters?

2. The surface temperature of the planet Venus is about 2.5 times that of the earth and its atmospheric pressure is 90 times that of the earth's at sea level. If one liter of gas from the earth's atmosphere at sea level were subject to the conditions on Venus, what volume would it have?

3. Ten liters of gas at 300 K and one atmosphere of pressure are taken to the center of the sun where the pressure is about 3.4×10^{11} atmospheres and the temperature is about 15×10^6 K. What volume will the 10 liters of gas now have?

9 VIBRATION AND SOUND

$\mathcal{O}$ne of the truly significant links that human beings have with each other and indeed with nature itself is through the phenomenon we call *sound.* For human beings the spoken word is perhaps our most important mode of communication and can range from the shouts of derision tendered at a referee or umpire for a poor call to the intimate whispers that pass between lovers. Each time we speak or are spoken to we employ sound. Nor are we alone in our use of sound. Our fellow creatures on the planet have fashioned some extraordinarily fascinating uses for this phenomenon. Bats employ a sound-echo system to navigate in the black of night, and ocean creatures find each other and their prey in the even blacker depths of the ocean by sounds emitted and received in that environment. Nature's nonliving constituents provide us with an impressive array of sounds as well. When we were children, who among us, for example, didn't scurry under a bed or into a parent's comforting arms on hearing a particularly loud clap of thunder? The howling of the wind, the roar of the surf pounding on an ocean shore, and the ''babbling'' of a stream or brook are but a few examples of how sound is not only a key item in the repertoire of the talents possessed by living creatures, but is also a phenomenon inherent to the physical world as well. Thus it is a prime candidate for our scientific investigations and understanding.

9.1 Vibrations

To develop a physical description of the phenomenon of sound, let us first consider a type of motion called vibratory motion, which was briefly alluded to in an earlier chapter. In a general sense, any type of repeating or periodic motion can be classified in physics as *vibratory motion* or simply as a *vibration.*

One of the simplest and most familiar examples of vibratory motion is that produced by a spring. Whenever a spring is stretched or compressed to a length that is different from its ''equilibrium'' length, the spring produces an internal restoring force, which acts in such a way as to attempt to ''restore'' the original length of the spring. Thus when the spring is stretched, the restoring force acts to reduce the length of the spring and when the spring is compressed, the restoring force acts to increase the length of the spring (Figure 9.1). Moreover, the magnitudes of these ''restoring'' forces increase in direct proportion to the magnitude of the stretching or deformation. Thus if a spring is stretched out to twice its equilibrium length by a given force, it will take about three times as much force to stretch the spring twice further or to a length four times as long as its equilibrium length. Similarly, if a spring is compressed to one-half of its equilibrium length by a given force, it will take about $\frac{3}{2}$ as much force to compress the spring to one-half of that length or $\frac{1}{4}$ of the original equilibrium length of the spring.

Suppose now that we attach a block to a spring having some equilibrium length as shown in Figure 9.2. If we pull sideways on the block so as to stretch out the spring and displace the block from its equilibrium position, the spring will exert a restoring force on the block in a direction opposite that of the stretching. If the block is released, and there are no other forces such as friction acting on the block, the block will, in response to the restoring force, accelerate to the left. As the block approaches its equilibrium position, however, there is no counter force to stop its motion and it ''overshoots'' its equilibrium position. As it passes to the left of that position, it is then moving in such a way as to compress the

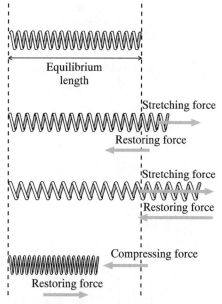

FIGURE 9.1 ....

Restoring Force   When a spring is stretched or compressed to a length different from its equilibrium length, the springs exerts a force proportional to the amount of the stretching or compression. Such forces are called restoring forces.

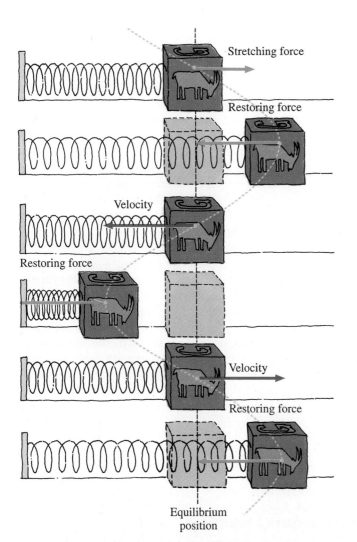

Stretching force

Restoring force

Velocity

Restoring force

Velocity

Restoring force

Equilibrium
position

FIGURE 9.2

Simple Harmonic Motion If the block is pulled from its
equilibrium position, the restoring force from the spring will
cause the block to move to the left past its equilibrium position,
at which time the spring will exert a restoring force that causes
the block to slow down, stop, and then overshoot to the right.
As the block overshoots to the right, the restoring force from
the spring causes the block to slow down, stop, and then move
to the left, thus beginning the overall process anew. The
resulting left and right motion of the block is called simple
harmonic motion.

spring. The spring responds to this attempt at compressing its length from the
equilibrium value by exerting a restoring force to regain its equilibrium length.
In response to this force the block now slows down, stops, and then begins to
accelerate to the right. The motion of the block to the right causes it to again
overshoot the equilibrium position, and once more the spring exerts a restoring
force back to the left. As a result, if no other forces are present, the block will
continue to oscillate back and forth, each time overshooting its equilibrium position and each time being yanked in the opposite direction by the spring's
restoring force when it does. The result is a to and fro type of motion that physicists
refer to as *simple harmonic motion.*

The concept of simple harmonic motion provides a most useful and versatile
mathematical model for analyzing a variety of motion phenomena in nature
including such seemingly disparate phenomena as orbital motion, the motion of
a pendulum, and some aspects of the behavior of solids (Figure 9.3). In each of
these instances, the motion can be thought of as an ongoing ''up and down'' or
''back and forth'' motion. This is easy to see for the case of a weight bouncing
up and down or back and forth at the end of a spring, but it is not as obvious for
such as the pendulum and an orbiting planet. Suppose, however, we observe the

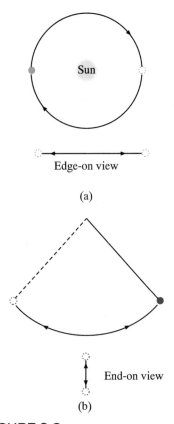

FIGURE 9.3

Examples of Simple Harmonic Motion (a) A planet in orbit about the sun; (b) a pendulum.

earth's motion about the sun from a position outside of the earth's orbit, yet in the plane of the earth's orbit. Under these circumstances the earth would appear to oscillate back and forth from one side of the sun to the other as it moved along in its orbit. Similarly, when a pendulum bob is viewed either "head on" or "end on," its motion also appears as an up and down motion.

Perhaps the most important aspect of simple harmonic motion, however, lies in its intimate relationship to a phenomenon called *wave motion*. If we attach a frictionless pen to the block described above and let it trace out the back and forth motion of the block on a moving strip chart as shown in Figure 9.4, the result is a "squiggly" looking curve, which is referred to as a *sine curve* or a *sine wave* because of its intimate relationship with the sine function in trigonometry. As we shall see, various types of oscillating systems produce oscillating types of disturbances, which in turn propagate outward in a manner and form not dissimilar to the wavy tracing produced on our hypothetical length of strip chart paper.

Simple harmonic motion can be characterized by several observable parameters. The first of these is the *frequency* or number of up and down or back and forth cycles that occurs in a given interval of time. Frequencies are usually measured in cycles per second or *hertz*, where one hertz (Hz) equals one cycle per second. A second observable characteristic of simple harmonic motion is the length of time it takes for one up and down or back and forth cycle to occur. This parameter is called the *period* of the simple harmonic motion and is most commonly used in describing the oscillations of a pendulum and the orbital motion of a celestial object. The period and frequency of a simple harmonic oscillator are inversely related to each other according to

$$\text{period} = \frac{1}{\text{frequency}}.$$

A simple harmonic oscillator having a frequency of 10 hertz, for example, would have a period of $\frac{1}{10}$ second. Conversely, a simple harmonic oscillator with a period of 2 seconds would exhibit a frequency of

$$2 \text{ seconds} = \frac{1 \text{ cycle}}{\text{frequency}}$$

or

$$\text{frequency} = \frac{1 \text{ cycle}}{2 \text{ seconds}}$$

FIGURE 9.4

Waves from a Simple Harmonic Oscillator If the back and forth motion of a simple harmonic oscillator is plotted in time, the result is a wave type of curve called a sine curve or sine wave.

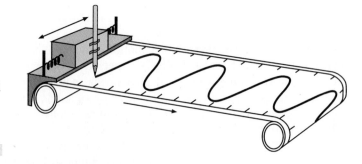

$$= \frac{1}{2} \text{ cycle/second}$$

$$= \frac{1}{2} \text{ hertz} \ .$$

Since the equilibrium position or equilibrium point lies halfway between the maximum displacement positions of the motion of a simple harmonic oscillator, we define the *amplitude* of the motion as the distance between the equilibrium point and either of the maximum displacement points. If the weight at the end of a spring oscillates between 5 cm and 15 cm about an equilibrium position of 10 cm, then the amplitude of that motion would be either 15 cm − 10 cm = 5 cm or 10 cm − 5 cm = 5 cm (Figure 9.5).

While the concept of simple harmonic motion finds considerable use in the analysis of many types of simple motion phenomena, its most powerful application arises from the fact that *any* periodic, oscillating, or vibrational motion, no matter how complex, can be represented as a set or "series" of simple harmonic oscillations, which have the proper set of frequencies and amplitudes.

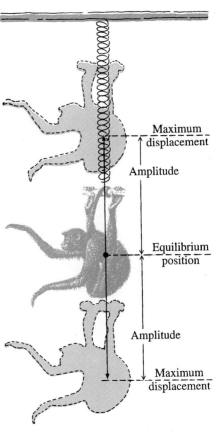

FIGURE 9.5 · · · · · · · · · · ·
Amplitude   The amplitude of a simple harmonic oscillator is the distance between the maximum displacement of the oscillation and the equilibrium position of the oscillation.

WORKED EXAMPLE

A certain tuning fork vibrates with a frequency of 500 hertz. What is the period of this vibration?

SOLUTION: Since the period of the vibration is given by

$$\text{period} = \frac{1}{\text{frequency}} \ ,$$

we have

$$\text{period} = \frac{1 \text{ cycle}}{500 \text{ cycles/sec}}$$

$$= \frac{1}{500} \text{ sec}$$

$$= 0.002 \text{ sec} \ .$$

1. What is simple harmonic motion? How is it related to wave motion?
2. Give some examples of simple harmonic motion.
3. Describe some of the parameters used to characterize simple harmonic motion.
4. Explain why the period of a simple harmonic oscillator is equal to 1/frequency.
5. Compare and contrast vibratory motion with simple harmonic motion.

REVIEW 9.1 QUESTIONS

1. A simple harmonic oscillator has a frequency of 1 hertz. How many cycles does the oscillator go through in
   a. one minute?   **b.** one hour?   **c.** one day?
2. The cycle of lunar phases occurs over a period of about 29.5 days. What is the frequency of this phenomenon?

9.2 Wave Motion

Suppose we now connect an oscillator to a very long spring as shown in Figure 9.6. If the oscillator is moving up and down, an up and down pattern is induced in the spring, which travels outward from the oscillator. This traveling pattern is referred to as a wave. Because the up and down motion of the wave pattern is perpendicular to or "transverse" to the direction in which the wave is moving, such waves are called *transverse waves.* The amount that the spring is displaced upward or downward from its straight line configuration is referred to as the *amplitude* of the wave. The distance between successive corresponding parts of the wave such as the distance between successive crests or successive troughs is called the *wavelength.* The *frequency* of the wave is defined as the number of wave crests that pass a given point in a given time. Notice the similarities between these parameters and those defined above for a simple harmonic oscillator. Also, the regular "wavy" pattern of a transverse wave is identical in format to the sine curve or sine wave described earlier.

If we now turn the oscillator at the end of the spring so that it is now moving back and forth along the axis of the spring we find that a second type of wave is generated along the axis of the spring. This wave consists of a series of compressions and stretchings of the coils in the spring, which also move outward from the oscillator (Figure 9.7). This type of wave is called a *longitudinal wave* due to the fact that for such waves the direction of the wave motion is the same as that for the wave pattern, or if you prefer, the wave motion pattern is "along" the direction of the wave's motion. Like transverse waves, longitudinal waves can be characterized by their frequency and wavelength. In the case of longitudinal waves, however, it is the successive higher density condensations and lower density rarifications rather than crests and troughs that are employed to define the wave's characteristic parameters of frequency and wavelength.

Let us now suppose that we are observing some sort of wave motion having a known frequency and wavelength. If we start a timer with the passage of a

FIGURE 9.6 ⋯⋯⋯⋯

The Transverse Wave   "Up and down" disturbances that are perpendicular to the direction in which the disturbance propagates are called transverse waves. In this case the transverse waves in the spring are being produced by the up and down motion of the oscillator on the left.

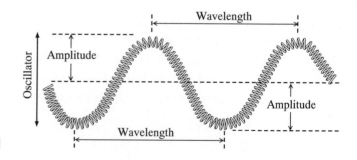

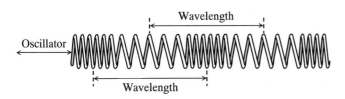

FIGURE 9.7 ···········

The Longitudinal Wave ''Back and forth'' disturbances that produce successive condensations and rarifications of the medium in which they are propagating and which are along the direction the disturbance is traveling are referred to as longitudinal waves.

given wave crest or compression and stop it some time later, the number of crests or compressions that will have passed by the point in that time will be the frequency times the elapsed time. Thus, if the frequency of a wave motion is 10 hertz, then the total number of crests or complete cycles that will have passed a given point in an elapsed time of 5 seconds is

$$\text{total cycles} = \text{frequency} \times \text{elapsed time}$$
$$= 10 \text{ cycles/second} \times 5 \text{ seconds}$$
$$= 50 \text{ cycles} .$$

In that same time the first crest will have moved a distance equal to the number of crests times the wavelength. Thus the distance traveled by first crest equals the number of wave crests × wavelength, and since the number of wave crests is equal to the frequency × elapsed time, we have the result that the distance traveled by first crest is equal to the frequency × elapsed time × wavelength. This relationship can be written as

$$\frac{\text{distance traveled by first crest}}{\text{elapsed time}} = \text{frequency} \times \text{wavelength} .$$

The distance traveled by the first crest/elapsed time is just the velocity of the first crest or, in effect, the velocity of the wave itself (Figure 9.8). Thus, for any

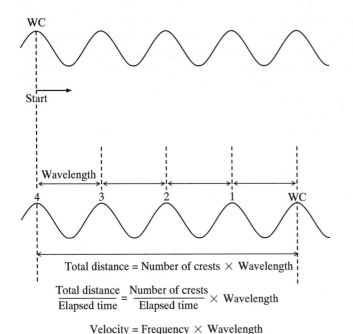

FIGURE 9.8 ···········

The Velocity of a Wave Disturbance As the wave crest WC moves to the right from the starting point, the distance it travels is equal to the wavelength × the number of crests that pass the starting point. If the distance is traveled in a given amount of elapsed time, the resulting wave velocity is equal to the product of the wave's frequency and its wavelength.

Wave A

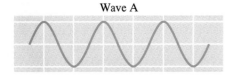

Wave B

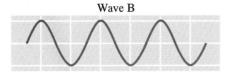

Enhanced wave

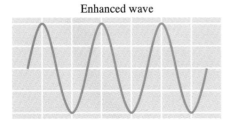

FIGURE 9.9 ···········
Constructive Interference If the crests and troughs of a first wave pattern are coincident with those of a second, the wave effect is reinforced and the waves are said to interfere constructively with each other.

Wave A

Wave B

No net wave

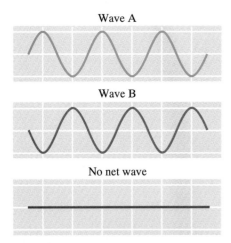

FIGURE 9.10 ···········
Destructive Interference If the crests of a first wave pattern are coincident with the troughs of a second and vice versa, the wave effect is canceled and the waves are said to interfere destructively with each other.

wave motion, we have the basic relationship that

$$\text{velocity of wave} = \text{frequency} \times \text{wavelength} \qquad \textbf{9.1}$$

or

$$v = f \times \lambda \ .$$

Note that the velocity described here is the velocity of the wave along the spring and not a velocity of the coils of the spring itself. If we were to number the spring coils and trace their motion as a wave passed by, we would find that the coils would jump up and down as a transverse wave passed by and oscillate back and forth as a longitudinal wave passed by. In both instances, however, there would be no net motion of the spring coils in any direction. In short, the coils undergo simple harmonic motion as they transmit the wave, but exhibit no net displacement after the wave passes.

Both transverse and longitudinal waves also exhibit a phenomenon called *interference.* Interference arises out of the fact that two or more waves can be at the same place at the same time. Thus if we toss a pebble into a pond, the disturbance caused when the pebble strikes the water will produce a series of concentric ripples or waves. If we toss a second pebble into the pond, a second set of waves will be produced that will "interfere" or interact with the first set of waves in the pond.

If the crest of a wave from the first pattern becomes coincident with the crest of a wave from the second pattern, then the two waves are said to exhibit *constructive interference* and display a wave effect that is enhanced or reinforced over the effects of the individual waves by themselves (Figure 9.9). On the other hand, if a crest from the first pattern becomes coincident with a trough from the second pattern, then the waves are said to exhibit *destructive interference* and in effect destroy each other's wave pattern at the coincidence point (Figure 9.10). Various intermediate levels exist in which constructive and destructive interference can occur in a given situation. Thus, the interaction of two or more waves will produce an interference "pattern" consisting of regions of wave enhancements, wave neutralizations, and all stages in between (Figure 9.11). Moreover, this interference pattern can change in time as the waves move past each other and alter the positions of the regions of constructive and destructive interference. A similar interference phenomenon is encountered with longitudinal waves. In this instance, the condensations in such waves will either be reinforced or canceled by the interference process.

One important manifestation of the interference process is the phenomenon of a *standing wave.* If we attach one end of a string to a wall and attach the other end to an oscillator and emit a single wave from the oscillator, the wave will travel down the string and, upon reaching the wall, will rebound or reflect off the wall back toward the oscillator. Suppose we now emit a series of waves along the string. The first wave as it reflects off the wall and begins its return now runs into and interferes with the oncoming waves. If the waves all have the same wavelength, the reflected waves will interfere with the incident waves in such a way as to create a wave disturbance that does not move along the string, but instead appears to oscillate in a fixed or "standing" position (Figure 9.12).

The positions along the standing wave that do not experience any displacement from the string's "undisturbed" or equilibrium position are referred to as *nodes* or *nodal points.* The other points along the string exhibit various amounts of displacement from the string's equilibrium position. The points along the string that experience the most displacement from the string's equilibrium position are

FIGURE 9.11 ·········
An Interference Pattern The two sets of concentric waves produce a pattern of light and dark areas where the waves constructively and destructively interfere.

called *antinodes*. If the distance between the oscillator and the wall is some length *L*, then standing waves will be created in the string for any wavelength that has a value equal to 2*L* divided by any positive integer. Hence, the simplest standing wave that can be created in a string of given length *L* is one having a wavelength of 2*L*/1 or 2*L*. The next possible standing wave occurs at a wavelength of

Wave A (incident)

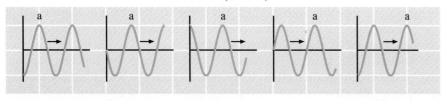

Wave B (reflected)

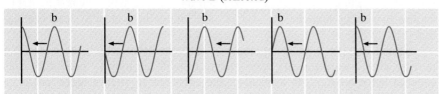

Resultant standing wave

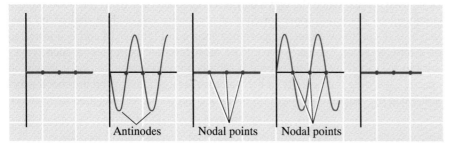

Antinodes   Nodal points   Nodal points

FIGURE 9.12 ·········
The Standing Wave As an incident wave passes a reflected wave moving in the opposite direction, the two waves interfere constructively and destructively to produce a ''standing'' wave that oscillates about a set of nodal points.

FIGURE 9.13 ···········

The Doppler Effect If a series of waves originates from a moving source, the waves "crowd up" in the direction of the motion and "thin out" in the opposite direction. An observer along the line of motion at point A will thus see a wavelength that is shorter than that emitted by the source. An observer at point B will see a wavelength that is longer than that emitted by the source.

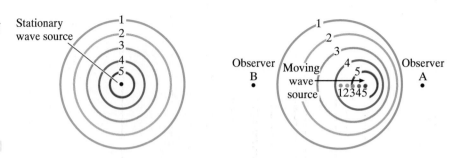

$2L/2 = L$, the next at $2L/3$, and so on. Standing waves can be set up whenever the wave medium is confined in some way. Among the more familiar examples of such standing waves are those induced in wind and string instruments. The "strings" of stringed instruments such as violins, pianos, and guitars are confined by being tied down at each end, and the air of the wind instruments such as the pipe organ, bassoon, or flute is confined in hollow tubes.

Suppose we now construct an experiment in which individual drops of water can be released into a large quiet pool from an eyedropper just above the water level in the pool. A single drop of water released into the pool will cause a disturbance or vibration that will propagate outward from the drop's impact point in the form of a single circular wave front along the surface of the water. If two drops are released successively into the pool, two concentric outwardly moving wave fronts will result, and so on.

Suppose, however, the eyedropper moves between the impact time of the first drop and that of the second. We now find that the second circular wave front is no longer concentric with the first, but rather is offset by a distance equal to the distance the eyedropper has moved between the impacts. The distance between the two wave crests along the direction of the eyedropper's motion is less than for the case of the stationary eyedropper and is larger in the direction opposite that of the eyedropper's motion. If we create a series of waves from a moving eyedropper, the effect is thus to create a "crowding" of waves in the direction of the eyedropper's motion and a "thinning out" effect in the opposite direction. This effect is called the *Doppler effect* and has a great many interesting applications in various branches of science (see Figure 9.13 and Physics Formulation 9.1).

As the dropper velocity approaches the velocity at which the waves are moving, the forward moving waves begin to build up into a "pile" of waves called a *wave barrier*. As a result of the existence of such a wave barrier, a considerable amount of additional effort is required to thrust the wave source through this "wave stack" in order for the wave source to exceed the velocity of its own waves. Once the wave barrier is breached, however, the wave pattern produced by the eyedropper moving at a velocity larger than that of its own waves takes on the V-shaped appearance shown in Figure 9.14 and is referred to as a bow wave or "shock" wave. The most familiar examples of such waves are the V-shaped wave left behind a rapidly moving speedboat as it moves at speeds exceeding the velocity of the water waves it generates and the shock wave generated by "supersonic" aircraft moving at speeds faster than their own sound.

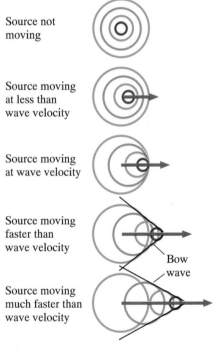

FIGURE 9.14 ···········

The Bow Wave As the velocity of the wave source approaches and exceeds the velocity of the waves it generates, a V-shaped "front" of waves develops which is called a bow wave.

PHYSICS FACET

9.1 Music

There is not a single human culture on the face of the planet that does not enjoy some version of the art form we call music. Whether it is a rock band in a London night club or an Australian aborigine playing his digeredoo, music in some form is being played by someone somewhere every minute of every day.

For the most part, music, as it is composed and played in Western culture, is based on the so-called "diatonic" scale of musical "notes," each note of which has a characteristic frequency or "pitch." The frequencies of the various musical notes are chosen so as to produce the greatest amount of musical "harmony" or "concordance," i.e., sounds that are pleasing to the human ear. The familiar musical notes of "do," "re," and "mi," in the C major scale for example, have frequencies, respectively, of 264 hertz, 297 hertz, and 330 hertz (see the table).

Musical notes are produced by three general types of musical instruments. The "stringed" instruments or "strings" create their musical sounds from vibrating strings. Examples of stringed instruments include the violin, banjo, and guitar. The "wind" instruments generate their musical sounds when an air column within the instrument is made to vibrate, as in the case of a pipe organ, flute, or trumpet. Musical sounds can also be produced by "percussion" instruments, which consist of a surface or membrane that is struck in an appropriate fashion by the musician. Drums and cymbals are examples of this type of instrument.

In each case, the various musical notes and their frequencies can be generated by altering some aspect of the instrument. A violinist, for example, can alter the pitch of the violin's sound by moving his or her fingers to various locations along the strings of the violin. A trumpeter creates various notes by altering the nature of the vibrating air column in the trumpet through the use of valves, a drummer can change the sound of the drum by striking different portions of the drum surface, and so on.

When a note is played on a given musical instrument, that instrument generates not only the sound frequency associated with that particular note, but also an entourage of higher frequency, lower intensity sounds called "harmonics" and "overtones." The harmonics and overtones are

⋯ The Basic Types of Musical Instruments (a) Percussion (Buddy Rich on drums); (b) string (Eric Clapton on guitar); (c) wind (Louis Armstrong on trumpet).

(a)   (b)   (c)

PHYSICS FACET

Continued

unique to each type of instrument and comprise what we refer to as the musical quality of a given instrument. It is this uniqueness of musical quality that enables us to tell the difference between music played on a guitar and a banjo, even though both are string instruments and both are sounding the same series of notes.

Musical instruments can be combined in a variety of ways, giving rise to the impressive array of musical art forms that have developed around the world throughout human history. The question of what constitutes "music to our ears," however, is one that is not readily answered. A meaningless assemblage of dissidence to one culture is another culture's Top Ten tune. What is beautiful music to one person may seem dull and boring to someone else. The choice as to whether we load Lawrence Welk or Iron Maiden into our compact disk player then becomes a matter of personal aesthetics, a realm that is decidedly an unscientific one.

The Diatonic C Major Scale

Note	Letter Designation	Frequency (hertz)
do	C	264
re	D	297
mi	E	330
fa	F	352
sol	G	396
la	A	440
ti	B	495
do	C′	528

WORKED EXAMPLE

One end of a string 0.5 meters long is attached to a wall. At the other end is an oscillator that is producing waves traveling down the string at a velocity of 2 meters/second. Find (a) the wavelength of the simplest standing wave that the oscillator can induce in the string and (b) the frequency at which the "free" end of the string should be oscillated in order to produce such a wave.

SOLUTION: (a) In general, the simplest standing wave that can be induced in a string of length L is one having a wavelength of $2L/1$ or $2L$. For $L = 0.5$ meters, the wavelength of the simplest standing wave that can be created is equal to 2×0.5 meters or 1 meter.
(b) For any wave motion

$$\text{velocity of wave} = \text{frequency} \times \text{wavelength} .$$

If the wavelength is 1 meter and the wave velocity is 2 meters/second, we have

Continued

> ## WORKED EXAMPLE
>
> *Continued*
>
> $$2 \text{ meters/second} = \text{frequency} \times 1 \text{ meter} .$$
>
> Solving for the frequency yields
>
> $$\text{frequency} = (2 \text{ meters/second})/1 \text{ meter}$$
> $$= 2 \text{ Hz} .$$
>
> The string should be oscillated at a rate of 2 cycles/second.

1. Sketch a wave and indicate on your drawing its wavelength.
2. Discuss the differences between a longitudinal wave and a transverse wave. How are these types of waves similar?
3. Compare and contrast a wave and the vibration that produces the wave.
4. Describe the phenomenon of interference. Under what conditions is interference constructive? Under what conditions is interference destructive?
5. What are standing waves? How are such waves produced? Give an everyday example of a standing wave.

1. A certain wave has a frequency of 50 Hz and a wavelength of 4 m. Find the velocity of this wave.
2. Sound waves travel in air at a rate of roughly 300 m/sec. If a given sound has a frequency of 400 Hz, what is the wavelength of this sound?

9.3 Sound

The phenomenon we refer to as "sound" is in effect the array of longitudinal waves generated by the wide variety of vibrating and oscillating objects existent in the physical world. The vibrations of these objects create longitudinal "pressure" waves in which the medium through which the sound is transmitted is alternately compressed and rarified as each sound wave passes. In this context, then, sound waves require the presence of a physical medium of some sort, such as water or air, in order to be propagated. For this reason, the phenomenon of sound is nonexistent in a vacuum chamber or on the surface of an airless world such as the moon.

The speed with which sound waves travel varies considerably with the characteristics of the medium in which they move. As a general rule, the speed of sound is most rapid in solid substances where the atoms are most closely packed and are hence able to pass a disturbance along to their neighbors more efficiently.

In a similar fashion, we find that the speed of sound is usually slower in liquids than in solids, and moves least rapidly of all in gases, where the atoms are largely disjoint from each other. Thus the speed of sound in air is about 300 m/sec, increases to nearly 1500 m/sec in water, and is over 5000 m/sec in solid iron and aluminum.

Because it takes energy to create the condensations and rarifications of a medium that characterizes sound waves, a given sound will have associated with it a certain amount of energy. If these sound waves are incident on some sort of surface at some given rate, then the intensity of the sound is defined as

$$\text{intensity} = \text{energy/(area} \times \text{elapsed time)} \ .$$

Recall that energy/elapsed time is equal to power; therefore, we have

$$\text{intensity} = \text{power/area} \ .$$

Thus the intensity of a sound wave is simply the power per unit area delivered by the wave and as such is measured in watts/meter^2.

Intensities of sound waves can range over many orders of magnitude. The human ear, for example, can respond to sound waves having power per unit area intensities ranging from about 10^{-12} watts/square meter for the faintest sounds we can detect up to 1 watt/square meter at the threshold of pain. Because of this very large ratio in intensities, scientists use a more compacted scale called the *decibel scale* in which a difference of ten decibels or one bel in sound level corresponds to a factor of 10 in actual intensity (see Physics Formulation 9.2). Thus a difference of 20 decibels corresponds to a factor of 10^2, 30 decibels to a factor of 10^3, and so on. The zero point of this scale is defined as a sound intensity of 10^{-12} watts/square meter, which is the approximate threshold level of human hearing. Some representative sounds and their associated intensities and decibel levels are presented in Figure 9.15.

Despite what our ears would tend to tell us, the amount of actual energy residing in even the most intense sounds that we hear is rather small when compared to the energy we receive from other sources. The sound level of 140 decibels which we would receive standing 30 meters behind a jet airliner at takeoff and which would very probably cause damage to our hearing has about the same intensity in watts/m^2 as the intensity of the light energy received from a 100-watt light bulb situated at a distance of about 30 cm.

As sound waves travel through a given medium, they can strike a surface and "bounce off" or be reflected from that surface. The most familiar example of such sound reflection is the echo effect. If we speak or yell in a relatively closed space such as an empty room, cave, or canyon, the sounds we utter travel outward, strike one of the confining surfaces, and bounce back toward us to be heard as an echo. If there are several reflecting surfaces present such as in an auditorium, for example, it is possible for a sound wave to be prolonged by multiple reflections. This phenomenon is called "reverberation" and a great deal of time and effort is spent designing lecture halls and auditoriums in such a way as to either minimize or altogether eliminate this undesirable sound effect.

Sound reflection is also the basis for a variety of so-called "sonar" devices, which are employed by both humans and other creatures alike. Very simply, in a sonar device, a pulse of sound is emitted. If that pulse strikes some object, the reflected sound will be picked up by the sonar detection gear. The distance, rate

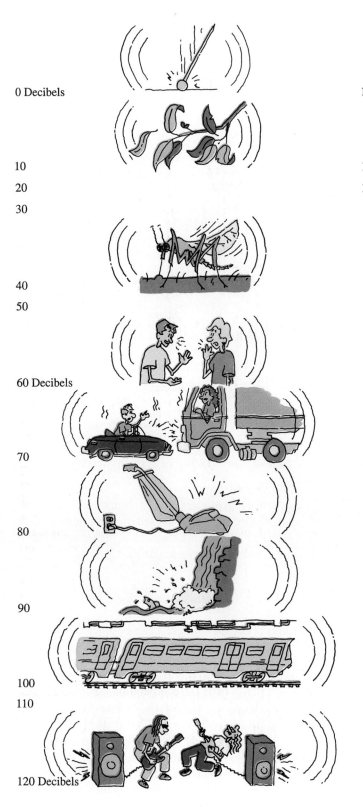

0 Decibels

10

20

30

40

50

60 Decibels

70

80

90

100

110

120 Decibels

Faintest
audible
sounds
1.0×10^{-12} W/m²

Rustling
of leaves
1.0×10^{-11} W/m²

1.0×10^{-10} W/m²

1.0×10^{-9} W/m²

Buzzing
mosquito
1.0×10^{-8} W/m²

1.0×10^{-7} W/m²

Ordinary
conversation
1.0×10^{-6} W/m²

Heavy
street
traffic
1.0×10^{-5} W/m²

Loud
vacuum
cleaner
1.0×10^{-4} W/m²

Base of
a large
waterfall
1.0×10^{-3} W/m²

Subway
train
1.0×10^{-2} W/m²

1.0×10^{-1} W/m²

Threshold
of pain
1.0 W/m²

FIGURE 9.15 · · · · · · · · · ·

The Decibel Scale   The span of 120
decibels between the detection threshold
of human hearing and the threshold of
pain for human hearing corresponds to a
ratio of 10^{12} in actual sound intensity or
power/unit area.

9.2 Speech and Hearing

One of the more amazing aspects of human beings is their ability to not only detect or "hear" sounds but to generate them as well.

Human speech is initiated by air rushing past a set of "vocal cords" located in the throat. As air is exhaled past these cords, they are set vibrating. These sound vibrations then move up the throat and out of the mouth. As they do so, they are modified by the mouth and nasal cavities in the head. The well-known versatility of the human voice in generating sound arises from two factors. First of all, we have the ability to vary the tension in the vocal cords, thus controlling the vibrations that result. The so-called "resonant" nasal and oral cavities can also be changed so as to exert a control over the final sound that emanates from our mouth. The movement of the tongue within the mouth is the most familiar example of the latter type of modulating process.

The human ear, of course, serves the function of sound receptor in the human body. The human ear (see the figure) consists first of the "outer ear," which collects the incoming sound and transmits it to a thin membrane called the eardrum. As the eardrum vibrates in response to these sound waves, bones in the middle ear then transmit those vibrations to the so-called "inner" ear. The inner ear consists of two canals filled with fluid, which are separated by a flexible partition. The inner ear is constructed so that different frequencies arriving from the bones in the middle ear

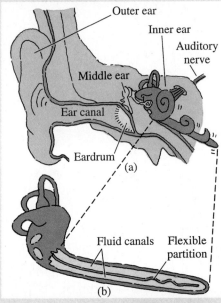

••• **The Human Ear** (a) Overview; (b) the uncoiled inner ear.

cause the partition to flex at different points along its length. This location of this flexing is sensed by nerve hairs located in the inner ear and the information is then passed on to the brain by means of nerve impulses.

of motion, and other information about the "target" can be inferred from the characteristics of the return echo. Sonar is perhaps most familiar to us for its role in warfare as a device to detect submarines. However, a large number of creatures employ similar devices in conducting their affairs. Bats, for example, use high-frequency sounds to navigate and locate their insect prey. Beneath the sea there is a plethora of sounds emanating from the inhabitants of that portion of our biosphere as they go about making a living.

Sound that has frequencies beyond the detectable limit of the human ear is called *ultrasound* and has found great use in medical science. When ultrasound pulses are directed into the human body, the pulses are reflected differently by different types of tissue and fluids within the body. By analyzing the reflected

PHYSICS FACET

Continued

The human ear is remarkable on a number of different counts. Unlike most other sound detectors such as microphones, human hearing does not seem to be appreciably affected by motions or vibrations experienced by the body or by sounds interior to the body, such as the beating of the heart, the rush of air in and out of the lungs, etc. Human hearing permits us to locate the position of a sound through a process called "auditory localization." The range of sound intensity to which a normal human ear can respond spans some 12 orders of magnitude, or a ratio of 10^{12} in actual intensity. The human ear also operates at a very low energy intensity. The sound energy intensity by which we are able to hear normal conversations easily, for example, corresponds to the intensity of the light energy an unaided human eye receives from a distant star. Moreover, we also have the ability to concentrate on specific sounds in the presence of many other sounds. It is this ability that allows us to "tune out" someone who is trying to talk to us while we are watching a game or favorite show on television and enables a parent to hear the cries of a lost child amidst the sounds of a large and noisy crowd of people.

The human ear is also capable of detecting a considerable range of sound frequencies as well. The lowest frequency sounds audible to the human ear have frequencies of about 20 Hz. The high-frequency cutoff for human sound detection occurs at about 20,000 Hz. By contrast, dogs can hear sounds up to 44,000 Hz in frequency and rats as high as 70,000 Hz. The human ear is most sensitive to sounds having frequencies of about 3000 Hz. Sounds that lie outside of this "audible range" of human hearing are referred to as *infrasound* if they have frequencies less than 20 Hz and *ultrasound* if they have frequencies larger than 20,000 Hz.

Many fascinating and important questions can be asked about our hearing. For example, to what extent is human behavior affected by continuous bombardment of sound or "noise pollution"? Can we be affected by so-called "subliminal" sounds—those we do not consciously hear, but which are nevertheless perceived by the subconscious portions of our minds? What is the mechanism by which we sort out and respond to the nuances, such as sarcasm, associated with the words we speak to one another? What makes two actors, each doing the same lines from the same play, so vastly different? Unfortunately, these and other topics can only be partially described by the physics of sound, as we once more enter a region of human experience that has thus far defied an accurate quantitative description.

sound in an appropriate manner, a "picture" can be generated of the conditions inside the body without having to resort to the more dangerous procedures of using x rays or performing surgery in a potentially traumatic medical operation (Figure 9.16). Sound wave reflection in a confined or closed medium can result in the generation of standing sound waves in that medium, and as such, the effect is basic to the musical sounds produced by wind and string instruments.

The wave nature of sound results in a number of other interesting "sound effects." For example, if two sound waves of unequal frequency interfere with one another, there will be constructive and destructive interference of the higher and lower density regions of the sound waves just as there is constructive and destructive interference for the crests and troughs of transverse waves. This

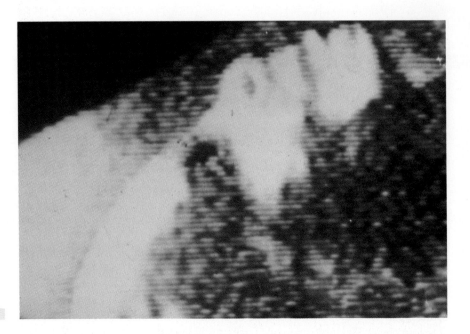

FIGURE 9.16 ⋯⋯⋯⋯

An Ultrasound ''Picture'' of an Unborn Child When pulses of high-frequency sound are directed into the human body, the pulses are reflected differently by different types of tissues and fluids within the body. By using appropriate detectors, a ''sonic picture'' of the region of interest can be produced without resorting to the more dangerous procedures of surgery or X ray use.

interference phenomenon results in a periodic series of variations in sound intensities, which are referred to as *beats*. The beat occurs when the condensation of two or more sound waves interfere constructively. The frequency of a beat generated by two intersecting sound waves is simply equal to the difference in their respective frequencies (Figure 9.17). Thus two sounds having frequencies of 300 and 350 hertz, for example, will produce a beat with a frequency of 350 hertz − 300 hertz = 50 hertz.

Associated with every object is a *natural frequency* or frequency at which the object will vibrate if it is struck by an outside blow. The difference in the sound produced by a silver bell when struck as opposed to that of a lead bell is

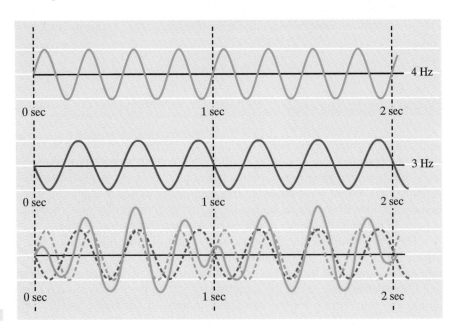

FIGURE 9.17 ⋯⋯⋯⋯

Beats If two waves having different frequencies interact with one another, a ''beat'' wave is produced that has a frequency equal to the difference in the frequencies of the two interacting waves. In this case the frequency of the beat wave is equal to 4 Hz − 3 Hz or 1 Hz.

FIGURE 9.18 **"Galloping Gertie"** At the time of its opening on July 1, 1940, the Tacoma (Washington) Narrows Bridge, nicknamed "Galloping Gertie", was the third largest suspension bridge in the world. Less than six months later, on the morning of November 7, 1940, winds caused the bridge to vibrate at its natural frequency. The resulting resonant oscillations continued violently for over four hours before the entire bridge span finally gave way and fell over 60 meters into Puget Sound.

due to the fact that the natural frequencies of the two bells are quite different. The natural frequency of a given object depends in a very complex way on the object's composition, size, shape, etc., but when an object is subjected to sound waves having the same frequency as the object's natural frequency, the incoming sound waves in effect constructively interfere with the natural frequency of the object. As a result the object will vibrate with an unusually large amplitude. The sound emanating from a stereo playing "heavy metal" rock music for example, can cause an entire house to vibrationally "shake, rattle, and roll" if the frequency of the stereo sound is equal to the natural frequency of the house. Aircraft can be torn apart by interactions with wind vibrations if care is not taken to avoid a resonance between the natural frequency of the plane and the frequency of the vibrations that the plane may encounter. In 1940 an entire bridge plunged into the Tacoma Narrows along the coast of Washington State largely because winds set the ill-fated bridge vibrating at its natural frequency (Figure 9.18).

There are a variety of devices by which sound waves can be detected or "heard." In each instance, the sound waves are first "picked up" by some sort of membrane that is free to vibrate. As the sound waves strike the membrane, the resulting vibrations are converted by the detector into some other form of energy, usually electrical, which can be more readily transmitted or analyzed. The human ear, for example, converts the sound incident on the eardrum into electric nerve impulses, which are then fed into the brain (Physics Facet 9.2). On the other hand, the membrane in a microphone is hooked into an electric circuit in such a way that the vibrations produced in the membrane by incident sound waves generate small variations in the electric current flowing through the circuit. These variations in current can then be amplified, recorded, and transmitted in a variety of ways that will be described in more detail in later chapters.

> WORKED EXAMPLE
>
> What is the beat frequency that would be observed for two sources of sound having frequencies of 510 Hz and 545 Hz?
>
> **SOLUTION:** The beat frequency is simply the difference between the frequencies of the two sources, which in this case is 545 Hz − 510 Hz = 35 Hz.

1. What is sound?
2. Why do sound waves tend to travel most rapidly in solid materials and least rapidly in gases?
3. Compare and contrast the measurement of sound intensity in watts/m^2 and in decibels.
4. Discuss some of the effects that can arise when sound waves are reflected.
5. Describe the physics of what happens as a jet plane approaches and then goes through the "sound barrier."

1. The energy from a certain sound strikes a wall 2 meters high and 4 meters long at a rate of 1500 J/sec. What is the intensity of this sound?
2. Sound A has an intensity that is 1000 times smaller than sound B. What is the difference in the intensities of these sounds expressed in decibels?

Chapter Review Problems

1. The planet Jupiter orbits the sun once every 12 years. What is the frequency of Jupiter's motion?
2. A weight at the end of a spring bounces up and down once every quarter-second. Find the frequency of this motion in hertz.
3. The Crab Nebula pulsar spins on its axis 30 times a second. What is the period of this motion? What is the frequency?
4. The weight at the end of a spring oscillates from 15 cm above the top of a table to 25 cm above. What is the amplitude of this motion?
5. A wave moves at 100 meters/second and has a frequency of 2 Hz. What is the wavelength for such a wave?

6. A certain wave has a frequency of 5 Hz and a wavelength of 3 m. What is the velocity of this wave?

7. A sound wave moves about 300 m/sec and has a frequency of 15 Hz. What is the wavelength of this sound?

8. How many wave crests will pass a given point in 10 sec as a wave moving at 100 m/sec and having a wavelength of 5 m passes by?

9. A wave has a frequency of 100 Hz. How many crests of this wave will pass by a given point in
   a. one second?   **b.** one minute?   **c.** one hour?

10. One end of a string 2 meters long is attached to a wall. At what frequency should the other end of the string be moved up and down in order to produce a standing wave with two antinodes? Assume the waves travel down the string at 1 m/sec.

11. What are the limits in the wavelength of sounds that can be heard by human beings if we can hear frequencies from 20 to 20,000 Hz? Assume that the speed of sound is 300 m/sec.

12. What is the beat frequency of two sound waves having frequencies of 30 Hz and 25 Hz?

13. Two sound waves interact to produce a beat frequency of 15 Hz. If one of the interacting sound waves has a frequency of 35 Hz, what are the possible frequencies of the second sound wave?

14. A sonar pulse emitted toward a whale returns as an echo 20 seconds later. If the speed of sound in water is 1500 m/sec, how far away is the whale?

15. The diameter of the earth's orbit about the sun is about 300 million kilometers. What is the amplitude of the earth's orbital motion as viewed by an observer located outside of the earth's orbit but still in its plane?

16. Find the ratio of intensities for two sound waves that have relative intensities of 35 and 55 decibels.

17. For a wave moving at 0.1 m/sec, find the wavelengths of the first three possible standing waves that can be induced in a wire 60 cm long.

18. What is the ratio of intensities for two sound waves that have relative intensities of 40 and 80 decibels?

19. A certain sound has a relative intensity of 50 decibels and is 10,000 times more intense than a second sound. What is the level in decibels of the second sound?

20. Sketch a transverse wave having an amplitude of 5 cm and a wavelength of 5 cm.

21. Sound waves at 500 Hz move through a certain medium and exhibit a wavelength of 70 cm. What is the speed of these sound waves?

22. What is the beat frequency that would be observed if the notes mi (E) and la (A) were sounded simultaneously?

23. What possible pair(s) of notes, if any, would produce a beat frequency of 33 Hz when sounded simultaneously?

1. Does a longitudinal wave have an amplitude? Explain.

2. How do you suppose ocean waves are produced?

3. Longitudinal shock waves have been observed to propagate in solids, liquids, and gases, while transverse shock waves will only propagate through solids. Explain why this is so.

4. If a tree collapsed and fell at the center of a forest many kilometers from any human or animal, would there be a sound from that event? Explain.

5. Compare and contrast the artistic and scientific aspects of the phenomenon of musical sound.

6. Where does the energy associated with sound waves come from?

7. Why do you suppose that marine creatures almost always have relatively poor eyesight but well-developed hearing?

8. Do percussion musical instruments make use of standing waves? Explain.

9. Discuss the possibility that waves exist in nature that are neither transverse nor longitudinal.

10. Is rotational motion a form of simple harmonic motion? Explain.

11. What factors limit the ability of the human ear to detect:
    a. very faint sounds?
    b. high-frequency sounds?
    c. low-frequency sounds?

12. Quite often at ballparks and sports arenas, the crowd will do ''the wave.'' How does such a ''wave'' compare to the waves we have described in this chapter?

13. How would the velocity of sound waves moving in a plasma compare with that for sound waves moving in a solid, gas, or liquid?

14. Describe the differences, if any, between a bow wave and a shock wave.

15. If a source of waves is moving perpendicular to the observer's line of sight, will the observer detect a Doppler effect in the wave motion? Explain.

16. List three classrooms on your campus that exhibit the phenomenon of reverberation.

17. Describe how you would employ the Doppler effect to determine the velocity of an object such as a speeding auto, a pitcher's fastball, or a hockey player's slapshot.

18. Can the Doppler effect be observed for waves other than sound waves? Explain.

19. ''Infrasound'' refers to sound waves that have frequencies less than those audible to the human ear. Can you think of a practical use for such sound waves?

20. Will a building or other structure resonate if exposed to sound having frequencies that are integral multiples or fractions of the structure's natural frequency? Explain.

21. How would you measure the speed of sound in a solid or liquid medium?

22. Do you think there could be other types of musical instruments besides wind, string, and percussion? Explain.

23. Two objects are moving directly toward an observer at the same speed. One is giving off its own sound while the other is giving off sound waves that have been emitted by the observer and then reflected off the approaching object back to the observer. Discuss the differences, if any, in the Doppler effects produced by the two objects.

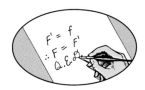

The Doppler Effect

Let us consider in greater detail the case of a wave source that is moving at some velocity less than that of the waves it emits. If the goose shown in Figure 9.19 emits a series of sound pulses in the form of "honk, honk, honk," the successive emitted pulses will be separated by some wavelength λ_{em} and will pass by a stationary observer at a frequency of f_{em}. The velocity v of the "honks" will be the product $f_{em} \times \lambda_{em}$. If, however, the goose is set in motion relative to the observer at some line of sight or "radial" velocity v_r, then during the time interval T between successive "honks," the goose has moved a distance d, which is given by

$$d = v_r \times T .$$

The time interval T, however, is simply the period between "honks," and is equal to the reciprocal of the emitted frequency or $1/f_{em}$. The frequency f_{em} in turn is equal to v/λ_{em}. Thus we have

$$T = 1/f_{em} = \frac{\lambda_{em}}{v}$$

and

$$d = v_r \times \frac{\lambda_{em}}{v} .$$

If the goose source is moving away from the fox observer, the wavelength λ_{obs}, which is perceived by the fox as coming from the goose, is equal to the wavelength actually emitted by the goose plus the distance d, or

$$\lambda_{obs} = \lambda_{em} + d .$$

Substituting the previous expression for d gives us

$$\lambda_{obs} = \lambda_{em} + v_r \frac{\lambda_{em}}{v} .$$

If we solve this equation for v_r, we obtain

$$v_r = \frac{\lambda_{obs}\, v - \lambda_{em}\, v}{\lambda_{em}}$$

or

$$v_r = \left(\frac{\lambda_{obs} - \lambda_{em}}{\lambda_{em}} \right) v .$$

FIGURE 9.19 ··········

The Doppler Effect   (a) If there is no relative motion between the goose source and the fox observer, the frequency and wavelength of the wave motion perceived by the fox will be the same as those emitted by the goose. (b) If the source is moving at some line of sight velocity v_r relative to the observer, the frequency and wavelength of the wave motion perceived by the observer will be different from those emitted by the source. In this case, since the goose is moving away from the fox, the fox will perceive a smaller frequency and larger wavelength than that actually emitted by the goose.

If the goose is moving toward the observer, the observed wavelength will be λ_{em} − d, and we can obtain this same expression for v_r, provided we regard the velocity of approach as having a negative value. A similar relationship can be derived in terms of the emitted frequency f_{em} and the observed frequency f_{obs}, which ends up as follows:

$$v_r = \left(\frac{f_{em} - f_{obs}}{f_{obs}} \right) v \, ,$$

where once again the velocity v is positive if the relative motion of the source is away from the observer and negative if the relative motion of the source is toward the observer. Any net relative motion along the line of sight between a source of any wave motion and an observer will produce changes in the observed frequencies and wavelengths governed by these equations.

QUESTIONS

1. Describe some practical applications of the Doppler effect.
2. Can the Doppler effect distinguish between a source moving toward a stationary observer and an observer moving toward a stationary source? Explain.
3. Do all forms of wave motion exhibit a Doppler effect? Explain.

PROBLEMS

1. Derive the expression for v_r in terms of frequency.
2. A musician is playing the musical note of B at 480 Hz aboard a train. As the train leaves the station at 5 m/sec, what frequency do the people on the station platform hear? Could this Doppler-shifted B note be mistaken for some other note?
3. A student pulled over for running a red light claims that because of the Doppler effect the light appeared to be green as he approached the intersection. If the wavelength of green light is about 5.3×10^{-7} m and that of red light is about 6.6×10^{-7} m, calculate the velocity at which the student would have to be moving for his story to be true. Assume that light waves travel at 300,000 km/sec. In light of your calculation, do you think the student's explanation is plausible?

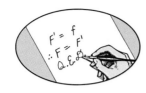

The Decibel Scale

Quite often in science a particular phenomenon must be observed over an extraordinarily large range of numerical values. Such is the case with sound. The faintest sounds we can detect with our human ears have intensities about 10^{-12} times those that bring our hearing to the pain threshold. To avoid working with such large ratios, scientists have devised a scale called the *decibel scale,* which in effect permits the rating of sound intensities using more convenient numbers.

The zero point of the decibel scale is, by definition, taken to be the faintest sounds that can be detected by the human ear. Although this level of intensity varies from human to human, scientists have set the power/unit area or *intensity* of a zero decibel sound to be 1.0×10^{-12} watts/m². In other words, the typical human ear can detect sounds with an associated intensity of 1.0×10^{-12} watts/m² or more. A tenfold increase in the power intensity of a given sound is defined to be an increase of one *bel,* named in honor of the American inventor of the telephone, Alexander Graham Bell. Each jump of one bel thus corresponds to a factor of 10 increase in the actual intensity of the sound. An increase of two bels corresponds to an increase of 10×10 or 100, three bels to an increase of $10 \times 10 \times 10$ or 1000, and so on. Acoustical scientists and engineers have found that the size of the bel is somewhat inconvenient and have introduced a smaller unit of relative sound intensity in which the bel is divided up into 10 equal ratios called decibels (dB). If I_0 is the sound intensity at zero decibels and I is the intensity at b decibels, then we may write that

$$I = I_0 \times 10^{b/10} .$$

Since the intensity I_0 of a zero decibel sound is 1.0×10^{-12} watts/m², we have

$$I = 1.0 \times 10^{-12} \times 10^{b/10}$$

$$= 10^{(b - 120)/10} \text{ watts/m}^2 .$$

Thus a sound having a relative intensity of 20 decibels will have a corresponding intensity of

$$I = 10^{(20 - 120)/10} \text{ watts/m}^2$$

$$= 10^{-100/10} \text{ watts/m}^2$$

or

$$= 10^{-10} \text{ watts/m}^2 .$$

The intensity of sound at which we begin to experience pain is about 1 watt/m² or 10^0 watts/m². The decibel level for such an intensity is given by

$$10^0 = 10^{(b - 120)/10} .$$

Since both quantities are equal, the exponents of 10 on both sides of the equation must be equal as well, and we have that

$$\frac{b - 120}{10} = 0 .$$

Solving for b we get

$$b - 120 = 0$$

or
$$b = 120 \text{ db} .$$

Thus the entire range of human hearing that covers 10^{12} W/m^2 in actual sound intensity is reduced to 120 decibels of relative sound intensity on the decibel scale and 12 bels on the bel scale of relative sound intensity.

QUESTIONS

1. Why do physicists employ the decibel scale?
2. Can you think of scales used in other areas of science that are similar to the decibel scale?
3. How would you measure the intensity of a given sound in decibels?

PROBLEMS

1. A certain sound has a relative intensity of 80 decibels. Find the intensity in watts/meter^2 of this sound.
2. What is the decibel level of sound waves having an intensity of 100 watts/meter^2?
3. Suppose that the quantity B were the sound intensity measured in bels. Derive a relationship between I, I_0, and B.

10 ELECTRICITY AND MAGNETISM

Nowhere in nature can one find a phenomenon as startling and spectacular as a bolt of lightning. Human beings have figuratively and sometimes even literally been struck by the awesome power contained in these arcing atmospheric events. In Greek mythology the power to hurl such bolts across the sky was reserved only for Zeus, the king of the gods. More modern times have seen renditions of lightning bolts often decorating the sides of our war machines, the logos of our light and power companies, and the chests of our superheroes, superheroines, and sci-fi spaceship commanders. In a partial foreshadowing of modern scientific discovery, Mary Shelley even employs lightning as the ''life energy'' by which Baron von Frankenstein is able to activate his infamous creation. On a far less grandiose scale, nearly all of us at one time or another have walked briskly across a carpet toward a door and, on reaching for the doorknob, gotten ''zapped'' by what was in effect a miniature lightning bolt. Such phenomena are manifestations of the existence of a fundamental quantity in nature we have called the electric charge.

Electrical charges, be they at rest or in motion, exhibit a wide variety of interesting, useful, and important effects. In this chapter, we will take a closer look at the entity that is responsible for lightning, ''static electricity,'' and even the neurological processes that permit you to read this book.

10.1 Electric Charge

If one performs an experiment by combing one's freshly cleaned hair very rapidly and then passing the comb just above several small bits of paper, one finds that the paper will ''jump up'' and cling to the comb (Figure 10.1). Some agent associated with either or both the comb and the paper has generated the observed upward motion in the paper. It is also obvious that this new motion-producing agent is not gravitational in nature. There is not enough mass in the comb for it to be able to yank on the paper with sufficient strength to allow the paper to ''defy'' the earth's gravity. It can be further shown that the action of combing one's hair plays a crucial role in the phenomenon as well. A comb that has not been run through hair and is passed over the bits of paper will not pick up the

FIGURE 10.1 ⋯⋯⋯⋯⋯

Electrical Force   The comb can pick up the bits of paper against the earth's gravity because a second force, called an electrical force, exists between the comb and bits of paper.

paper, nor will a comb that has been run through greasy, unclean hair. To describe this new form of behavior, scientists have developed the concept of *electric charge.* As the comb is run through our hair, the comb somehow becomes "charged." This charged state somehow allows the comb to pick up the bits of paper. A variety of objects can be charged by rubbing them with a variety of different materials. The Greek Thales recorded the first instance of such charging more than 2500 years ago when he rubbed amber with a wool-like fabric and described the results.

If we take two similar objects, such as a pair of glass rods, and charge them by rubbing them with a piece of silk, we find that the rods will, unlike the comb and paper, repel each other if suspended as shown in Figure 10.2. If we repeat the procedure using rubber rods instead, which have been charged by wiping them with wool, we find that the rubber rods repel one another. But if we take one of the charged rubber rods and bring it close to a suspended charged glass rod, we find that the rods attract one another. Any other charged object when brought close to the charged rubber or glass rods will repel either one or the other but not both. Moreover, if the object repels one of the charged rods, it will always attract the other.

From this set of simple experiments, we may conclude that there are two types of electrical charge in the universe, the charge that resides on the glass rod or so-called positive or (+) charge and the charge that resides on the rubber rod or so-called negative or (−) charge. A given electrical charge will repel other electrical charges like itself and will attract other electrical charges which are unlike itself. Thus, two negative charges will exert a repulsive force on each other as will two positive charges. On the other hand, a positive and a negative charge will attract each other.

In general, objects in the universe possess a mixture of positive and negative electrical charges, which in effect balance each other out and make the object as a whole electrically "neutral." If we rub a rod, walk across a rug, or comb our hair briskly, the frictional forces between the surfaces involved can "strip off"

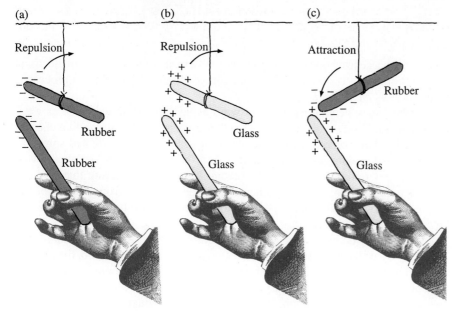

FIGURE 10.2 · · · · · · · · · · ·
Electric Charge Interactions (a) If two rubber rods are charged in exactly the same way, the rods will repel one another. (b) If two glass rods are charged in exactly the same way, the rods will once more repel one another. (c) A charged rubber rod, however, will attract a charged glass rod. We can thus conclude that there are two types of electrical charge, positive and negative, and that like charges exert a repulsive force on each other while unlike charges exert an attractive force on each other.

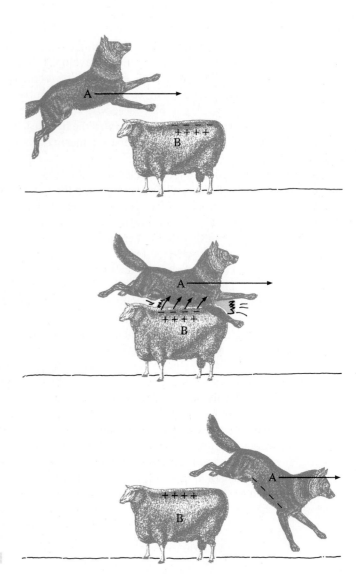

FIGURE 10.3 ⋯⋯⋯⋯⋯

Charging by Friction As object A rubs over object B, the frictional forces between the objects cause some of the electrical charge to be stripped from B, thus leaving both objects with a net electrical charge.

electrical charge from one object to another. As a result of this transfer of electrical charge, both objects now possess a net charge (Figure 10.3). Such a process is referred to as *charging by friction.*

Objects may be charged in other ways as well. Suppose we touch an electrically neutral body with one that possesses a charge. Because the like charges repel each other, some of them will ''run'' onto the neutral body and spread as far away from each other as possible. When the charged body is taken away, these charges remain behind on the second body, thereby giving that object a net electrical charge. Since the charging process resulted from actual contact between the two bodies, this type of charging process is referred to as *charging by contact* (Figure 10.4).

One may also charge objects by ''induction.'' If we frictionally charge ourselves up by walking across a rug and then reach toward a doorknob, the charges in the doorknob will divide up into those charges which are repelled to the far side of the doorknob by the charge on our hand and those which are attracted to the near side of the doorknob. If an uncharged individual now touches the far

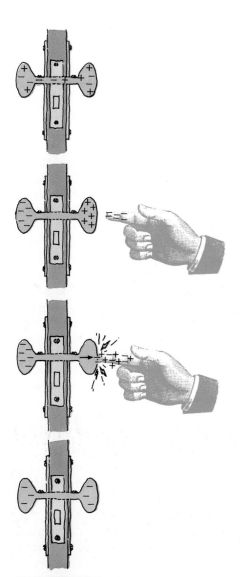

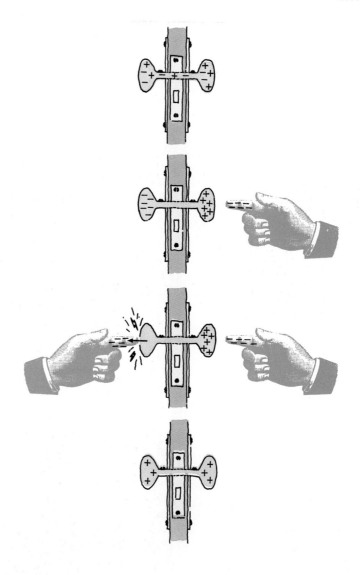

FIGURE 10.4 ··········
Charging by Contact If an electric charge is brought into contact with an electrically neutral object, such as a doorknob, the unlike charges in the doorknob will migrate onto the charged object. When the object is removed, the unlike charges are taken with it, thus leaving a net charge on the doorknob. Such a process is called charging by contact.

FIGURE 10.5 ··········
Charging by Induction If an electric charge is brought close to an electrically neutral object such as a doorknob, the unlike charges will migrate toward the charge and the like charges will be repelled. If the opposite side of the doorknob is touched, the repelled charges will be repelled off the doorknob onto the touching agent. When contact with that agent is removed, the repelled charges are taken with it and the doorknob now has a net charge. Such a process is called charging by induction.

side of the doorknob, the charges on that side can be repelled even farther away from our hand by moving along the arm of the second person. When that hand is taken off the doorknob, a net charge is left on the doorknob. This process of charging is said to be *charging by induction* (Figure 10.5).

By performing additional experiments on electrically charged objects, we can also deduce that "unit" positive and negative electrical charges exist in nature which are equal in magnitude and are incapable of further division* (Physics Formulation 10.1). Every electric charge observed on any object in nature consists of an integral multiple of these unit charges. Thus, we can observe an electric charge that is 122 times this unit charge but not a charge that is, say, 122.5 times this unit charge. Interestingly the ultimate indivisible unit of positive electrical charge resides on each and every proton in the universe and the equal magnitude ultimate unit of negative charge on each and every electron. Neutral atoms are thus composed of an equal number of positive charges, which reside on the protons in the nucleus, and negative charges, which reside on the electrons out in the "halo" of the atom. This structure, as we will see later on, was deduced from experimental results.

This structure also raises a number of interesting questions regarding atoms. If, for example, like charges repel each other, how do the protons in the nucleus remain "stuck" together in the nucleus? In an extreme case, the element uranium has 92 such protons in its nucleus. What is it that keeps the dozens of protons in a uranium atom nucleus compacted into a comparatively small volume of space? In a similar vein, because unlike charges attract, we may also ask what is it that prevents the negatively charged electrons from falling into the positively charged nucleus? These very important questions carry us directly into the realm of the atom itself and as such will be deferred to a later chapter. For now, we will simply note that atoms have somehow figured out a way to overcome these difficulties.

As a consequence of the comparatively low-mass, randomly moving, negatively charged electrons being located in the outer regions of atoms, it is the electron that is the most important particle in terms of electrical phenomena. For example, when we rub the aforementioned glass rod, we are in effect stripping electrons from the glass, leaving behind a net positive charge. On the other hand, the rubber rod strips electrons from the wool, thereby picking up a net negative charge. The number and location of the outer or valence electrons and their associated electric charges controls the location of a given element in the periodic table. When such electrons become involved in a chemical reaction, their fate is of considerable significance in determining both the nature of that reaction as well as the types of compounds that will result. In addition, electric charge can also move or flow from one atom to another in a given substance, thus giving rise to an important phenomenon called *electric current*.

1. Describe how you would prove to someone that electric charges exist in nature.

2. How are we able to determine the existence of two fundamental types of electric charge?

3. Compare and contrast charging by friction, contact, and induction. Give an example of each type of process.

---

*The possible existence of subatomic particles called "quarks," which possess simple fractions of these unit charges, will be discussed in Chapter 13.

4. A helium atom consists of two protons in its nucleus surrounded by two electrons. Discuss the various *electrical* forces that exist between these charged particles.

5. Why is the electron generally more important in electrical phenomena than the proton?

10.2 Coulomb's Law

As we have already seen, electrical charges can, depending on the circumstances, exert either attractive or repulsive forces on each other. In 1787 the French physicist Charles Coulomb was able to determine that the magnitude of the force exerted between two charges at rest separated by a known distance could be written as

$$\text{force between two charges} = \frac{\text{constant} \times (\text{charge 1}) \times (\text{charge 2})}{(\text{distance})^2}$$

or

$$F_{\text{electrostatic}} = \text{constant}\, \frac{Q_1 Q_2}{d^2}. \qquad \textbf{10.1}$$

Because the charges are at rest, this force is often called the electrostatic force and the above relationship is called *Coulomb's law*.

The mathematical form of Coulomb's law is thus almost identical to that of Newton's universal law of gravitation. There is, however, a very important difference. In gravitational interactions we only observe attractive forces, while in electrostatic interactions, we can observe both attractive and repulsive forces. This situation is represented mathematically by the use of $+$ and $-$ signs on the charges. The product of two positive charges or two negative charges mathematically results in a positive-valued electrostatic force, while the product of a positive and a negative charge results in a negative-valued force. In this mathematical representation, repulsive electrostatic forces are positive-valued and attractive electrostatic forces have negative values.

The unit of electric charge most commonly used in science is called the coulomb (C) and will be more formally defined in the next chapter. For now we will take the coulomb to be the total electric charge residing on either 6.25×10^{18} electrons or 6.25×10^{18} protons. Suppose we set up a pair of electrical charges at a given distance from one another. Then we simultaneously measure the magnitude of the electrostatic force between the two charges in newtons, the distance between the two charges in meters, and the size of the electric charges in coulombs. We find that the numerical value of the constant in Coulomb's law is 9×10^9 N-m^2/C^2. Coulomb's law can thus be rewritten as

$$\frac{\text{electrostatic}}{\text{force}} = \frac{(9 \times 10^9) \times (\text{charge 1}) \times (\text{charge 2})}{(\text{distance})^2}.$$

Thus, the electrostatic force between two charges of $+4$ and -3 coulombs, which are separated by 2 meters, will be

$$\begin{aligned}
\text{electrostatic force} &= \frac{9 \times 10^9 \text{ N-m}^2/\text{C}^2 \times (+4 \text{ C}) \times (-3 \text{ C})}{(2 \text{ m})^2} \\
&= \frac{9 \times 10^9 \times (-12)}{4} \text{ N} \\
&= -27 \times 10^9 \text{ N} .
\end{aligned}$$

The negative sign in the final result tells us that the electrostatic force between the two charges is an attractive force. At the atomic level, electrostatic forces are far larger and far more important than gravitational forces. Thus, atoms and molecules, as we have already implied, are held together not by gravity, but by electrical and nuclear forces.

WORKED EXAMPLE

An electric charge of $+2$ C is placed 0.5 m from a charge of -4 C. What is the magnitude and direction of the electrostatic force between these charges?

SOLUTION: The force between two electric charges is given by Coulomb's law:

$$\text{electrostatic force} = 9 \times 10^9 \frac{(\text{charge 1}) (\text{charge 2})}{(\text{distance})^2} .$$

Charge 1 $= +2$ C, charge 2 $= -4$ C, and the distance is 0.5 m. The electrostatic force is thus given by

$$\begin{aligned}
\text{electrostatic force} &= 9 \times 10^9 \frac{(+2) (-4)}{(0.5)^2} \text{ N} \\
&= 9 \times 10^2 \frac{-8}{0.25} \text{ N} \\
&= \frac{-72}{0.25} \times 10^9 \text{ N} \\
&= -288 \times 10^9 \text{ N} \\
&= -2.8 \times 10^{11} \text{ N} .
\end{aligned}$$

Thus, the electrostatic force between these two charges has a magnitude of 2.88×10^{11} N, and because the force has a negative value, it is an attractive force.

1. What is Coulomb's law?

2. Compare and contrast Coulomb's law with the universal law of gravitation.

3. Discuss how the concept of two different types of electrical charge is represented mathematically.

4. What is meant when a given force is said to be attractive or repulsive?

5. Show that the units on the constant in Eq. 10.1 are N-m²/C².

1. A +5-C charge is placed 10 meters from a +15-C charge. What is the magnitude and the direction of the electrostatic force between these charges?

2. How far should a −3-C charge be placed from a +4-C charge in order for the force between them to have a magnitude of 7.5 × 10⁸ N? Is this force attractive or repulsive?

10.3 Electric Field and Potential

In describing electrostatic and electrodynamic phenomena, the concept of the *electric field* is useful. Suppose we have an object that possesses an electrical charge. If we place that object at some point in space and it experiences an electrostatic force, then the electric field at that point in space is defined as

$$\text{electric field} = \frac{\text{electrostatic force on object}}{\text{total charge on object}} \qquad \textbf{10.2}$$

or

$$E = \frac{F_{\text{electrostatic}}}{Q} \; .$$

The electric field strength at a certain point in space for an object having a charge of 5 coulombs and experiencing a force of 20 newtons is

$$\text{electric field} = \frac{20\text{N}}{5 \text{ C}}$$

$$= 4 \text{ N/C} \; .$$

The direction of the electric field is the direction in which a positive charge would move if placed in that field. In essence, the electric field tells us the magnitude and direction of the electrostatic force that a charged object would experience at any given location in that field. Quite often an electric field is represented by a set of so-called ''field lines'' in which the arrows on the lines denote the direction of the force that would be exerted on a positive charge at a given point. The degree to which the field lines are crowded provides an indication of how strong the force will be at a given point. Examples of electric field line diagrams are presented in Figure 10.6 for a variety of simple charge configurations.

Suppose now we wish to move a charged particle from one position to another in the presence of an electric field. If we do so, then the work done either on the particle or by the particle will be, by definition,

$$\text{work} = \text{force} \times \text{distance moved} \; .$$

But the electrostatic force due to an electric field, as we have seen, is related to the electric field by

(a)   (b)

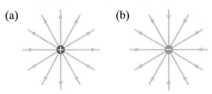

(c)

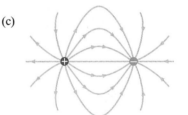

(d)

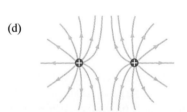

(e)

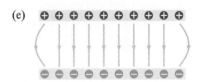

FIGURE 10.6 ··········
Electric Field Lines   Diagrams of the electric field lines that exist about (a) a positive charge, (b) a negative charge, and between (c) unlike charges, (d) like charges, and (e) charged parallel plates. The arrows denote the direction of the force that would be exerted on a positive charge at the given point. The degree to which the field lines are crowded provides a rough indication of the magnitude of the force that will be exerted on that charge.

PHYSICS FACET

10.1 An Atmosphere of Electricity

The thin layer of gaseous atmosphere that blankets our earth is a dynamic and complex entity in which warm and cold air masses sweep and swirl across the earth's surface, interacting with land, water, and each other in a number of interesting and important ways. In the course of these interactions, air masses can quite often create electrically charged regions in a fashion not unlike that in which rubbing motions can create electrically charged glass or rubber rods.

The most common manifestation of such electrical charging occurs during a thunderstorm. As the thunderstorm develops, large-scale vertical convective motions occur that can carry moist warm air from the earth's surface to altitudes in excess of 15 kilometers. In the process, positive electric charge is carried to the cloud top while the negative electric charge is left behind at the bottom of the cloud. Potential differences between these charge arrays can be as high as 300,000 volts. Similar electrical potential differences can build up between the earth's surface and the negatively charged cloud base. At electrical potentials of such extreme values, the chemical bonding between charges in the atoms and molecules making up the earth's atmosphere is overridden and a large number of ions and free electrons are created. At this point, the earth's atmosphere, which is normally a strong electrical insulator, now takes on the characteristics of a strong electrical conductor, and an electrical discharge in the form of a lightning bolt occurs.

A single lightning bolt can transfer over 100 coulombs of charge in about 2×10^{-4} sec, which corresponds to a current of some 500,000 amperes. The power developed in such a bolt can be as high as 10^{12} watts! The energy released in such a discharge heats up the atmosphere along the path of the bolt to temperatures in excess of 25,000 K and creates a localized pressure of hundreds of atmospheres along the discharge path. This pressure disturbance then propagates outward from the bolt path as an acoustical shock wave, which can be heard as the "sonic boom" of thunder.

At any given time, some 2000 to 3000 lightning-producing thunderstorms are estimated to be active somewhere on the earth's surface, and electrical discharge phenomena have also been observed in the atmospheres of other planets besides the earth. Photographs taken of the night side of the planet Jupiter, for example, recorded enormous lightning bolts in which the power developed is orders of magnitude larger than what we experience here on earth. Such electrical activity raises some intriguing possibilities. When mixtures of simple gases such as methane, ammonia, carbon dioxide, and water vapor are subjected to repeated electrical discharges in the laboratory over extended periods of time, the result is the formation of long-chain hydrocarbon molecules whose structural complexity begins to rival that of life itself. In response to such experimental findings, scientists have speculated that millions of years of similar electrical discharges in the earth's primordial atmosphere might well have literally sparked into existence the beginnings of life on our planet. If that is indeed the case, then careful studies of planets such as Jupiter take on special significance. It is on these worlds that primordial-like processes seem to be occurring at the present time and it is on these worlds, therefore, that the secrets of how living matter came to exist in the physical world may await our discovery.

••• **Lightning Bolts** The atmospheric motions that occur during a thunderstorm serve to separate electric charges within the clouds, thus creating large electric potentials between the clouds and the ground and between various regions of the clouds themselves. When the electric potential is sufficiently large, an electrical discharge occurs in the form of a bolt of lightning.

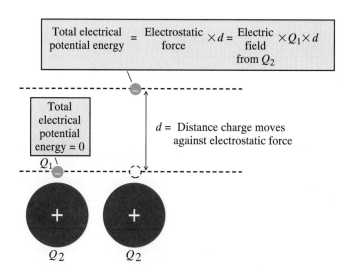

$$\boxed{\begin{array}{ccc}
\text{Total electrical} \\
\text{potential energy}
\end{array} = \begin{array}{c}\text{Electrostatic}\\ \text{force}\end{array} \times d = \begin{array}{c}\text{Electric}\\ \text{field}\\ \text{from } Q_2\end{array} \times Q_1 \times d}$$

Total electrical potential energy = 0

Q_1

d = Distance charge moves against electrostatic force

Q_2 Q_2

FIGURE 10.7 ···········

Electric Potential Energy   As the positive charge is moved farther away from the negative charge, the work done moving the positive charge against the electrostatic force of attraction is acquired by the positive charge as electrical potential energy. The total electrical potential energy acquired by the positive charge is equal to the electric field from the negative charge × the total positive charge × the distance the positive charge moves.

$$\text{electric field} = \frac{\text{electrostatic force}}{\text{total charge}}$$

from which we obtain:

$$\text{electrostatic force} = \text{electric field} \times \text{total charge} .$$

Thus, the work done to move a charged particle a given distance in an electric field is

$$\text{work} = \text{electric field} \times \text{total charge} \times \text{distance moved} .$$

Depending on the circumstances, movement of a charged particle in an electric field can result in energy being absorbed or released. The situation is very much analogous to that of a mass in a gravitational field. When we lift an object, it takes energy or work to perform the task. If we drop the object, the energy is recovered in the form of the object's kinetic energy. Similarly, if we try to push a charged object toward a like charge, we must expend work or energy to do so. If a charged object is released, the energy is recovered in the form of kinetic energy as the charged object is repelled by the like charge. Thus a charged particle in an electric field can possess stored energy in the form of *electrical potential energy,* the magnitude of which depends on its position within an electric field as well as the nature of that field (Figure 10.7). In dealing with electrical potential energy, the *electric potential* is defined as follows:

$$\text{electric potential} = \text{voltage} = \frac{\text{total electrical potential energy}}{\text{total charge}} .$$

In effect, the electric potential is the electric potential energy per charge and is measured in joules/coulomb. The most common unit of electrical potential is the *volt* where one volt equals one joule/coulomb. As a result of this definition, the term "voltage" has crept into the technological lexicon and is now widely used interchangeably with electric potential (Figure 10.8). The difference between the voltages or electrical potentials at two points is called the *electrical potential difference* and is of considerable importance in the analysis of charge flow, as we shall see in the next section.

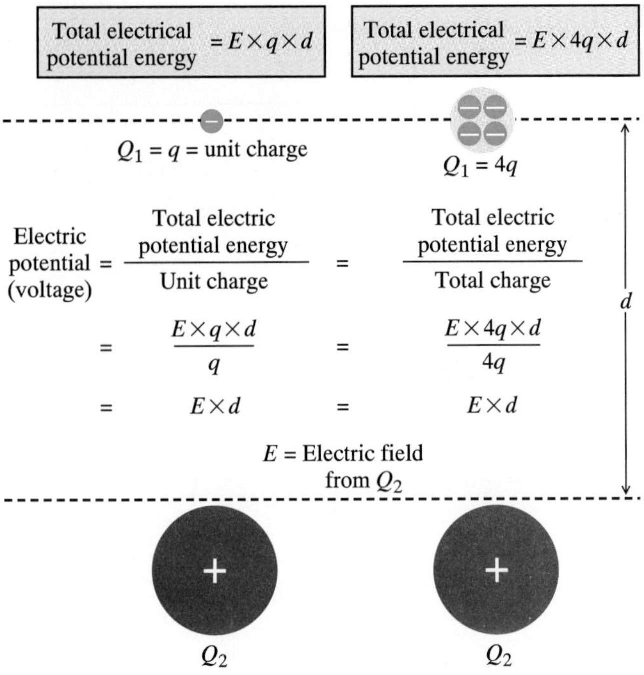

$$\boxed{\text{Total electrical potential energy} = E \times q \times d} \qquad \boxed{\text{Total electrical potential energy} = E \times 4q \times d}$$

$Q_1 = q =$ unit charge $Q_1 = 4q$

$$\text{Electric potential} = \frac{\text{Total electric potential energy}}{\text{Unit charge}} = \frac{\text{Total electric potential energy}}{\text{Total charge}}$$
(voltage)

$$= \frac{E \times q \times d}{q} \qquad = \frac{E \times 4q \times d}{4q}$$

$$= E \times d \qquad = E \times d$$

$E =$ Electric field from Q_2

d

Q_2 Q_2

FIGURE 10.8 ···········

Voltage The electric potential or voltage at a given point is the electrical potential energy of a given total charge located at the point divided by the magnitude of that total charge. Voltage is thus the electrical potential energy of a single unit charge. In this case the total electric potential energy of the charge array on the right is four times that of the charge on the left. The voltage or electric potential of both configurations, however, is the same.

WORKED EXAMPLE

An object with an electric charge of $+5$ C is moved a total of 1.5 meters against a field of 50 N/C. Find (a) the force that was employed, (b) the work that was done in this process, and (c) the change in the electric potential of the object.

SOLUTION: (a) In general the electric field is given by

$$\text{electric field} = \frac{\text{electrostatic force}}{\text{total charge}}$$

For this case, the electric field is 50 N/C and the total charge is $+5$ C. We thus have

$$50 \text{ N/C} = \frac{\text{electrostatic force}}{+5 \text{ C}}$$

or

$$\text{electrostatic force} = 50 \text{ N/C} \times +5 \text{ C}$$
$$= 250 \text{ N} .$$

(b) The work done in this process is

$$\text{work} = \text{electrostatic force} \times \text{distance moved} .$$

Continued

WORKED EXAMPLE

Continued

For this case,

$$\text{work} = 250 \text{ N} \times 1.5 \text{ m}$$
$$= 375 \text{ N-m} .$$

(c) Assuming that all of the work done on the charged object goes into changing the electric potential energy of the object, we have

$$\text{work} = \frac{\text{change in electric}}{\text{potential energy}} .$$

If the work done is 375 N-m, then

$$\frac{\text{change in electric}}{\text{potential energy}} = 375 \text{ N-m} .$$

The change in electric potential is

$$\frac{\text{change in electric}}{\text{potential}} = \frac{\dfrac{\text{change in electric}}{\text{potential energy}}}{\text{total change}} .$$

For the present situation,

$$\frac{\text{change in electric}}{\text{potential}} = \frac{375 \text{ N-m}}{+5 \text{ C}}$$
$$= 75 \text{ J/C} .$$

1. What is meant by an electric field?
2. How are electric fields represented in a field line diagram?
3. Describe the electric field about a pair of unlike charges that have equal magnitudes.
4. Compare and contrast electric field, electric potential, and electric potential difference.
5. Which way would a negatively charged particle move if placed in each of the fields shown in Figure 10.6?

1. The electrostatic force on a charge located in an electrostatic field of 500 N/C is −20 N. Find the sign and value of the charge.
2. What force will a 300 N/C electric field exert on a +15-C electric charge?

10.4 Electric Current

Just as objects in a gravitational field will move from a position of higher gravitational potential energy to a region of lower gravitational potential energy (books falling to the floor, etc.) and thermal energy will flow from a higher temperature reservoir to a lower temperature reservoir, so too do positive electrical charges flow from a higher electrical potential or voltage to a lower one. Such a flow of electrical charge is referred to as *electric current.* Because electrons are located in the more readily accessible outer regions of atoms and molecules, electric currents almost always involve a flow of the negative charges of electrons. The size of an electric current is defined as the rate at which electrical charges pass a given point per unit time, or

$$\text{electric current} = \frac{\text{total charge flow}}{\text{total elapsed time}}.$$

Electric currents are measured in amperes. A current flow of 1 ampere (A) corresponds to 1 coulomb of charge passing a given point in 1 second of elapsed time. This in turn means that if a current of 1 ampere is flowing past a given point, 6.25×10^{18} unit electrical charges are passing by that point each and every second!

Such electrical currents can be made to perform a variety of useful tasks. In one simple example, suppose we set up a device that will allow electric current to flow from a region of high voltage to a region of low voltage in such a way that it passes through the filament of a light bulb. As the charges move through the filament, they encounter friction or resistance from the filament, which in turn causes the filament to heat up to the point where it will give off light that is useful for studying, illuminating the darkness, etc.

In constructing such a simple electrical device, however, several problems arise. Suppose we have tediously separated a large number of electrical charges from their corresponding atoms and separately stored these two sets of charges at a positive reservoir or terminal and a negative reservoir or terminal (Figure 10.9). Despite the fact that there is a potential difference between the terminals, the charges will not flow from one terminal to another. This is because the air between the terminals does not readily permit the flow of charges from one terminal to another just as it did not readily permit the flow of thermal energy from the coffee spoon to our hand in Chapter 8. As a result, the air is said to be an electrical insulator and hence a poor electrical conductor.

The ability of a substance to permit the flow of electric current is called its electrical conductivity. Electrical conductivities of substances can range from very high values for the "conducting" substances, or *conductors,* such as copper, silver, and other metals, to very low values for the poor conductors, or *insulators,* such as rubber, glass, and air. Typically the ability of a metal to conduct an electric current is some 10^{24} times better than that of a good insulator! Roughly halfway between these two extremes are the *semiconductors,* whose electrical conductivities can be greatly altered by the addition of tiny amounts of impurities to their crystalline structure. Other substances, usually metals, or metal alloys, exhibit a phenomenon called *superconductivity,* in which their ability to resist the flow of electrical currents essentially vanishes as their temperatures approach absolute zero. Clearly, all of these phenomena are intimately linked to the mobility of electric charge in a given substance. For the conductors, the charges are metallically bonded and as such are quite free to wander about from atom to

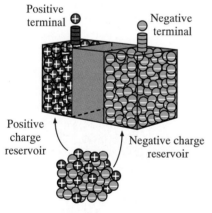

Positive terminal

Negative terminal

Positive charge reservoir

Negative charge reservoir

FIGURE 10.9 ············

Creating a Voltage Supply If an array of positive and negative charges is separated into positive and negative charge reservoirs, the work performed in accomplishing such a charge separation is now available as electric potential energy. If an electrical device is hooked up to the terminals of the voltage supply, negative electric charge will flow from the negative terminal through the device to the positive terminal.

PHYSICS FACET

10.2 Bioelectricity

Hungarian physiologist Albert Szent-Györgyi, a Nobel laureate, once stated that "what drives life is a little electric current." Although Szent-Györgyi was referring to the sunlight-induced electric currents that occur in plants during the process of photosynthesis, his words can be readily extended to include the electrical processes that occur in virtually every life form on the face of the earth.

Perhaps the most dramatic example of the importance of electrical processes in living creatures are those that occur in human beings. Virtually all of the information that is transmitted within the human body is done so by a vast and complex network of cells called the nervous system. The building blocks of the nervous system are the single nerve cells or neurons. The typical neuron, as shown in the figure, consists of a cell body surrounded by a branching array of fibers called dendrites. Extending out of one side of the cell body is a long cord-like appendage, which is called the axon. In human beings and higher animals, the axon is surrounded by a layer of fatty material called the myelin sheath. The myelin sheath in turn is segmented sausage-style into lengths of about one millimeter by minute spaces roughly 10^{-6} meters across, which are called the nodes. The axon then branches out into terminal fibers, which are collectively referred to as the axonal tree.

When a neuron is in its "resting" or "ready" state and not transmitting information, a steady electric potential of about 70 millivolts exists across the cell membrane between the membrane's positively charged outer surface and its negatively charged inner surface. When certain stimulant chemicals such as hormones or neurotransmitters come in contact with the dendrites or the cell body, the permeability of the cell membrane briefly changes in the area of contact. Positively charged sodium ions in the outside interstitial fluid now rush toward the negative inside portion of the membrane, causing a localized change in the electric potential of about 100 millivolts in a matter of only a few milliseconds. This shift of electric potential is called an *action potential* and in the course of the ion rush across the cell membrane, similar responses are triggered in the adjacent areas of the cell membrane. The result is a pulse of electric potential that propagates out of the neuron's cell body and along the axon as a nerve impulse where it is

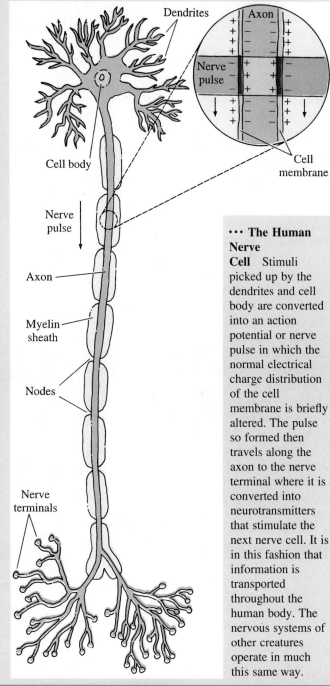

··· The Human Nerve Cell Stimuli picked up by the dendrites and cell body are converted into an action potential or nerve pulse in which the normal electrical charge distribution of the cell membrane is briefly altered. The pulse so formed then travels along the axon to the nerve terminal where it is converted into neurotransmitters that stimulate the next nerve cell. It is in this fashion that information is transported throughout the human body. The nervous systems of other creatures operate in much this same way.

PHYSICS FACET

Continued

reinforced at each of the nodes to prevent it from becoming too weak.

As the nerve impulse passes, the rest state charge distribution on the cell membrane's inner and outer walls is quickly reestablished in preparation for the next nerve impulse. When the nerve impulse reaches the axon tree, it stimulates the release of molecules called neurotransmitters, which move across a junction called a synapse to strike and stimulate the next nerve cell in the chain, thus beginning the process anew in that neuron. It is in this fashion that nervous impulses are transmitted throughout the human body. Nerve centers exist that coordinate and evaluate all of the incoming pulsed information and in a similar electrochemical fashion formulate some sort of pulsed response. The most magnificent of these coordinating centers is, of course, the human brain.

The patterns of electrical activity within a human being can be used to determine the health of that individual. For example, unusual electrical behavior indicative of various illnesses can be detected through the use of instruments such as the electroencephalograph (EEG), which is used to monitor electrical activity in the brain, and the electrocardiograph (EKG), which is used to monitor electrical activity in the heart.

Human beings are not the only creatures that make extensive use of electrical phenomena in their existence. Nervous systems and the electrical processes by which they operate appear in animals as diverse as jellyfish, worms, and insects. Several species of fish, most notably the electric eel, the electric catfish, and the electric skate, have nerve endings located in modified muscle cells called electroplaques. When one of these cells receives a nerve impulse, it will not contract as does ordinary muscle tissue, but instead discharges the pulse into the surrounding water. Amazingly, the individual nerve impulses transmitted to the thousands of electric plaques are sent in such a way that they all arrive at their respective electroplaque destinations at exactly the same time! The result is a coordinated pulsed electrical discharge with a potential as high as 650 volts.

The rate at which these pulses are produced varies from one per second up to more than 1000 per second, depending on the species involved. These electrical discharges surround the fish with an electrical field, the magnitude and direction of which the fish can detect by means of highly specialized electrosensory cells in the outer skin. If the electric field is distorted by something having a different electrical conductivity from that of the surrounding water, the sensory cells are stimulated and the fish is alerted. Thus, through the use of electric fields, these ''electric fish'' can navigate, detect prey, avoid predators, and communicate with each other, even in the muddy waters in which they are most often found.

Electric phenomena have been found to play an indispensable role even in that most fundamental of life processes, photosynthesis. When sunlight strikes the chlorophyll molecules in plant tissue, electrons are liberated from the ''reaction center'' of the chlorophyll. Once freed, the electrons and their associated electric charges participate in a variety of chemical reactions associated with the overall photosynthesis process. The charge flow can be cyclic, in which case the electrons are returned to their initial reaction center, or noncyclic when the electrons end up elsewhere. Although there are still some details to be learned about the complex reactions that go into the process of photosynthesis, it is clear that electric charge flow is both present and important. So it is with the plants and animals of our earth.

atom. For insulators, however, the charges are much more rigidly locked into their respective chemical bonds, thereby inhibiting the flow of electric currents.

If we develop enough electrical potential or a sufficiently high voltage between the terminals, then it is possible to override the effects of chemical bonding even within an insulating substance. In doing so, large numbers of ions and free electrons are created. At this point the insulator is said to ''break down'' and briefly becomes an electrical conductor, with the result that the stored charges almost instantly flow to the lower electrical potential. Familiar examples of this type of electrical discharge include the ''zap'' we get when we walk across a rug and touch a doorknob and the ''striking'' of lightning bolts from cloud to cloud or from cloud to ground during an ''electrical'' storm.

Clearly such a mode of electrical discharge is not suitable for operating our light bulb in a useful way. Any such discharges are exceedingly brief and occur in a random fashion. Our light bulb would thus only be lit for a brief instant if it were ''struck'' by such a discharge. Somehow, then, we need to channel the current flow through our light bulb as well as maintain that flow through the light bulb for a period of time that will permit us to study for the next day's midterm exam.

The channeling of the charge flow in an electrical device is accomplished through the use of wires made of materials having a high electrical conductivity. Given a choice, electrical charges much prefer the path of least resistance and will therefore journey along a wire made of conducting material rather than ''zap'' their way across an insulator such as air or glass. As a result, electrical currents can be ''conducted'' along such wires to and from whatever electrical devices we wish.

Unless we have a way to maintain the electric potential, however, the flow of current will drop off rather quickly as the charges race to equalize the electrical potentials of the two reservoirs. Electrical potential differences are maintained by devices that are variously referred to as ''voltage supplies,'' ''seats of electromagnetic force or emf,'' or ''electron pumps.'' Regardless of the terminology, as each charge flows from a high-potential reservoir to a low-potential reservoir, a voltage supply will immediately replace that charge so as to maintain an ongoing potential difference. One familiar example of such a voltage supply is the chemical battery in an automobile. These batteries separate electric charges through the use of chemical reactions, usually involving two different metals immersed in an acidic bath. If this charge separation is not maintained by periodically ''recharging'' the battery, then the electrical potential between the terminals falls to zero and the battery goes ''dead.'' In automobiles this recharging process is usually accomplished by the use of a generator in which rotational energy is used to reseparate the charges in the battery, thereby maintaining the battery's electrical potential. We will describe generators in more detail in the next chapter.

WORKED EXAMPLE

A lightning bolt transfers 120 C of charge to the ground in one-thousandth of a second. What was the current flow during this strike?

SOLUTION: A total of 120 C of electric charge was transferred to the ground in one-thousandth or 1×10^{-3} seconds. The current flow was then

$$\text{current} = \frac{\text{charge transferred}}{\text{time interval}}$$

Continued

WORKED EXAMPLE

Continued

$$= \frac{120 \text{ C}}{1 \times 10^{-3} \text{ sec}}$$
$$= 120 \times 10^3 \text{ amperes}$$
$$= 1.2 \times 10^5 \text{ amperes} .$$

The current on the lightning bolt was 120,000 amperes!

1. Why does electric charge flow from one place to another?
2. Which type of electric charge is most often involved in electric current? Why?
3. What is an ampere?
4. Explain how an electric light bulb works.
5. Discuss the uses of substances that are (a) good electrical conductors, (b) good electrical insulators.

1. A total of 5000 coulombs of charge passes a certain point in 20 seconds. Find the current in amperes.
2. A current of 25 amperes flows through a certain wire for an hour. How many coulombs passed through the wire? How many unit electrical charges?

10.5 Electrical Circuits

If a source of emf is hooked up by conducting wires to some sort of electrical device such as a light bulb, the entire array is said to be an electrical circuit, since the charges will flow from the high-potential reservoir and then get "pumped" by the voltage supply back to their original position in the high-potential reservoir. The process is much like the one that occurs when water is cycled through an artificial decorative waterfall. The water flows downward from a reservoir of higher gravitational potential energy to one of lower gravitational potential energy. A pump then provides the energy to lift the water back to the higher potential energy reservoir.

The ability of a given circuit to "resist" the flow of charge is referred to as that circuit's *electrical resistance*. The electrical resistance of a given circuit depends on a variety of factors. For example, a longer conducting path throughout the circuit will result in a larger resistance to the flow of the charges, while a larger wire diameter and hence a larger wire cross-sectional area will permit a larger flow of charge. The type of conducting material present in the circuit can also play a significant role. Wires made of a blend of iron, chromium, and nickel called Nichrome can conduct electrical charges nearly 60 times better than wires having the same dimensions, but made of copper.

Overall, however, if a given voltage is applied to a given circuit, the flow of electric current through that circuit depends inversely on the overall amount of resistance present in the circuit. Thus, for a given voltage, a large overall resistance in the circuit results in a smaller current flow and a smaller overall resistance in the circuit results in a larger current flow. At the same time, if a circuit has a fixed electrical resistance, an increase in the voltage will result in a proportional increase in the current flow through the circuit. These observed facts can be summarized in a relationship between voltage, current, and electrical resistance called *Ohm's law,* which states:

$$\text{current} = \frac{\text{voltage}}{\text{resistance}} \qquad 10.3$$

or

$$I = \frac{V}{R} .$$

The unit of electrical resistance is called the *ohm.* An ohm is defined as the electrical resistance that will permit one ampere of current to flow when a potential of one volt is placed across the resistance.

The very simplest circuit consists of a voltage supply, a resistance, and a connecting wire as shown in Figure 10.10. As current flows through the resistance, it will cause the resistance to heat up because of the friction between the resistance and the passing charges. This heat can then be employed in a variety of appliances such as a toaster, stove, iron, or blow dryer. The amount of power developed in such an appliance depends on both the resistance of the appliance as well as the amount of current flowing through it in the following way (see Physics Formulation 10.1):

$$\text{electrical power} = (\text{current})^2 \times (\text{resistance})$$

or, in an alternate fashion,

$$\text{electric power} = \text{current} \times \text{voltage} .$$

Thus, the electrical power developed in an appliance having a resistance of 5 ohms and through which a current of 3 amperes is flowing will be

$$\begin{aligned}\text{electrical power} &= (3 \text{ amperes})^2 \times 5 \text{ ohms} \\ &= 9 \text{ (amperes)}^2 \times 5 \text{ ohms} \\ &= 45 \text{ watts} .\end{aligned}$$

The electrical power developed in an appliance operating at 120 volts and through which a current of 0.5 amperes is flowing will be

$$\begin{aligned}\text{electrical power} &= 0.5 \text{ amperes} \times 120 \text{ volts} \\ &= 60 \text{ watts} .\end{aligned}$$

When electric charges flow from a location of high electric potential to one of lower potential, they do so along the path of least resistance. On occasion we inadvertently provide such a path of least resistance for the current in an electric circuit and the result is called a *short circuit.* Perhaps the simplest example of a

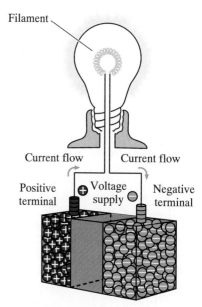

FIGURE 10.10 · · · · · · · · · ·
A Simple Electrical Circuit   As the charge current flows from the negative to the positive terminals of the voltage, the friction developed in the light bulb filament causes it to heat up, thereby giving off useful illumination.

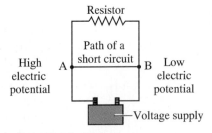

FIGURE 10.11 ···········
A Simple Short Circuit   If there is no other conducting path from the point of high potential (A) to that of low potential (B), the current will flow through the resistor. If, however, a much lower resistance wire is connected to A and B as shown, the bulk of the current will flow along the wire bypassing the resistor and a short circuit will result.

short circuit is illustrated in Figure 10.11. In this instance, a resistance is connected to a voltage supply and, under normal circumstances, the current will flow around the circuit through the resistance. If, however, we connect a wire to points A and B in that circuit, the electrons now see a far easier way to get to the low-voltage terminal and will tend to flow along the wire directly from A to B. In effect, the bulk of the electrons no longer flow through the resistance, but rather move along the short circuit. The difficulties presented by such a short circuit arise from the rapid heat buildup that can occur and the attendant danger of fire and other related problems.

It is also possible for a human being to provide a short circuit path of least resistance for current flow in a circuit as shown in Figure 10.12. When an electric current passes through us we experience an electric ''shock.'' Usually, we are shocked when we touch a source of voltage while standing on the ground. The earth is very efficient at absorbing electric currents and is always in effect at zero electrical potential. When we contact a wire or other voltage source, we provide a path for the current to flow through us to the earth's zero potential or ''ground.'' The damage to the human body that results from an electric shock is due entirely to the current that passes through the body during the course of that shock. In turn, the amount of current is, from Ohm's law, dependent on the body's resistance and the voltage involved. These conditions obviously can vary considerably. The resistance of the human body, for example, can change by factors of several thousand depending on whether the skin is wet or dry, etc. Currents in the range of 0.2 ampere are usually fatal as a result of their impact on the human heart and nervous system. A variety of safety precautions can be taken to protect against either a short circuit or electric shock. These include proper insulation of wires and the use of ''ground'' wires, which provide a current path from an unwanted or unexpected voltage in an electrical appliance or some other electrical device to the earth or ground.

Quite often we wish to operate a number of appliances off the same voltage supply. There are basically two ways in which multiple resistances can be hooked up to a voltage supply. In the *series circuit,* the resistances are connected in such a way that the current can flow along one single route, which carries it through each of the resistors in the circuit (Figure 10.13). In this type of circuit the current flowing through each resistance is the same and the voltage drops off across each

FIGURE 10.12 ···········
Electric Shock   If a human being provides a low-resistance path between a high and low electric potential, current will flow through that person and that individual experiences an electric shock.

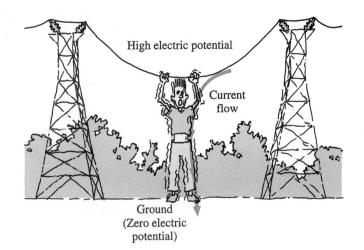

resistance in accordance with Ohm's law. If one of the resistances breaks the flow of electrons through it, the entire circuit will also be broken.

A *parallel circuit,* on the other hand, is set up so that each resistance in the circuit is hooked up directly across the terminals of the voltage supply (Figure 10.14). By connecting resistances in this way, the current now has several possible paths through the circuit even if one or more of the resistances in the circuit breaks the current flow. The failure of a single element in a parallel circuit thus does not result in a shutdown of the entire circuit as for a series circuit. For this reason most wiring in homes and offices is done in parallel circuits.

Problems can arise in parallel circuits, however, if an attempt is made to run too many appliances off a single voltage supply. As each additional resistance is added to a parallel circuit, the total current drawn from the voltage supply increases. At some point, the connecting wires are called on to carry an "overload" of current and heat up dangerously to the point where a fire could be started. To prevent such occurrences, safety fuses or circuit breakers are installed at key locations in the circuit. If an overload occurs, such devices respond by breaking the circuit. In a fuse, for example, a small ribbon of metal within the fuse melts down and breaks when subjected to a current overload.

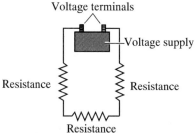

FIGURE 10.13 ··········
A Simple Series Circuit   In a series circuit, the resistances are connected in such a way that the electric current flowing through each resistance element is the same.

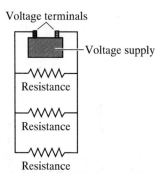

FIGURE 10.14 ··········
A Simple Parallel Circuit   In a parallel circuit, the resistances are connected in such a way that the voltage across each resistance element is the same.

WORKED EXAMPLE

An appliance is plugged into a 220-volt outlet and draws a total of 5 amperes of current. What is the resistance of the appliance?

SOLUTION: From Ohm's law, we have that

$$\text{voltage} = \text{current} \times \text{resistance} .$$

For a voltage of 220 V and a current of 5 amperes, this equation becomes

$$220 = 5 \times \text{resistance} .$$

Solving this equation for the resistance, we obtain:

$$\text{resistance} = \frac{220}{5} \text{ ohms}$$

$$= 44 \text{ ohms} .$$

---

1. Describe some of the factors that affect electrical resistance.
2. State Ohm's law. Of what value is this law?
3. Compare and contrast a series electrical circuit and a parallel electrical circuit. What are the advantages and disadvantages of each type of circuit?
4. Discuss the conditions under which a person might receive an electric shock.
5. Show that units of [(amperes)^2 × ohms] and [amperes × volts] are both equivalent to watts of power.

REVIEW
10.5
QUESTIONS

REVIEW 10.5 PROBLEMS

1. A current of 0.2 amperes flows through a circuit when a voltage of 100 V is applied to the circuit. What is the overall electrical resistance in the circuit?

2. A drier is plugged into a 220-volt outlet. If the drier draws 1.5 amperes of current, what power is developed in the drier?

10.6 Magnetism

For centuries humans have thought magnetism to be one of the basic forces of nature. In our first recorded encounters with this phenomenon over 2000 years ago, it was noticed that certain stones, called "lodestones," could attract objects made of iron. It was also recognized that while this force, called magnetism, was in many ways similar to electrical force, there were also some pronounced differences.

If we support an elongated lodestone or some other form of a "bar" magnet in such a way that it is free to rotate, we can observe several interesting properties of magnetism (Figure 10.15). First of all, the magnet will always align itself in exactly the same way, with one end of the magnet pointing roughly in the direction north and the other pointing south. If we define the "north-seeking" end of the magnet as the "north pole" of the magnet and the "south-seeking" pole of the magnet as the magnet's "south pole," we find that the north pole of the magnet will repel the north poles of other magnets and attract their south poles. In like fashion the south pole of the magnet will repel the south poles of other magnets and attract their north poles. Thus, like electric charges, there are two basic types of magnetic poles, north poles and south poles. Also similar to electric charges, like magnetic poles exert a repulsion or positive force on each other, while unlike magnetic poles exert an attractive or negative force on each other. Interestingly, the mathematical form for the magnitude of the force exerted between two magnetic poles was discovered by Coulomb in 1785, two years before he proposed his more famous Coulomb's law for electric charges. The mathematical form for "Coulomb's law of magnetic forces" is virtually identical to those for gravitational and electrical forces and can be stated as follows:

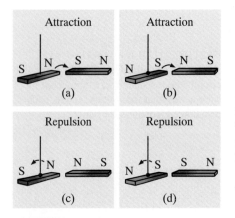

FIGURE 10.15 ··········

Magnetic Interactions When the pole of one magnet is brought close to the pole of a second magnet, we find that unlike poles exert an attractive force on each other, (a) and (b); while like poles exert a repulsive force on each other, (c) and (d). The end of a magnet that "seeks" or points toward the magnetic north pole of the earth's magnetic field is designated as N and the end that points toward the earth's south magnetic pole is denoted by S.

$$\text{magnetic force} = \text{constant} \, \frac{\left(\begin{array}{c}\text{pole strength} \\ 1\end{array}\right)\left(\begin{array}{c}\text{pole strength} \\ 2\end{array}\right)}{(\text{distance})^2},$$

where pole strengths are expressed in units of ampere-meters and the value of the constant in this equation is 1.0×10^{-7} N/ampere². Like electrical charge, magnetic pole strengths have an associated magnetic field, which can be described both mathematically and through the use of field diagrams. Some of these are shown in Figure 10.16.

As similar as magnetic forces are to those of electrostatic forces, however, one finds some interesting contrasts as well. For example, if we cut a magnet in two in order to isolate the north and south poles of the magnet, we find instead that we now have two smaller magnets, each of which now has both a north and a south pole (Figure 10.17). Despite the best efforts to date of some of the world's best scientists, no magnetic "monopole" has been isolated. Every known magnet, regardless of its size, always has a paired north and south pole. This "side-by-side" existence of magnetic north and south poles within the same magnet tells us that such poles are bound up in the magnet in a very intrinsic way and are unable to "flow" in the way that electric charges move. We can "magnetize" a given substance by stroking it with a magnet and "demagnetize" it by either heating it up or dropping it on the floor. Of all the substances in nature, however, only a few are capable of being magnetized. Such materials are said to be "ferromagnetic" and include iron, nickel, cobalt, the rare-earth metals gadolinium and dysprosium, and many alloys of these elements.

By the start of the nineteenth century, it appeared that these and other observations had firmly established magnetism as an independent force of nature, similar in many ways to gravity and electricity, but nevertheless a force in its own right. It was thus with a considerable amount of surprise and irony that in 1820 a Danish physics teacher named Hans Øersted, during a lecture demonstration specifically designed to show that electricity and magnetism were completely independent of one another, found that quite the opposite was true. Thus came the beginning of the realization that electricity and magnetism are intimately related to one another in a number of important and fundamental ways.

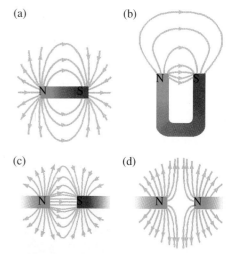

FIGURE 10.16 · · · · · · · · · · ·
Magnetic Field Lines Diagrams of the magnetic field lines that exist about (a) a bar magnet, (b) a "horseshoe" magnet, (c) like magnetic poles, and (d) unlike magnetic poles. The arrows denote the direction toward which the north pole of a magnet would point if it were situated at that position in the field.

WORKED EXAMPLE

What is the direction and magnitude of the force between the north pole of a bar magnet, which has a pole strength of 120 ampere-meters, and that of the south pole of a bar magnet having a pole strength of 350 ampere-meters located 0.5 m away?

SOLUTION: The magnetic force between two magnetic poles is given in general by

$$\text{magnetic force} = 1.0 \times 10^{-7} \frac{\left(\begin{array}{c}\text{pole strength} \\ 1\end{array}\right)\left(\begin{array}{c}\text{pole strength} \\ 2\end{array}\right)}{(\text{distance})^2}.$$

For this case, we have

$$\text{magnetic force} = 1.0 \times 10^{-7} \text{ N/amp}^2$$
$$\times \frac{120 \text{ (amp-m)} \times 350 \text{ (amp-m)}}{(0.5 \text{ m})^2}$$
$$= 1.0 \times 10^{-7} \text{ N/amp}^2$$
$$\times \frac{42{,}000 \text{ amp}^2\text{-m}^2}{0.25 \text{ m}^2}$$
$$= 168{,}000 \times 10^{-7} \text{ N}$$
$$= 1.68 \times 10^5 \times 10^{-7} \text{ N}$$
$$= 1.68 \times 10^{-2} \text{ N}.$$

FIGURE 10.17 · · · · · · · · · ·
Duality of Magnetic Poles If a magnet is cut in two, the result is not an isolated north and south magnetic pole, but two complete magnets each having both north and south magnetic poles.

1. How are magnetic forces similar to electric forces? How are they different?
2. Describe some of the observed properties of magnets.
3. What is a magnetic monopole? Has one ever been observed?
4. Discuss how north and south magnetic poles interact.
5. Sketch the magnetic field lines associated with (a) a bar magnet and (b) a horseshoe-shaped magnet.

1. Two magnetic poles with identical pole strengths of 1000 ampere-meters each are placed 0.4 meters apart. What is the magnitude of the magnetic force between these magnetic poles?
2. If the distance between two magnets is increased by a factor of 5, how do the magnetic forces between the two magnets compare before and after the increase?

Chapter Review Problems

1. An electric charge of $+3$ C is placed 2 m from a charge of -0.5 C. What is the magnitude and direction of the electrostatic force between these charges?
2. Suppose the distance in Problem 1 were tripled. How much larger (or smaller) would the magnitude of the electrostatic force now be?
3. How much charge must be placed 4 meters from a $+5$-C charge in order to obtain a repulsive force of exactly 1 N?
4. An electric charge of $+5$ C is placed one meter to the right of a -2-C charge. A third charge is to be placed to the left of the -2-C charge at a distance of 2 meters. What sign and magnitude must this third charge have if the net electrostatic force on the -2-C charge is to equal zero?
5. A scientist finds that when two positive charges of 6×10^{-4} C each are placed 2 meters apart, a repulsive force of 450 N is observed. What is the value of the constant in Coulomb's law for these data?
6. A total of 300 C of electric charge is observed to pass a certain point in one minute. What is the current flow in amperes?
7. A current of 5 amperes is flowing in a certain wire. How long will it take for a total of one million coulombs of charge to flow past a given point?
8. A current of 3 amperes flows past a point for a total of 4 minutes. What was the total charge that passed the point?
9. If the unit of electrical charge is 1.6×10^{-19} C, how many unit electric charges pass a given point in one minute if the current flow is 10 amperes?
10. What voltage would be required to produce a current of 3 amperes through a resistance of 10 ohms?
11. A current of 5 amperes flows through a given resistance when the resistance is hooked up to a 120-volt battery. What is the value of the resistance?

12. A 10-ohm resistor is plugged into a 110-volt outlet. What current will flow through the resistor?

13. If a person is shocked by a 120-volt electrical outlet, how much resistance must the person have in order to keep the current passing through that individual under 0.2 ampere?

14. An iron with a resistance of 8 ohms draws a current of 4 amperes. What is the total power developed by the iron?

15. A 75-watt light bulb draws a current of 0.5 ampere. What is the resistance of the light bulb? What is the applied voltage?

16. A 100-watt light bulb has a resistance of 4 ohms. What current will such a bulb draw? What is the applied voltage?

17. How much power is developed in an appliance having a resistance of 100 ohms when a current of 2 amps flows through the appliance?

18. An appliance is plugged into a 120-volt outlet and draws a current of 0.8 amps. What power does the appliance develop?

19. A 250-watt light bulb is observed to draw a current of 2 amperes when plugged into a certain outlet. What is the voltage of the outlet?

20. What resistance will be needed for an appliance to develop 1000 watts of power when it is plugged into a 220-volt outlet?

21. Describe what happens to the magnetic force between two magnets A and B when

   a. the pole strength of A is tripled.
   b. the pole strength of B is cut in half.
   c. the distance between A and B is increased by 5 times.
   d. all of the above occur at once.

22. Two magnets separated by 1 meter have the same pole strengths. What magnitude should the pole strengths be if the force between the two magnets is 1 N?

23. Two magnets having respective pole strengths of 50,000 amp-m and 8 million amp-m experience a magnetic force of 100 N. How far apart are these magnets?

24. The mean pole strength of the magnetic field of the planet Saturn is about $7/10$ that of the Earth and the mean pole strength of the magnetic field of the planet Mercury is about $1/100$ that of the earth. If Saturn is 18 times farther away than Mercury, find the ratio of the Earth-Saturn magnetic force to the Earth-Mercury magnetic force.

THOUGHT QUESTIONS

1. Joe Cool slides across the sofa and as he leans over to give Patti Party-Time a kiss, a spark jumps between their lips. Joe Cool says it must be true love. If you were Patti, would you believe Joe? Explain.

2. What is "static cling" in a clothes drier? How do you think it arises? How might you prevent static cling?

3. Is it possible that a third kind of electrical charge can exist in nature? Explain.

4. What are the differences, if any, between a negative electric charge and an electron?

5. Sketch the electric field lines between two charges having a like sign but differing in magnitude by a factor of 4.

6. Discuss whether lighting can occur when there is no stormy weather present. Can lightning occur without thunder? Thunder without lightning?

7. If a substance is a good thermal conductor, does that mean it will be a good electrical conductor as well? Explain.

8. Look up the diagrams of the nervous systems of three types of creatures other than human beings. Compare and contrast these nervous systems.

9. What do you think will happen if a progressively larger number of appliances are hooked into a series circuit? A parallel circuit? Are these desirable results? Why?

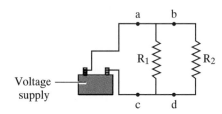

10. A person falls from a certain height and grabs onto a high-voltage wire in mid-air. As the person hangs on the wire, does he or she experience an electric shock?

11. In the parallel circuit shown at the left (upper), R_1 and R_2 are resistance. Resistance R_2 is disconnected from the circuit at points b and d. Describe how you would connect R_2 back into the remaining circuit so that you would have a series circuit involving R_1 and R_2.

12. In the series circuit shown at the left (lower), R_1 and R_2 are resistances. The voltage supply is disconnected from the circuit at points a and b. Describe how you would reconnect the voltage supply to R_1 and R_2 so that you would have a parallel circuit involving R_1 and R_2.

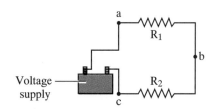

13. Where in the circuit shown in Figure 10.14 would you place a safety fuse or circuit breaker? Why?

14. Given a choice of purchasing a set of Christmas tree lights that were hooked up in series or a set of lights that were hooked up in a parallel, which string of lights would you purchase? Why?

15. Can you think of an alternate set of units for magnetic pole strength from the ones given in the text?

16. If a pile of iron filings is placed on a piece of paper and a magnet is brought up underneath the paper, the iron filings will appear to "stand up." Why does this happen?

17. In what ways are magnetic pole strengths similar to electric charges? How do they differ?

18. Devise an experiment to verify Ohm's law.

19. Show that the expression for electric power given in the text has the units of power, i.e., joules/sec or watts.

20. Why do you suppose that "horseshoe" magnets are more useful in many instances than "bar" magnets?

21. Is it possible to transform a nonmagnetic material into one that is magnetic? If so, how?

22. Discuss how you would experimentally determine the constant present in the magnetic force equation.

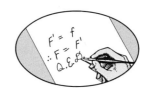

PHYSICS FORMULATION 10.1

The Fundamental Unit of Electrical Charge

Less than two decades after the discovery of the electron by Sir Joseph Thompson in 1897, the American physicist Robert Millikan in 1913 was able to determine experimentally the magnitude of the electrical charge residing on Thompson's newly discovered particle. This measurement was of crucial importance, because, quarks notwithstanding, the charge on the electron continues to this day to be the fundamental unit of electric charge in the universe.

Millikan's "oil drop" experiment, as it is called, is a classic of elegance and simplicity. The basic apparatus consists of two parallel plates as shown in Figure 10.18. Before any electric charge is placed on the plates, a fine mist of oil drops illuminated by sidelight is sprayed into the space between the plates. As the oil drops fall, they are acted on by a downward force of gravity, which seeks to accelerate them to higher velocities, and the frictional force resulting from air resistance, which seeks to inhibit the downward motion and increases with velocity. As the velocity of a given drop increases due to the acceleration of the earth's gravity, the velocity-dependent air resistance force also increases. Eventually the two forces on the drop become equal and the drop's downward velocity reaches a constant "terminal" velocity, whose value can be measured. At this point we have

$$\text{force of gravity} = \text{force of air resistance} .$$

If m is equal to the mass of the drop and v is the drop's terminal velocity, we may write that

$$mg = k_f v ,$$

where g is the acceleration of the earth's gravity and k_f is the "friction constant" between the drop and the air. Solving for the mass of the drop yields

$$m = \frac{k_f v}{g} .$$

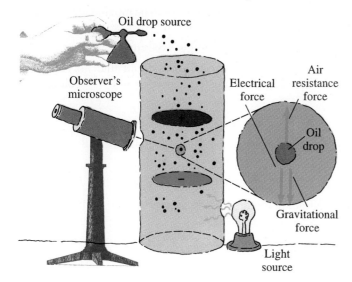

FIGURE 10.18 ·········

The Millikan Oil-Drop Experiment If oil drops carrying an electric charge are sprayed between two charged parallel plates, an air resistance force, a gravitational force, and an electrical force are all acting on each individual drop. If the velocity of a given drop is constant, then the net force acting on the drop must be zero. From this fact, one can deduce the value of the unit of electric charge.

Thus the mass of the drop can be determined, assuming we know the value of the friction constant associated with the frictional "drag" of the air on the downward fall of the oil drop.*

Once the mass of the drop has been determined, the plates are then charged so that a uniform electric field of known magnitude E exists between the plates. A source of ionizing radiation such as X rays is directed at the space between the plates and a fine mist of oil drops is once more introduced into the apparatus. From time to time one of the ions produced in the air by the X rays will attach itself to one of the falling drops. At this point, the drop becomes subject to a third force from the charged plates. This force has a magnitude equal to the product of the total charge on the drop and the electric field. The presence of this third force causes the terminal velocity of the "charge-infected" drop to change its magnitude and, depending on the sign of the attached charge, perhaps its direction as well. Such charged drops can easily be distinguished from electrically neutral ones and the values of their new terminal velocities measured. Now we have

force of air resistance = force of gravity + electric field force, thus

$$mg = k_f v + QE .$$

If we solve this equation for Q, the total charge on the drop, we obtain,

$$Q = \frac{mg - k_f v}{E} .$$

Since all of the quantities on the right-hand side of the equation are either known or have been measured, the total charge on the drop can be readily calculated.

In performing such experiments and calculations, Millikan found that the values of the electrical charges residing on the oil drops were always integral multiples of 1.60×10^{-19} coulomb and, in fact, never took on any values smaller than this. He thus concluded that the fundamental unit of electric charge was equal to 1.60×10^{-19} coulomb.

QUESTIONS

1. Why is it necessary in this experiment for the drops to be moving at a constant velocity?

2. Describe the forces that act on the drops in this experiment.

3. Discuss the significance of the fact that Millikan could find no electrical charges in his experiments that were smaller than 1.60×10^{-19} coulomb.

PROBLEM

1. A physicist operating a Millikan oil drop apparatus obtains the following data for a certain charged droplet: $m = 5.6 \times 10^{-12}$ kg

   $k_f = 2 \times 10^{-10}$ kg/sec $v = 0.04$ m/sec $E = 1.0 \times 10^8$ V/m

   What is the electric charge on the droplet? Is this an integral multiple of the fundamental unit of electric charge?

---

*Interestingly, much of Millikan's Nobel prize winning work on the unit of electric charge went into the determination of the exact mathematical law describing how the fall of small liquid spheres is affected by the presence of the air resistance.

PHYSICS FORMULATION 10.2

Electric Power

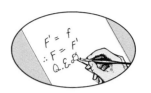

Since electrical currents are employed to drive a wide range of electrical devices, it is useful to be able to calculate the amount of power a given electrical device can develop. As we have seen, for a particle with a total charge Q in an electric field E, the work W is given by

$$W = E \times Q \times d .$$

The voltage V is equal to the product of the electric field and the distance the charge moves, thus

$$W = V \times Q .$$

By definition, the power P developed in a given system is

$$P = \frac{W}{\Delta t} ,$$

where Δt is equal to the elapsed time. Hence,

$$P = \frac{V \times Q}{\Delta t} .$$

The total charge per elapsed time is merely the electric current I and we have

$$P = V \times I .$$

If a given device has a certain resistance R, then from Ohm's law we may write

$$V = I \times R$$

and

$$\begin{aligned} P &= (I \times R) \times I \\ &= I^2 \times R . \end{aligned}$$

This equation may also be written

$$P = V^2/R .$$

By making use of these equations one can determine, for example, how much current a given appliance or device will draw from a given voltage supply. A 100-watt light bulb operating on a 120-volt line thus draws 100 watts/120 volts or ⅚ of an ampere of current. An appliance having a resistance of 50 ohms operating on a 120-volt line will consume power at a rate of (120 volts)²/50 ohms or 288 watts, and so on.

QUESTIONS

1. Show that the units of the product (voltage)²/resistance are those of power, i.e., joule/sec or watts.

2. Show that the equation

    electrical power = (current)² × resistance)

    can be written as

    electrical power = (voltage)²/resistance .

3. Sketch graphs of each of the following:
   a. current versus electric power with constant resistance
   b. resistance versus electric power with constant current
   c. voltage versus electric power with constant resistance.

   Do your results suggest anything concerning the design of electrical appliances? Explain.

PROBLEM

1. A 60-watt light bulb is plugged into a 120-volt outlet. Find
   a. the current that will flow through the light bulb and
   b. the resistance of the light bulb.

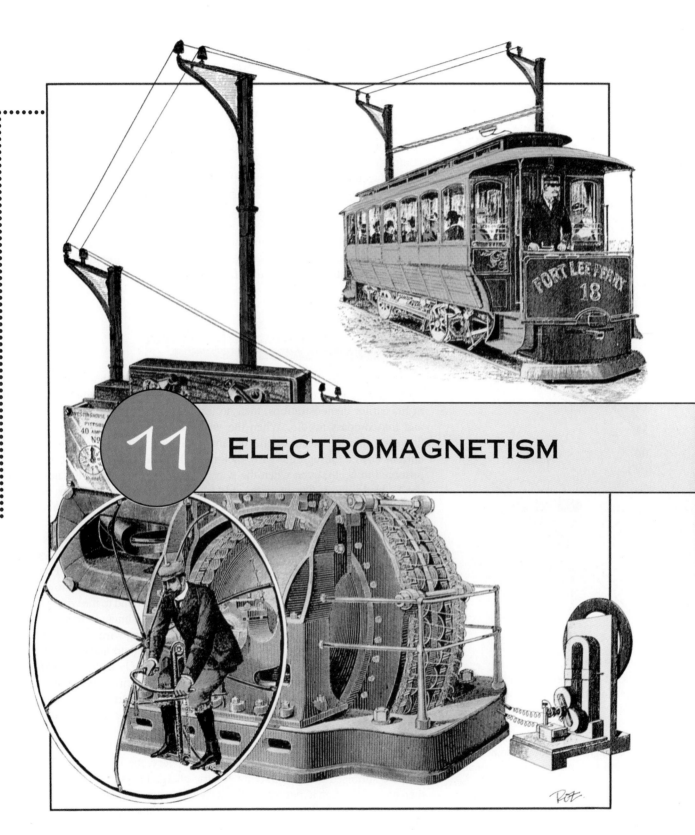

11 ELECTROMAGNETISM

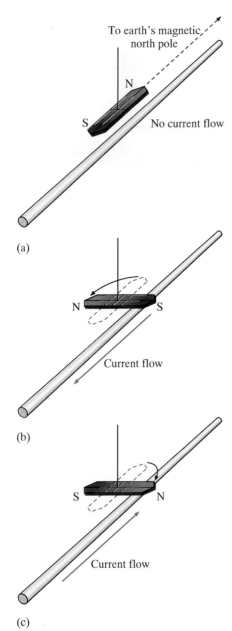

FIGURE 11.1 ··········
**Øersted's Magnetic Effect
Experiment**   (a) If a magnet aligned
along the earth's magnetic field lines is
placed over a wire through which no
electric current is flowing, the north pole
of the magnet continues to point toward
the earth's magnetic north pole. (b and
c) If a current is passed through the wire,
the magnet will deflect as shown in
response to the magnetic field created
about the wire by the flow of electric
charge.

$\bigvee$ ery often important discoveries in science come about in a most surprising
and unexpected fashion. Such was certainly the case when in 1820 a Danish
professor of physics at the University of Copenhagen named Hans Christian
Øersted began a demonstration lecture on electricity and magnetism. As an integral part of this lecture, Øersted would hook a wire up to the ends of a battery
and then place the current-carrying wire directly over a magnetic compass in
such a way that the wire was at right angles to the compass needle. When the
compass needle failed to respond to the proximity of an electric charge flow,
Øersted would proudly hail the result as a clear-cut demonstration that magnetism
and electricity were no more related to each other than gravity and electricity.
On this particular day, however, much to Øersted's horror and no doubt to the
delight of his students, the compass needle *did* respond to the presence of a
current-carrying wire by lurching in a sideways fashion. Øersted had inadvertently
stumbled onto an intimate relationship between two seemingly independent interactions in nature. This relationship is now known as *electromagnetism*.

11.1 Electromagnetic Force

In pursuing his discovery further, Øersted was able to show that when the current-
carrying wire was perpendicular to the magnetic compass needle as it had been
in his earlier demonstrations, then no interaction occurred between the current
and the compass needle. When the wire was lined up parallel to the needle,
however, then the needle deflected in a sideways direction. When no electric
current flowed through the wire, then no deflection of the magnetic needle was
observed to occur, even when the wire was aligned parallel to the needle (Figure 11.1). In short, Øersted was able to demonstrate that the flow of *electric*
charge produced a *magnetic* field that was capable of affecting another magnet.

Having established the existence of a magnetic field about a current-carrying
wire, one would next like to somehow map out such a field. To do this, we first
mount a wire vertically through a flat plate of insulating material. As current
flows through the wire, the associated magnetic field can be mapped by placing
a small magnetic compass at various positions on the plate, as shown in Figure 11.2. At each position, the compass needle will respond to the wire's magnetic
field by aligning itself along the wire's magnetic field lines. In this fashion, one
can deduce that the magnetic field of a current-carrying wire can be represented
by a series of circular field lines. All of these circular field lines have cylindrical
symmetry with the wire and their strengths vary inversely with the perpendicular
distance to the wire (Figure 11.3). The direction of the wire's magnetic field lines
can be readily obtained through the use of the so-called ''right-hand rule'' shown
in Figure 11.4. If one points the thumb of one's right hand in the direction of the
positive charge flow along the wire, then the fingers of the right hand tend to
curl in the direction of the surrounding circular magnetic field lines.

Within a year after Øersted's initial discovery, the French physicist Andre
Ampere surmised that if a moving electric charge behaved as a magnet, and it
then moved through a magnetic field, it should experience an ''electromagnetic''
force just as surely as a bar magnet would experience a magnetic force when
placed in that same field. This hypothesis was in fact confirmed by Ampere in
a series of imaginative experiments involving current-carrying wires. More detailed investigation revealed that when an electric charge is moving through a

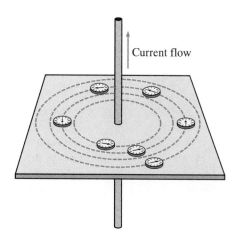

FIGURE 11.2 ··········
Mapping the Magnetic Field Around a Current-Carrying Wire   If a small magnetic compass is placed on a flat insulating plate, the current in the wire will cause the compass needle to align with the magnetic field lines produced by the current flow. In this way, we can map the direction and shape of the wire's magnetic field lines.

magnetic field parallel to the magnetic field lines, then this electromagnetic force is zero. The value of the force increases as the charge moves at progressively larger angles to the magnetic field lines. Finally, when the charge is moving at some velocity perpendicular to the magnetic field lines, then the magnitude of the electromagnetic force exerted on the charge is given by

$$\begin{matrix} \text{electromagnetic} \\ \text{force} \end{matrix} = \text{charge} \times \begin{pmatrix} \text{velocity} \\ \text{of} \\ \text{charge} \end{pmatrix} \times \begin{pmatrix} \text{strength of} \\ \text{magnetic} \\ \text{field} \end{pmatrix} \quad \textbf{11.1}$$

or

$$F_{\text{electromagnetic}} = Q \times v \times B \ .$$

Magnetic field strengths are measured in units called *teslas*. One tesla is defined as 1 newton per coulomb-meter/sec (N/C-m/sec) or 1 newton per ampere-meter (N/A-m). The force is exerted in a direction perpendicular to the plane determined by the directions of the velocity of the charge and the magnetic field lines. This direction can be obtained through the use of a variation of the right-hand rule, as shown in Figure 11.5. If the fingers of one's right hand are initially pointed in the direction of the velocity of a moving positive charge and then "curled" toward the direction of the magnetic field lines, the thumb of the right hand will point in the direction of the electromagnetic force on the charge. If the charge

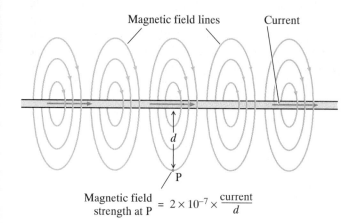

$$\begin{matrix} \text{Magnetic field} \\ \text{strength at P} \end{matrix} = 2 \times 10^{-7} \times \frac{\text{current}}{d}$$

FIGURE 11.3 ··········
The Magnetic Field About a Current-Carrying Wire   The magnetic field lines about a current-carrying wire consist of a set of rings that have cylindrical symmetry with the wire. The strength of the field decreases as the perpendicular distance from the wire increases.

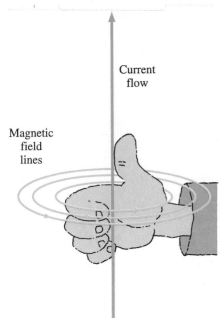

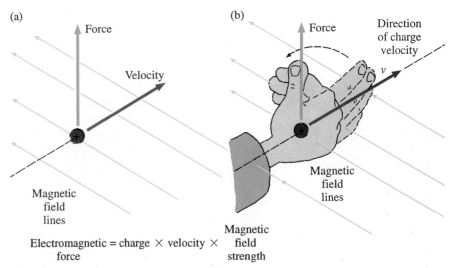

Electromagnetic = charge × velocity × field
force   strength

FIGURE 11.5 ···········
The Electromagnetic Force   (a) If a charged particle is moving perpendicular to the field lines of a magnetic field, it will experience a force having a magnitude equal to the product of the mass, charge, and magnetic field strength. (b) The direction of the electromagnetic force is determined by a version of the right-hand rule in which the fingers of one's right hand initially point in the direction of the velocity of a positive electric charge. When the fingers are curled toward the direction of the magnetic field lines, the right-hand thumb will point in the direction of the electromagnetic force on the charge.

FIGURE 11.4 ···········
The Right-hand Rule   If the thumb of the right hand is pointing in the direction of the current flow of positive charges, then the fingers of the right hand will curl along the direction of the magnetic field lines associated with the current flow.

is negative, then the electromagnetic force will be exerted in the exact opposite direction from which the right-hand thumb is pointing.

Suppose that the single moving electric charge discussed above is now replaced by a current-carrying wire as shown in Figure 11.6. As each electric charge moves down the wire, it experiences an electromagnetic force that is perpendicular to the plane determined by the direction of the charge velocity and the direction

FIGURE 11.6 ···········
The Force on a Current-Carrying Wire in a Magnetic Field   If electric charges are moving along a wire in a magnetic field, then each charge in the wire will experience a force due to its motion in the magnetic field. The total of all of these individual forces is the total force on the wire.

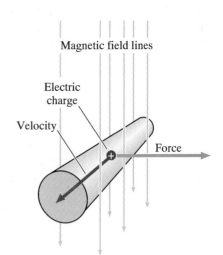

of the magnetic field lines. If we calculate the total effect of all of the individual electromagnetic forces acting on the individual charges moving along the wire, we obtain (Physics Formulation 11.1) a total force whose magnitude is

$$\frac{\text{force}}{\text{length}} = \text{current} \times \begin{array}{c} \text{magnetic} \\ \text{field} \\ \text{strength} \end{array}$$

and whose direction is the same as that for the individual forces.

This result provides us with the basis for the formal definition of the ampere and the coulomb. Suppose two current-carrying wires are set parallel to each other as shown in Figure 11.7. As current flows through both wires, each wire will have a magnetic field associated with the motion of its respective electric charges. These magnetic fields then exert electromagnetic forces on the other wire's moving charges, thereby producing a mutual force of attraction if the currents in the wires are in the same direction. However, a mutual force of repulsion is produced if the currents in the wires are in opposite directions. From Physics Formulation 11.1, the magnitude of this mutual force per unit length is found to be

$$\frac{\text{force}}{\text{length}} = (2 \times 10^{-7}) \frac{\left(\begin{array}{c} \text{current in} \\ \text{first wire} \end{array} \right) \left(\begin{array}{c} \text{current in} \\ \text{second wire} \end{array} \right)}{\text{distance between wires}}.$$

Armed with this equation, physicists have formally defined the ampere as the amount of current flowing in each of two, long parallel wires that will produce a force of 2×10^{-7} newtons per meter of wire length when the wires are separated by a distance of exactly 1 meter (Figure 11.8). This is precisely equivalent to the "preliminary" definition of the ampere set forth earlier. In the present definition, however, the "standard" ampere is expressed in terms of the far more easily measured quantities of force and distance.

The coulomb can now be formally defined as that quantity of electric charge which passes by a given point in a wire in 1 second when a current of exactly 1 ampere is flowing through that wire. In other words, a current of 1 ampere corresponds to a charge flow of 1 coulomb per second. Since we have already defined the ampere in terms of the observed force between two parallel current-carrying wires at a known distance from each other, the coulomb is now formally defined in terms of these same parameters as well.

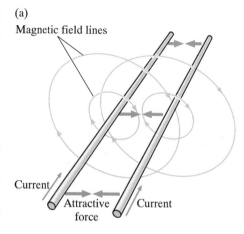

(a)

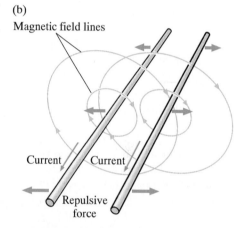

(b)

FIGURE 11.7 ··········
Forces on Parallel Current-Carrying Wires If two parallel wires are carrying electric current, each will produce a magnetic field. The motion of each wire's charges in the other wire's magnetic field produces a mutual force between the wires that is (a) attractive if the currents in the wires are in the same direction or (b) repulsive if the currents in the wires are in opposite directions.

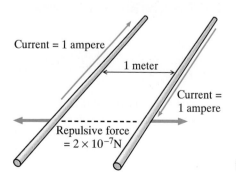

FIGURE 11.8 ··········
The Ampere If equal electrical currents flowing through two wires separated by a distance of 1 meter produce an attractive or repulsive force between the wires of 2 $\times 10^{-7}$ N per meter of wire, then the current in the wires has, by definition, a magnitude of one ampere.

11.1 The Measurement of Electric Current

In the course of dealing with electrical phenomena, it is absolutely necessary to be able to measure electric currents with relative ease and precision. An important device in this regard is an offshoot of the electric motor and electric generator called a *galvanometer*.

Like the electric motor and generator, the galvanometer consists of a rotatable wire coil mounted in a magnetic field. In the case of the galvanometer, however, a small light-weight pointer is attached to the coil as shown in the figure. The entire coil-pointer ensemble is mounted on two "control" springs, which serve as electrical terminals for the coil and align the coil pointer to the zero point on the galvanometer scale when no current is flowing through the coil.

As current passes through the coil, the moving charges in the coil experience an electromagnetic force that, like the coil in an electric motor, causes the coil to rotate. In the galvanometer, however, this rotation is opposed by the restoring force exerted by the control springs. Since the restoring force of the spring increases by the amount it is displaced from its equilibrium or "zero" position, the coil's rotation will continue only so far before coming into equilibrium with the springs' restoring force. At this point, the coil will have rotated to an equilibrium position for the given current, and the pointer will point to a given point on the galvanometer scale for that current.

Values for the galvanometer scale can thus be set by passing known amounts of electric current as determined from a current balance, etc. (Physics Formulation 11.1), and noting the position of the pointer on the scale for each such

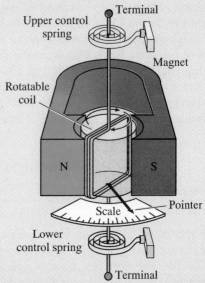

••• The Galvanometer   As current flows through the coil, an electromagnetic force causes the coil to rotate. The restoring force from the control springs balances this force, and the coil takes on an equilibrium orientation, which is indicated by the pointer on the galvanometer scale.

current. Once this calibration process is completed, unknown electric currents can be measured by reading from the galvanometer scale how much of a deflection they produce in the coil orientation compared to currents having known values.

WORKED EXAMPLE

A force of 0.005 newtons per meter is exerted between two parallel wires when the wires carry a current of 10 amperes each. How far apart are the two wires?

Continued

WORKED EXAMPLE

Continued

SOLUTION: The mutual force per unit length between two parallel current-carrying wires is given by

$$\frac{\text{force on wire}}{\text{unit length}} = 2 \times 10^{-7} \frac{\left(\begin{array}{c}\text{current in} \\ \text{first wire}\end{array}\right) \left(\begin{array}{c}\text{current in} \\ \text{second wire}\end{array}\right)}{\text{distance between wires}}.$$

In this case, we have

$$0.005 = 2 \times 10^{-7} \frac{10 \times 10}{\text{distance between wires}}.$$

Solving this equation for the distance between the wires, we obtain

$$\begin{aligned}
\text{distance between wires} &= \frac{2 \times 10^{-7} \times 10 \times 10}{0.005} \text{ m} \\
&= \frac{2 \times 10^{-7} \times 10 \times 10}{5 \times 10^{-3}} \text{ m} \\
&= 0.4 \times 10^{-7+1+1+3} \text{ m} \\
&= 0.4 \times 10^{-2} \text{ m} \\
&= 4 \times 10^{-3} \text{ m}
\end{aligned}$$

or about 4 millimeters.

---

1. Explain how Øersted demonstrated that electricity and magnetism are related phenomena.

2. Describe the electromagnetic force that is exerted on a charged particle moving in a magnetic field. How does this force compare with magnetic and electrostatic forces?

3. What is the right-hand rule? How is it used to describe electromagnetic interactions?

4. Discuss the electromagnetic forces exerted between two current-carrying wires.

5. How are the ampere and coulomb formally defined? What is the advantage to the scientist in defining these quantities in such a fashion?

REVIEW
11.1
QUESTIONS

---

1. Find the magnitude and direction of the electromagnetic force exerted on a charge of 2×10^{-6} C, which is moving directly away from you along your desktop at a speed of 5 m/sec through a magnetic field of 1.5×10^{-3} teslas whose field lines are pointing perpendicularly upward from the desktop.

REVIEW
11.1
PROBLEMS

2. Currents of 3 amperes and 5 amperes run through a pair of parallel wires located 0.1 meters from each other. Find the magnitude and direction of the electromagnetic force between the wires if the currents are moving in opposite directions.

11.2 Electromagnetic Induction

The discovery that an electromagnetic force is exerted on an electric charge as it moves in a magnetic field raises an intriguing question. If a moving electric charge has a force exerted on it by a stationary magnetic field, could the reverse situation be true as well? Could stationary electric charges be set in motion from an electromagnetic force delivered by a "moving" magnetic field? In short, could electric voltages and electric currents somehow be generated from magnetic fields? Little more than a decade after Øersted's original discoveries, an American physicist, Joseph Henry, and an English scientist, Michael Faraday, simultaneously but independently discovered in 1831 that electric charges can indeed be set in motion and a voltage "induced" by a "moving" magnetic field.

There are several possible ways in which a "moving" magnetic field can be created. First of all, one can move the electric charges through a stationary magnetic field, thus causing the charges to be subject to an electromagnetic force with magnitude of

$$\text{force} = \text{charge} \times \text{velocity} \times \genfrac{}{}{0pt}{}{\text{magnetic}}{\genfrac{}{}{0pt}{}{\text{field}}{\text{strength}}} \ .$$

For simplicity, we will assume for the time being that the motion of the electric charges is perpendicular to the magnetic field lines. If the charges are located in a segment of wire having a length L and the wire segment is pushed through a magnetic field as shown in Figure 11.9, the electric charges in the wire segment will respond to the electromagnetic force exerted on them in an interesting and important way. The positive charges will experience an electromagnetic force which pushes them to one end of the wire segment, while the negative charges will experience a similar force which pushes them toward the opposite end of the wire segment. The result is a separation of electrical charge that is identical to that found in a battery or voltage supply. The magnitude of the voltage associated with such a separation of charges can be obtained by recalling the basic definition of work

$$\text{work} = \text{force} \times \text{distance} \ .$$

In this case the force is the electromagnetic force on the charges and the distance is the length of the wire segment. We thus have that the work done in moving a charge along the length of the wire segment is

$$\text{work} = \text{charge} \times \text{velocity} \times \genfrac{}{}{0pt}{}{\text{magnetic}}{\genfrac{}{}{0pt}{}{\text{field}}{\text{strength}}} \times \genfrac{}{}{0pt}{}{\text{length}}{\genfrac{}{}{0pt}{}{\text{of}}{\text{wire}}} \ .$$

The work required per unit charge to move charges the length of the segment is

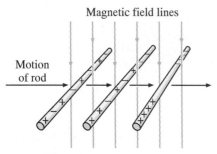

Magnetic field lines

Motion of rod

FIGURE 11.9 ··········

Induced Voltage As a rod moves through a magnetic field perpendicular to the field lines, the electromagnetic forces on the charges in the rod lead to a separation of charge, and hence an electrical potential difference or voltage develops between the ends of the rod.

$$\frac{\text{work}}{\text{charge}} = \text{velocity} \times \begin{array}{c}\text{magnetic}\\ \text{field}\\ \text{strength}\end{array} \times \begin{array}{c}\text{length}\\ \text{of}\\ \text{wire}\end{array}.$$

Because energy must be conserved, the work put into a given system must show up as some form of energy. In this case the work per unit charge needed to separate the electrical charges reappears in the form of an electrical potential energy difference per unit charge or a voltage between the ends of the wire segment. Thus we have

$$\text{voltage} = \text{velocity} \times \begin{array}{c}\text{magnetic}\\ \text{field}\\ \text{strength}\end{array} \times \begin{array}{c}\text{length}\\ \text{of}\\ \text{wire}\end{array} \qquad \textbf{11.2}$$

or

$$V = v \times B \times L.$$

The polarity of the induced voltage and the direction of any current flow produced by it are determined by a statement initially proposed by the Russian scientist Emil Lenz and is referred to as *Lenz's law*. Lenz's law can be stated as follows:

> The polarity of an induced voltage and the direction of any current flow it produces will always be such as to oppose the change which initially produced them.

For example, if we were to hook up the ends of the wire segment in Figure 11.9 to an outside electrical circuit, a current would flow through the wire. Such a current flow will produce a magnetic field around the wire segment. Lenz's law tells us that this induced current and its associated magnetic field must oppose the change that initially produced them. In the case we are currently considering, that initial change was the motion of the wire segment through the magnetic field. Thus the induced current resulting from the induced voltage will flow in such a way that its interactions with the surrounding magnetic field will tend to stop that motion of the wire. If this were not true, then the induced current would actually accelerate the motion of the wire, which in turn would create a larger induced voltage, leading to a larger induced current, etc. In short, we would have a perpetual motion device that would be a clear-cut violation of the law of the conservation of energy.

Similar effects can be observed when a stationary array of charges interacts with a magnetic field that is moving at some velocity relative to the charges. One simple example of such a situation is illustrated in Figure 11.10. The charges are located along a wire conductor that has been shaped into a spring-like configuration called a *coil*. When a bar magnet is pushed through the coil, an electromagnetic force is exerted on the charges in the coil, causing them to separate. This charge separation results in the development of an electric potential difference between the ends of the coiled wire. If some sort of device such as a current-measuring ammeter is attached to these endpoints, current will flow through the meter.

Faraday recognized that these and other similar phenomena could be neatly summarized in terms of a quantity called *magnetic flux*. The magnetic flux associated with a given surface area is equal to the product of the surface area and the component of the magnetic field strength perpendicular to that area. Thus if the magnetic field lines are parallel to the surface area, no part of the magnetic field strength is perpendicular to the area, and the magnetic flux is zero. If the

(a)

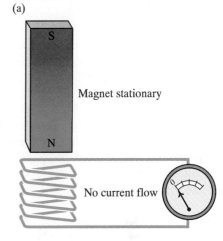

Magnet stationary

No current flow

(b)

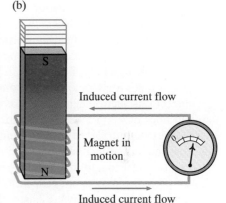

Induced current flow

Magnet in motion

Induced current flow

FIGURE 11.10 ··········

Electric Current Induced by a Moving Magnetic Field   (a) When the magnet is stationary relative to the coil, no current flows. (b) If the magnet is set in motion through the coil, then the moving magnetic field will exert a force on the charges in the coil and an induced electric current will flow through the coil and the connected current meter.

11.2 Geomagnetism

For centuries, sailors from cultures the world over have taken advantage of the fact that a freely rotating bar magnet will align itself in a roughly north-south orientation. It was not until the turn of the seventeenth century, however, that a viable explanation for this phenomenon was advanced by the Englishman Sir William Gilbert, court physician to Queen Elizabeth I. Gilbert fashioned a lodestone into a sphere and demonstrated that a small compass placed anywhere on the surface of the sphere would always align in a north-south direction. Thus, he reasoned, Earth must be a gigantic magnet in which the magnetic poles are located roughly at Earth's geographic poles.

Since then we have found that Earth's magnetic field is far more complex and interesting than anything Gilbert could have imagined. In a general sort of way, Earth's magnetic field can be approximated by imagining an enormous bar magnet buried inside of the earth at a slight angle to Earth's axis of rotation. The region of magnetic field about the Earth is called Earth's *magnetosphere,* and interaction between Earth's magnetic field and particle emissions from both the Sun and deep space has given rise to a number of interesting phenomena. For example, as high-energy charged particles impinge on Earth from outer space, they are subject to an electromagnetic force as they move through the field lines of Earth's magnetic field. As a result, these charged particles are forced to spiral along Earth's magnetic field lines and are effectively trapped in zones called *Van Allen belts.* Occasionally the Sun will emit a tremendous burst of particles in an event called a solar flare. As these particles reach Earth's magnetosphere, they too are forced to spiral along Earth's magnetic field lines, except in this case they have enough kinetic energy to violently collide with the molecules in the outer layers of Earth's atmosphere and cause the electrons in these molecules either to be stripped from the molecule or relocated within the molecule. As the electrons ''recombine'' or return to their former positions, they give off the energy picked up from the collision in the form of a shimmering display of light which is called an *aurora.*

The origin of Earth's magnetic field is something of a mystery. The field is somehow generated by the looping

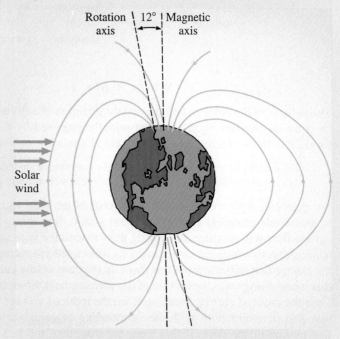

··· The Earth's Magnetic Field The pattern of magnetic field lines for the earth is similar to that of a gigantic bar magnet embedded in the earth at an angle of about 12° relative to the earth's rotational axis. The distortions in the earth's magnetic field line pattern are due to interactions with a flow of charged particles from the sun called the solar wind.

motions of large numbers of charged particles in Earth's interior, but understanding the exact nature of that process has thus far proved to be most elusive. If we assume, for example, that the required current loops are produced by Earth's rotation, then we are faced with explaining why Earth's magnetic poles are situated more than 1700 kilometers away from its rotational or geographic poles.

Most fascinating of all, however, has been the discovery in recent years that Earth's magnetic field has changed dramatically throughout its geological history. By studying ''fossilized'' magnetic properties of various rock strata in its crust, scientists have deduced that Earth's magnetic field has gone through several ''reversals'' in which the north and south poles of Earth's magnetic fields reverse them-

PHYSICS FACET

Continued

selves. As these reversals proceed, there are periods of time, sometimes lasting as long as thousands of years, during which Earth's magnetic field strength is near zero. It is suspected that these periods of zero magnetic field strength in turn have had a great impact on the manner and direction in which life has developed on our planet, but for the present one can only speculate on such possibilities.

Well within the realm of scientific fact, however, is the observation that magnetic fields are not the sole province of Earth. The mighty planet Jupiter, for example, is in possession of a magnetic field that is some 20 thousand times that of Earth's, and even the lowly planet Mercury exhibits a small magnetic field that is most likely a fossil remnant

of a magnetically more powerful era in that planet's geological past. On an even larger scale, we now know that stars such as our Sun possess magnetic fields that can be thousands of times stronger than Earth's field and can engulf billions of square kilometers of space. On what is perhaps the most ostentatious scale of all, the Milky Way, the vast system of stars to which our solar system belongs, has an overall magnetic field, as do the hundred remote star systems we have named the galaxies. Magnetic fields thus seem to exist across the entire span of our perception of the physical universe, from the tiny current loop magnetic fields of atoms to the gigantic galactic magnetic fields that engulf hundreds of billions of stars.

surface area is perpendicular to the magnetic field lines, then all of the magnetic field strength is perpendicular to the surface area and we have

$$\begin{matrix} \text{magnetic} \\ \text{flux} \end{matrix} = \begin{matrix} \text{magnetic} \\ \text{field} \\ \text{strength} \end{matrix} \times \begin{matrix} \text{surface} \\ \text{area} \end{matrix} .$$

If the surface area–field lines angle is at some intermediate angle between 0° and 90°, then the magnetic flux takes on corresponding values, which are intermediate between zero and the product of the surface area and magnetic field strength (Figure 11.11). Because magnetic flux is essentially a product of magnetic field strength and area, this quantity is expressed in units of tesla-m^2.

Having defined the magnetic flux, let us now consider the system shown in Figure 11.12 in which a square-shaped conducting wire is placed in a magnetic field such that its plane is perpendicular to the magnetic field lines. Assume further that one of the wire sides is movable. If the sliding wire is moved at some velocity, then the magnitude of the induced voltage at the ends of the sliding wire will be

$$\begin{matrix} \text{induced} \\ \text{voltage} \end{matrix} = \begin{matrix} \text{magnetic} \\ \text{field} \\ \text{strength} \end{matrix} \times \begin{matrix} \text{length} \\ \text{of} \\ \text{wire} \end{matrix} \times \begin{matrix} \text{velocity} \\ \text{of} \\ \text{wire} \end{matrix} .$$

FIGURE 11.11 ··········

Magnetic Flux The magnetic flux passing through a given area is equal to the product of the magnetic field strength and that part of the area which projects perpendicular to the magnetic field lines.

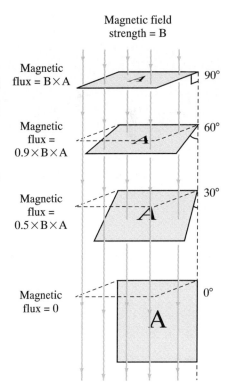

Magnetic field strength = B

Magnetic flux = B × A   90°

Magnetic flux = 0.9 × B × A   60°

Magnetic flux = 0.5 × B × A   30°

Magnetic flux = 0   0°

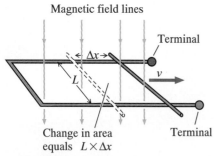

Magnetic field lines

Terminal

Δx

L

v

Change in area equals $L \times \Delta x$

Terminal

FIGURE 11.12 ··········

Faraday's Law As the wire segment moves perpendicular to the magnetic field lines, the area of the loop is increasing at a rate of $L \times \Delta x / \Delta t$. The induced voltage is equal to the product of this time rate of area change and the magnetic field strength, or simply the time rate of change of the magnetic flux.

The velocity of the wire is equal to

$$\text{velocity} = \frac{\text{distance wire travels}}{\text{elapsed time}},$$

and thus

$$\begin{matrix}\text{induced}\\\text{voltage}\end{matrix} = \begin{matrix}\text{magnetic}\\\text{field}\\\text{strength}\end{matrix} \times \begin{matrix}\text{length}\\\text{of}\\\text{wire}\end{matrix} \times \frac{\text{distance wire travels}}{\text{elapsed time}}.$$

The product of the length of the wire and the distance the wire travels is equal to the change in the area of the wire loop as the sliding wire moves through the field. Thus we have

$$\begin{matrix}\text{induced}\\\text{voltage}\end{matrix} = \begin{matrix}\text{magnetic}\\\text{field}\\\text{strength}\end{matrix} \times \frac{\text{change in area}}{\text{elapsed time}}.$$

The product of the magnetic field strength and the change in area is equal to the change in the magnetic flux. As a result, the magnitude of the induced voltage is

$$\begin{matrix}\text{induced}\\\text{voltage}\end{matrix} = \frac{\text{change in magnetic flux}}{\text{elapsed time}}.$$

The induced voltage will, in accordance with Lenz's law, always develop in such a way as to oppose the change that produced it. This fact is denoted mathematically by placing a negative sign in front of the expression on the right-hand side of the previous equation. The equation now reads as follows:

$$\begin{matrix}\text{induced}\\\text{voltage}\end{matrix} = -\frac{\text{change in magnetic flux}}{\text{elapsed time}} \qquad \text{11.3}$$

or

$$V = -\frac{\Delta \phi}{\Delta t}.$$

This very important equation was first formulated by Michael Faraday and is known as *Faraday's law*. This physical principle tells us that a voltage is induced whenever a change in the magnetic flux occurs. Such flux changes can occur as a result of either or both a changing area or a changing magnetic intensity. This relatively simple principle has provided the basis for some of the most extraordinarily important technological devices of the past 150 years.

WORKED EXAMPLE

A wire shaped into a rectangular loop 0.4 m $\times$ 0.2 m is placed in a magnetic field having a field strength of 0.3 tesla. Find the maximum voltage that can be induced in the loop if (a) the orientation of the loop in the field is altered over a 0.02-sec time interval and (b) the field strength drops to zero in a 0.2-sec time interval.

Continued

WORKED EXAMPLE

Continued

SOLUTION: (a) The maximum induced voltage occurs when the plane of the loop changes from being parallel to the field lines to an orientation that is perpendicular to them. For such a rotation, the area of the loop perpendicular to the field lines changes by

$$\text{change in area} = 0.4 \text{ m} \times 0.2 \text{ m}$$

$$= 0.08 \text{ m}^2 .$$

If this change in projected area occurs in a magnetic field strength of 0.3 tesla, then the change in magnetic flux is

$$\text{change in magnetic flux} = 0.3 \text{ tesla} \times 0.08 \text{ m}^2$$

$$= 0.024 \text{ tesla-m}^2 .$$

From Faraday's law we know, in general, that

$$\frac{\text{induced}}{\text{voltage}} = -\frac{\text{change in magnetic flux}}{\text{elapsed time}} .$$

For the present case, then

$$\frac{\text{induced}}{\text{voltage}} = -\frac{0.024 \text{ tesla-m}^2}{0.02 \text{ seconds}}$$

$$= -1.20 \text{ volts}$$

(b) If the magnetic field drops to zero, then the field strength changes by an amount equal to 0.3 tesla. For the maximum effect to occur, the plane of the wire loop must be oriented perpendicular to the magnetic field lines when this change occurs. If it is, then the change in magnetic flux is

$$\text{change in magnetic flux} = \frac{\text{change in magnetic field strength}}{} \times \text{area} .$$

In this case,

$$\text{change in magnetic flux} = 0.3 \text{ tesla} \times 0.4 \text{ m} \times 0.2 \text{ m}$$

$$= 0.024 \text{ tesla-m}^2 .$$

Since this change occurs over an elapsed time of 0.20 seconds, we know from Faraday's law that

Continued

WORKED EXAMPLE

Continued

$$\frac{\text{induced}}{\text{voltage}} = \frac{0.024 \text{ tesla-m}^2}{0.20 \text{ seconds}}$$

$$= -0.12 \text{ volts} .$$

The negative signs in each result indicate that the voltage will develop in such a way as to oppose the orientation change of the loop or the drop in the magnetic field strength.

REVIEW 11.2 QUESTIONS

1. Explain how a voltage can be produced by the principles of electromagnetism.
2. What are Lenz's and Faraday's laws? Discuss their importance.
3. In what ways, if any, are Lenz's law and Faraday's law related to each other?
4. Define what is meant by magnetic flux. How is magnetic flux related to magnetic field strength?
5. How can the magnetic flux be changed?

REVIEW 11.2 PROBLEMS

1. At what speed must a 1.0-m wire be moved through a magnetic field having a field strength of 1.0 tesla in order to produce an induced voltage of 1.0 volts? Assume the wire moves in a direction perpendicular to that of the field lines. Does your result suggest a possible way to define the volt? Explain.
2. A wire shaped into a circular loop 0.5 m in diameter is placed in a magnetic field having a field strength of 0.01 tesla. If the magnetic field strength remains constant, what is the maximum voltage that can be obtained by changing the loop's orientation in the field at a rate of one complete rotation per second? The area of a circle is equal to $\pi \times (\text{radius})^2$.

11.3 Electromagnetic Devices

The discovery that voltages could be produced by electromagnetic induction led to an almost immediate recognition that such processes literally and figuratively offered enormous potential for human technology. Scientific legend has it that a young William Gladstone, who eventually became a four-time prime minister of England, is supposed to have asked after seeing Faraday demonstrate such induced voltages, ''But, Mr. Faraday, of what use is this?'' Faraday reportedly replied, ''Sir, in twenty years, you will be taxing it!''

Initially, the difficulty in making practical use of electromagnetically induced voltages lay in the fact that they could not be sustained for a significant length

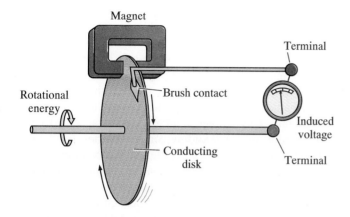

FIGURE 11.13 ··········
Faraday's Electric Generator   As the conducting disk is rotated between the poles of a magnet, a voltage is induced between the rapidly moving outer edges of the disk and the stationary shaft.

of time. If a magnet is shoved into a wire coil, for example, the induced voltage will occur only during the relatively small time interval when the magnetic fluid lines are moving past the coil. Similarly, it is not practical or convenient to have a wire segment linearly move through a magnetic field over a very large distance for a very long interval of time.

Within two months of his discovery of electromagnetic induction, Faraday was able to solve this difficulty and constructed a device in which an electromagnetically induced voltage could be sustained. A schematic of Faraday's apparatus is shown in Figure 11.13. A thin copper disk is mounted on a shaft in such away that the outer edge of the disk passes through the poles of a very strong magnet. As the shaft is turned, the outer edge of the disk passes through the magnetic field at a relatively high speed, thus inducing a voltage in the disk. Since the disk regions close to the shaft are neither moving at high speed nor through magnetic field lines, a voltage difference develops between the central shaft and the outer edge of the disk. As long as the disk is rotating, the induced voltage is maintained and can be used as a voltage source in an electrical circuit. Faraday's device, in effect, was able to transform mechanical energy into electrical energy on a continuing basis. It was the world's first electric generator.

Since Faraday's time, electric generators have become far larger and more sophisticated, but the basic concept remains the same. In modern electric generators, a wire coil consisting of a large number of insulated loops is placed between the poles of a magnet as shown in Figure 11.14. The ends of the wire coil are connected to two rings, which are called *commutator rings*. Each of these rings touches a conducting sliding contact "brush," which in turn is connected to the generator's output terminals. If the coil is rotated by means of input

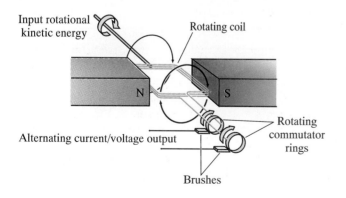

FIGURE 11.14 ··········
The Operation of an Electrical Generator   Rotational kinetic energy is supplied to a wire coil placed between the poles of a magnet. The ends of the wire coil are connected to two commutator rings, each of which touches a conducting brush connected to one of the generator's output terminals. As the coil rotates in the magnetic field, an electric current is produced at the generator's output terminals, which alternates its direction as the plane of the coil changes its orientation in the magnetic field.

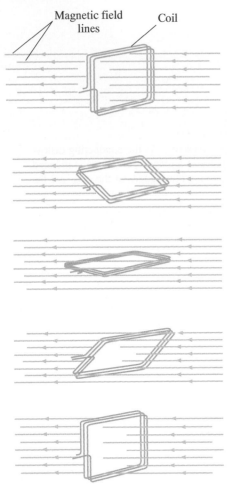

Magnetic field lines   Coil

FIGURE 11.15 ··········

The Rotating Coil   When a coil rotates in a magnetic field, it presents various orientations of its area relative to the magnetic field lines. The projected area can thus range from zero when the coil is parallel to the field lines to the full area of the coil when the coil is perpendicular to the field lines.

FIGURE 11.16 ··········

The Operation of an Electric Motor   An alternating current supply connected to a coil placed between the poles of a magnet produces a continuous rotation of the coil. If a shaft is connected to the rotating coil, it becomes a source of rotational kinetic energy.

mechanical energy, then the plane of the coil will cyclically display various orientations relative to the magnetic field lines in which it is located (Figure 11.15). As these changes in orientation occur, the magnetic flux passing through the coil varies from zero when the plane of the coil is parallel to the magnetic field lines to its maximum value when the plane of the coil is perpendicular to the magnetic field lines. The magnetic flux through the coil is thus changing in time and, from Faraday's law, an induced voltage must therefore develop across the coil terminals. This voltage source is then "tapped" by an outside circuit via the commutator rings and sliding contact brushes.

Note that the output voltage of such a device does not have a constant value, but rather varies cyclically from zero to alternating positive and negative maximum voltages with a period equal to the rotation period of the coil. Such an alternating voltage will produce a similarly varying electrical current in the circuits to which they are connected. Because of this cyclic nature, such voltages and currents are referred to as alternating voltages and currents. Almost all of the electrical energy made available commercially by power companies comes in this format. This is due not only to the fact that alternating voltages and currents are the logical products of a rotating coil electric generator, but, as we shall see, electrical energy in this format can also be readily transmitted over very large distances.

Electromagnetic induction can also be called on to perform the important "reverse" task of converting electrical energy into mechanical energy. The most familiar device by which this is accomplished is the electric motor. The electric motor can be thought of as a sort of "reverse generator," which produces output rotational energy from input electrical energy. The key to that output is once more electromagnetism.

The basic design of an electric motor is very similar to that of the electric generator (Figure 11.16). A conducting coil of wire is placed in a magnetic field and voltage is applied to the terminals of the coil. As the electric charges flow through the sides of the coil, which are perpendicular to the magnetic field lines, they are subjected to an electromagnetic force. Because there are two such current-carrying directions, one side of the coil is subject to an upward force, the other to a downward force. As a result, a net torque is exerted on the axis of rotation of the coil or the motor's "drive shaft." This ongoing torque causes the drive shaft to continuously rotate (Figure 11.17), thus providing a source of rotational energy.

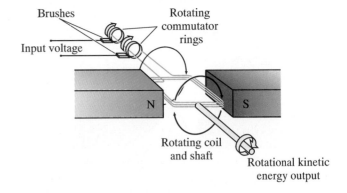

Brushes   Rotating commutator rings

Input voltage

N   S

Rotating coil and shaft

Rotational kinetic energy output

Let us now reconsider the simple current-carrying straight wire shown in Figure 11.18(a). As we saw earlier, the magnetic field lines about the wire are in the form of concentric circles, which are centered on the wire and whose planes are perpendicular to the wire. If the wire is now made into a loop as shown in Figure 11.18(b), the field lines tend to ''crowd together'' at the center of the loop, thereby creating a much higher magnetic field strength at that position. The effect can be enhanced by ''piling up'' a number of such ''current loops'' to create an electric coil or *solenoid* as shown in Figure 11.18(c). In this arrangement, each current loop in the coil adds to the overall field strength of the coil. It can be shown that for a given solenoid length the magnitude of the magnetic field strength along the axis of symmetry of the coil would be twice as large for two loops or ''turns'' compared to a single loop or turn, three times as large for three loops, and, in general,

$$\begin{matrix} \text{magnetic field} \\ \text{strength} \\ \text{(center of coil)} \end{matrix} = \text{number of loops} \times \begin{matrix} \text{magnetic field} \\ \text{strength} \\ \left(\begin{matrix}\text{center of a}\\\text{single turn}\end{matrix}\right) \end{matrix} .$$

A close examination of the magnetic field associated with this type of helical arrangement reveals that while the magnetic field strength is relatively uniform along the solenoid's axis of symmetry, nonetheless a tendency exists for some

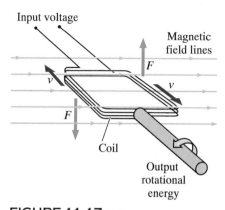

FIGURE 11.17 ··········
The Armature of an Electric Motor   As current flows through the coil, electromagnetic forces are exerted on the parts of the coil that are perpendicular to the magnetic field lines. The resulting torques cause the armature to rotate.

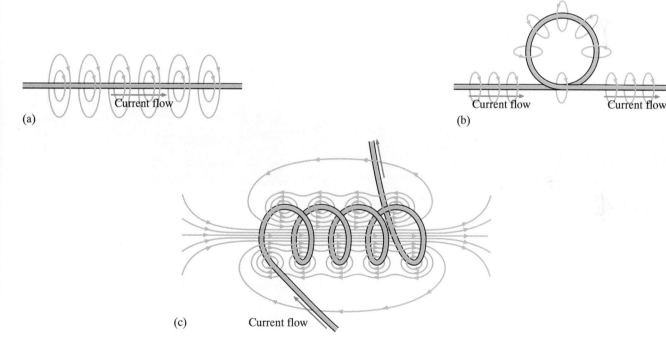

FIGURE 11.18 ··········
Magnetic Fields Produced by Current Flow   Magnetic field line diagrams for (a) a wire carrying an electric current flow; (b) a wire loop carrying an electric current flow; (c) a series of wire loops or helix carrying an electric current flow.

of the magnetic field lines to ''stray'' and ''scatter'' out between the loops [see Figure 11.18(c)]. This situation can be corrected by placing a piece of magnetic material, usually iron, at the core of the solenoid. When the flow of electric current through the coils produces a magnetic field, the iron core becomes magnetized and acts as a sort of ''conductor'' for the magnetic field lines. As a result, the magnetic field within the core is almost completely confined to the iron core and is highly uniform in its strength throughout the core. The field lines at the ends of the iron core are identical to those emanating from a ''permanent'' magnet of similar shape, regardless of whether it is a bar, horseshoe, etc. As a result, a current-carrying helix with an iron core is referred to as an *electromagnet*.

(a)

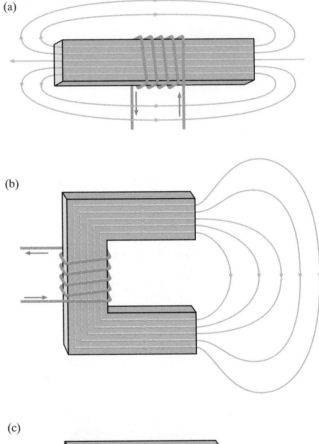

(b)

(c)

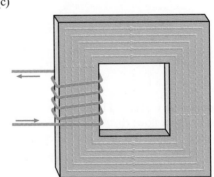

FIGURE 11.19 ··········

Electromagnets (a) A bar electromagnet; (b) a horseshoe electromagnet; (c) a closed-core electromagnet.

Interestingly, if the iron core is in the shape of some sort of closed loop such as a ring or square, then the magnetic field lines will permeate the volume of the entire core, even though most of the core may lie outside of the current-carrying coil. Such an electromagnet is said to have a closed core.

With these results in mind, consider the arrangement shown in Figure 11.19 in which two coils, each having a different number of loops, are wrapped about different sections of a closed core. If an alternating voltage is hooked to the terminals of the "primary" coil on the left, the resulting ac current will produce an alternating magnetic field strength throughout the closed core. Since the magnitude of the magnetic field strength is varying, a corresponding variation in the magnetic flux also occurs throughout the closed core. From Faraday's law, this variation in magnetic flux will induce a voltage at the terminals of the "secondary" coil, which, from Physics Formulation 11.2, is given by

$$\frac{\text{number of loops}}{\text{voltage}} = \frac{\text{number of loops}}{\text{voltage}} .$$
$$\text{(primary coil)} \qquad \text{(secondary coil)}$$

This equation provides the basis for an electromagnetic device called a *transformer* (Figure 11.20). Transformers are able to "transform" voltage levels from lower to higher values (step-up transformers) or from higher to lower values (step-down transformers), depending on the number of loops in the primary and secondary coils. For example, if we wished to step down the voltage from a coil having 1000 loops and an input voltage of 1200 volts to a secondary coil voltage of 120 volts, the number of loops in the secondary coil would be given by

$$\frac{\text{number of loops in}}{\text{secondary coil}} = \frac{1000 \text{ loops}}{1200 \text{ volts}}$$

$$\frac{\text{number of loops in}}{\text{secondary coil}} = \frac{1000 \text{ loops}}{1200 \text{ volts}} \times 120 \text{ volts}$$

$$= 100 \text{ loops} .$$

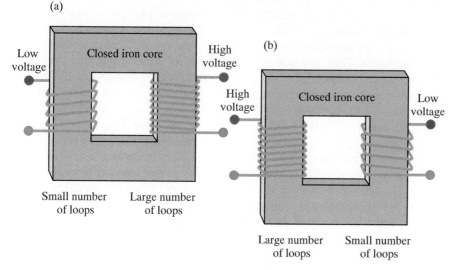

(a)

Low voltage

Closed iron core

High voltage

Small number of loops

Large number of loops

(b)

High voltage

Closed iron core

Low voltage

Large number of loops

Small number of loops

FIGURE 11.20

The Transformer A voltage applied to the terminals of the primary coil on the left can be (a) "stepped-up" or (b) "stepped-down" to a desired output voltage at the terminals of the secondary coil on the right by adjusting the number of loops in the two coils.

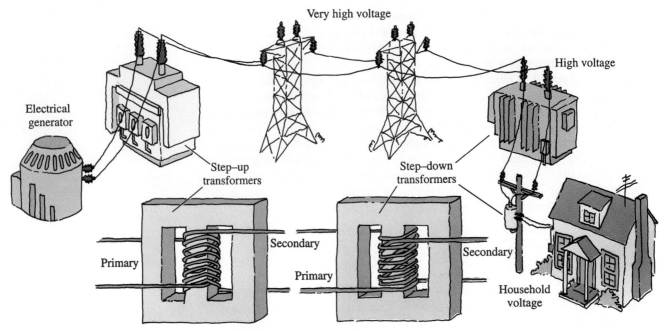

Very high voltage

High voltage

Electrical generator

Step–up transformers

Step–down transformers

Secondary

Primary

Secondary

Primary

Household voltage

FIGURE 11.21 ··········
Transformers and the Transmission of Electrical Energy   An electrical transformer can ''step-up'' voltages to high values for transmission of electrical energy over large distances and then ''step-down'' to lower voltages for consumer use.

The ability to alter voltages in this fashion permits electrical energy to be transmitted over very large distances via ''power'' lines or ''transmission'' lines. A typical delivery system for electrical energy is illustrated schematically in Figure 11.21. The electrical energy is first produced at a power plant by electric generators that can be run in several possible ways (see Chapter 17).

To transmit this electrical energy over long distances via transmission lines, the heat loss from the flow of the electric current through the lines must be reduced as far as possible. Recall that the heat loss in any given electrical resistance varies as the square of the current passing through that resistance. Also, from Ohm's law, we know that for a given resistance, the voltage/current ratio has a constant value. Thus, the heat loss in a given resistance equals the product of the current and the voltage. If we seek to minimize the heat loss in an electrical transmission line, we must minimize the current. To do this, we must in turn maximize the voltage. As a result of all this, power companies transmit their electrical energy over large distances by means of power lines that are operated at voltages as high as several hundred thousand volts. The step-up process from the comparatively low-voltage output of an electric generator (a few kilovolts) to the several hundred thousand volts at which power lines are normally maintained is accomplished by a step-up transformer. When the electrical energy arrives at its destination, it is distributed to consumers at lower voltages by means of step-down transformers. In this way electrical energy can be delivered over long distances from electric power plants to consumers in the form of relatively low-voltage, high-current output, which can be readily used by a wide range of electrical tools, appliances, and other devices.

WORKED EXAMPLE

A transformer primary coil has a total of 150 turns and a secondary coil with 15,000 turns. If the input voltage to the primary coil is 2200 volts, what output voltage will develop at the terminals of the secondary coil?

SOLUTION: In general we have for a transformer that

$$\frac{\text{number of loops}}{\text{voltage}} = \frac{\text{number of loops}}{\text{voltage}} .$$
$$\text{(primary coil)} \qquad \text{(secondary coil)}$$

For this case,

$$\frac{150 \text{ loops}}{2200 \text{ volts}} = \frac{15,000 \text{ loops}}{\text{voltage}} .$$

Solving for the voltage in the secondary coil, yields

$$\text{voltage} = \frac{15,000 \text{ loops} \times 2200 \text{ volts}}{150 \text{ loops}}$$

$$= 220,000 \text{ volts} .$$

---

1. Explain how an electric generator operates.
2. Compare and contrast an electric motor and an electric generator.
3. How are alternating currents and voltages produced?
4. Describe the operation of a transformer.
5. Write a short description of the physics associated with the production and delivery of electric power.

---

1. What is the magnetic field strength that will result from a 15-turn coil if the field strength is 0.05 tesla for a single loop? If the area of each loop is 0.04 m², what is the magnetic flux at the center of this coil?
2. An engineer wishes to design a transformer that will produce an output voltage of 600 volts from an input voltage of 100 volts. If the input voltage primary coil is wrapped with a total of 40 turns, how many turns should the secondary coil have?

11.4 Microscopic Electromagnetism

In the preceding sections, we described the intimate relationships that exist between moving electrical charges and magnetic fields. In light of such electromagnetic phenomena, how then can objects such as lodestones and "permanent"

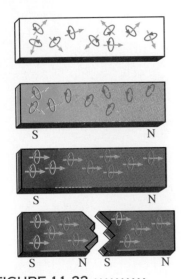

FIGURE 11.22 ·········

Magnetism and Atomic Theory Atoms can be thought of as tiny current loops produced by revolving and rotating electrons and protons. As the degree of magnetization increases, these atoms and their collected aggregates, called domains, become more and more aligned, thus producing an overall magnetic field of increasing magnitude. Because of this alignment, when a magnet is broken in two, the individual pieces have both north and south poles and thus function as two complete magnets.

magnets exist that possess magnetic fields in the apparent absence of electric charge flow? To answer this question, we must descend to the scale of the atom. Recall that an atom can be thought of as a system of negatively charged electrons orbiting about a positively charged nucleus of protons and neutrons. As these charged electrons orbit, they create tiny loops of electrical current, which in turn create tiny magnetic fields. Moreover, the electrons and protons in an atom are also spinning about their axes of rotation, thereby creating a second type of current loop and hence a second type of magnetic field.

For most substances, these tiny magnetic fields cancel out due to the variety of spin and orbital directions for the various electrons and protons. In magnetic materials such as iron and nickel, however, there is a significant magnetic field at the atomic level that goes uncancelled. Each iron and nickel atom is in effect an atomic-sized magnet that can collect into aligned groups called *magnetic domains*. When a piece of iron is magnetized, the magnetic domains are all aligned in the same way, thus producing the overall field of the magnet. When the iron is not magnetized, the domains are randomly aligned and tend to cancel one another out. When a piece of magnetic material that is not magnetized is placed in a magnetic field, the domains in the object now align to produce a magnetic field in that object. If the object is dropped or heated, the alignment of the magnetic domain can be destroyed and along with it the aligned strength of the magnet itself. If a magnetized object is cut up in a fairly gentle way, the alignment of the domains is not disturbed, and two magnets, each complete with a set of north and south poles, are created (Figure 11.22).

Perhaps the most amazing electromagnetic aspect of all was that deduced by the Scotsman James Clerk Maxwell in the middle of the last century. Maxwell theorized that if an electron were made to oscillate, it will, because of the motion of the charge, create a variable electric field and induce a variable magnetic field in the region of space surrounding the oscillating electron. This variable electric and magnetic field combination will then propagate outward from the vibrating charge in the form of ''electromagnetic waves'' or ''electromagnetic radiation.'' Maxwell's work not only revolutionized our understanding of the light we see, but opened up to us incredible vistas of energy that have since ''come to light'' in the present century.

1. Discuss how an atom can have a magnetic field.
2. Why aren't all substances magnetic in nature?
3. What is a magnetic domain?
4. How are electromagnetic phenomena responsible for the properties of permanent magnets?
5. Explain why heating a magnet can destroy its magnetic properties.

Chapter Review Problems

1. An electric charge of 0.2 C is moving through a magnetic field at a velocity of 10 m/sec perpendicular to the field and experiences a force of 5 N. What is the strength of the magnetic field?

2. An electric charge of 0.05 C is moving through a magnetic field having a strength of 0.08 T at a velocity of 3 m/sec perpendicular to the field. Find the magnitude of the electromagnetic force on this charge. If the magnetic field lines are perpendicular to this open page and are pointing down and the charge is moving directly toward you, find the direction of the force.

3. An electric charge is moving through a 0.01-T magnetic field at a velocity of 0.3 m/sec perpendicular to the field. If the charge experiences a force of 5 N, what is the magnitude of the electric charge?

4. An object with a charge of 4 C is moving at 8 m/sec perpendicular to a magnetic field having a strength of 25 T. What is the magnitude of the electromagnetic force exerted on the object as it moves through this field?

5. An object with a charge of 2×10^{-2} C is moving perpendicular to the earth's magnetic field at a velocity of 5×10^7 m/sec and experiences a force of 10 N. What is the strength of the earth's magnetic field?

6. A current of 0.4 ampere is passed through a wire segment 0.2 m in length perpendicular to a magnetic field of 2 T. What is the magnitude of the electromagnetic force experienced by the wire segment?

7. A wire segment 0.4 m in length and oriented perpendicular to a magnetic field experiences a force of 8 N when a current of 2.0 amperes flows through the segment. What is the strength of the magnetic field?

8. How long should a wire segment be in order for it to experience a force of 1 N when placed perpendicular to a magnetic field of 1 T. Assume a current flow of 1 ampere through the wire segment. Does your result suggest a possible way for defining a standard tesla or ampere? Explain.

9. How much current will be required to produce an electromagnetic force of 4 N on a 0.3-m wire segment oriented perpendicular to a magnetic field of 0.8 T?

10. An electron having a charge of -1.6×10^{-19} C is moving horizontally in a northern direction at a velocity of 2×10^6 m/sec through a magnetic field having a field strength of 1.5 T and whose field lines are pointing directly to the west. Find the magnitude and direction of the electromagnetic force exerted on the electron.

11. A current of 5 amperes is passed through two parallel wires separated by a distance of 0.08 m. Find the magnitude and the direction of the force/meter of wire exerted between the two wires if the currents in the wires are:
    a. moving in the same direction
    b. moving in opposite directions.

12. How much current should be run through two parallel wires separated by 0.01 m in order to produce a force/meter between the wires that is exactly equal to one newton?

13. A force of 2×10^{-5} newtons/meter is exerted between two parallel wires when the wires are each carrying 2 amperes of current. How far apart are the two wires?

14. Find the magnetic field associated with a wire in which a current of 1 ampere is flowing at a perpendicular distance from the wire of
    a. one centimeter   **b.** one meter   **c.** one kilometer.

15. The poles of a certain magnet face each other at a distance of 2 cm and are in the shape of squares 10 cm on a side. The strength of the magnetic field

between the facing poles is uniformly 0.5 T. What is the magnetic flux through an area 10 cm × 10 cm between the poles oriented

a. perpendicular to the magnetic field lines?

b. parallel to the magnetic field lines?

16. What would the magnetic flux be for a square area 0.05 m on a side which is placed between the poles of the magnet in Problem 15 perpendicular to the magnetic field lines?

17. A coil having an area of 0.1 m^2 is placed in a 2.5-T magnetic field so that its area is perpendicular to the field lines. The coil is suddenly turned parallel to the field lines in a total time of 0.1 second. What voltage develops at the terminals of the coil?

18. A coil is to be turned from a parallel orientation to a perpendicular orientation to the field lines of a 1.5-T magnetic field. If the rotation is to occur over a period of 0.05 seconds, how large an area should the coil have in order to produce an induced voltage of 2 volts?

19. A voltage of 0.4 volts is induced in a coil when it is suddenly pulled out of a magnetic field. If the coil is pulled out in a period of 0.15 seconds and has an area of 0.5 m^2, find the strength of the magnetic field. Assume that the area of the coil was perpendicular to the field lines as it was pulled from the field.

20. The primary coil of a certain transformer has a total of 60 wire loops and its secondary has a total of 15 wire loops. If the voltage across the transformer's primary coil is 120 volts, what is the voltage across the transformer's secondary coil?

21. A power company wishes to step the 2000-volt output from an electrical generator up to 200,000 volts for transmission lines. If there are 350 wire loops in the primary coil of a transformer, how many wire loops must the secondary coil in this transformer have in order to obtain the desired transmission line voltage?

22. A coil having an area A is placed so that its plane is oriented perpendicular to a magnetic field having a strength of B. Find the total change in magnetic flux in terms of A and B as the coil is rotated through

a. ¼ rotation   **b.** ½ rotation   **c.** ¾ rotation   **d.** one full rotation

THOUGHT QUESTIONS

1. Using Figures 11.1 and 11.2 for reference, explain Øersted's observations regarding the behavior of a magnetic compass needle in the vicinity of a current-carrying wire.

2. Sketch the magnetic field lines for a current loop that is shaped like a figure eight.

3. What do you think would happen to the compasses in Figure 11.2 if the plate were made of conducting material rather than insulating material?

4. What sort of electromagnetic force, if any, will exist between two current-carrying wires that are at right angles to each other?

5. Discuss how you would develop a series of ''left-hand'' rules to describe electromagnetic phenomena.

6. Two wires are aligned parallel to each other and one is carrying an electric current, the other is not. Describe the electromagnetic forces, if any, that exist between the wires.

7. Is there another convenient way that the ampere could be defined? Explain.

8. How might you measure the strength of the earth's magnetic field?

9. Discuss how you would obtain direct current and direct voltage from an electrical generator.

10. Look up the positions of nickel and iron in the periodic table and speculate on what other elements might have magnetic properties.

11. Suppose that the magnet in an electric generator were rotated about a stationary coil. Would such an arrangement produce a voltage at the coil's terminals? If so, what would be the advantage, if any, in such a design for an electric generator?

12. Compare and contrast a galvanometer, an electric motor, and an electric generator.

13. Is Lenz's law a consequence of action = reaction? Explain.

14. In this chapter we have discussed alternating voltages and currents. Is it possible to have an alternating electrical resistor? Explain.

15. Do you think that magnetic fields are vector quantities? Explain.

16. Why do you suppose that the electrical energy furnished by power companies is made available in the form of alternating voltages and currents?

17. Verify that the units of $\Delta\phi/\Delta t$ are the units of voltage.

18. Why is Faraday's law such an important principle?

19. Discuss all the possible ways that one could change magnetic flux.

20. What effect on the output voltage of an electric generator will each of the following have?
    a. a larger coil area
    b. a faster rotation rate of the coil
    c. both (a) and (b).

21. Why are transformers important in modern technology?

22. Describe how you might use a galvanometer to measure voltages and resistances.

23. Explain how an electromagnet works from an atomic view point.

24. It has been stated that the discovery of electromagnetism is the most important discovery in the history of human technology. Comment on this claim.

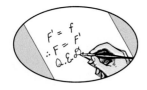

The Standard Ampere

The existence of electromagnetic phenomena provides the physicist with the opportunity to define precisely a measurable standard for the flow of electric charges and indeed for the quantity of electric charge as well.

To see how such standards are developed, let us consider a charge Q moving down a long thin wire, which is located in a magnetic field perpendicular to the wire having a strength of B. The electromagnetic force F exerted on this charge as it moves along the wire is

$$F = Q \times v \times B ,$$

where v is the velocity of the charge. If there are a total of N charges moving in the wire, then the magnitude of the electromagnetic force on these charges is N times larger or

$$F_{total} = N \times Q \times v \times B .$$

If the charges move a distance L in a time interval t, then

$$v = \frac{L}{t}$$

and

$$F_{total} = N \times Q \times \frac{L}{t} \times B .$$

Now the quantity NQ/t is simply the total charge flow past a given point per unit time. This quantity, by definition, is equal to the magnitude I of the current flowing in the wire. We may therefore write that

$$F_{total} = I \times L \times B .$$

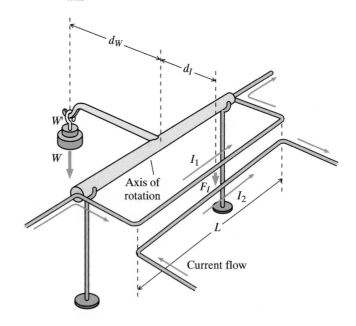

FIGURE 11.23 ···········

The Current Balance   As current flows through the two wire segments L, they exert a mutual force on each other. The torque exerted by F_I over the lever arm d_I is balanced by the added weight w' acting over the lever arm d_w.

The strength of the magnetic field surrounding the wire may also be evaluated in terms of I. In the 1820s, the French physicists Jean Biot and Felix Savart experimentally measured the characteristics of the magnetic field produced when electric charges flowed along a long straight wire. They found that the pattern of the magnetic field lines over a plane perpendicular to the wire consists of a series of circles concentric to the wire and having a direction given by the right-hand rule. They also found that the magnetic field strength B varies with the perpendicular distance d to the wire as follows:

$$B = 2 \times 10^{-7} \frac{I}{d} ,$$

where I is the magnitude of the current flowing through the wire. This equation is often called the Biot-Savart law.

Suppose now that we have two parallel current-carrying wires, each with a length L as shown in the apparatus depicted in Figure 11.23. As current flows through the wires, the force on wire 1 due to the magnetic field from wire 2 is

$$F_{1 \text{ from } 2} = I_1 \times L \times B_2 ,$$

where B_2 is the magnitude of the magnetic field strength from wire 2. The magnitude of B_2 is given by the Biot-Savart law as

$$B_2 = 2 \times 10^{-7} \frac{I_2}{d} ,$$

where d is now the perpendicular separation between the wires.

Substitution of this expression for B_2 into the previous equation yields

$$F_{1 \text{ from } 2} = I_1 \times L \times \frac{2 \times 10^{-7} I_2}{d} .$$

If we calculate the force on wire 2 due to the magnetic field of wire 1, we obtain an identical expression. Thus, there is a mutual force F between the wires whose magnitude is

$$F = 2 \times 10^{-7} \frac{I_1 I_2 L}{d}$$

and which is repulsive if the current flow in the wires is in opposite directions and attractive if the current flow in the wires is in the same direction. This equation is often written in the form of an expression for a force per unit length of wire, F/L, as follows:

$$\frac{F}{L} = 2 \times 10^{-7} \frac{I_1 I_2}{d} .$$

Let us now consider the apparatus shown in Figure 11.23 in more detail. Known as a current balance, this instrument consists of a wire segment of known length L that is attached to a voltage supply by connecting wires at right angles to L. The wire segment is mounted at a known distance d_l from a parallel rod, which is located along the axis of rotation of the instrument. The wire segment is balanced by placing a weight w on an arm attached to the rotation axis at a known distance d_w. A second wire of length L is then placed parallel to the first segment at a known distance d. Identical currents are then run through the two wire segments in such a way that a force of attraction is created between the segments. This force is balanced by adding an additional weight w' to the opposite side of the balance. The torque τ_I on the current side of the balance is equal to the electromagnetic force F_I between the wires acting over the distance between the wire segment and the axis of rotation of the balance, or

$$\tau_I = F_I \times d_I \ .$$

The force F_I between the wire segments is

$$F_I = 2 \times 10^{-7} \frac{I_1 I_2 L}{d}$$

and τ_I is then

$$\tau_I = 2 \times 10^{-7} \frac{I_1 I_2 L}{d} \times d_I \ .$$

The torque τ_w exerted by the additional weight w' on the opposite side of the balance is given by

$$\tau_w = w' \times d_w \ .$$

If the weight w' is adjusted so that no rotation occurs, the two torques balance each other and we can then write

$$\frac{2 \times 10^{-7} I_1 I_2 L}{d} \times d_I = w' \times d_w \ .$$

If we adjusted the current in both line segments so that $I_1 = I_2 = I$, then

$$\frac{2 \times 10^{-7} I^2 L}{d} \times d_I = w' \times d_w \ .$$

Solving for the current, we obtain

$$I^2 = \frac{w' \times d_w \times d}{2 \times 10^{-7} \times L \times d_I}$$

and

$$I = \sqrt{\frac{w' \times d_w \times d}{2 \times 10^{-7} \times L \times d_I}} \ .$$

Thus, the current flowing through the wire segments is expressed in terms of the measured quantities w', L, d, d_I, and d_w. We can now define an ampere of current as that charge flow that will produce a force of 2×10^{-7} N between two wires separated in a current balance by exactly one meter.

Since one ampere is also defined as a flow of one coulomb of charge per second, the current balance enables us to define a standard coulomb in terms of measurable quantities as well.

QUESTIONS

1. Describe the physical principles involved in a current balance.

2. Why is it necessary to have the contacts to the two wire segments perpendicular to the segments?
3. What adjustments would you make, if any, in the current balance if current were to flow in opposite directions through the wire segments?

PROBLEMS

1. Find the magnetic field strength 0.4 m from a wire in which a current of 1.5 amperes is flowing.

2. Calculate the magnitude and direction of the force between two parallel wires 1.5 meters in length separated by a distance of 0.5 meters in which a current of 2 amperes is flowing in the two wires in opposite directions.

3. A current of 2 amperes is flowing through a pair of 0.2-m wire segments separated in a current balance by 0.1 m. If the wire segment is 0.25 m from the rotation axis of the balance, how much weight would be needed to balance the torque from the electromagnetic force if the weight is to be hung 1 meter from the rotation axis?

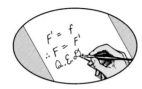

Transformers

One of the more useful devices based on electromagnetic phenomena is the electrical transformer. A typical transformer is shown in Figure 11.24 and consists of a closed iron core, which is wrapped with two coils of conducting wire whose loops are electrically insulated from the core and from each other. The first coil or primary coil has a total of N_1 turns of wire and the secondary coil has a total of N_2 turns. If an alternating voltage V_1 is applied to the terminals of the primary coil, then the changing current flowing in the primary coil will induce a change per unit time in the magnetic flux of $-\Delta\phi/\Delta t$ at the center of the coil for each loop in the coil. The total change in the magnetic flux per unit time in a primary coil having N_1 turns is equal to $N_1 \times -\Delta\phi/\Delta t$. From Faraday's law, we have in the primary coil that

$$V_1 = -N_1 \frac{\Delta\phi}{\Delta t}$$

or

$$-\frac{\Delta\phi}{\Delta t} = \frac{V_1}{N_1} .$$

If the coils are wrapped about a closed iron core, then the value of $\Delta\phi/\Delta t$ is nearly constant throughout the core and the total change in the magnetic flux in a secondary coil having N_2 turns would be equal to $N_2 \times -\Delta\phi/\Delta t$. However, from Faraday's law, such a change in magnetic flux must induce a voltage V_2 at the terminals of the secondary coil, which is given by

$$V_2 = -N_2 \frac{\Delta\phi}{\Delta t} .$$

Solving for $-\Delta\phi/\Delta t$ yields

$$-\frac{\Delta\phi}{\Delta t} = \frac{V_2}{N_2} .$$

Equating the two expressions for the quantity $-\Delta\phi/\Delta t$, we have that

$$\frac{V_1}{N_1} = \frac{V_2}{N_2} .$$

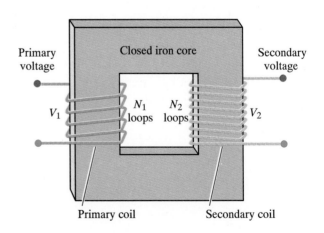

FIGURE 11.24 ·········

The Closed-Core Transformer The magnetic field lines produced by the current flow in the primary coil are "conducted" to the secondary coil by the iron core. The changes in the magnetic field produced by the primary then induce a voltage at the terminals of the secondary coil. The respective voltages are determined by the number of loops in the primary and secondary coils.

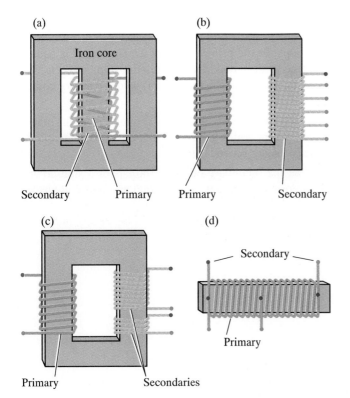

(a)

Iron core

Secondary   Primary

(b)

Primary   Secondary

(c)

Primary   Secondaries

(d)

Secondary

Primary

FIGURE 11.25 · · · · · · · · · ·

Various Transformer Designs   (a) The ''shell'' transformer;
(b) the tapped secondary transformer; (c) the two secondary
transformer; (d) the autotransformer.

The transformer thus provides us with the opportunity to alter voltages for
a variety of uses. For example, electrical power is most efficiently transmitted
at high voltages and low currents, but is most safely and conveniently used for
appliances at low voltages and higher currents. The transformer permits a given
voltage to be ''stepped up'' to a higher voltage or ''stepped down'' to a lower
voltage by simply adjusting the number of loops in the primary and secondary
coils.

Transformers can take on a variety of designs, depending on their use. Transformers, for example, can have their primary and secondary coils superimposed
in the ''shell'' type of arrangement shown in Figure 11.25, or one can have a
''tapped'' secondary in which the position of our electrical contact can be varied
along the length of the secondary coil. In this latter arrangement, the voltage to
the secondary coil terminals can be adjusted to a desired value by moving the
contact to a position where the number of individual loops included in the secondary circuit will result in the desired output voltage. Some transformers have
two secondary coils so that two different secondary voltages can be produced
from one transformer, and the autotransformer consists of a single coil that is
hooked up to both the primary and secondary terminals.

QUESTIONS
· · · · · · · · ·

1. Is it possible to have a direct current (dc) transformer? Explain.

2. Does a step-up transformer violate the principle of the conservation of energy?
   Explain.

3. Discuss possible uses for each of the transformers illustrated in Figure 11.25.

PROBLEMS

1. Assuming that the power input to the primary coil of a transformer is equal to the power output to the secondary coil, derive a relationship between N_1, N_2, and the currents I_1 and I_2, which flow in the primary and secondary coils, respectively.

2. An electrical engineer desires an output voltage of 120 volts from an input voltage of 600 volts. If the primary coil is to have 1000 loops, find the number of loops the engineer needs for her secondary coil.

3. Find the lowest voltage that the secondary output terminals of a step-down transformer can have if the primary coil has 50 turns and an input voltage of 100 volts.

In an earlier chapter we described how, in the seventeenth century, the Englishman Isaac Newton was able to describe successfully a wide variety of motion-related phenomena with his single, comparatively simple theory of classical mechanics. Less than two centuries later, another British Islander, Scotsman James Clerk Maxwell performed what is perhaps an even more impressive feat of theoretical elegance when he was able to capture phenomena as diverse as x rays and moonbeams within the framework of a single physical description. Although at the time Maxwell was concerned with describing the nature of the light we see as human beings, as a result of his ideas, physicists now recognize the fact that visible light is but one aspect of a much more comprehensive phenomenon we now refer to as electromagnetic radiation.

12.1 Electromagnetic Waves

Suppose we place an electric charge on the end of a spring and then cause the charge to oscillate as shown in Figure 12.1. Maxwell argued that because the electric charge was oscillating it would induce a similarly oscillating magnetic field in the space around the electric charge. The changing magnetic field, however, would now, in turn, induce an oscillating electric field about the vibrating charge as well. Maxwell was able to demonstrate mathematically that these induced electric and magnetic fields would not remain confined to the space in the vicinity of the vibrating charge, but rather would radiate outward from it.

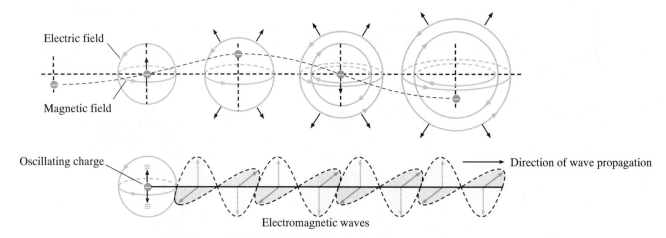

FIGURE 12.1 • • • • • • • • • •

The Generation of an Electromagnetic Wave As the negative charge moves upward, it creates the magnetic field denoted by the light blue field lines. Since the magnetic field is changing, that change induces the electric field denoted by the light green field lines. As the charge reverses its direction, the old electric and magnetic fields decrease to zero and a new set of electric and magnetic field lines is created, but in the opposite sense. Meanwhile, the first set of variable field lines expands outward from the oscillating charge. As the process continues, an electromagnetic wave having variable electric and magnetic field components propagates outward as an electromagnetic wave front.

12.1 James Clerk Maxwell (1831–1879)

The front of a T-shirt that is very popular among physics and engineering students reads:

"... and God said

$$\vec{\nabla} \times \vec{E} = -\frac{\partial \vec{B}}{\partial t}$$

$$\vec{\nabla} \times \vec{H} = \vec{J} + \frac{\partial \vec{D}}{\partial t}$$

$$\vec{\nabla} \cdot \vec{D} = \rho$$

$$\vec{\nabla} \cdot \vec{B} = 0$$

... and there was light."

The person for whom this very important and fundamental set of equations is named was not omnipotent, but was born quite human on June 13, 1831, in the Scottish city of Edinburgh as James Clerk Maxwell.

After surviving the death of his mother when he was eight years old and a grade school tutor who pronounced him to be "slow at learning," Maxwell was enrolled at the Edinburgh Academy where, at the age of 14, he published his first scientific paper on a mechanical method for tracing nonelliptical types of oval curves. Two years later, Maxwell entered the university at Edinburgh and promptly published two more scientific papers, one on rolling curves in mathematics and the other on equilibrium in elastic solids. His developing talent then carried him to Cambridge University in 1850, where, among other accomplishments, he wrote a prize-winning essay in which he mathematically demonstrated that the rings of material around the planet Saturn could not be rigidly solid, but instead consisted of a multitude of relatively small particles. The correctness of these assertions would be ultimately verified decades later, first by earth-based astronomical measurements of particle motions within the rings and then in a most dramatic fashion from direct close-up photographs of the Saturnian ring system taken by space probe flybys.

The late 1850s saw turbulent and changing times for Maxwell, times that included a marriage, the deaths of his

beloved father and a close friend, and two changes of employment, one occurring at the University of Aberdeen in Scotland when the university was reorganized in 1860 and Maxwell's professorship was declared "redundant." After being turned down for a position at his alma mater in Edinburgh, Maxwell finally obtained stable employment as professor of natural philosophy at King's College in London. With a difficult period of his life behind him, Maxwell proceeded in subsequent years to establish himself as one of the preeminent physicists of human history.

In his five short years at King's College, Maxwell investigated topics as diverse as the geometric optics of a "fish-eye" lens, molecular motions in gases, electricity and magnetism, and color phenomenon. Even in his whimsical

••• **James Clerk Maxwell (1831–1879)**

PHYSICS FACET

Continued

moments, Maxwell could be thought-provoking. His proposed "Maxwell's Demon," a creature (or device) that could presumably separate the fast- and slow-moving molecules of a gas into high and low temperature reservoirs spurred a lively discussion of the laws of thermodynamics and has since come to play an important role in the development of modern information theory. Although he would become renowned for his theoretical work, Maxwell was also quite skilled in the laboratory as well. In 1861, for example, he demonstrated with a device of his own design the feasibility of making color photographs at a time when black-and-white photography had been on the technological scene for less than a generation.

After five years at King's College, Maxwell resigned and retired to the family estate in Scotland. There he collected and galvanized his thoughts into his two key works in physics. In the first, *Theory of Heat,* Maxwell applied the methods of probability and statistics to obtain a solid mathematical formulation for the then qualitative kinetic theory of gases. His most important and enduring contribution, however, came in 1873 when he published his famed *Treatise on Electricity and Magnetism.* Drawing heavily from the earlier work of scientists such as Gauss, Ampere, and especially Faraday, Maxwell developed the first overall

mathematical description of the phenomenon we now know as electromagnetism. In Maxwell's view, all electric and magnetic phenomena could be described by the four Maxwell's equations that adorn the T-shirt described earlier.

Maxwell further deduced that it was possible to create waves consisting of electric and magnetic fields that could propagate through space at a speed equal to that of light. From this result, Maxwell wrote that he could "scarcely avoid the inference that light consists . . . of electric and magnetic phenomena." Thus, did Maxwell not only produce a synthesis of electricity and magnetism, but was also able to include light itself within the framework of that synthesis.

Maxwell returned to Cambridge in 1871 where he took on the task of designing and supervising the construction of the now-famous Cavendish Laboratory. As a result of his efforts, he was appointed the laboratory's first director, a position he held until struck down in 1879 by the very same type of abdominal cancer that had spelled his mother's doom four decades earlier. Thus came the premature end to a man whose work Albert Einstein would describe a half-century later as "the most profound and the most fruitful that physics has experienced since the time of Newton."

Moreover, Maxwell found that these propagating electric and magnetic fields had to leave their source in the space at a particular speed. If the outward speed of the fields were too fast, then one could easily demonstrate that an energy buildup would have to occur in both the electric and magnetic fields, resulting in what would in effect be a creation of energy. On the other hand, if the outward speed of the fields were too slow, then the energy contained in the electric and magnetic fields would "die out" and energy would in effect be destroyed. Such creation or destruction of energy is, of course, in direct violation of the principle of the conservation of energy. If the fields move outward at just the right speed, however, there is neither an energy buildup or an energy loss, and energy imparted to the electric and magnetic fields by the vibrating electric charge propagates in a fashion compatible with the principle of energy conservation. Maxwell was able to calculate this critical outward speed for an oscillating charge in free space and found it to be about 300,000 km/sec, the same as the value for the speed of light in free space.

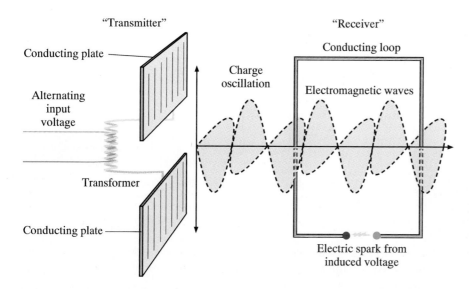

"Receiver"

Conducting plate

Conducting loop

Charge oscillation

Electromagnetic waves

Alternating input voltage

Transformer

Conducting plate

Electric spark from induced voltage

FIGURE 12.2 ⋯⋯⋯⋯

The Hertz Experiment   An alternating voltage produces an oscillating charge between the two plates. The oscillating charge in turn produces electromagnetic waves, which induce enough voltage in the receiver to cause an electric spark to arc across the receiver gap.

In 1887, the German physicist Henrich Hertz was able to verify Maxwell's hypothesis experimentally by creating just such waves in his laboratory. A schematic diagram of Hertz's apparatus is shown in Figure 12.2 and consists of a "transmitter" and a "receiver." The transmitter is made up of two large conducting plates hooked up to an alternating voltage. The receiver is a circular conducting wire, which is broken by a small gap. When Hertz applied an alternating voltage to the plate terminals, an alternating electric charge flow was produced between the plates, in effect resulting in an oscillating electric charge. When the transmitter was turned on, a spark of electricity could be observed periodically arcing across the gap in the receiver wire, thus demonstrating that electromagnetic waves had been produced, propagated to the ring, and induced the charge flow necessary for the spark to occur.

Hertz was further able to demonstrate that these electromagnetic waves had a wavelength of about 5 meters, yet traveled at Maxwell's predicted speed of 300,000 km/sec, the same as that of visible light. Thus by the end of the nineteenth century, physicists recognized the existence of "electromagnetic" waves which, like sound waves, are generated by vibrations and oscillations.

When an electromagnetic disturbance is created, its propagation can be represented as a spherical shell of energy whose radius is increasing at the same rate in all directions or *isotropically*. As the shell expands, the total energy in the shell remains the same, but the total surface area of the shell at any given time increases as the square of the shell's radius. As a result, if we define the energy intensity or apparent brightness of a source as the power per unit area received from a given source of electromagnetic waves at a given distance, we find (Physics Formulation 12.1) that the intensity of the source is given by

$$\frac{\text{intensity}}{\text{of source}} = \frac{\text{power output of source}}{4\pi \times (\text{distance to source})^2} .$$

Notice that the intensity of the source decreases as the square of the distance to the source. In other words, the intensity of a light source can be described with an inverse square law (Figure 12.3). Thus, the energy intensity of a streetlight is larger when you are standing directly beneath it than when you are viewing it

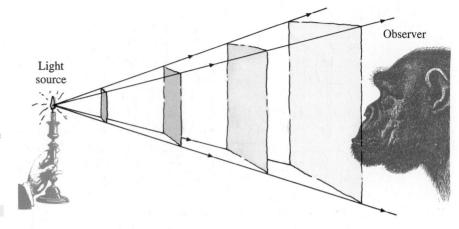

FIGURE 12.3 ··········
The Inverse Square Law for Light Intensity As the distance between an observer and a source of light increases, the intensity or apparent brightness of the radiant energy received from the source decreases with $1/(\text{distance})^2$ as the source's power spreads out over progressively larger areas of space.

from several blocks away. The dimming-with-distance effect can be mathematically expressed in the following way (see Physics Formulation 12.1):

$$\frac{\text{intensity at distance 1}}{\text{intensity at distance 2}} = \left(\frac{\text{distance 2}}{\text{distance 1}}\right)^2 .$$

Thus, the apparent brightness or intensity of a source of electromagnetic wave energy obeys an inverse square law similar to those found for gravitational, electrostatic, and magnetic forces.

Although Maxwell was the first to deduce its electromagnetic nature, the wave properties of light were proposed and theoretically described nearly two centuries earlier by English physicist Robert Hooke and Dutch theorist Christian Huygens. These hypotheses concerning the wave nature of light were then experimentally confirmed through the work of the English physicist Thomas Young early in the nineteenth century. Huygens in particular proposed the idea that waves spreading out from a source could be thought of as being composed of a series of "miniaturized waves" or wavelets. Moreover, any point on a given wave front could itself be regarded as a generator of these secondary wavelets (Figure 12.4). This idea is called Huygens's principle and leads to the idea that such wavelets can be combined into wave "fronts" according to "Huygens's construction" as shown in Figure 12.5. Although it is only an approximation, such a representation of light as wavelets permits us to describe many of the

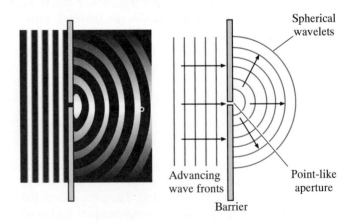

FIGURE 12.4 ··········
Huygens's Principle Each point on a wave front can be considered a source of small secondary spherical wavelets. In this illustration, the part of the advancing wave front that passes through the point-like opening in the barrier gives rise to the spherical wavelet shown. All of the other points along the wave front are stopped by the barrier and hence do not contribute their wavelets.

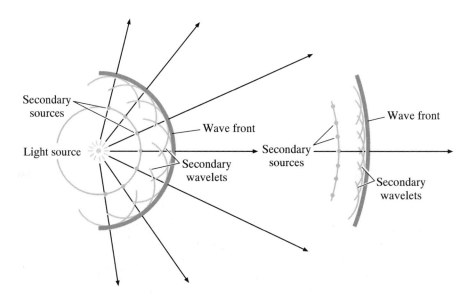

Secondary sources

Light source

Wave front

Secondary wavelets

Secondary sources

Secondary wavelets

Wave front

Secondary wavelets

FIGURE 12.5 ··········
The Wave Front As waves spread out from a point source, secondary wavelets are formed. The envelope of these and succeeding wavelets forms a traveling wave front for the wave. At large distances from the source, the wave front has a nearly planar geometry.

commonly observed properties of light, such as its refractive and reflective properties, in a way consistent with the view of light as an electromagnetic wave. At comparatively large distances from the source, the surface of a given spherical wave front can be approximated by a plane wave front or plane wave. Thus the light we receive from very distant celestial objects can be approximated to a high degree of accuracy by assuming such a plane wave model.

WORKED EXAMPLE

A streetlight has an observed intensity of 24 watts/m^2 when viewed from a line-of-sight distance of 5 meters. At what distance will the streetlight appear to have an intensity of 6 watts/m^2?

SOLUTION: The relationship between the intensity of a light source and the distance is

$$\frac{\text{intensity at distance 1}}{\text{intensity at distance 2}} = \frac{(\text{distance 2})^2}{(5 \text{ m})^2}.$$

Solving for (distance 2)^2, we obtain

$$(\text{distance 2})^2 = \frac{24 \text{ watts/m}^2}{6 \text{ watts/m}^2} \times 25 \text{ m}^2$$

$$(\text{distance 2})^2 = 4 \times 25 \text{ m}^2 = 100 \text{ m}^2$$

$$\text{distance 2} = 10 \text{ m}.$$

The streetlight will appear to have an intensity of 6 watts/m^2 when viewed from a distance of 10 meters.

1. Explain how electromagnetic waves are produced.
2. Describe how Hertz was able to demonstrate the existence of electromagnetic waves.
3. Compare and contrast an electromagnetic wave and a sound wave.
4. Why does a light source appear to get fainter as the distance to the light source increases?
5. What is Huygens's principle? How is this concept used to approximate the propagation of light?

1. A 120-watt light bulb is emitting its radiant energy uniformly in all directions. Calculate the intensity of this source at a distance of
   a. 1 meter   b. 100 meters   c. 1 kilometer.
2. A light source known to have a power output of 100 watts is observed to have an intensity of 0.32 watts/m^2. How far is the light source from the observer?

12.2 The Speed of Light

For over two centuries prior to Maxwell's ponderings regarding the nature of light, scientists struggled with the not so trivial task of measuring the speed at which light travels through space. Even with the crude techniques available early in the Scientific Revolution, it was recognized that light moves at enormous speeds compared to what we encounter in our "everyday" lives.

Sometime around 1625, Galileo cleverly attempted to measure the speed of light by employing a method that was used at the time to determine successfully the speed of sound. The experiment was set up by first carrying a covered lantern to the top of a hill and placing an assistant with a similar lantern on top of a second hill at a known distance away. Galileo then produced a flash of light by rapidly uncovering and covering his lantern. On seeing Galileo's flash, the assistant would similarly generate a flash of light from the second lantern. Galileo would note the total interval between the time he uncovered his lantern and the time he saw the return flash from his assistant's lantern. Galileo repeated this experiment many times at many distances using several assistants, and found that unlike sound, this measurement was thwarted by the fact that the amount of time his assistants needed to react to the initial flash and uncover their own lanterns was much larger than the light travel time back and forth between the lanterns. Based on this result, Galileo concluded that light travels at a speed that is essentially infinite.

In 1675, the Danish astronomer Olaus Roemer was able to show that the "winking out" of the light from Jupiter's satellites as they entered the shadow of the planet took roughly 22 minutes longer to reach his telescope when Earth was at its farthest point from Jupiter than when it was at its nearest point (Figure 12.6). Roemer concluded from this *light*-time effect that light did not travel at an infinitely large speed, but rather traversed space at a rate of one earth orbital diameter every 22 minutes or at a speed of some 250,000 km/sec. In retrospect, we now know that Roemer's measurements were somewhat in error, but they

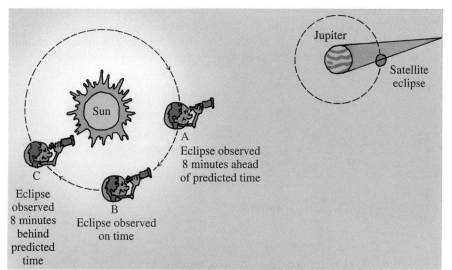

The Light-Time Effect When Danish astronomer Olaus Roemer predicted the times when Jupiter's satellites would pass into that planet's shadow and become eclipsed, he found that the satellite eclipses occurred 11 minutes ahead of their predicted times when Earth was closest to Jupiter (A); right at their predicted times when Earth was at its mean distance from Jupiter (B); and eleven minutes behind their predicted times when Earth was farthest from Jupiter (C). The modern value of this light time interval is approximately 8 minutes.

nonetheless provided the first unambiguous observational evidence that light traveled at a finite, albeit very large, velocity.

In 1727, the English astronomer James Bradley demonstrated that the velocity of light from a distant star vectorially adds to the earth's orbital velocity to produce an apparent angular displacement in the star's observed position. By measuring the value of this angle, called the *angle of aberration* of starlight, and knowing the velocity of the earth in its orbit, Bradley deduced that light travels at a rate of some 300,000 km/sec, a value that within the accuracy of the observations agrees with the currently accepted value for this quantity.

Since the eighteenth century, the speed of light has been measured with ever increasing accuracy in a variety of most ingenious ways. In one technique, first devised by the Frenchman Hippolyte Fizeau, and later refined by the American Albert Michelson, a beam of light is "chopped up" into a series of pulses by means of a toothed wheel or rotating mirror. The time required for a given pulse of light to travel a known distance can be accurately determined by adjusting the rate at which the wheel or mirror turns, thereby permitting the determination of the speed of light (see Physics Formulation 12.2). Such techniques are in essence "assistantless" versions of Galileo's earlier attempts to measure the speed of light.

Even more modern accurate determinations of the speed of light have come from experiments in which the values of the wavelength and frequency of a given beam of light are simultaneously measured to a very high degree of accuracy. Since light behaves as a wave, the speed of light is then the product of these two measurements.

The most accurate measurements of the speed of light in a vacuum to date yield a value of 2.99792458×10^8 m/sec, but for the purposes of this text we will assume a rounded off value for the speed of light equal to 300,000 km/sec or 3×10^8 m/sec, and will designate this quantity as c. Further, measurements of the speed of light reveal that its value does not change with either the wavelength or frequency of the light involved and is in fact experimentally found to be the same for all other forms of electromagnetic waves as well, just as Maxwell had predicted.

The speed of light does, however, vary with the medium in which it is moving. The speed of light in water, for example, is about 225,000 km/sec, and 120,000 km/sec as it travels through a diamond. The ratio of the speed of light in a vacuum to that in a given substance is defined as the *index of refraction* of that substance, or

$$\begin{matrix} \text{index of refraction} \\ \text{of substance} \end{matrix} = \dfrac{\begin{matrix}\text{speed of light} \\ \text{in vacuum}\end{matrix}}{\begin{matrix}\text{speed of light} \\ \text{in substance}\end{matrix}} .$$

The index of refraction for water, for example, would be

$$\text{index of refraction} = \frac{300,000 \text{ km/sec}}{220,000 \text{ km/sec}}$$
$$= 1.33 .$$

That for diamond is

$$\text{index of refraction} = \frac{300,000 \text{ km/sec}}{120,000 \text{ km/sec}}$$
$$= 2.50 .$$

Interestingly, no form of particle or energy has thus far been discovered that can travel faster than the speed of light. The speed of light in a vacuum, thus, appears to be the ultimate speed limit in the universe. This fact has given rise to a bit of graffiti that from time to time appears on the walls of the campus science or engineering buildings and states: ''300,000 km/sec: It's not just a good idea, it's the law!''

The most amazing aspect of all concerning the speed of light in a vacuum, however, is the observationally verifiable fact that regardless of how an observer is moving with respect to a given light ray, that observer will always measure the same value, 300,000 km/sec, for the speed of light. Such a result is contrary to the way speeds and velocities are predicted to behave in the Newtonian view of motion, and is, in fact, as we shall see, one of the key underpinnings of the more modern view of motion called relativity theory.

WORKED EXAMPLE

The speed of light in the earth's atmosphere is measured to be 299,913 km/sec. If the speed of light in a complete vacuum is 300,000 km/sec, what is the index of refraction of the earth's atmosphere?

SOLUTION: The index of refraction is defined as

$$\begin{matrix}\text{index of refraction} \\ \text{of substance}\end{matrix} = \frac{\text{speed of light in vacuum}}{\text{speed of light in substance}} .$$

For the earth's atmosphere

Continued

WORKED EXAMPLE

Continued

$$\frac{\text{index of refraction}}{\text{of earth's atmosphere}} = \frac{300,000 \text{ km/sec}}{299,913 \text{ km/sec}}$$

$$= 1.00029$$

which is very nearly 1.00.

1. How fast does light travel? How does this value compare to the speed of the following?
   a. a sprinter   **b.** an automobile
   c. a jet plane   **d.** a rocket
2. Was Galileo's attempt to measure the speed of light based on correct thinking? If so, why did his attempts fail?
3. Describe one of the successful techniques for measuring the speed of light.
4. What is the index of refraction of a substance?
5. Discuss some of the characteristics of the speed of light.

1. The distance between New York and Los Angeles is about 5800 km. How long will it take a light beam to make a round trip from New York to Los Angeles?
2. The speed of light in a certain substance is found to be 30,000 km/sec. What is the index of refraction for such a substance?

12.3 The Electromagnetic Spectrum

One consequence of electromagnetic wave theory is that it is possible for an electric charge to oscillate at any frequency larger than zero, thereby allowing for the possible existence of an infinite variety of electromagnetic waves. This range of possible electromagnetic frequencies and wavelengths is referred to as the *electromagnetic spectrum* (Figure 12.7). If we measure the wavelengths of the electromagnetic waves visible to us as light, we find that the human eye can detect wavelengths as small or as "short" as 4×10^{-7} m and as large or as "long" as 7×10^{-7} m. This range of electromagnetic wave observable with the human eye is referred to as the *visible* region of the electromagnetic spectrum and has a corresponding range of frequencies between 7.5×10^{14} Hz and 4.0×10^{14} Hz. Beyond these limits, the human eye is not able to detect the presence

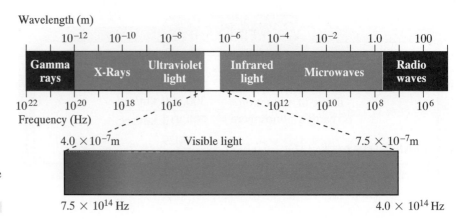

FIGURE 12.7 ⋯⋯⋯⋯

The Electromagnetic Spectrum The radiant energy we see as visible light is but one small portion of the entire range of frequencies and wavelengths that make up the electromagnetic spectrum.

of electromagnetic waves. There is, of course, no reason whatsoever to assume that, just because we as human beings cannot directly ''see'' radiant energy beyond these limits, no such energy exists. In fact, our bodies can, to some extent, detect such energy. The thermal energy radiating from the hot spoon handle described in an earlier chapter, for example, can be felt by our hand as ''heat,'' and numerous creatures, such as pit vipers, are equipped with detectors that operate quite superbly at nonvisual frequencies and wavelengths.

One of the supreme achievements of human technology over the past century has been the exploration of this electromagnetic spectrum, particularly in our development of the ability to detect and, in many instances, make use of its many types of radiant energy. As a result, we are now aware of a wide variety of electromagnetic waves that extend outward from the high- and low-frequency ends of the visible spectrum. The thermal energy mentioned above, for example, has frequencies smaller than 4.3×10^{14} Hz, which marks the red end of the visible spectrum, and is hence said to be a part of the *infrared* region. The *microwave* region of the electromagnetic spectrum extends over a frequency range of about 10^9 to 10^{11} Hz, followed by the *radio* region, whose frequencies are lower than 10^9 Hz. Included in this region of the electromagnetic spectrum are those frequencies over which commercial AM radio broadcasts are made (535–1605 kilohertz), FM radio broadcasts (88–108 megahertz), and several channels of television broadcasting (54–890 megahertz).

On the other side of the visible region, extending from 7×10^{14} Hz to roughly 3×10^{15} Hz, is the region we call the *ultraviolet* region. It is primarily the ultraviolet waves we receive from the sun that provide us with our summertime tans and sunburns. A yet higher frequency region for electromagnetic waves extends from about 3×10^{15} Hz to 10^{18} Hz and is called the *X-ray* region. X rays are perhaps most famous for their ability to penetrate matter. In fact, their use as a diagnostic tool in the medical sciences is due to their ability to pass through human tissue readily. The most energetic and probably the most feared types of electromagnetic waves are the *gamma rays*. Gamma rays have frequencies in excess of 10^{18} Hz and are the result of incredibly high charge vibration rates generated within atomic nuclei by various nuclear processes. Because gamma rays are a by-product of nuclear reactions and because they have an extraordinary ability to penetrate and disrupt matter, they are a major factor to be contended with in the production of nuclear power (Chapter 17).

WORKED EXAMPLE

What is the frequency of orange colored light, if its wavelength is about 6.1×10^{-7} m?

SOLUTION: The product of the frequency and wavelength of an electromagnetic wave is equal to the speed of the wave, or

$$\text{wavelength} \times \text{frequency} = \text{speed of light} .$$

For the orange light in question, we have

$$6.1 \times 10^{-7} \text{ m} \times \text{frequency} = 3 \times 10^{8} \text{ m/sec} .$$

Solving for the frequency yields

$$\text{frequency} = \frac{3 \times 10^{8} \text{m/sec}}{6.1 \times 10^{-7} \text{ m}}$$
$$= 0.49 \times 10^{15} \text{ Hz} = 4.9 \times 10^{14} \text{Hz} .$$

The frequency of orange light is about 4.9×10^{14} Hz.

---

1. What is meant by the electromagnetic "spectrum"?
2. Which wavelengths and frequencies of the electromagnetic spectrum can be seen by the human eye?
3. Summarize the properties of the various types of nonvisual electromagnetic radiation.
4. Briefly describe a practical use for
   a. radio waves   **b.** microwaves   **c.** x rays
   d. infrared rays   **e.** gamma rays.
5. Compare and contrast radio waves and gamma rays.

---

1. Write down the frequency of your favorite radio station and calculate the wavelength at which it is broadcasting electromagnetic waves.
2. Alternating current from commercial sources typically has a frequency of 60 hertz. What wavelength will the electromagnetic waves have that emanate from such a charge oscillation?

12.4 Reflection and Refraction

When an electromagnetic wave front strikes a given surface it is possible for a new wave front to be created that leaves the surface symmetrically with respect to the incoming wave front. The wave front in effect "bounces" off the surface

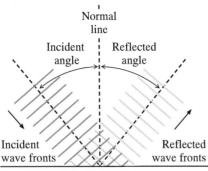

FIGURE 12.8 ··········

The Reflection of Light When parallel wave fronts strike a reflecting surface, they "bounce" off the surface at a reflected angle to the normal line that has the same value as the incident angle to the normal.

and is said to be *reflected* off the surface. If we define the *normal* to a given point on a surface as the line that is perpendicular to the surface at that point, then the *angle of incidence* is said to be the angle between the direction of the incoming wave front and the normal (Figure 12.8). The angle between the reflected wave front and the normal is called the *reflected angle* and the reflection always occurs in such a way that the reflected angle is equal to the angle of incidence. This statement is often called the law of reflection and can be stated as

$$\text{angle of reflection } = \text{ angle of incidence .}$$

If the surface is smooth, the reflected angles of the various portions of an incoming wave front will be the same. The surface is then said to be "polished" or "highly reflective." Perhaps the most familiar example of a polished surface is a common mirror. If on the other hand the surface is not smooth, then the various portions of this "rough" surface will reflect various portions of an incoming wave front in different angles. The result is a phenomenon called *diffuse reflection,* which permits the reflecting object to be viewed equally well from a variety of directions (Figure 12.9). For example, the light reflected from a page in your notebook or text is diffusely reflected and permits you to read your notes or text equally well from a variety of different angles.

The direction in which electromagnetic wave fronts move can also be changed by a process called *refraction*. Refraction arises from the fact that the speed at which electromagnetic waves travel depends on the medium in which they find themselves. Contrary to what happens in free space, the transmission of light through a material medium consists of a high-speed series of processes in which light waves incident on the medium are absorbed by the electrons at the boundary, causing them to vibrate at the same frequency as the incoming light. The oscillating electrons in turn reradiate electromagnetic waves at the same frequency, which then induce the next set of neighboring charges down the line of propagation in the medium to oscillate, and so on. The time it takes for such absorptions and reradiations of electromagnetic waves to occur has the effect of reducing the speed at which the waves are able to travel through the medium.

As we have already seen, each transparent medium can be characterized by its index of refraction, or the ratio of the speed of light in free space to that in the given medium. If a wave front strikes a boundary between media having differing indices of refraction as shown in Figure 12.10, the change in speed of the various parts of the wave front as they cross the boundary produces a "bending" or refraction of the direction of the wave front. The situation is analogous to that of an automobile that leaves the pavement on a highway at some angle.

FIGURE 12.9 ··········

Reflection from Smooth and Rough Surfaces For a smooth or polished surface, the angle of reflection is the same for all parts of the wavefront. However, a wavefront reflected from an uneven or rough surface can have many different angles of reflection, giving rise to a phenomenon called diffuse reflection.

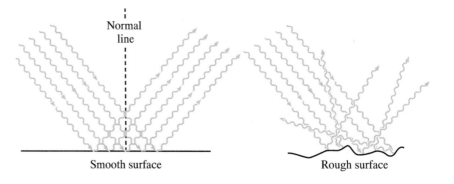

As one of the front wheels passes across the boundary from pavement to gravel, it slows down, thus causing the front axle of the car to twist to a different angle. As a result, the motion of the car itself then changes into a different direction.

The actual relationship between the incident and refracted angles as a wave front crosses a boundary is somewhat complicated for the purposes of this text. Basically, the relationship states that the wave front will bend in such a way that the largest angle between the wave front and the normal will always occur in the medium having the smaller index of refraction (Figure 12.11).

If an electromagnetic wave in a medium with a higher index of refraction is incident on a boundary with a lower index medium at a sufficiently large angle of incidence, the wave may not be refracted into the second medium at all, but rather will be completely reflected at the medium interface. This phenomenon is called *total internal reflection* and is perhaps most familiar to us in the realm of ''fiber optics.'' In fiber optics, light is transmitted from one location to another by means of internal reflectors along a small-diameter fiber of transparent material (Figure 12.12).

It has also been observed that in a given medium electromagnetic waves having different wavelengths will travel at slightly different speeds. For example, red light travels about one percent faster in a given transparent medium such as water or glass than does violet light. This, of course, means that the index of refraction of a given substance depends on the wavelength of the wave involved. Wave fronts of differing wavelength will, therefore, be refracted differently in a given medium. A most familiar example of this property, called *dispersion,* is the ''rainbow'' effect seen when sunlight strikes droplets of water. In this instance, white sunlight strikes a given water droplet and because the different colors making up the white sunlight are refracted differently, the light that emerges from the droplet is spread out into all of its component colors. We see the result as a spectrum of colors or a rainbow (Figure 12.13).

Although we have primarily described reflection and refraction phenomena in the context of the more familiar visible light examples, it should be emphasized

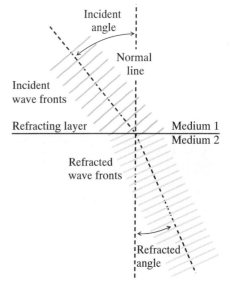

FIGURE 12.10 ⋯⋯⋯⋯

The Refraction of Light   When parallel wavefronts pass from one material to another, the speed of the wavefronts is either increased or decreased, thus causing the wavefronts to be ''bent'' or refracted. In this case, the wave fronts are passing from a region of lower index of refraction to one of higher index of refraction.

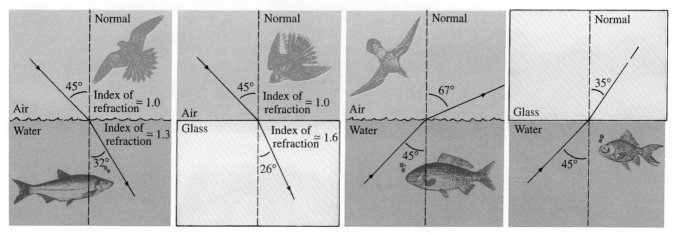

FIGURE 12.11 ⋯⋯⋯⋯

Refraction Across Various Media Boundaries   In each case, the angle between the normal line and the path of the wavefronts is larger for the medium having a lower index of refraction.

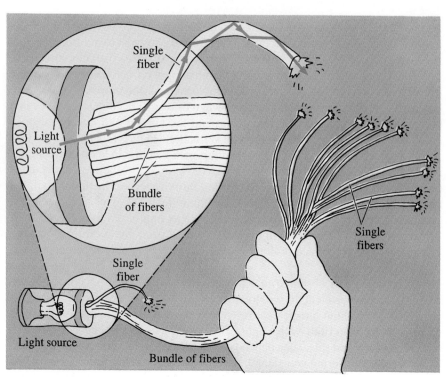

FIGURE 12.12
Fiber Optics By using the phenomenon of total internal reflection, it is possible to transmit beams of light along small-diameter "light pipes."

that other types of electromagnetic waves can also be reflected and refracted, although not necessarily by the same materials that refract and reflect visible light. The wire-mesh disk of a television or radio antenna, for example, can collect and focus radio waves by reflection just as surely as a metal-coated concave glass mirror can collect and focus visible light waves. On the other hand, that same wire mesh would fail miserably were we to attempt to use it as a collecting device for visible light waves.

Using the principles of refraction and reflection, scientists have fashioned a variety of instruments that, because of their ability to collect and form light rays into images and then resolve detail in those images, have dramatically extended

FIGURE 12.13
Dispersion of Light While light passes across a boundary between substances having different indices of refraction, the different wavelengths composing the white light are refracted at different angles, thus, giving rise to a spectrum of colors. Rainbows are the most familiar examples of dispersion effects.

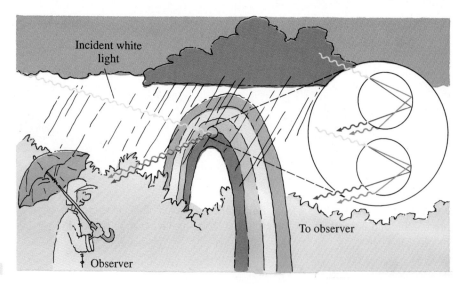

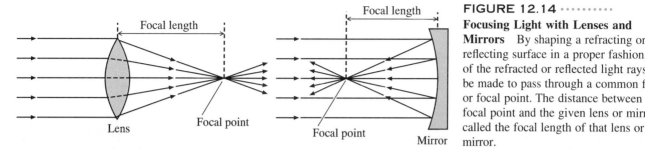

FIGURE 12.14 ·········
Focusing Light with Lenses and Mirrors By shaping a refracting or reflecting surface in a proper fashion, all of the refracted or reflected light rays can be made to pass through a common focus or focal point. The distance between the focal point and the given lens or mirror is called the focal length of that lens or mirror.

our perception of the physical world. With microscopes, we have gained access to a microbial world whose inhabitants are so tiny as to be invisible to our unaided or ''naked'' eye. With telescopes, we have extended our vision to objects in space so distant that the light by which we view them began its journey toward the earth long before the earth and solar system even existed.

The principles underlying such instruments are relatively simple. If we fashion the surface of a thin glass disk in just the right way, every light ray normal to the plane of the disk passing through and refracted by that ''figured'' piece of glass or lens can be made to pass through a single point called the *focal point* of the lens (Figure 12.14). If the light happens to be coming from an extended source, then the lens will form the light rays into a complete image of that object. Mirrors can be similarly figured so that all of the light normal to the plane of the mirror reflected from the mirror's surface passes through a single focal point. Any given optical instrument consists of some sort of combination of lenses and/ or mirrors.

In the simple refracting telescope, for example, two lenses of unequal focal lengths are placed in a line with the shorter focal length (eyepiece) lens at the back of the instrument. The forward, longer focal length lens forms the incoming light into a magnified image, which is then further enhanced by the eyepiece lens (Figure 12.15). A simple microscope works basically the same way, but due to the difference in the location of the object relative to the lens system, the path of the light rays in a microscope is somewhat different (Figure 12.16). In both

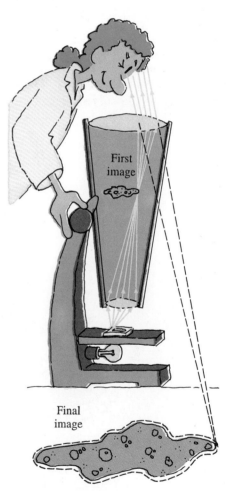

FIGURE 12.16 ·········
The Simple Compound Microscope By using two lenses as shown, it is possible to obtain an image of enhanced size for a very small object close to one of the lenses.

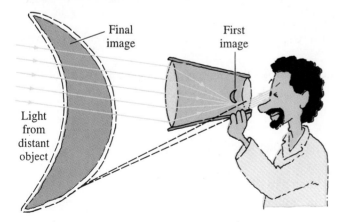

FIGURE 12.15 ·········
The Simple Refracting Telescope By using two lenses as shown, we can obtain an image of enhanced size for a distant object.

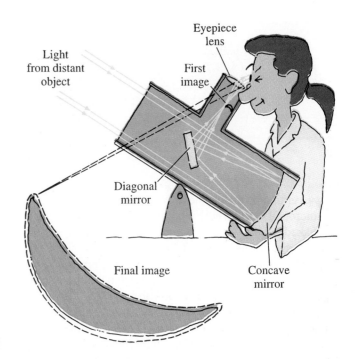

FIGURE 12.17 ···········
The Newtonian Reflector This type of telescope employs two mirrors and a lens as shown to obtain an image of enhanced size for a distant object.

instances, however, the instrument forms an image that is far more enhanced than what could be seen with our unaided eye.

Enhanced imaging can also be produced by reflection processes as well. One of the most common optical systems along these lines is the Newtonian reflecting telescope (Figure 12.17), named for its inventor, Sir Isaac Newton. In this type of telescope, light rays coming from a distant object strike a concave mirror. Before the rays can come to focus, however, they are intercepted by a flat ''diagonal'' mirror and reflected to a position outside of the path of the incoming light. Once again, the system forms the incoming light into a magnified image, this time off to the side of the telescope, which can be further enhanced by an eyepiece lens.

REVIEW
12.4
QUESTIONS

1. Describe the physics of what happens when a beam of light is either reflected or refracted.

2. Discuss the relationship between the angle of incidence for a beam of light and its
   a. angle of reflection
   b. angle of refraction.

3. What is meant by total internal reflection? Of what practical use is such a phenomenon?

4. Give some everyday examples of the phenomenon of dispersion.

5. Explain how each of the following instruments is able to produce magnified images by the principles of refraction and reflection.
   a. microscope   **b.** reflecting telescope   **c.** refracting telescope

12.2 The Human Eye

The human eye is one of the most amazing sensing devices imaginable. The eye basically operates as a simple refracting telescope without the eyepiece. Instead of focusing on an eyepiece lens, the lens of the eye focuses the incoming light onto a sandwiched, spherically shaped set of cells called the retina. Cells in the retina, referred to as rods and cones, respond to the light that strikes them by generating small electrical nerve pulses, which are assembled by nerve cells in the retina and then passed to the brain for interpretation.

The fluid lens of the eye is surrounded by a set of muscles called the ciliary muscles, which serve to change the shape of the lens, thereby permitting the eye to change focus from a distant subject, such as the homework assignment on the front blackboard of a large lecture hall, to one that is nearby, such as the notes you are taking in class. The eye also has an iris, which can adjust the aperture through which the light enters the eye. When it is dark and we need as much

light as we can to try to see, the iris opens wide, while on a bright sunny day, when there is an abundance of light, the iris closes down to a smaller area. Surrounding the eyeball itself is a dark layer of material called the choroid, which serves to prevent multiple reflection of light within the eyeball.

The human eye has excellent visual acuity and can resolve detail nearly down to the theoretical limit imposed by diffraction effects. It can also respond to a range of light intensities that extends over some nine orders of magnitude or a ratio of 10^9 from the brightest to faintest intensities we can detect. The electromagnetic radiation to which the eye responds ranges in wavelength from about 4×10^{-7} m in the violet to about 7.5×10^{-7} m in the red. In other words, all of the color the eye perceives in our everyday world is sorted and interpreted from an interval of wavelengths that is scarcely $\frac{1}{25,000}$ of a centimeter in extent! The human eye

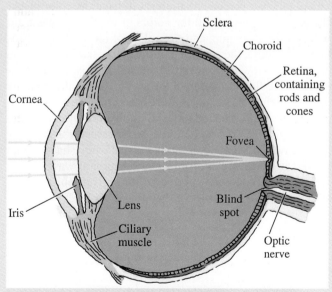

··· **The Human Eye**   Light from an object is refracted and focused onto the retina where it is concerted into electrical impulses for interpretation by the brain.

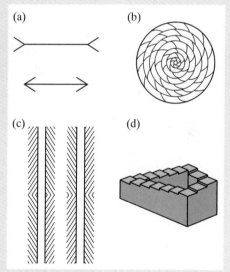

··· **Optical Illusions**   As marvelous a sensing device as it is, the human eye can be "fooled" by a variety of optical illusions such as: (a) the "unequal" line segments; (b) the "spiral"; (c) the "warped" lines; (d) the "stairway to nowhere."

PHYSICS FACET

Continued

even has tear ducts to cleanse it and a "lens cap" to cover and protect it when not in use.

Because it is so intimately tied to the human psyche, the human eye often "sees" only what the brain expects or wants it to see. For this reason, and because the eye can also be fooled by optical illusions, scientists have, in recent decades, tended to place a low weight on the scientific value of visual observations. Instead, there has been more reliance on "eye-like" instruments, such as the camera or phototube, which, like the eye, convert incoming light into some sort of chemical or electronic images, but are not plagued by the psychological interaction between the eye and the human mind.

12.5 Color

Within the interval of electromagnetic wavelengths detectable by the human eye, it has been found that different wavelengths of light generate different physiological and psychological perceptions. This variety of perceptions is the phenomenon we call *color*. Light having a wavelength of about 4×10^{-7} m, for example, is perceived as having the color violet, light as a wavelength of 5.3×10^{-7} m appears green and so on (see Figure 12.18). If there is no visible light emanating from a given object, we perceive that object to have the color black, while an object simultaneously giving off all the colors of the visible spectrum appears to be white. Clearly, black, white, and the colors of the visible electromagnetic spectrum do not constitute the sum total of the colors that we see. One can verify this fact simply by viewing a beautiful sunset, a shopping mall at Christmas time, or an extensive flower garden in the spring. Yet if these colors are visible to us, they must somehow be related to those basic colors that constitute the visible electromagnetic spectrum. That relationship is obtained by examining a phenomenon called *color mixing*, a concept familiar to any artist who has ever worked in color media. The basic idea of color mixing is that every shade or hue of coloration we perceive can be generated from a combination of wavelengths and intensities of visible light. The color we perceive for a given object is the result of the color composites of light that leave that object.

FIGURE 12.18 ··········

The Visible Spectrum All colors we perceive as human beings are combinations of wavelengths of electromagnetic radiant energy ranging in value from about 4×10^{-7} m to 7.5×10^{-7} m.

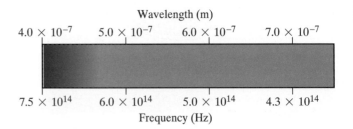

The color of the light departing the object depends first of all on the nature of the incident light. Suppose we have a surface such as a white piece of paper, which reflects equally all of the colors in the color spectrum. If we now shine monochromatic light (light having only a single color) onto the paper, the paper will, upon reflecting that light, appear to take on the color of the incident light. Thus, the paper appears blue in blue light, red in red light, and so on. Suppose, however, we shine two or more monochromatic beams of light onto the paper. The resulting color that we perceive is said to be *additively mixed.* Through the process of additive mixing, we can impart any color we wish to the paper simply by superimposing light beams from the proper number of monochromatic light sources operating at the proper wavelengths and intensities. If, for example, we shine a red beam of light onto a green beam, the resulting spot of light will be yellow. Every perceived color can in fact be generated by shining various combinations of intensities of only three colors—blue, green, and red—which are called the *additive primary colors* (Figure 12.19). Additive color mixing is one of the important principles employed in obtaining the colors we see in a color television picture (Chapter 16).

In addition to the color of the incident light, the absorptive properties of a given object also play an important role in determining the perceived color of that object. A white piece of paper appears white in the sunlight because it is reflecting all of the visible wavelengths of the light incident from the sun. On the other hand, a dandelion appears yellow in the same sunlight because it is reflecting only the yellow light and is absorbing all of the other colors. Such processes are called *selective absorption and reflection,* and the perceived reflected color is created by subtracting one or more colors from the incident light. Similar subtractive efforts can be observed for objects or media that transmit light. Such objects or media in effect filter out or otherwise subtract all color from the incoming beam except that which makes it through. The mechanism by which various materials can subtract wavelengths out of a given light ray are very much dependent on an understanding of atomic structure. As such this mechanism will be described more fully in that context in the next chapter.

Analogous to the phenomenon of color addition, three subtractive primary colors exist: cyan, which is composed of a band of blue and green wavelengths, which give rise to a sky blue color; magenta, a reddish purple color composed of bands of red and blue wavelengths; and an "expanded yellow" consisting of red, orange, and green wavelengths. Like their additive counterparts, cyan, magenta, and yellow can be combined to create any color we wish (Figure 12.20). In this instance, however, the color mixing is accomplished by the selective absorptions on the part of the material making up the object. For example, many color films are made of three-layered emulsions, each of which absorbs one of the subtractive primaries. The combination of the various responses of each of the layers in the emulsion is thus able to reproduce whatever color existed in the image the camera initially recorded.

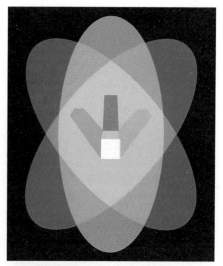

FIGURE 12.19

Additive Color Mixing   Any color imaginable can be obtained by proper combinations of the additive primary colors of blue, green, and red.

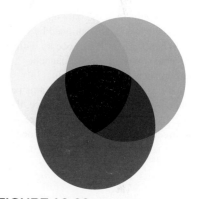

FIGURE 12.20

Subtractive Color Mixing   Any color imaginable can also be obtained by the proper combination of the subtractive primary colors of cyan, expanded yellow, and magenta.

1. What is meant by color? Estimate the wavelength of your favorite color.
2. Explain in terms of reflection and absorption phenomena why the pages in this text are white and the printed type is black.
3. Compare and contrast additive and subtractive color mixing. Give an everyday example of each.

REVIEW
12.5
QUESTIONS

4. Discuss some practical uses of additive and subtractive color mixing.

5. Describe the sets of additive and subtractive primary colors. In what ways are they similar? How are they different?

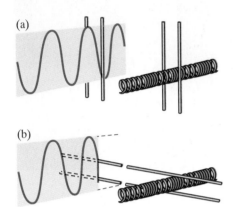

FIGURE 12.21 ··········

Polarization (a) If transverse and longitudinal waves are incident on a pair of vertical bars, both waves will pass through the bars. (b) If the bars are rotated 90°, the longitudinal wave is still able to pass through the bars but the transverse wave is now stopped by the bars. When such an effect occurs with the transverse waves of electromagnetic radiation such that only certain orientations of the wave's vibrations pass through a given object or substance, the emergent wave is said to be polarized. The direction of the "bars" is defined as the polarization axis of the object or substance.

12.6 Polarization

Recall for a moment the spring-oscillator system described in Chapter 9 and imagine that spring passing through a "grating" made up a pair of vertical parallel bars as shown in Figure 12.21. If we generate a longitudinal wave along the spring it will pass through the grating undisturbed. In a similar fashion, a vertical transverse wave will also pass undisturbed through the vertical bars of the grating. Suppose we rotate the grating 90° so that the bars are now situated horizontally rather than vertically. Any longitudinal wave generated along the spring will still pass undisturbed through the rotated bars of the grating. The vertical transverse wave, however, will be stopped by the now horizontally oriented bars. In fact, the vertical transverse wave will be stopped in varying degrees by any orientation of the grating other than a vertical one. Similarly, if we could generate a series of transverse waves in the spring which had different angular orientations relative to a vertical grating, we would find that the only transverse waves which would pass through the grating would be those whose vibrational directions were aligned parallel to the bars of the grating. When such waves have their plane of vibration restricted in such a way to a single plane of orientation they are said to be plane polarized or *polarized*. If we place a second grating whose bars are aligned perpendicular to those of the first, we can stop any and all transverse waves moving along the spring.

Since sound waves are longitudinal waves, we do not expect them to exhibit this polarization phenomenon, and they don't. Light waves, on the other hand, can be polarized. If we shine light through a pair of layers of a transparent polarizing substance such as calcite, we can reduce the intensity of the emerging light to nearly zero by rotating one of the calcite layers at right angles to the other. In this case the calcite crystals act as tiny gratings for the incoming light. The first layer polarizes the light into a single plane and the second stops that emergent light. From such observations, as well as the theoretical consequences of Maxwell's equations, we may conclude that light as well as other forms of electromagnetic radiation behaves as transverse waves.

FIGURE 12.22 ··········

Polarized Sunglasses Most of the light that reflects as "glare" off sand, snow, or clouds is horizontally polarized. By wearing glasses with a vertical polarization axis, most of the glare can be reduced or eliminated.

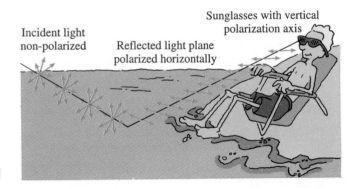

Incident light non-polarized

Reflected light plane polarized horizontally

Sunglasses with vertical polarization axis

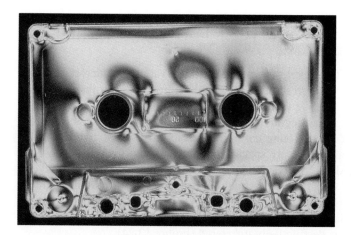

FIGURE 12.23 ··········

Optical Activity Stresses in a transparent object cause the plane of polarization of a light beam to be rotated differently in different regions of the object as the light passes through. This property is called the optical activity of the object and is greatest where the stresses in the object are the greatest.

Light waves can be polarized by a variety of other processes in nature as well. Most notably, light reflected off nonmetallic horizontal surfaces tends to be highly plane-polarized in a horizontal direction. Thus a pair of sunglasses made of a polarizing material which has its "grating" aligned in a vertical direction is very effective at cutting down the intensity of the horizontally polarized light that "glares" off, for example, sand and water at the beach, and snow at the ski resorts (Figure 12.22).

The phenomenon of polarization has a number of interesting uses and applications. In nature, bees and other insects have eyes that are composed in part of layers of "grating-like" material. This characteristic enables the insect to use the angle of polarization of blue skylight to orient themselves and to navigate relative to the sun's position in the sky.

A number of transparent materials have a property called *optical activity* in which the plane of polarization of the light passing through such a substance will experience a certain amount of rotation (Figure 12.23). Measurement of the optical activity of transparent solutions containing sugars and other organic molecules can provide important information regarding the structure and composition of the material present. Many transparent solid substances become optically active when subjected to stress. The degree of optical activity is directly proportional to the stress involved. Thus, a polaroid picture of a given object will pinpoint the regions of greatest optical activity, hence, the regions of the object subject to the largest stress. Such regions, or "stress points," represent the parts of the object with the largest probability of experiencing mechanical failure. The ability to locate stress points in a given object is of considerable value to the engineer who is constantly seeking designs that minimize the stress levels of such points or eliminate them altogether.

1. Discuss the phenomenon of polarization.
2. What types of waves, if any, cannot be polarized?
3. Explain how a transparent crystal can cause polarization to occur.
4. How do polaroid sunglasses work?
5. Discuss some of the practical applications of the phenomenon of polarization.

REVIEW
12.6
QUESTIONS

12.7 Interference and Diffraction

As we have seen in Chapter 9, waves can exhibit a variety of interference phenomena in which the wave crests and troughs can reinforce or destroy one another. Perhaps the most impressive demonstration of this aspect of the wave nature of light was provided at the beginning of the nineteenth century by the English physicist Thomas Young. Young carefully directed a beam of light so that it passed through a pair of closely spaced pinholes and then onto a background screen. Young found that the resultant illumination on the screen consisted of a pattern of bright and dark fringes that could only be interpreted as an interference pattern. The bright fringes were those regions on the screen where the light waves constructively interfered and the dark fringes were those regions on the screen where the light waves destructively interfered (Figure 12.24).

The intensity of light observed at a given point on the screen thus depends on the difference in path length between the two openings and that given point. The two arriving waves will constructively interfere to produce a bright fringe if the path lengths differ by integral multiples of the wavelength of the light. They will destructively interfere to produce a dark fringe if the path lengths differ by half-integral numbers of the wavelengths (Figure 12.25). The interference pattern obviously changes with the wavelength of the light involved. For this reason, interference phenomena of various kinds can be used in spectrographs to spread a beam of light into an array or spectrum of its component wavelengths.

Interference of light waves is also observed to occur in very thin layers of transparent materials or thin films. In this case, a beam of light incident on the thin film is split into a reflected portion and a refracted portion at the film's surface as shown in Figure 12.26. The refracted portion of the beam then reflects off the lower surface back to the upper surface where it reemerges parallel to the reflected portion of the beam. In this case, the refracted portion of the beam has traveled further than the reflected portion, and the relationship between the crests and troughs of the two portions of the beam as they recombine and hence the type of interference exhibited depends on the thickness of the film. The array of colors

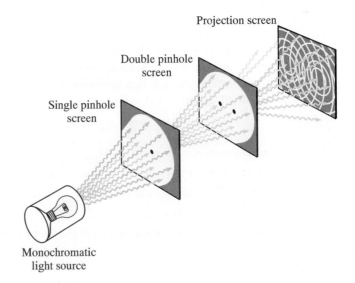

FIGURE 12.24 ··········

Young's Experiment In 1801, the English physicist Thomas Young directed a beam of monochromatic light toward three colinear screens. The first screen had a single pinhole, followed by a screen with a pair of pinholes, and then a projection screen. The light emerging from the double-pinhole screen was found to interfere with itself to produce an interference pattern of light and dark regions on the projection screen.

(a)

Interference pattern

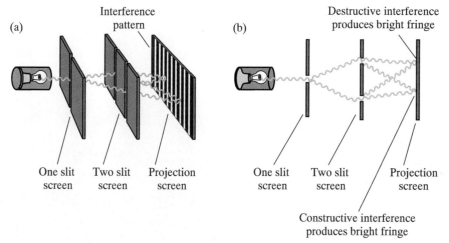

One slit screen   Two slit screen   Projection screen

(b)

Destructive interference produces bright fringe

One slit screen   Two slit screen   Projection screen

Constructive interference produces bright fringe

FIGURE 12.25 ············

Constructive and Destructive Interference in Young's Experiment   If the pathlength difference between the two-slit screen and the projection screen is an integral multiple of the wavelength of the incident light, the two light waves constructively interfere to produce a bright region on the projection screen. If the pathlength is a half-integral multiple of the wavelength of incident light, the two light waves destructively interfere to produce a dark region on the projection screen. Part (a) shows a side view; part (b) shows a top view.

exhibited by light reflected from a thin layer of gasoline or oil floating on top of a body of water is perhaps the most familiar example of thin-film interference.

Interference phenomena can also occur as light passes by an opaque object. Huygens's principle, as we have seen, assumes that a wave front of light can be regarded as being composed of tiny wavelets, each of which propagates from a point on the wave front. If no obstacles are placed in the path of the wave front, it will travel in a straight line. Suppose, however, that the wave front encounters an obstruction of some sort. As the wave front passes the obstruction, the points on the wave front close to the edge of the object give rise to spherical wavelets that not only spread out beyond the "geometric" shadow of the object, but can interfere with each other as they do it. The result is a phenomenon called *diffraction* in which a pattern of light fringes is produced at the boundary of the object's geometric shadow (Figure 12.27).

The degree to which diffraction occurs depends on the size of the wavelength of the light involved compared to the size of the obstructing object. The diffraction

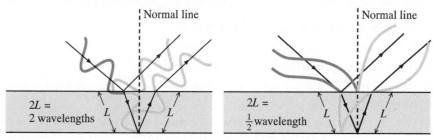

FIGURE 12.26 ············

Interference in a Thin Film   As a beam of light strikes a transparent thin film, part of the beam is reflected at the film surface. The other part is refracted at the film's surface, reflected by its lower boundary, and then refracted by the film surface before reemerging from the film. The remaining part of the beam thus has a pathlength difference equal to $2L$. When $2L$ is an integral multiple of the wavelength of the incident light, construction interference occurs. When $2L$ is an odd half-integral multiple ($\frac{1}{2}$, $\frac{3}{2}$, $\frac{5}{2}$, etc.) of the wavelength of the incident light, destructive interference occurs.

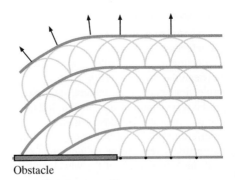

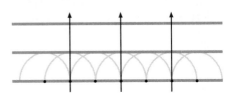

Obstacle

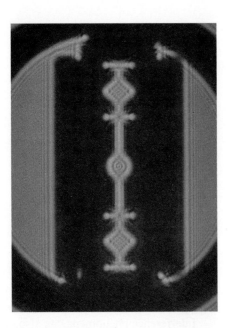

FIGURE 12.28 ·············

A Diffraction Pattern   The edges of the shadow from this razor blade are not sharp, but consist of a series of light and dark fringes. These fringes are a result of one region of the passing wave front constructively and destructively interfering with another region of the passing wavefront.

FIGURE 12.27 ·············

Diffraction   As a wave front passes an obstacle, Huygens's principle tells us that spherical waves will propagate outward from each point on the wavefront as shown. The light thus appears to bend around the obstacle to create an interference pattern.

associated with a wave front of light passing the edges of a large-aperture window is much smaller than for portions of that same wave front passing through a pinhole-sized aperture. In all cases, however, a ''blurring'' effect occurs at the boundary of the geometric shadow, which places a fundamental restriction on the degree of accuracy to which we can measure the dimensions of that shadow (Figure 12.28). This effect can perhaps best be illustrated by trying to measure the length of a shadow cast by one of your friends on a bright sunny day. Diffraction is, in a certain sense, the bane of experimental and observational scientists in that it places an ultimate limit on the degree to which optical instruments such as microscopes and telescopes are able to resolve detail.

REVIEW 12.7 QUESTIONS

1. Describe the phenomenon of interference.
2. Discuss several ways in which one can cause light waves to interfere.
3. What is diffraction? How is it related to interference?
4. How do interference and diffraction demonstrate that light can behave as a wave motion?
5. Why do diffraction phenomena restrict the ability of optical instruments such as microscopes and telescopes to detect image detail?

12.8 Scattering

Because the electrons within atoms and molecules are free to vibrate, they may possess one or more ''natural'' or ''resonant'' frequencies which are electromagnetic analogues to the natural frequencies described for acoustical phenom-

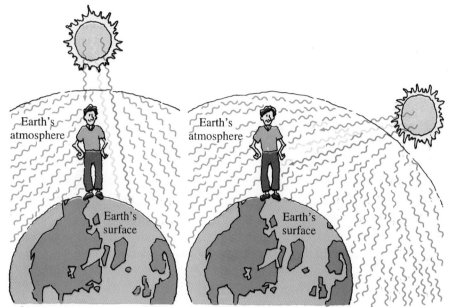

FIGURE 12.29 ⸱⸱⸱⸱⸱⸱⸱⸱⸱⸱
Scattering in the Earth's Atmosphere As light rays enter the earth's atmosphere, the shorter wavelength experiences a greater degree of scattering. We thus observe a blue sky, yellow sun, and red sunsets.

ena. In a gaseous medium, where the atoms and molecules are relatively far apart, electromagnetic waves with frequencies at or near the natural frequencies of the atoms in the gas can interact with these atoms in several ways.

The energy from such waves, for example, can be absorbed by the electrical charges in the atoms and molecules of the gas causing these charges to vibrate. As they do so, they generate electromagnetic waves having the same frequency as the incident wave, but because of the random orientation of the atoms in the gas, these "regenerated" waves are emitted in a random fashion. A given atom is thus just as likely to radiate electromagnetic waves back toward the original source as it is to send them along their original direction of propagation. As a result of this type of interaction, some of the wavelengths and frequencies contained in the original incident electromagnetic radiation are in effect subtracted out of the incident beam. This subtraction process is often referred to as *scattering* and is the mechanism underlying the fact that we observe a blue sky, a yellow sun, and red sunsets (Figure 12.29).

As white sunlight strikes the earth's atmosphere, the violet and blue wavelengths are fairly close to the natural frequencies of the nitrogen and oxygen molecules making up the bulk of the earth's atmosphere. Hence these wavelengths are scattered out of the sunbeam more effectively than the lower frequency waves of the visible spectrum. For every ten violet waves scattered out of the beam, for example, there will be roughly five green waves, two orange waves, and one red wave. The scattered waves are then absorbed and randomly reradiated by other nitrogen and oxygen molecules in the earth's atmosphere in these same proportions. Some of these randomly reradiated waves make it to the surface of the earth where they give the sky a bluish color when observed by the human eye.

Meanwhile, the rays we perceive as coming directly from the sun now have much of their blue and violet wavelengths scattered out of the beam. When the sun is high in the sky and the layer of atmosphere through which the sun's rays must pass in small, this color subtraction results in a yellowish-colored sun. As the sun begins to set, however, the sunlight must traverse ever-thicker layers of the earth's atmosphere, resulting in ever-greater degrees of scattering for the frequencies all along the visible portion of the electromagnetic spectrum. The ratios of the waves scattered at various colors however remain the same. Thus the low-frequency red waves are the least affected by the scattering processes in the earth's atmosphere. Therefore, at sunset it is the sun's red waves that by and large make it through the atmosphere for us to see.

1. Explain what is meant by scattering.
2. Explain why the sky is blue and the sun appears yellow at high noon and red at sundown.
3. Other than the earth's atmosphere, can you describe an example of scattering?
4. What effect do you think atmospheric pollution would have on our perception of the colors of the sky and the sun?
5. What effect do you think the altitude of a person above sea level would have on our perception of the colors of the sky and of the sun?

12.9 Light as a "Particle of Energy"

In the last decade of the nineteenth century, Maxwell's description of light as an electromagnetic wave enjoyed enormous success in providing an explanation for the behavior of both visual light and other forms of radiant energy. It was with considerable surprise and consternation that around the turn of the century a number of light-related phenomena were observed that could simply not be accounted for by Maxwell's electromagnetic wave theory.

If we carefully examine the radiant energy given off by a nonreflecting, self-glowing or "incandescent" solid, we would find that a plot of the wavelength of radiant energy versus the intensity of the energy radiated at the wavelength would have the hump-shaped appearance shown in Figure 12.30. At the end of the last century the British physicist Lord Rayleigh attempted to explain theoretically the characteristics of this so-called blackbody curve by assuming that the radiant energy emitted from an incandescent solid was due to electronic charges vibrating within the solid over a continuous distribution of energies of oscillation. Rayleigh was able to construct a theoretical plot of wavelength versus energy intensity and found to his dismay that instead of passing through a maximum value and returning toward zero as the experimental curve did, his theoretical curve "took off" toward infinitely large values of the energy intensity as the wavelength decreased in value. Not losing their sense of humor in this matter, the physicists of the day referred to this discrepancy as the "ultraviolet catastrophe."

In the scientific method, whenever there is a disagreement between theory and replicable observations, it is the theory that must be altered to fit the observations. So it was in 1900 when a German physicist named Max Planck found that he could theoretically replicate the shape of the experimental wavelength-energy intensity plot if he assumed that the energies of oscillation in the solid were not continuous as Rayleigh has supposed, but were instead limited to discrete

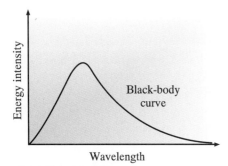

FIGURE 12.30

The Blackbody Curve If one observes an incandescent source of electromagnetic radiation and constructs a plot of the wavelength versus the energy intensity observed at that wavelength, the result is this hump-shaped curve, which is referred to as a "blackbody" curve because of the fact that such objects are radiating their own energy as opposed to energy reflected from some other source.

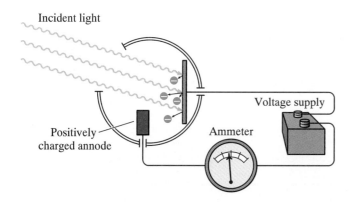

Incident light

Positively charged annode

Voltage supply

Ammeter

FIGURE 12.31 ···········

The Experimental Apparatus for the Photoelectric Effect When light shines on the negatively charged plate, electrons are liberated and flow across the gap to the positively charged plate, and a photocurrent flows through the circuit.

"quantized" values, which were integral multiples of a basic, indivisible "energy quantum." Planck further surmised that this energy quantum was equal to the product of the oscillator frequency and a mathematical constant, which is now known as Planck's constant and has a value equal to 6.63×10^{-34} joule-seconds (see Physics Formulation 13.1). Planck was appalled at the idea that energy of any form in the universe had to come in discrete packages, and would later refer to his calculations as an "act of desperation." Nonetheless Planck's calculations offered a theoretical explanation for the behavior of wavelength-energy intensity plots.

About this same time, Hertz had discovered that under certain circumstances some metals could exhibit a *photoelectric effect* in which light incident on a given metal would cause electrons to be liberated from that metal. Such a phenomenon was not in itself surprising, since in theory waves could cause electrons to vibrate in such a way as to be ejected from the surface of the metal. More detailed investigation of the photoelectric effect, however, revealed characteristics of this phenomenon that could not be explained in terms of a wave-matter interaction.

In one experiment, two oppositely charged plates are hooked up to a voltage supply and an ammeter in the circuit shown in Figure 12.31. If no light shines on the plates, the circuit behaves as if it were an open circuit and no current flows. If, however, the plates are exposed to light, electrons, called *photoelectrons,* are ejected from the negatively charged cathode plates and flow across the plate gap to the positively charged anode plate. The circuit is thus completed. The rate at which the photoelectrons are excited from the cathode plate can be measured from the observed rate of current flow through the ammeter. In performing such an experiment, the rate at which electrons were ejected from the cathode was found to be directly proportional to the intensity of the incident light when photoelectrons were produced. Surprisingly, however, a *threshold frequency* was also found to exist for the incident light below which no electrons were ejected from the cathode surface, regardless of how intense the light became.

By adjusting the electrical potential between the plates to an appropriate value called the *stopping potential* it is also possible to stop the flow of the photoelectrons across the plate gap. Since the kinetic energy of the photoelectrons arriving at the anode is the product of the electron charge and the electric potential between the plates, a measurement of the stopping potential will permit the calculation of the kinetic energy of the liberated electrons. Measurements of the photoelectron kinetic energies made in this fashion produced additional surprises. It was found that the kinetic energy of the photoelectrons increased linearly with

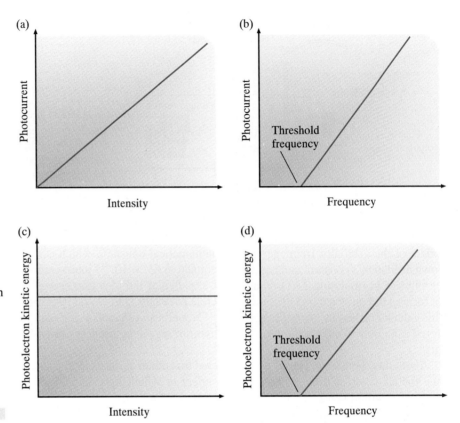

FIGURE 12.32 ···········

Observed Characteristics of the Photoelectric Effects (a) Photocurrent varies linearly with intensity of incident light. (b) Photocurrent varies linearly with frequency of incident light and has a threshold frequency. (c) Photoelectron kinetic energy does not depend on intensity of incident light.
(d) Photoelectron kinetic energy varies linearly with frequency and has a threshold frequency.

the frequency of the incident light but the intensity of that light had no effect whatsoever on the kinetic energies (Figure 12.32).

Such observations are at odds with the idea that light acts as a wave. If electrons were being ''vibrated'' out of the metal as the wave model suggested, then we would expect the absorption of the higher energy associated with higher intensity waves to manifest itself in higher kinetic energies for the electrons. Instead it manifests itself in more electrons being liberated. Also puzzling was the fact that some frequencies of light were simply unable to liberate photoelectrons regardless of the incident light intensity. Once again, the energy associated with a high-intensity wave should be able to liberate electrons from the metal surface independently of the wave's frequency—if the light behaved as a wave.

In 1905, the legendary German physicist Albert Einstein (Physics Facet 15.1) offered an ingenious explanation for these observed properties. Einstein claimed that all of the difficulties encountered by the wave theory of light in attempting to account for the observed properties of the photoelectric effect could be resolved by simply assuming that light does not behave as a continuous wave motion. Instead light is composed of discrete packets of energy called *photons,* which exhibit particle-like behavior. Thus, in Einstein's mind, Planck's explanation for the incandescent radiation curve was backward. Instead of the oscillations of the light-producing electrons being restricted within the structure of the solid to discrete values and hence generating discrete values of wave energies, it was the light itself that was quantized into ''packets'' of energy having a value equal to Planck's constant × frequency, or

$$\begin{matrix} \text{energy of} \\ \text{a photon} \end{matrix} = \begin{matrix} \text{Planck's} \\ \text{constant} \end{matrix} \times \begin{matrix} \text{frequency of} \\ \text{the photon} \end{matrix},$$ 12.1

$$E = h \times f.$$

The light striking the metal cathode in the photoelectric effect is composed of these photons, and a given electron will completely absorb one photon and the energy associated with that photon. If the energy absorbed by the electron is less than the metal's *work function* or the energy needed for the electron to break free of the metal, then the electron cannot escape the cathode and no current flows. An increase in the light intensity will increase the number of photons incident on the cathode and hence the number of electrons that will have absorbed these photons. Regardless of their total number, however, each and every electron that has absorbed a photon still lacks the energy to clear the work function of the cathode and, despite the increase in light intensity, no photocurrent is generated. On the other hand, if the energy of an incident photon is higher than the work function, any electron absorbing that photon will be ejected from the surface of the cathode with a kinetic energy equal to the difference between the energy of the incident photon and the work function of the cathode. An increase in the intensity of the incident light will once more lead to an increase in the number of electrons that have absorbed incident photons, but the ejected electrons will have the same kinetic energies. Thus an increase in the intensity of the incident light will produce an increase in the photocurrent, but will not cause the electrons' kinetic energy to increase. This result is, as we have seen, in accordance with experimental results.

There is no doubt, then, that light can behave as if it were a particle. On the other hand, we have spent a good bit of this chapter describing the impressive scientific success achieved by thinking of light as a wave. In fact, the phenomena of the interference and diffraction can be accounted for *only* by assuming that light behaves as a continuous wave motion. Light thus exhibits a sort of "wave-particle" duality in which it can somehow behave as both a particle and as a wave, but not simultaneously! Neither of these comparatively simple mainstay representations from classical physics is adequate for the cause. Therefore, suggestions have been made that the photons making up light beams are perhaps some sort of new entity, which defies description in terms of analogues with objects and phenomena in the "everyday" world. Photons, however, are only the beginning. In the next chapter as we pass from the macroscopic everyday world into the microscopic realm of the atom, we will find that our explorations will be marked with even more amazing and profound surprises.

WORKED EXAMPLE

A yellow light source with a frequency of 5.2×10^{14} Hz has an observed intensity of one watt/m^2. How many yellow photons per second will strike a square one meter on a side under these circumstances?

Continued

> WORKED EXAMPLE
>
> *Continued*
>
> **SOLUTION:** The energy of a photon is given by
>
> $$\text{energy of photon} = h \times \text{frequency of photon} .$$
>
> Each yellow photon thus has an energy given by
>
> $$\text{energy of photon} = 6.63 \times 10^{-34} \times 5.2 \times 10^{14} \text{ J}$$
> $$= 34.5 \times 10^{-20} = 3.45 \times 10^{-19} \text{ J} .$$
>
> Since one watt equals one joule/sec, the number of yellow photons striking the square is
>
> $$1 \text{ joule/sec} \times \frac{1 \text{ photon}}{3.45 \times 10^{-19} \text{ J/photon}}$$
>
> or
>
> $$(1/3.45) \times 10^{19} \text{ photons/sec} = 2.9 \times 10^{18} \text{ photons/sec} .$$
>
> Thus, 2.9×10^{18} yellow photons are striking the square each second.

1. What is a blackbody curve?
2. Describe the photoelectric effect. In what way(s) does the wave model fail to explain the observed characteristics of the photoelectric effect.
3. Can you think of any practical applications of the photoelectric effect? Describe them.
4. What is a photon? What evidence do we have for the existence of photons?
5. Compare and contrast the various representations of light (ray, wave, particle, etc.) discussed in this chapter and comment on how each has been successful and how each has failed in describing the various phenomena exhibited by light.

1. The green, yellow, and red lights of a traffic signal have approximate wavelengths, respectively, of 5.5×10^{-7} m, 5.8×10^{-7} m, and 6.6×10^{-7} m. Find the energy of one photon of each color of light coming from a traffic signal.
2. The energy of a certain photon is 2×10^{-20} joules. Is this photon visible to the naked eye? Explain.

Chapter Review Problems

1. What is the intensity of a 75-watt bulb when viewed from a distance of 3 meters if it has an intensity of 6 watts/m^2 when viewed from a distance of 1 meter?

2. A certain light source is observed to have an intensity of 2 watts/m^2 when viewed from a distance of 100 meters. At what distance will the source appear to have an intensity of 50 watts/m^2?

3. The planet Jupiter is approximately five times farther from the sun than Earth. How much more (or less) intense is the Sun's light as seen from Jupiter compared to the Sun's light as seen from Earth?

4. Moving at a speed of 300,000 km/sec, how far can light travel in

   a. one minute?   **b.** one day?   **c.** one year?   **d.** one decade?

5. How long does it take light to travel

   a. one kilometer?

   b. once around the circumference of the earth (40,000 km)?

   c. to the moon (distance = 384,000 km)?

   d. to the sun (distance = 150 million km)?

6. The largest known value for the index of refraction is about 4.0. What is the speed of light in such a substance?

7. A physicist measures the speed of light through a certain transparent solid and obtains a value of 275,000 km/sec. What is the index of refraction for this solid?

8. The lowest frequency of electromagnetic radiation thus far detected has a value of about 0.01 Hz. What is the wavelength of such radiant energy?

9. A certain form of electromagnetic energy coming from interstellar hydrogen gas has a wavelength of 21 cm. At what frequency are the oscillations in the hydrogen gas occurring in order to produce this particular form of energy?

10. A physicist measures the wavelength and frequency of the radiant energy emitted by a microwave oven and finds that they have respective values of 0.03 m and 1.0×10^{10} Hz. At what speed do microwaves travel through space?

11. What is the wavelength of electromagnetic radiation having a frequency of 1 Hz? What problems, if any, would arise when trying to detect such radiation?

12. A scientist discovers a new form of wave energy with a wavelength of 30 km and a frequency of 100 Hz. Can this energy be electromagnetic in nature? Explain.

13. A light wave strikes a plane mirror at an angle of 20° to the surface of the mirror. Find:

   a. the angle of incidence.

   b. the angle of reflection.

14. A light ray strikes the right-angle array of mirrors shown at the right. With the use of a protractor, trace the ray's path as it reflects off each of the three mirrors.

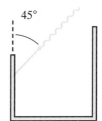

15. A certain type of glass has an index of refraction of 1.5. At what speed will light travel through this glass?

16. A physicist measures the speed of light in a certain diamond to be 130,000 km/sec. What is the index of refraction of the diamond?

17. The scattering of light by the earth's atmosphere varies as 1/(wavelength)^4. How would you expect the scattering of light by the earth's atmosphere to depend on the frequency of light?

18. The highest frequency photons observed in the universe thus far have frequencies of about 2×10^{22} Hz. What is the wavelength and energy of one such photon?

19. How many photons of red light, each having a frequency of 3.5×10^{14} Hz, are required to produce 20 joules of energy?

20. The wavelength of a certain photon measured in meters has the same numerical value as its frequency measured in Hz. Find the wavelength and frequency of this photon.

21. Express the energy of a photon in terms of its wavelength, if the energy of a photon is equal to $h \times$ frequency.

22. A photon of blue light has a wavelength of 4.7×10^{-7} m and an energy of 4.0×10^{-19} J. From these data, find the value of the Planck's constant h.

23. A radio photon is found to have a wavelength of 40 cm and an energy of 4.5×10^{-25} joules. Using this data, find the speed of the radio photon.

THOUGHT QUESTIONS

1. It is possible to have just an "electric" wave or "magnetic" wave by itself? Explain.

2. Describe how the presence of fog, smoke, or air pollution would affect the relationship between the observed intensity of a light source and the distance to the source.

3. Roemer's discovery of the light-time effect for the satellites of Jupiter was hailed at the time as proof that the Earth orbits the Sun. Explain why this effect constitutes such a proof.

4. How would you show that the electromagnetic waves of Hertz's demonstration were not the result of direct electromagnetic induction?

5. Discuss how the production of electromagnetic waves is a consequence of the principle of the conservation of energy.

6. Does aberration of starlight demonstrate the earth's orbital motions? Explain.

7. Is it possible for a substance to have an index of refraction that is less than 1.0? Explain.

8. Suppose you found a new form of electromagnetic energy that had wavelengths shorter than the gamma rays. Discuss what sort of properties you would expect this new form of radiant energy to have.

9. Discuss how you would set up an apparatus for producing very low frequency, very long wavelength forms of electromagnetic waves. Why might you want to do this?

10. Summarize the various types of electromagnetic waves that the human body can detect in one way or other.

11. What are the properties of the gamma rays that make them so dangerous to human beings?

12. In what way(s) is Huygens's principle *not* a good representation of the behavior of light waves?

13. Discuss how you might define a "smooth" surface and a "rough" surface in terms of measurable quantities.

14. Describe some examples of the use of refraction and reflection processes by nonhuman creatures.

15. Would you expect sound waves to exhibit refraction or dispersion phenomena? Explain.

16. Give examples of refraction and reflection for nonvisual forms of electromagnetic waves.

17. Does the velocity of a light ray change when it is reflected? Explain.

18. Describe one possible application of fiber optics in the medical sciences.

19. Is it possible to have refraction without dispersion? Dispersion without refraction? Explain.

20. How might you use its index of refraction as a means of identifying an unknown substance?

21. How does color differ from wavelength?

22. Look up the structure and functions of the eyes of three different nonmammalian creatures. Compare each with one another and with those of the human eye.

23. In what ways does the human eye make a good scientific instrument? In what ways does it not?

24. Describe in detail how you might set up an experiment to decide whether or not human behavior is affected by color.

25. How might light be polarized by processes other than reflection which occur naturally in the physical world?

26. Can ocean waves be polarized? If so, is there any reason we would want to do such a thing?

27. What role might the phenomenon of diffraction play in designing an optical device?

28. Describe how the phenomenon of interference could be used to determine the thickness of very thin film layers.

29. When the Viking space probes landed on the surface of the planet Mars, they found that the Martian sky was pink. Propose some possible explanations for why this is so.

30. What are some of the properties of light that cannot be accounted for if one assumes that light behaves as

    a. a ray.   **b.** a wave.   **c.** a particle.

31. How should the scientific method proceed in order to resolve the contradiction outlined in Question 30 concerning the "true" nature of light?

32. Discuss how the "ultraviolet catastrophe" is an illustration of the scientific method in action.

33. Why do you suppose that Max Planck referred to his explanations for the behavior of the blackbody curve as an ''act of desperation''?

34. Does a photon carry any electrical charge? How would you demonstrate your claim?

35. Is there a phenomenon analogous to the photoelectric effect for nonvisual forms of electromagnetic radiation? Explain.

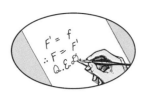

Light Intensity

Whenever a source of electromagnetic waves is observed, the detector, whether it is a human eye, camera film, or a phototube, responds in some fashion to the amount of energy received from the source. Because of its importance, this quantity is referred to as the *apparent brightness* or *intensity* of the source. Intensity is defined as the amount of energy per unit time passing through a unit area perpendicular to the line of sight between the source and the observer.

It is obvious to even the most casual observer that the apparent brightness of a light source decreases with increasing distance, but for a variety of reasons, we would like to know in precisely what fashion. To deduce the formal mathematical relationship between distance and intensity, consider a source of electromagnetic waves that has a power output of P watts. This power output in watts or joules/sec is sometimes also referred to as the *luminosity* of the source.

As the electromagnetic waves are produced by the source, suppose they propagate outward equally in all directions from the source. As a result, a given electromagnetic disturbance can be thought of as a sort of expanding spherical shell of energy. As this energy shell propagates outward, the total power associated with the disturbance remains the same, but the total area of space over which it is spread increases as its distance from the source increases (Figure 12.33). From solid geometry we know that the surface area of a sphere is given by

$$\text{surface area of a sphere} = 4\pi \left(\text{radius of sphere}\right)^2.$$

The perceived intensity of the source at some distance d from the source will then have a value equal to the total power of the source divided by the surface area of a sphere having a radius equal to the distance to the source or

$$\frac{\text{apparent brightness}}{\text{of source}} = \frac{\text{intensity}}{\text{of source}} = \frac{\text{power of source}}{4\pi\,(\text{distance})^2}.$$

Thus, if a given source is observed from a distance d_1, the perceived intensity I_1 of the source at that distance will be

$$I_1 = \frac{\text{power of source}}{4\pi(d_1)^2}.$$

Similarly, if the same source is observed from a second distance d_2, the observed intensity at that distance will be

$$I_2 = \frac{\text{power of source}}{4\pi(d_2)^2}.$$

If we divide these two equations and recall that the quantity power of source/4π remains the same for both distances, then we have

$$\frac{I_1}{I_2} = \frac{(d_2)^2}{(d_1)^2} = \left(\frac{d_2}{d_1}\right)^2.$$

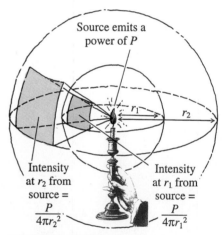

FIGURE 12.33 ··········
The Inverse Square Law for Light Intensity   As light propagates outward from a given source, it spreads out over spheres of progressively larger radii and hence progressively larger surface areas. The surface area of a sphere is equal to $4\pi(\text{radius})^2$. Since the power of the source is constant, the light intensity that is observed at some distance r from the source is equal to (power of source) ÷ (surface area of sphere), which has a radius equal to the distance to the source, or intensity = power/$4\pi r^2$. This is the inverse square law for light intensity.

If, for example, the distance to a certain light source is doubled such that $d_1/d_2 = 2$, then the ratio of the intensity I_1 at the new distance to I_2 at the old distance will be

$$\frac{I_1}{I_2} = \left(\frac{d_2}{d_1}\right)^2 = \left(\frac{1}{2}\right)^2$$

or ¼ as much. Thus, if the distance to a given source is doubled, the received intensity will be ¼ what it was at the old distance.

In a similar fashion, if the source is brought three times closer, then the value of $d_1/d_2 = \frac{1}{3}$ and the intensity ratio I_1/I_2 at the new distance is now given by

$$\frac{I_1}{I_2} = \left(\frac{d_2}{d_1}\right)^2 = \left(\frac{3}{1}\right)^2 = 9 \ .$$

or I_1 is 9 times larger than I_2. Thus, if the distance to the source is three times smaller, the source will appear to be 9 times brighter when viewed from this reduced distance.

The expression for light intensity can also be used to deduce the distances to light sources. Suppose we have measured the intensity I_0 of a certain light source from a known distance d_0. If we observe that same light source or one identical to it from some unknown distance d_{unk} and find that it now has an observed intensity I_{obs}, the unknown distance is then

$$d_{unk} = d_0 \sqrt{\frac{I_0}{I_{obs}}} \ .$$

As we shall see, this relationship will be of fundamental importance when we discuss the physics of the remote and inaccessible celestial objects that populate the outer space of our physical universe.

QUESTIONS

1. In what ways, if any, has the principle of the conservation of energy been used in deriving the equation for the apparent brightness of a source?
2. What is the difference between the power and intensity of a light source?
3. Compare and contrast the mathematical expression for the intensity of a light source with Coulomb's law and the law of gravity.

PROBLEMS

1. A student standing 2 m from a desk lamp observes that the intensity of the lamp from that distance is 2 watts/m². What is the power output of the lamp?
2. A certain plant can withstand a light intensity of 80 watts/m² without being damaged. How close can the plant be placed to a 200-watt light bulb and not suffer any adverse effects from the light bulb?
3. A 5000-watt landing light for a jet aircraft is observed to have an intensity of 1.0×10^{-5} watt/m² as the plane approaches the end of an airport runway. How far is the aircraft from the runway?

"Terrestrial" Determinations of the Speed of Light

Despite the fact that the first determination of the speed of light came as early as 1670 with Roemer's investigations of the light-time effect displayed by Jupiter's satellites, it was not until nearly two centuries later that a purely "terrestrial" determination of this fundamental quantity of nature was successfully made. It is little wonder when we consider that in one second light traverses a straight-line distance equivalent to some 7½ times around the world. Even though the experimental difficulties involved in accurately measuring such a large speed were obviously immense, scientists felt that an earth-based determination of the speed of light would not only serve to confirm the astronomical or "extraterrestrial" measurements, but would make possible much more accurate determinations, because of the carefully controlled conditions that could be created in an all-earth or terrestrial setting.

The first successful attempts in this regard came out of France and were developed by a wealthy Parisian named Hippolyte Fizeau in the middle of the last century. In Fizeau's experiment (Figure 12.34), a light beam is directed at a mirror mounted a known distance away. Before the light gets to the distant mirror, however, it must traverse a partially silvered, diagonally oriented mirror and then pass through the teeth on the rim of a rotating cogwheel. When the light beam strikes the distant mirror, it is reflected back through the teeth of the cogwheel before being reflected 90° by the diagonal mirror to the observer. If the cogwheel is rotating, each time the wheel is oriented in such a way that the light beam can pass through one of the gaps between the cogs, a flash of light is sent toward the distant mirror. As the flash travels to and from the distant mirror, the cogwheel continues to rotate. When the flash returns to the cogwheel rim, if it encounters a gap, it will pass onto the diagonal mirror and will be reflected to the observer. If, however, the light flash encounters a cog, it will be stopped at the cogwheel rim and not make it to the observer. Fizeau found that as the angular speed of the cogwheel was slowly increased, there came a point at which each flash produced by a given gap would be stopped on its return trip by the adjacent cog, thus causing the beam to appear to the observer to be totally blotted out. By

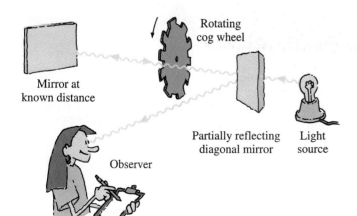

Rotating cog wheel

Mirror at known distance

Partially reflecting diagonal mirror

Light source

Observer

FIGURE 12.34 ··········

Fizeau's Determination of the Speed of Light Light shines through a partially reflecting diagonal mirror and a rotating cogwheel before being reflected back to the observer by a mirror at a known distance. If the wheel is rotating at just the right rate, light passing through a given gap will be stopped on its return by the very next cog and the observer will see no light reflecting off the diagonal mirror. By knowing the distance to the far mirror and measuring the rotation rate of the wheel, which will produce a total "eclipse" of the source for the observer, the speed of light can be determined.

knowing the rotation rate of the cogwheel in order to ''eclipse'' the light beam, Fizeau could determine the light travel time to and from the distant mirror. Thus, Fizeau's experiment was an ''assistantless'' version of Galileo's theoretically correct but observationally naive attempt to determine the speed of light.

In Fizeau's original experiment, the light flashes traveled a total round trip distance of 17.25 km and he observed that the first total eclipse of the light beam occurred when the cogwheel was rotating at 12.6 rev/sec. Since Fizeau's cogwheel had 720 cogs and 720 gaps, this result meant that the wheel had turned through $1/2 \times 720$ or $1/1440$ of the revolution in the time it took the light to pass through one gap in the cogwheel, travel the 17.25 km, and then get stopped by the gap's adjacent cog. The time it took the wheel to turn through $1/1440$ of a revolution was $1/1440$ rev $\div$ 12.6 rev/sec or 5.5×10^{-5} seconds. For this experiment then, light traveled 17.25 km in 5.5×10^{-5} seconds or at a rate of 313,000 km/sec. Although Fizeau's value for the speed of light is appreciably higher than the currently accepted value, his represents the first successful terrestrial determination of this quantity.

Since Fizeau's time, researchers have not only improved and refined his experiment through the use of rotating mirrors, ''electro-optic'' shutters, etc., but have developed alternate techniques as well. One of the simplest recalls that for any wave motion

$$\frac{\text{speed of}}{\text{wave}} = \frac{\text{frequency}}{\text{of wave}} \times \text{wavelength} .$$

Thus, if we can simultaneously measure the frequency and wavelength of an electromagnetic wave to a very high degree of precision, the product of these two quantities, the speed of the electromagnetic waves, will have a high degree of precision as well. As a result of these sorts of efforts, what was once a seemingly inaccessible quantity is now one of the most accurately determined constants in all of science.

QUESTIONS

1. What effect, if any, does the fact that light is moving through air in Fizeau's experiment have on his measurements?
2. In one variation of Fizeau's experiment, a rotating multifaced mirror replaces the cogwheel. Describe how such an apparatus could be used to measure the speed of light.
3. How might you employ a space probe such as the Mariner, Pioneer, or Voyager to measure the speed of light?

PROBLEMS

1. How long will it take light moving 3×10^8 m/sec to travel back and forth between two hilltops separated by 2 km? How does this travel time compare with the human reaction time of about 0.2 sec?
2. Suppose Fizeau's cogwheel had 840 cogs and the first eclipse had occurred at 8.7 rev/sec. If the mirror distance was 19.75 km, what value for the speed of light would Fizeau have obtained from these data?
3. A radio wave has a frequency of 104.8×10^6 Hz and an observed wavelength of 2.861 cm. Find the speed of this wave, and comment on the limits these frequency and wavelength data place on the accuracy of your result.

13

THE REALM OF THE ATOM

𝒥n earlier chapters we alluded to the atom as a sort of billiard-ball building block of matter, which can rattle off walls, participate in chemical reactions, and otherwise exhibit a number of interesting phenomena. In the course of those brief sketches, the description of atomic structure was limited to the context of the discussion then at hand and few details were offered. We now return to scrutinize the atom in a much more detailed fashion.

13.1 Atomic Structure

The oldest and most simplified view of the atom, indeed the view of Democritus himself, is that of a tiny impenetrable, indivisible billiard ball. By the end of the nineteenth century, this billiard-ball model had evolved into the ''plum pudding'' model of the English physicist Sir Joseph Thompson. After discovering the electron to be a negatively charged particle that was distinct from an atom, Thompson envisioned the atom as a semi-solid, positively charged sphere in which these negatively charged electrons were embedded, much as raisins are embedded in plum pudding. At almost the exact same time, three French physicists, Henri Becquerel, Marie Curie, and Pierre Curie, found that certain elements exhibited an effect that Marie Curie labeled *radioactivity.* It was clear from the outset, for reasons we will soon discuss, that the phenomenon of radioactivity was not associated with the electrons that participated in chemical bonding and chemical reactions, but rather arose from somewhere deeper inside the atom. Was it possible that Thompson's ''plum pudding spheres'' possessed structures of their own? The man who provided the first answer to this question was the English physicist, Sir Ernest Rutherford.

By 1910, it had been established that one of the forms of radioactive emissions, the *alpha particles,* possessed a positive electrical charge and, to a certain degree, the ability to penetrate matter. Rutherford hit on the ingenious idea of probing the Thompson plum pudding spheres by ''shooting'' at them with alpha particles. Rutherford directed a beam of alpha particles into a thin layer of atoms, in this case gold foil, as shown in Figure 13.1. The foil was almost completely surrounded by a detecting screen made of zinc sulfide and was capable of registering any impact made by an alpha particle at any angle to the incident particle

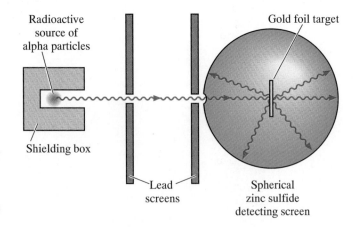

FIGURE 13.1 ··········

The Rutherford Experiment   A beam of alpha particles from a radioactive source is directed at a gold foil target, which is surrounded by a spherical detector made of zinc sulfide. The spherical detector is capable of detecting an alpha particle scattered in any direction from the target.

13.1 Marie Sklodowska Curie (1867–1934)

Radioactivity is one of those twentieth century scientific terms that has managed to strike a considerable amount of fear, apprehension, and anxiety in the human psyche as modern civilization continues its ongoing struggle with the ''genie'' of nuclear energy. Ironically, the woman who coined the term was hardly a person to be feared.

Marie Sklodowska was born in Warsaw on November 7, 1867. Even as a child she exhibited the prodigious memory and talent that were to serve her so very well later in her career. At the age of 16 she completed her secondary education at the Russian lycée in Warsaw and received a gold medal for an impressive inventory of academic achievements. Family financial difficulties interrupted her career for several years, during which time she held positions as a high school teacher and a governess.

Even in this relatively mundane period of her life, Marie Sklodowska managed to carry out some remarkable activities. The Poland to which she was born had been ruthlessly and callously partitioned out of existence as an independent state decades earlier by the Russians, Austrians, and Prussians with Sklodowska's beloved Warsaw falling to the control of czarist Russia. It was thus with no small amount of danger involved that in her off-hours she clandestinely taught reading and writing of the surpressed Polish language to women workers at the outlawed Polish nationalist ''free university'' in Warsaw.

After financing her sister's medical education in Paris, Sklodowska herself journeyed to ''The City of Light'' in 1891, where she enrolled in physics at Paris's famed Sorbonne. There she met and married one of the university's noted physicists, Pierre Curie, in 1895. Over the next decade, the Curie partnership became the most scientifically productive husband and wife team of all time. In the summer of 1898 the Curies isolated a new chemical element which Marie named polonium for her imprisoned native land. Later that year a second element, radium, was also isolated. Both of these new elements exhibited a then recently discovered radiation-like phenomenon, which Marie called ''radioactivity'' owing to the penetrating and ionizing nature of the radioactive ''rays.'' So eminent was Marie's work that in 1903, the same year she completed her doctorate at Sorbonne, she shared the Nobel Prize in physics along with her husband and another Frenchman, Henri Becquerel, for their discovery and investigations of radioactivity.

••• **Marie Sklodowska Curie (1867–1934)**

Continued

A tragedy of ghastly proportions was visited on Marie Curie when in April 1906, Pierre was accidentally hit and instantly killed by a runaway horse-drawn cart on the Rue Dauphine in Paris. Devastated to the depths of her soul, Marie Curie nonetheless continued on with her brilliant career. Less than a month after her husband's death, Marie Curie was elected unanimously by the faculty at Sorbonne to fill Pierre's professorship, and with that appointment, became the first woman ever to teach at the Sorbonne. In 1910, she published her famous *Treatise on Radioactivity,* which stands as one of the fundamental works of modern physics. When her candidacy for admission to the prestigious, but all-male French Acadèmie des Sciences was defeated by one vote in 1911, Marie Curie responded as only a Marie Curie could—she won the 1911 Nobel Prize in Chemistry! With that award, which was in recognition for her work on polonium and radium, Marie Curie is to this day the only scientist to win two Nobel Prizes in two different areas of scientific endeavor.

The war years of 1914–1918 saw Marie Curie, along with her daughter Irene, formulate the beginning technology for radiography and other medical applications of radioactive substances. In 1932 it was with the greatest satisfaction that she attended the inauguration, with her sister as director, of the Radium Institute in her native Warsaw, a city that had emerged from World War I as the capital of a free and independent Poland.

Perhaps the greatest contribution by the "Mother of Nuclear Chemistry," however, was her insistence that high-intensity radioactive sources be accumulated and stockpiled for research purposes. Such farsightedness provided the preacceleration sources of particle beams, which led, among other things, to Chadwick's discovery of the neutron and Irene Joliot-Curie's creation of artificial radioactivity, an achievement for which she won the 1935 Nobel Prize in chemistry. Unfortunately, Marie Curie did not live to see her daughter claim her award. On July 4, 1934, Marie Sklodowska Curie died of leukemia, which, in retrospect, was almost certainly caused by prolonged exposure to the very radioactivity which she had named and which had played such an important role in her rise to scientific immortality.

beam. As the beam of alpha particles struck the gold foil, Rutherford found that most of the alpha particles passed through the gold foil and struck the detector behind the foil undisturbed and undeflected. Most of Thompson's plum pudding spheres were empty space! A certain small percentage of the alpha particles, however, were deflected by the foil, in some instances by so much that they appeared to almost "rebound" backward off the foil.

From these results, Rutherford concluded that a small compact and massive nucleus existed at the center of the atom that contained all of the positive charge existent in the atom and virtually all of the atom's mass, hence the deflection of the positively charged alpha particles. Rutherford was even able to determine that the size of the nucleus was approximately 10^{-15} meters in diameter or about $\frac{1}{100,000}$ the size of the atom itself. On such a scale, if the nucleus of an atom were the size of a basketball, the atom itself would be about 25 kilometers in diameter!

Since atoms are observed to have no net electric charge, however, Rutherford was forced to conclude that the small, compact atomic nuclei are surrounded by the appropriate number of electrons required to balance the total positive charge on the nucleus. Furthermore, since these surrounding electrons seemed to have no effect on the motion of the alpha particles, they must necessarily have a low mass compared to that of the alpha particles. Rutherford's "nuclear" model for the atom then was somewhat similar to a tiny solar system with the nucleus taking on the role of a "sun" and the electrons the surrounding "planets" (Figure 13.2). On the basis of our Chapter 10 discussion of electric charge interactions, the reader should perceive some serious electrostatic difficulties with Rutherford's portrait of the atom. One of these difficulties, the question of how the positive charges in the nucleus are kept together when they should fly apart from electrostatic repulsion, will be considered in the more appropriate context of nuclear structure (Section 13.2). For the time being, let us instead consider the electrons, which in Rutherford's picture, must somehow surround the nucleus.

The experimental laws of electrostatic forces dictate that unlike charges attract one another. On that basis we would thus expect the negatively charged electrons surrounding the positively charged nucleus to be drawn into that nucleus, but somehow the electrons in an atom are able to escape such a fate. One suggestion for resolving the dilemma was an extension of the idea of a "solar system atom" in which the electrons electrostatistically "fall" around the nucleus just as planets gravitationally fall about the sun. Electrons orbiting in such a fashion, however, are essentially oscillating charges, which must radiate electromagnetic energy. That energy must, in turn, be drawn from somewhere. The only viable "somewhere" was the kinetic energy of the orbiting electron. But if energy was radiated in this fashion, the electron ought to lose its kinetic energy rapidly and spiral into an eagerly waiting, positively charged nucleus. This was not happening.

The mystery was compounded even further by several effects exhibited in gases. These effects had been known throughout most of the nineteenth century but had never been satisfactorily explained. In 1815, the German optician Joseph Fraunhofer noticed that when light from the sun was broken up into a spectrum, the result was not a continuous distribution of colors but an array of colors, which contained several hundred dark gaps or "lines" (Figure 13.3). A similar effect

FIGURE 13.2

Rutherford's Nuclear Model for the Atom On the basis of the results obtained in his alpha particle probes of the atom, Rutherford concluded that atoms consisted of a very tiny positively charged nucleus surrounded by a large volume of electron-occupied space.

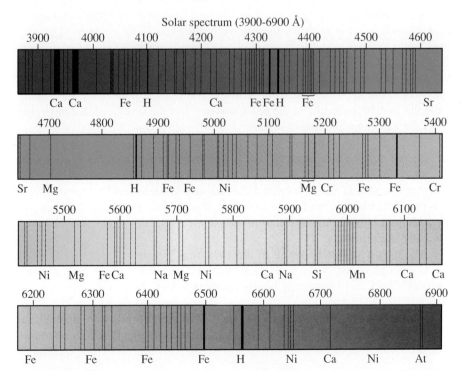

Solar spectrum (3900-6900 Å)

FIGURE 13.3

The Solar Spectrum in the Visible Region When sunlight is broken up into its component wavelengths by a spectrograph, one finds a multitude of dark lines superimposed on the background continuum of energy intensity. In this illustration, some of the more prominent lines are identified by the elements that produce them.

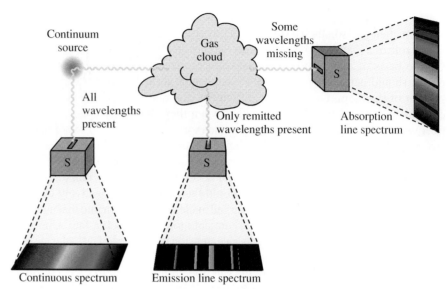

FIGURE 13.4 ··········

The Basic Types of Spectra In a continuum source of light that is radiating at all wavelengths, the spectrograph S will detect a continuous spectrum from the source, which also contains all wavelengths. If a gas medium is placed between the source and the spectrograph, then the gas will absorb some of the wavelengths from the light beam. A dark line or absorption line spectrum will result. If only the gas medium is observed by the spectrograph, then only those wavelengths reemitted by the gas will be observed and a bright line or emission line spectrum will result.

was observed in the spectra of distant stars, but the number, wavelengths, and intensities of the lines were generally different from the sun's "Fraunhofer" lines.

A terrestrial version of this effect could be produced by shining a beam of light from an incandescent light source through a low-pressure gas as shown in Figure 13.4. The spectrum observed for the light coming directly from the source exhibited a continuous array of color, while the spectrum of the light that passed through the gas was found to contain dark lines. Somehow the gas had imparted the dark lines onto the emergent beam of light. This conclusion was confirmed when it was found that the patterns of the dark lines changed from gas to gas. The line pattern for hydrogen gas was different from that for helium and so on. Moreover, it was found that if the gas were observed along a line of sight perpendicular to the incident light beam, the gas would appear to glow and give off light of its own. When this light was broken up into a spectrum, it was found to exhibit only a few bright lines of light whose spacing exactly matched that of the lines in the dark line spectrum. The spectrum of the light observed coming from the gas perpendicular to the beam was in effect the reverse of that emerging from the gas along the line of sight.

Although these effects were not understood at the time, it was nevertheless recognized that the uniqueness of the bright and dark line patterns of each substance in nature provided a potentially powerful tool for chemical analysis (see Chapter 16). It was also recognized that the key to understanding these effects was in the electrons surrounding the nucleus. If an atom were ionized, for example, by adding or subtracting an electron, the pattern of spectral lines was observed to change significantly.

The man who finally put all of these "atomic puzzle" pieces together was the Danish physicist Neils Bohr. Bohr in effect expanded the "quantization" ideas of Planck and Einstein by boldly claiming in 1913 that, at the atomic level, not only was light quantized into discrete values, but also the radii of the orbits of electrons about the nucleus as well. The "permitted" orbits were defined and characterized by integral values of the orbital angular momentum of the electron. The orbital angular momentum in turn was an integral multiple of the "quantum"

13.2 Atomic Clocks

For many decades the most accurate time on the earth was the sidereal time that astronomers kept at various observatories around the world such as the Royal Observatory at Greenwich in England and the U.S. Naval Observatory in Washington, D.C. The basic unit of time in all systems of measurement is the second, which, as we have seen, is defined as $1/86,400$ of the mean interval of time it takes the earth to spin once on its axis relative to the sun.

With the development of radio and radar in concert with World War II, physicists gained the ability to "tune in" to and determine frequencies in the radio region of the electromagnetic spectrum to the extraordinary accuracy of one part in 10^9.

If the electron within the atoms of a given substance can be made to vibrate between two energy states, they will emit electromagnetic waves having a frequency that corresponds to the energy of the electronic transition. In particular, the alkali metals such as sodium, lithium, and cesium have internal energy levels that are very close to one another. As a result, it is possible for electrons in these atoms to make very low energy transitions, which result in electromagnetic waves being emitted that have low energies and relatively low frequencies. One of the alkali metals, the element cesium, has an electronic transition that emits radiation at a frequency measured to be precisely 9.192631770×10^9 hertz. Because this frequency is so accurately known, scientists have adopted it as the standard of frequency and time, and the second is now defined as the interval of time required for 9,192,631,770 vibrations of a cesium atom to occur. The most accurate time in science is now kept by these so-called "atomic clocks" in which the "tick-tock" of the clock corresponds to an upward and downward transition of a cesium atom electron, an event that takes less than one billionth of a second to occur!

Microwave photon with:

Energy $= E_2 - E_1$

Frequency $= \dfrac{E_2 - E_1}{h}$

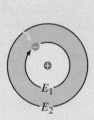

"Tick"

Photon absorbed

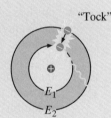

"Tock"

Photon reemitted

Microwave photon with:

Energy $= E_2 - E_1$

Frequency $= \dfrac{E_2 - E_1}{h}$

••• **The Atomic Clock**   The outermost electron of a cesium-133 atom can absorb a photon having a frequency of precisely 9.192631770×10^9 Hz to make an upward transition and then reemit a photon having precisely that same frequency when it makes a downward transition to its initial energy state. The successive up and down oscillations constitute the "tick-tock" of the cesium atomic clock, with the interval between successive "tocks" being equal to $1/9,192,631,770$ sec. Since the reemitted photons are in the microwave region, their frequencies can be electronically measured to an accuracy of one part in 10^9. This frequency is also almost completely unaffected by the cesium atom environment.

of orbital angular momentum, which Bohr found to be equal to the Planck's constant h divided by 2π.

From this basic assumption, Bohr was then able to demonstrate that the radii of the electron orbits can occur *only* at discrete distances from the nucleus, and that the total energy associated with an orbiting electron can take on *only* certain discrete values. Furthermore, as long as an electron stays in one of these orbits or "energy levels" it does not have to radiate electromagnetic energy. The electron, however, is no longer able to move about the nucleus in a continuous fashion, but must change its position by making discrete "jumps" between energy levels. To make an upward jump or "transition" to a higher energy level, the electron must absorb an energy equal to the difference in the energy values of the levels involved. On the other hand, if an electron makes a downward transition, it must give up a discrete amount of energy in the form of a radiated photon. In essence, Bohr claimed that along with the angular momentum of the electron, the electron orbits and electron energies within the atom were quantized as well.

The Bohr model for the atom enjoyed a qualitative success that was immediate and extensive. Virtually all of the observed characteristics of line spectra could now be *qualitatively* explained. If a beam of light containing a continuum of light frequencies passes through a relatively cool, tenuous layer of gas, the electrons present in the atoms comprising the gas will tap whatever energy from the radiation field they need to move to a higher energy level in the atom. Since a discrete amount of energy is involved in such a transition, only those photons having the proper frequency and corresponding energy will be absorbed by the electrons. As a result, the beam that emerges from the gas will be missing at least some of the radiation at a number of discrete frequencies. If this emergent beam is broken up into a spectrum, the observer will see a set of missing frequencies from the continuous spectrum that made it through the gas unaffected (Figure 13.5).

Electrons excited in this fashion generally do not remain in their upper levels for very long, but drop back down to their initial level. In doing so, the electron must, according to the Bohr model, reemit the energy it absorbed in jumping to its excited state. This energy is usually reemitted in the form of one or more photons. These reemitted photons, however, can move off in any direction and are just as likely to go back toward the incident light source as to proceed along the original direction. Thus a certain number of photons absorbed in the gas are returned to the original beam, but most are reemitted in other directions. If radiation observed from such a source of reemitted photons were broken up into a spectrum, no continuum of frequencies would be observed. Instead, a set of bright lines would be observed whose frequencies would correspond to the photon energies reemitted by electrons making downward transitions within the gas.

The pattern of spectral lines emerging from such a gas depends on the spacing of the energy levels of the electrons surrounding the nucleus, as well as on the location of an electron within those energy levels. If an electron is in the lowest possible energy level it can occupy within a given atom, it is said to be in its *ground state.* Any other electron energy levels are said to be *excited states.* The transitions an electron can make out of its ground state may vary considerably from transitions out of an excited state. In the very simple case of the element hydrogen, for example, an electron in the ground state can make a transition into any of the energy levels above the ground-state energy level, thus giving rise to a series of spectral lines at high-energy ultraviolet frequencies called the Lyman

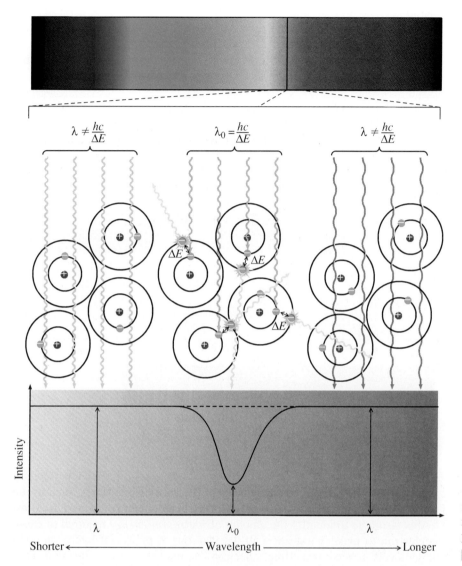

$$\lambda \neq \frac{hc}{\Delta E} \qquad \lambda_0 = \frac{hc}{\Delta E} \qquad \lambda \neq \frac{hc}{\Delta E}$$

Intensity

$\lambda \qquad \lambda_0 \qquad \lambda$

Shorter ⟵ ——————— Wavelength ——————— ⟶ Longer

FIGURE 13.5

The Formation of an Absorption Line If an electron requires a certain amount of energy ΔE to jump to a higher energy level, it can obtain that energy by absorbing a photon having a wavelength $\lambda_0 = hc/\Delta E$. Thus, wavelengths having values not equal to $hc/\Delta E$ pass through the atoms, while those having the wavelength λ_0 are absorbed and reemitted in random directions. The result is a reduced intensity at λ_0, which we perceive as a dark line.

series. On the other hand, if an electron is excited up to the second energy level in hydrogen, it can make a transition into any of the energy levels above it, thus giving rise to a second possible series of lines, called the Balmer series of lines, which are observed at visible and near ultraviolet frequencies, and so on (Figure 13.6).

For incandescent solids, liquids, and high-density gases, Bohr argued that the energy levels of the atoms are crowded to such an extent that their electron energy levels are no longer discretely separated. Under such circumstances, electrons can make transitions having any amount of energy. Hence, the frequency of the corresponding photon that is emitted can take on any value as well. The result is a "continuous" spectrum from such sources.

The Bohr model also enjoyed a considerable amount of quantitative success in dealing with the hydrogen atom and "hydrogen-like" atoms, such as singly ionized helium, which were ionized to the point of having only one electron remaining in orbit about the nucleus. Bohr, for example, was able to account

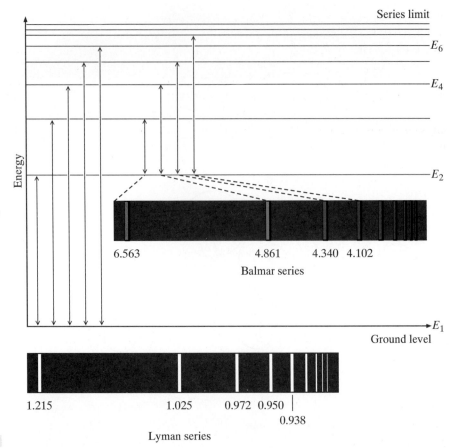

FIGURE 13.6 ··········

Electronic Transitions in Hydrogen If an electron is in the ground level E_1, it can make upward transitions to levels E_2, E_3, etc., giving rise to a set of spectral lines called the Lyman series. If the electron is in the E_2 level, then it can make upward transitions to levels E_3, E_4, etc., which results in the generation of a set of spectral lines called the Balmer series. The longer transition arrows on this diagram denote more energetic transitions and, hence, shorter wavelength, higher frequency transitions. All wavelengths indicated are in meters $\times 10^{-7}$.

successfully for the observed frequencies of the various line series, such as the Lyman and Balmer series, which were observed in the spectra of hydrogen atoms. He was also able to identify the amount of *ionization energy* required to cause the electron to make a transition all the way out of the set of hydrogen atom energy levels to become a "free" electron.

Despite its successes, however, the Bohr model was essentially a "transitional" theory, which simultaneously clings to some of the ideas of Newton's

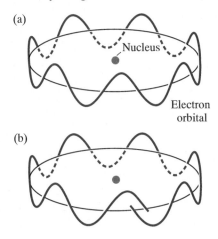

FIGURE 13.7 ··········

The de Broglie Wavelength The French physicist Louis de Broglie hypothesized that electrons can have wave-like properties, with the electron's associated wavelength or de Broglie wavelength equal to *h*/momentum. (a) For an electron to take on a stable orbital about an atomic nucleus, the circumference of that orbital must be equal to an integral member of de Broglie wavelengths. The result is the formation of a stable standing wave in the orbital. (b) If this condition is not met, the electron's de Broglie waves destructively interfere, the standing wave cannot maintain itself, and the electron orbital is unstable and cannot exist.

classical mechanics while furthering the idea of quantization in nature. As a result, the Bohr theory leaves a number of questions unanswered regarding atomic structure. In particular, what is it about electrons that would cause them to be restricted to certain orbits in the space surrounding the nucleus? If the "solar system" model of Bohr were totally correct, then the surrounding electrons, like the planets and asteroids of the solar system, should in theory be free to take on orbits of any size and shape. The dilemma was solved in part by the French physicist Louis de Broglie. De Broglie took a "what's good for the goose is good for the gander" approach by theorizing in 1924 that if light waves could have particle-like properties, then particles such as electrons could have wave-like properties. Maxwell had demonstrated that the momentum associated with an electromagnetic wave is given by

$$\text{energy} = \text{momentum} \times \text{speed of light} .$$

De Broglie reasoned that if electrons have wave-like properties, their associated energies would be equal to the Planck's constant $h \times$ frequency or

$$\text{energy} = h \times \text{frequency} = \frac{h \times \text{speed of light}}{\text{wavelength}} .$$

Using these two equations, de Broglie was able to show that the "particle wavelength" or *de Broglie wavelength* (Figure 13.7) of an electron or any other particle in nature is given quite simply by

$$\frac{\text{de Broglie}}{\text{wavelength}} = \frac{h}{\text{momentum}} .$$

De Broglie further demonstrated that such particle waves could maintain their stability about the nucleus only at values of the orbital circumferences that were integral multiples of the electron wavelengths. At other orbital circumferences, the electron waves would destructively interfere and the waves would cancel. If the de Broglie wavelength is confined to quantized integral values, then so must be the associated momentum. Bohr's quantization of the electron's orbital angular momentum was thus a direct consequence of the quantization of electron waves locating themselves in stable "standing wave" configurations about the nucleus. Less than two years after de Broglie's theoretical work, two American physicists, Clinton Davisson and Lester Germer, demonstrated experimentally that electrons can exhibit the phenomenon of diffraction and hence possessed the wave-like behavior predicted by de Broglie.

De Broglie, however, had addressed only the question of the wavelength of his "particle waves." The final details were worked out a bit later in the 1920s by an Austrian-German physicist, Erwin Schrödinger, and a German physicist, Werner Heisenberg, both of whom developed theoretical descriptions of *all* of the properties of such particle waves. In particular, Schrödinger hypothesized that at the atomic level a given particle does not occupy a specific place at a specific time, but must be viewed as being spread out over space and time in what he termed a "wave function." Such a wave function can perhaps be best thought of as the distribution of the probability of finding a given particle at a given place at a given time if that particle were a discrete entity. Thus, in Schrödinger's view, an electron does not orbit the nucleus as Bohr envisioned, nor is it the standing wave de Broglie proposed, but rather is a "cloud" of charge surrounding the nucleus (Figure 13.8). The electron's density of probability can

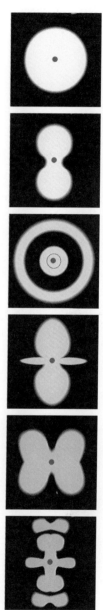

FIGURE 13.8 ··········

Wave Functions for Various Electron Energy States The levels of "brightness" indicate the degree of probability of finding a discrete entity electron in that particular location relative to the nucleus (red dot).

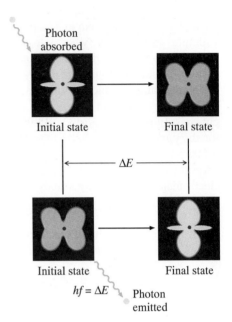

Photon absorbed

Initial state   Final state

←— ΔE —→

FIGURE 13.9 ··········

Wave Function Transitions   The absorption or emission of energy in an electron energy level transition results in a change of shape in the wave function characterizing the electron energy state.

Initial state   Final state

$hf = \Delta E$ Photon emitted

be completely described by solving a quantum mechanical "equation of motion" called the Schrödinger equation, just as the motion of a particle at the macroscopic or "everyday" level of nature can be completely described by solving Newton's equations of motion. For Schrödinger, the energy absorbed or emitted by an electron manifests itself not by making an upward or downward "jump" from one energy level to another, but rather by a change in the electron's wave function from one wave form to another (Figure 13.9).

Unfortunately, no visual analogue to Schrödinger's wave functions exists at the macroscopic level of our perception of the physical world. A particle approximates this behavior to some degree, as does a wave motion, but as we have seen, neither of these macroscopic models provides a complete description of atomic structure and the behavior of electrons therein. Schrödinger's wave functions, on the other hand, have been most successful in this regard. As a result, nearly all of the last vestiges of classical physics were swept from the world of the atom. In less than three short decades, the classical view of atoms had been completely revised by Schrödinger, Heisenberg, and others into a strange, new theory called *quantum mechanics,* whose properties, successes, and implications are described more fully later in this chapter.

WORKED EXAMPLE

A 0.005-kg bullet is fired at a velocity of 600 m/sec. Find the de Broglie wavelength for the bullet.

SOLUTION: The momentum of the bullet is the product of its mass and velocity, or the bullet's momentum is

$$\text{momentum} = 0.005 \text{ kg} \times 600 \text{ m/sec}$$
$$= 3.0 \text{ kg-m/sec} .$$

Continued

WORKED EXAMPLE

Continued

The expression for the de Broglie wavelength is

$$\frac{\text{de Broglie}}{\text{wavelength}} = \frac{h}{\text{momentum}}.$$

In this case, we have for the bullet

$$\frac{\text{de Broglie}}{\text{wavelength}} = \frac{6.63 \times 10^{-34}}{3} \text{ m}$$

$$= 2.21 \times 10^{-34} \text{m}.$$

Thus, the de Broglie wavelength for the bullet is 2.21×10^{-34} m! By contrast, the shortest known wavelength of gamma rays is about 10^{-13} m.

1. Compare and contrast the models of the atoms as put forth by Democritus, Thompson, Rutherford, and Bohr.

2. Explain why the existence of clustered positive charges in the nuclei of atoms and of electrons occupying the outer regions of atoms present difficulties if one attempts to understand atomic structure in terms of electrostatic forces alone.

3. What are Fraunhofer lines? Why are they important in our understanding of the structure of atoms?

4. How did Neils Bohr explain spectra line formation? Continuous spectra?

5. Describe what is meant by a de Broglie wavelength. Why is it important in our attempts to describe atomic structure?

1. Protons from the solar wind travel with a velocity of about 450,000 m/sec in the vicinity of the earth. If a proton has a mass of 1.7×10^{-27} kg, find the de Broglie wavelength for solar wind protons in the vicinity of the earth.

2. What are the limits on the value of the de Broglie wavelength of an electron having a mass of 9.1×10^{-31} kg if the electron can go no faster than 3×10^8 m/sec?

13.2 Nuclear Structure

During the beginning of the twentieth century, the ''electronic'' outer structure of the atom was found to be both extensive and complex. Left at the center of this array of electronic wave functions, however, was Rutherford's tiny, but massive nuclear "kernel," another billiard ball standing at the limits of our scientific perception. In light of the fact that Democritus's billiard ball was found to have structure, it was thus only reasonable to assume that Rutherford's nuclear billiard ball, although 100,000 times smaller, could also possess a structure of its own.

The first clues that such structure might exist came with Becquerel's aforementioned discovery of radioactivity. Since the level of radioactivity was not affected by chemical and physical changes in the radiative material, these "rays" could not be related to the electronic structures of atoms, which are deeply affected by such changes. The only remaining possible origin was the nucleus. Further investigation of radioactivity revealed that if a "beam" of radioactive rays is passed through a magnetic field or between a pair of parallel, electrically charged plates, the beam will split up into three component beams. One beam is deflected as if it had a positive electric charge (alpha "rays"), one is deflected as if it had a negative electric charge (beta "rays"), and one can pass through an electric or magnetic field without being deflected (gamma "rays") (Figure 13.10).

It was quickly established that the beta "rays" were not rays at all, but particles, in particular, high-velocity *electrons,* which were somehow being ejected from a positively charged nucleus. The alpha "rays" were also found to be particles as well, but possessing a much larger mass than the beta particles. Rutherford demonstrated in 1909 that the alpha particles were identical to helium atoms that had been stripped of both of their electrons. In other words, the alpha particles were the same as nuclei of helium atoms. It was further discovered that if a chemically pure sample of a radioactive substance were kept carefully isolated,

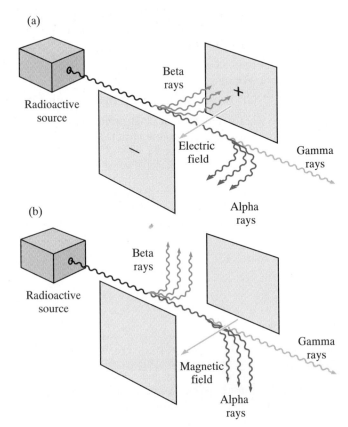

FIGURE 13.10 ··········

Radioactive Rays   When a beam of radioactive "rays" is passed through (a) an electric or (b) a magnetic field as shown, the beam splits up into three component parts. The alpha rays deflect the fields as if they had a positive electric charge, the beta rays deflect as if they had a negative electric charge, and the gamma rays exhibit no deflection in the presence of either an electric or magnetic field.

it would nevertheless become chemically contaminated by a different substance, which appeared at the same rate at which alpha or beta particles were being emitted. It was most difficult to reconcile such observed phenomena with structureless, billiard-ball nuclei.

The only true "rays" of the lot were the gamma rays, which were found to be a very high frequency, short-wavelength form of electromagnetic radiation. Gamma rays were found to be far more energetic and have far greater penetrating power than any other forms of such energy, including the then newly discovered X rays. Unlike the X rays, however, which Bohr correctly attributed to transitions of electrons occupying the innermost energy states of atoms, no such electronic transitions could be conjured up to account for the energies associated with the gamma rays. One logical explanation was that the gamma rays arise from high-energy transitions *within* the nucleus, which once more implied that Rutherford's nuclear kernels possessed structure in their own right.

If the nucleus did indeed possess the structure implied by the above phenomena, what then was the actual nature of that structure? Rutherford already provided some initial answers when, in the course of his aforementioned scattering experiments with alpha particles on gold foil, he found that the nucleus possessed a positive electrical charge. In the course of "shooting" alpha particles at nuclei, he found that if he could accelerate these particles to high enough velocities, he could split or "smash" his "target" nuclei into smaller pieces. The smallest nuclear "piece" that Rutherford could create in this fashion was a particle that had a positive electric charge. The charge on this particle was numerically equal to the charge on a single electron. Its mass was virtually the same as that of the hydrogen atom, and it possessed a radius of about 1.2×10^{-15} m. This positively charged nuclear piece was called a *proton* and represents one of the fundamental "building blocks" of the nucleus. Thus the simplest atom in nature, the hydrogen atom, consists of a single proton and a single electron.

If we then yield to temptation and hypothesize that additional elements are created by simply adding protons to the nucleus and electrons to the surrounding energy levels, we get no further than the next most complex element, helium, before running into problems. Helium atoms are observed to have two electrons and a nucleus containing two protons. We would therefore expect the mass of helium atoms to be twice that of hydrogen atoms, but in fact helium atoms are four times more massive than hydrogen atoms. The most reasonable explanation is that the helium nucleus consists of two protons and one or more additional particles that have no electrical charge.

In 1932 the English physicist James Chadwick bombarded a thin foil of the metallic element beryllium with a beam of alpha particles and found that one of the pieces from the "smashed" beryllium nuclei was a particle having a size and mass slightly larger than those of a proton, but which did not possess an electric charge. The long sought-after particle was called a *neutron* and, at the time, was thought to be the last of the nuclear building blocks. Each chemical element could be uniquely characterized by its atomic number or the number of protons in its nucleus. The atomic weight, on the other hand, is the combined total of protons and neutrons within the nucleus. It is thus possible for different "versions" or isotopes of the same element to exist that have the same characteristic number of protons but a different number of neutrons. A "carbon-13" nucleus, for example, consisting of six protons and seven neutrons, has an atomic weight of 13 and is an isotope of a "carbon-12" nucleus, which has six protons, six neutrons, and an atomic weight of 12.

Such a view of the nucleus, however, is at considerable odds with our perceptions of electrostatics, one of whose principles is that like charges will repel each other. Despite this fact, the nuclei of every chemical element except hydrogen are composed of positively charged protons packed into a close-quarters sphere, which does not fly apart from the expected electrostatic repulsion. Somehow the protons in the nucleus have found a way to "disobey" Coulomb's law. Moreover, when a proton and a neutron form a *deuteron,* or an isotope of the hydrogen atom nucleus, they are attracted and held together by some sort of attractive force that far exceeds any gravitational force these two particles could possibly exert on each other.

From these and other phenomena observed in the nucleus, physicists have concluded that an interaction exists within the nucleus that is as fundamental to nature as are the gravitational and electromagnetic interactions. It is variously referred to as the *strong nuclear interaction,* strong nuclear force, or simply the strong force. The ability to exert a strong nuclear force seems to be a characteristic of a family of nuclear particles called *hadrons,* which include protons and neutrons, but exclude electrons. This force does not depend on electric charge or gravitational mass. It does, however, depend on distance in a very important way. Whereas gravitational and electrostatic forces decrease with the square of the distance or (distance)^2, strong nuclear forces decrease much faster with distance. As a result, the strong nuclear forces are quickly overpowered by the electrostatic forces as the particle distance increases. For distances larger than about 2×10^{-15} m, the strong nuclear force becomes negligible for all intents and purposes. For distances smaller than about 1×10^{-15} m, however, the strong nuclear force becomes repulsive, as if the protons and neutrons have some sort of a hard core or kernel that resists further compression. As a result, the strong nuclear force exerts its effectiveness over a somewhat limited range of distances extending from 1×10^{-15} m to 2×10^{-15} m. Thus the repulsive electrostatic forces between protons are held in check over this range by the strong nuclear force.

The presence of neutrons in the nucleus enhances this stabilizing effect because they exert strong nuclear forces of their own. They also spatially increase the average separation between the protons, thereby weakening the electrostatic repulsion. As a result, every stable nucleus, has at least as many neutrons as protons, except for hydrogen and an isotope of helium, helium-3, which has two protons and one neutron in its nucleus. The heavier elements need increasingly larger numbers of neutrons to provide the "glue" to hold the protons together. One of the stable isotopes of the element lead, for example, has 82 protons and 124 neutrons in its nucleus. At atomic numbers beyond 82, however, even the neutron glue cannot maintain the stability of the nucleus against all of the electrostatic repulsive forces acting between all of the pair combinations of protons, and the nucleus will "decay" to a more stable configuration by some form of radioactivity.

Interestingly, theoretical nuclear physicists have predicted the possible existence of "islands of stability" in atomic nuclei having atomic numbers in the vicinity of 114, 126, and 164. Such nuclei are presumably stable against radioactive decay and hence might exist somewhere in nature as new, "superheavy" elements. Thus far, however, no evidence has been observed to indicate the existence of such materials, either on the earth or anywhere else in the cosmos.

For the stable nuclei, a given proton or neutron located deep inside a nucleus will experience the strong nuclear forces from all sides and will, in effect, be in a force-free or "free-fall" state much as the molecules away from the surface of a liquid are free to float and wander through the volume occupied by that liquid. The protons and neutrons at or near the surface of the nucleus, on the other hand, will experience the strong nuclear forces only from their neighbors on the "inboard" side of the nucleus, much as the intermolecular forces within a liquid provide a cohesive inward force at the liquid's surface boundary. Expanding on these ideas, Neils Bohr proposed in 1937 that the nucleus could be thought of as a droplet of incompressible "nuclear fluid" that exhibited a great many similarities to a drop of liquid at the macroscopic level of nature (Figure 13.11). Bohr's "waterdrop" model was able to account successfully for some of the overall properties of the nucleus. The spherical shapes of most nuclei, for example, can be accounted for in this model by a sort of "nuclear surface tension," which draws the protons and neutrons contained in a given nucleus into a minimum surface configuration for their given total volume.

It has also been discovered that the protons and neutrons within the nucleus, like electrons and photons, exhibit a wave-particle duality. A number of experiments can be conducted, for example, to demonstrate that neutrons exhibit the wave-like phenomenon of diffraction. As a result, the behavior of the protons and neutrons within a given nucleus are most appropriately described by assuming that they behave as wave functions, which can take on a variety of energy states or configurations within the nucleus, just as do the electronic wave functions outside of it. Whenever an incoming particle or photon is absorbed by the nucleus, the wave functions within the nucleus take on an "excited" energy configuration. Gamma rays, for example, can be absorbed and emitted by the nucleus in a fashion that is exactly analogous to the absorption and emission of lower frequency photons from the electrons surrounding the nucleus. In fact, a given nucleus can be characterized by the wavelength pattern of the gamma-ray spectral lines that it absorbs or emits.

In some cases the elevation of a nucleus to an excited state is caused by the capture of an incoming particle. Such interactions will produce a *transmutation* of the nucleus into a different element. Thus, if an alpha particle is captured by a carbon nucleus having six protons and six neutrons, the carbon nucleus will now contain two more protons and two more neutrons for a total of eight each. A nucleus with eight protons however is no longer a carbon atom nucleus, but rather is an oxygen atom nucleus. As a result, we say that the carbon atom nucleus has been transmutated into an oxygen atom nucleus.

Sometimes such captures result in the formation of a stable isotope, but most often particle captures produce a nucleus with an unstable energy configuration. Such nuclei will decay to a less energetic state by emitting gamma rays and/or particle radiation. This decay process can be described for an individual nucleus only in terms of a "decay probability" in which the nucleus might decay relatively quickly or it might take a much longer time to decay. The decay process can, however, be very accurately described for aggregates of atoms using the concept of a *radioactive half-life*. The half-life of a radioactive material is defined as the time it takes for half of the initial number of unstable nuclei to decay to a less energetic state. Uranium-230, an unstable isotope of the element uranium, which has 92 protons and 138 neutrons, for example, has a half-life of 21 days. Thus, if we isolate a sample of pure uranium-230, after 21 days we would find that half of the nuclei in the sample would have emitted alpha particles to transmute into the element thorium while the other half would still be uranium-230. After

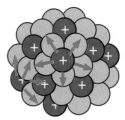

FIGURE 13.11 ··········
The Water-Drop Model for the Nucleus Deep within the nucleus of this silicon atom the net force on a given proton or neutron is essentially zero. At the surface of the nucleus, however, a strong inward force is exerted that prevents the proton or neutron from escaping the nucleus. Because of this observed similarity between a liquid and the atomic nucleus, Danish physicist Neils Bohr proposed that the nucleus can be thought of as a droplet of incompressible "nuclear fluid," which is composed of protons and neutrons.

13.3 Radioactive Age-Dating

One of the most powerful techniques for obtaining ages of objects in nature is a procedure called radioactive age-dating or simply *radioactive dating*.

Radioactive dating is predicated on the fact that a given radioactive isotope will decay to other substances at a particular rate, namely, the half-life of that isotope. If a given isotope begins to decay at a certain time, then one-half of the number of atoms will have decayed into their radioactive decay products after a time of one half-life of the original isotope, three-fourths of the atoms will have decayed after a time of two half-lives, $7/8$ of the atoms after three half-lives, and so on. If we could somehow ascertain the percentage of atoms that have undergone radioactive decay we could, by knowing the half-life of the isotope, determine how much time has elapsed from the time the radioactive decays began and thereby ''date'' the start of those decays.

Perhaps the most familiar example of radioactive dating is a technique employed in archaeology called carbon-14 dating. Carbon-14 is an isotope of the element carbon with a half-life of 5692 years and a relatively stable abundance because of the ongoing interactions between cosmic rays and the nitrogen atoms in the earth's atmosphere. During the course of its lifetime, any living creature will possess a carbon-14 to carbon-12 ratio that is the same as that of the environment. Upon that creature's death, however, the carbon-14 input ceases, and the ratio of carbon-14 to carbon-12 begins to decrease as the carbon-14 experiences radioactive decay. By measuring the carbon-14 to carbon-12 ratio for a given archaeological artifact, and com-

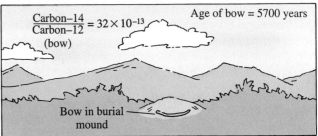

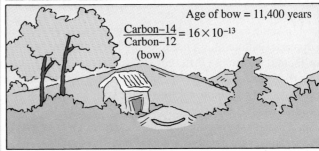

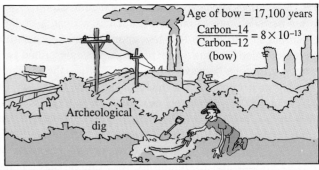

••• **Carbon-14 Dating** Living tissue has a carbon-14 to carbon-12 abundance ratio of 64×10^{-13}. When the tissue dies, the carbon-14 decays into nitrogen with a half-life of 5700 years. Since the abundance of carbon-12 remains the same, a measurement of the carbon-14 to carbon-12 abundance ratio indicates the extent to which the carbon-14 has decayed since the tissue died and, hence, the age of the dead tissue.

PHYSICS FACET

Continued

paring that ratio with the constant and ongoing carbon-14 to carbon-12 abundance ratio that occurs in the environment, an estimate can be had for the amount of carbon-14 radioactive decay that has occurred. From this we can obtain a value for the number of carbon-14 half-lives which have transpired and hence the age of the artifact. Ages obtained for artifacts using carbon-14 dating compare very favorably with ages for those same artifacts obtained by other means such as tree-ring counts.

Similar techniques have been applied to determine the geological ages of extremely ancient rocks such as those brought back from the moon during the Apollo lunar exploration program. In this case, isotopes having much longer half-lives, such as the decay of uranium-238 into lead-206, whose half-life is roughly 4.5 billion years, must now be employed. To date a given rock, the abundance ratio between a given "parent" isotope and a "nonradiogenic" isotope of its decay product is obtained and compared to the abundance ratio between the "radiogenic" and "non-radiogenic" isotopes of the decay product. Since the radiogenic isotope is produced by the decay of the given parent isotope and the nonradiogenic isotope is not, the age of the rock can be calculated from the half-life of the parent decay and a comparison of these two abundance ratios.

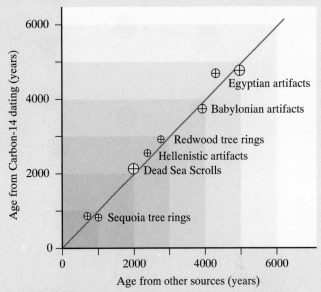

··· **The Calibration of Carbon-14 Dating** If a plot is constructed of known ages of objects and artifacts versus ages determined from carbon-14 dating, the result is a straight line, which indicates agreement within experimental error between the carbon-14 ages and ages obtained in other ways. The size of the crosses indicates the approximate uncertainties in the data points.

another 21 days, half of the remaining uranium would have decayed to thorium, leaving one-fourth of the original uranium-230 atoms, and so on. Half-lives of unstable nuclei range from very small fractions of a second up to billions of years. The decay of platinum-190 into osmium-186 has a half-life of about 6×10^{11} years, while the half-life of the decay of polonium-212 into lead occurs with a half-life of about 3×10^{-7} sec.

Whenever a particle such as a neutron strikes a nucleus and is absorbed, the impact causes a deformation of the shape of the nucleus. If the deformation exceeds a certain "critical" amount, then it is possible for the electrostatic repulsive forces within the nucleus to split that nucleus into two smaller nuclei of roughly equal size. Such a process is called *nuclear fission* and generally occurs only in the heavier nuclei such as uranium. Under certain circumstances, a substance can sustain such a fission process by means of a *chain reaction* in which

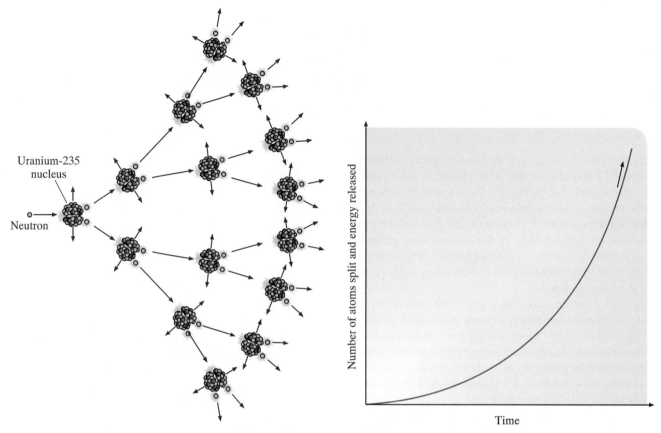

FIGURE 13.12 ··········
Schematic Diagram of a Uranium-235 Chain Reaction   When the incident
neutron strikes a uranium-235 nucleus, the nucleus breaks apart and releases two more
neutrons, which in turn strike two more uranium-235 nuclei, which break apart and
release two neutrons, each of which strikes four more uranium-235 nuclei, and so on.
As a result, the number of nuclei that are split and the energy released from each
splitting can build to very large values in a very short time.

the particles generated in the initial fission process are able to produce additional
fission processes. Perhaps the most famous example of a chain reaction involves
the uranium isotope uranium-235, which, on being struck by a neutron, will split
in two and release two or three neutrons, which in turn strike other uranium-235
nuclei, and so on (Figure 13.12). With each fission, some of the mass in the
nucleus is converted into its equivalent energy according to the principles of
special relativity theory (see Chapter 14). As a result, such "fissionable" materials constitute a potential source of impressive amounts of energy for either
constructive or destructive purposes.

At conditions of extraordinarily high temperatures and pressures, two or
more nuclei can collide and "fuse" together into a single nucleus having a larger
atomic number and atomic mass. The most familiar example of a fusion process
is the *thermonuclear fusion* of four protons into a helium nucleus or alpha particle
with an attendant conversion of mass into its equivalent energy. Such fusion
reactions are capable of releasing thousands of times more energy than the fission
chain reactions and are known to be the source of the energy produced by the

sun and stars. Sadly, the only fusion reactions artificially produced here on the earth to date have been in the form of explosive detonations of "hydrogen" bombs.

Up to this point, we have been able to account for the behavior of nuclei in terms of electromagnetic and strong nuclear interactions between the two "basic" building blocks of the nucleus, the proton and the neutron. Nowhere in such a nuclear portrait do we expect to encounter the electron. Yet, as we have seen, the beta particles emitted from radioactive nuclei are generally high-velocity electrons. Moreover, when such *beta decay* occurs within a nucleus, the atomic number of the nucleus left behind by the emitted electron increases by one while the atomic mass of that nucleus remains the same (Figure 13.13). The beta decay has in effect somehow changed one of the neutrons present in the nucleus into a proton! Certainly none of the interactions thus far described—gravitational, electromagnetic, or strong nuclear—can account for such behavior. Thus physicists have postulated a fourth basic interaction in nature, which has been variously termed the *weak nuclear interaction,* the weak nuclear force, or simply the weak interaction.

The weak nuclear force is a force whose strength decreases so rapidly with distance that its essential range is less than 10^{-17} m, or less than the size of a proton or neutron. Thus the weak nuclear force acts *within* a proton or neutron. The implication, of course, is that protons and neutrons are not tiny billiard balls, but are themselves composed of some sort of particle structure at an even more fundamental level, which under the right circumstances, can be affected by the weak nuclear force. Beta decay in its usual form is then viewed as a disruption of a neutron into its component parts, a proton and an electron, by a weak nuclear force.

In the course of such decays, however, the energy of the emitted electron was found to be less than the total energy lost by the nucleus. Moreover, it could be shown experimentally that neither the linear momentum of the system nor the angular momentum associated with the spins of the neutron, proton, and electron involved in the decay were conserved. Thus, at first examination, the process of beta decay seemed to violate three of the most cherished conservation principles in physics: the conservation of energy, the conservation of linear momentum, and the conservation of angular momentum. An ingenious theoretical solution was proposed in 1930 by German physicist Wolfgang Pauli. It was later refined by the Italian physicist Enrico Fermi, who hypothesized the existence of a "subatomic" particle called a *neutrino.* The neutrino was, in this view, created in a beta decay along with a proton and electron and accounted for all of the missing energy, linear momentum, and angular momentum observed in a beta decay.

Two years before Pauli's work, a British physicist, Paul Dirac, found that a number of seeming dilemmas within the framework of the new quantum mechanics could be resolved if for each basic type of particle in nature there existed an "antiparticle," which possessed the property that whenever it collided with its particle counterpart, both particle and antiparticle would be totally annihilated and their masses converted into the equivalent energy predicted by relativity theory. Hence, even before the actual experimental verification of the existence of the neutron, the third "fundamental" building block of the atom, some very compelling theoretical arguments were being advanced in favor of the existence of a considerable array of particles, some of which are even more elementary than the proton, neutron, and electron. Thus did physics make its beginning forays into a yet tinier realm of perception, the world of the "subatomic" or the "micro-microscopic."

Hydrogen-3 nucleus
(1 proton, 2 neutrons)

Hydrogen-3 nucleus
emitting beta particle

Beta particle

Now a helium-3 nucleus
(2 protons, 1 neutron)

Neutrino

FIGURE 13.13 ·········

Beta Decay Whenever a beta particle is emitted from a nucleus, such as the isotope hydrogen-3, a neutron is changed into a proton. To preserve the conservation of energy and linear and angular momentum, a "subatomic" neutrino must also be emitted along with the beta particle. The remaining nucleus then has one more proton and one less neutron, but almost the same atomic mass. In this case, the beta particle emission changes the nucleus from that of hydrogen-3 to helium-3.

WORKED EXAMPLE

A pure sample of the isotope polonium-210 is observed to decay to $\frac{1}{16}$ polonium-210 and $\frac{15}{16}$ lead-206 over a period of 560 days. What is the half-life of the polonium-210 isotope?

SOLUTION: During each half-life of the polonium-210, $\frac{1}{2}$ of the polonium decays into lead. Thus, after one half-life period, $\frac{1}{2}$ of the polonium remains; after two half-life periods, $\frac{1}{2} \times \frac{1}{2}$ or $\frac{1}{4}$ of the polonium remains. After three half-life periods, $\frac{1}{2} \times \frac{1}{2} \times \frac{1}{2} = \frac{1}{8}$ of the polonium remains; and after four half-life periods, $\frac{1}{2} \times \frac{1}{2} \times \frac{1}{2} \times \frac{1}{2} = \frac{1}{16}$ of the polonium remains. Thus, for the original sample to decay to $\frac{1}{16}$ polonium and $\frac{15}{16}$ lead, four polonium-210 half-life periods must have elapsed. Since the total time interval is 560 days, this means that the polonium half-life period is 560 days/4 = 140 days.

REVIEW 13.2 QUESTIONS

1. Discuss the evidence in favor of the existence of structure within the nucleus.
2. Describe the phenomenon of radioactive "rays."
3. What is meant by the radioactive half-life of a substance?
4. Describe Bohr's "water drop" model for the nucleus. In what ways was this model successful in accounting for the observed properties of the nucleus? In what ways did it fail?
5. Compare and contrast strong and weak nuclear interactions. What sorts of nuclear processes do they produce?

REVIEW 13.2 PROBLEMS

1. The isotope nitrogen-17 decays by emitting a beta particle. If the half-life of this decay process is about 4 seconds, find
   a. the substance into which nitrogen-17 transmutes
   b. the time it takes for a sample of 100% nitrogen-17 to decay into a sample that is $\frac{1}{16}$ nitrogen-17.
2. The isotope magnesium-27 emits a beta particle and decays into aluminum-27. If an initially pure sample of magnesium-27 is found to be $\frac{63}{64}$ aluminum-27 and $\frac{1}{64}$ magnesium-27 after a period of 57 minutes, find the half-life of this nuclear reaction.

13.3 "Elementary" Particles

News of Chadwick's experimental discovery of the neutron in 1932 had scarcely been announced when the American physicist Carl Anderson stumbled on the existence of a new particle he called the *positron*. Anderson's new particle pos-

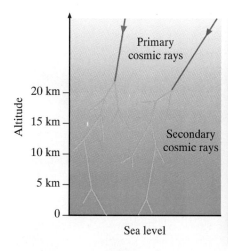

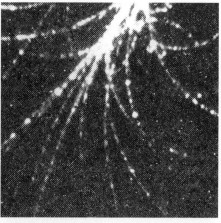

FIGURE 13.14 **A Cosmic Ray Shower** As high-energy "primary" cosmic ray particles strike the outer layers of the earth's atmosphere, they collide with the nuclei of the atmospheric atoms and produce showers of secondary particles, (left) which can be recorded by photographic plates or other types of detectors (right).

sessed the same mass as an electron, but whenever it collided with an electron, both particles would instantly be annihilated and converted into their equivalent energies. The positron was thus the antielectron whose existence Dirac had hypothesized four years earlier. Moreover, it was found that nuclei that were "proton-rich" exhibited a form of beta decay in which a proton decayed into a neutron while emitting a positron. Once more a discrepancy in energy and momentum was observed that could only be accounted for by theorizing the existence of an "antineutrino" being ejected from the nucleus as well. The neutrinos proposed by Pauli and Fermi had no mass, no charge, and could only interact with other particles by experiencing a weak nuclear interaction, which meant that they had to pass within 10^{-17} m of the "target" particle. The probability of a detectable interaction between a neutrino and another particle is thus extraordinarily low, yet the existence of these elusive phantom particles was confirmed experimentally in the 1950s by two American physicists, Frederich Reines and Clyde Cowan.

Since 1930, experimental studies of the interaction between high-energy particles and various nuclei have led to the discovery of some 300 "elementary" particles of which the neutrino, electrons, and their antiparticles are examples. The "sources" of such high-energy particles are found naturally in the form of *cosmic rays* or can be artificially produced here on the earth in large and sophisticated particle accelerators such as the cyclotron (see Chapter 15). The cosmic rays are in reality high-energy interstellar particles consisting primarily of protons, some electrons and alpha particles, and lesser abundances of the lighter atomic nuclei. As these particles strike the nuclei of the atoms in the earth's atmosphere, they produce showers of "secondary" particles, many of which have properties far different from the protons, neutrons, electrons, and neutrinos discussed thus far (Figure 13.14).

One class of such shower particles has masses that are several hundred times that of an electron, but are substantially less massive than either a proton or a neutron. One such particle called the π^+ *meson,* for example, has a mass about 273 times that of an electron and an electric charge equal to that of the proton. Its antiparticle, the π^- meson, has the same mass, but the charge of the electron. Each of these π-mesons or *pions* have been observed to further disintegrate into the μ-mesons or *muons,* which have masses of about 210 electron masses and the electric charge of either a proton or an electron. The terrestrial equivalent of

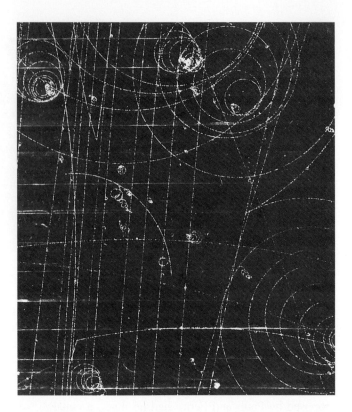

FIGURE 13.15 ··········
Terrestrially Produced Elementary Particles   In this photograph taken in a laboratory on earth, each trail represents a record of an elementary particle interacting with matter in the presence of electric and magnetic fields.

such nuclear "splattering" can be achieved through the use of particle accelerators, which are able to "shoot" beams of particles having extraordinarily high energies at "target" nuclei. The resulting nuclear "debris," like that from the cosmic rays, contains a multitude of particles having a wide variety of characteristics (Figure 13.15). Although these particles are referred to as "elementary particles," their complexity and numbers contradict such simplistic terminology.

In an attempt to lend some semblance of order to this "zoo" of elementary particles, as it has sometimes been called, physicists have, with some success, sought to classify these particles. They have tried to classify them according to which types of interaction—strong nuclear, weak nuclear, or electromagnetic—the particles are capable of experiencing* and by means of a characteristic set of "quantum numbers," which describes a given particle in terms of fundamental units of quantized parameters such as charge, "spin" momentum, etc. Using this type of classification, physicists have found that in addition to energy, linear momentum, and angular momentum, some of the more abstract and esoteric quantities defined by physicists, such as "isospin" and "baryon number," are also conserved in elementary particle reactions.

Current beliefs are that some of the elementary particles themselves are made up of a set of six subelementary building blocks, called quarks (after a term appearing in the James Joyce novel *Finnegan's Wake*).† The quarks are characterized by a fractional electric charge of either ⅔ of the charge on a proton or ⅓ of the charge on an electron. They are believed to possess quantized properties that have been labeled, with a completely straight face, charm, beauty, truth, and

---

*At the atomic level, the gravitational interaction is not important.

†Some physicists have gleefully noted that quark is also the German word for "rubbish."

color. To date such particles with fractional electronic charges have not been directly observed experimentally, but there is a fair amount of circumstantial empirical evidence for their existence. Some particle physicists are of the opinion that the ''color'' force that holds the quarks together is so strong that it may never be possible to isolate a single quark from an elementary particle, thus preventing their direct detection. Were it possible to perform such an isolation, there would, of course, be no guarantees that the quark itself might not possess structure of its own. So it goes in the world of the atom, seemingly *ad infinitum!*

1. Discuss the evidence in favor of the existence of:
   a. antiparticles
   b. particles that are more elementary than protons, neutrons, and electrons.
2. Describe three different elementary particles that are not protons, neutrons, or electrons.
3. Summarize the quantities that are *always* conserved in any type of particle interaction.
4. What are cosmic rays? Why are they important in the study of elementary particles?
5. What are quarks? Discuss the difficulties in trying to observe them.

13.4 Quantum Theory

Throughout this chapter we have described effects and concepts at the atomic level, which simply cannot be represented within the framework of Newton's particles or Maxwell's waves. As a result, physicists have had to formulate a description of natural phenomena at the so-called ''microscopic'' level of the atom and subatom. The bulk of that endeavor was accomplished over the three decades that opened the twentieth century, and the names of the theoretical physicists involved—Bohr, Born, Dirac, Heisenberg, Pauli, Planck, and Schrö-dinger—read like a list of inductees to a Physics Hall of Fame. Their resulting achievement is a theory we called *quantum theory* or quantum mechanics.

By its very name, quantum theory views the world of the atom as a world composed of indivisible building blocks of one sort of another. We have already come into contact with such ''quantized'' physical entities. Electric charge, for example, comes in ''packets'' or ''quanta'' of 1.6×10^{-19} coulombs each. The quark theory would cut this value by a factor of three, but nonetheless retains the idea of an electric charge quantum in nature. The work of Planck and Einstein established the fact that the energy associated with electromagnetic waves also comes in packets, each with a value equal to the Planck's constant h times the frequency of the wave. Einstein's demonstration of the equivalence of mass and energy requires that since energy is quantized, so are mass and, for that matter, linear momentum. Bohr recognized that, at the atomic level, orbital angular momentum is also a physical quantity that existed in packets, each of which was equal to the Planck's constant divided by 2π. This ratio of $h/2\pi$ is of sufficient importance in quantum mechanics to have its own special designation—$\hbar$, which is pronounced ''h-bar'' and has a value

$$\frac{h}{2\pi} = \quad = 1.05 \times 10^{-34} \text{ joule-seconds} .$$

Interestingly, the so-called ''spin'' angular momentum associated with particles such as electrons can occur in half-integral multiples of $\hbar$.

Such a ''packet picture'' of nature is, of course, at seeming odds with our perception of nature at the macroscopic level of our everyday existence. To the person in the street, light beams behave as if they were continuous waves. Matter appears to be continuously solid, and spinning tops do not, as they slow down, appear to ''hop'' from one value of spin angular momentum to another. Such hops do occur, but in such tiny increments as to appear to be continuous to a naked-eye observer. A spinning top that changes its spin angular momentum by one joule-second, for example, must traverse some 10^{34} quantum mechanical packets or states of spin angular momentum. To detect one single such transition out of so many would require amazing perception indeed!

In addition to continuity, one of the most cherished and satisfying concepts of classical physics was the idea that the fundamental laws of nature are precise, causal, and deterministic. In this view the universe can be regarded as a gigantic machine whose workings follow exact, empirically obtainable laws, which in turn can be used to recall the machine's past or predict its future. The laws of celestial mechanics, for example, can be used to recall an eclipse from antiquity for archaeological purposes or to predict a planetary alignment for the launch of a future space probe. Quantum theory replaces this comforting view of nature with a probabilistic one whose fundamental laws are, at the microscopic level, not deterministic.

No less a personage than Albert Einstein reacted to this abandonment of exactness and causality by lamenting that ''God does not play dice with the

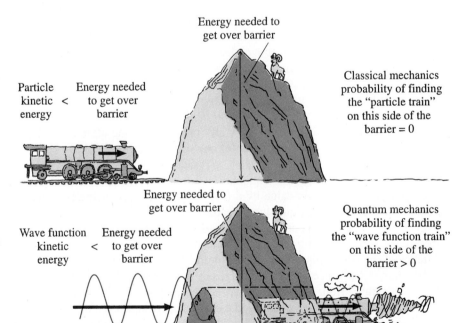

FIGURE 13.16

Quantum Mechanical Tunneling The classical mechanical train does not have enough energy to surmount the barrier and therefore, according to classical mechanics, the probability of finding the train to the right of the barrier is equal to zero. A quantum mechanical ''wave function train'' can, however, ''tunnel'' through the barrier so that there exists a non-zero probability of finding the train to the right of the barrier.

universe.'' At the atomic level, however, that is precisely what seems to happen. As we have already seen, the time-honored deterministic concepts of waves and particles from classical physics are not adequate descriptions of phenomena occurring at the atomic level. Instead, the interacting entities must be viewed as wave functions spread out over space and time. In attempting to describe the intensity of the wave function at various positions in space and time, probability is of paramount importance, for it is the ''probability amplitude'' of the wave function that determines its perceived behavioral properties. One fascinating aspect of this view is that quantum mechanical wave functions can exist at locations in space and time where classical waves and particles would be forbidden (Figure 13.16). A classical alpha particle within a nucleus, for example, does not have sufficient energy to overcome the potential barrier of the strong nuclear forces binding it to the nucleus. Therefore we would predict, from the principles of classical physics, that the alpha particle would forever be trapped within the confines of the nucleus. If, on the other hand, we view the alpha particle as a superposition of proton and neutron wave functions, the solution of the Schrö-dinger equation for this system predicts that a small but nonzero part of the wave function will lie outside of the potential barrier at any given time. Thus the alpha particle wave function can ''tunnel'' outside of the nucleus, and we observe it as the alpha particle emission for which classical physics was unable to account (Figure 13.17).

Probability also plays a most important role in the creation and annihilation of particles and mass. If a photon is incident on an array of atoms in a gas, for example, we can only specify the probability that a given photon will be absorbed and in effect ''annihilated'' by an atom in the gas. When an absorbing atom ''deexcites,'' we have no way of predicting the precise time the energy will be reemitted, the direction in which it will be reemitted, or what energy states the

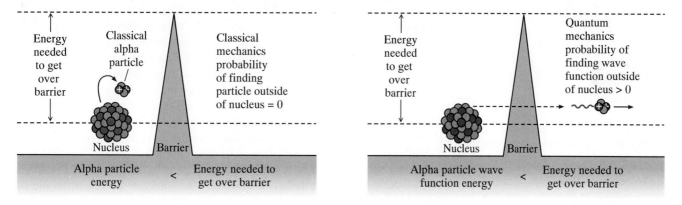

FIGURE 13.17 ···········

Alpha Particle Decay   One of the most important examples of quantum mechanical tunneling is the process of alpha particle decay. The energy of the alpha particle is not large enough to surmount the potential energy barrier that exists at the surface of an atomic nucleus. Therefore, the classical mechanics probability of an alpha particle escaping from an atomic nucleus is zero. Quantum mechanically, however, there is a non-zero probability of finding the alpha particle wave function outside of this potential energy barrier, hence allowing for a possible alpha particle escape. Since alpha particles are observed to be emitted from atomic nuclei, we deduce that the quantum mechanical explanation is the correct one.

atom will take on, and hence what wavelength photons will be reemitted as the atom deexcites. All of these events can be described for a single atom or photon only in terms of probabilities. In like fashion, the phenomenon of beta decay and its attendant creation of a proton, neutrino, and electron at the expense of a neutron annihilation is, as we have seen, quite probabilistic in its nature.

Obviously, probabilistic phenomena play an important role in the macroscopic world as well, whether present in human and animal behavior, "games of chance," or multiple-choice examinations. The crucial theoretical question is whether or not probability is a fundamental aspect of the physical universe. Quantum theory has enjoyed over a half-century of scientific victories by saying that it is. Within this probabilistic universe, however, are large numbers of events that have probabilities so close to zero or one that, for all intents and purposes, we can state with "certainty" a given event will occur (probability is essentially one) or will not occur (probability is essentially zero). The probabilities that a proton or electron will spontaneously decay into other particles and/or energy is essentially zero. Therefore, we expect to find stable matter in the universe that is composed of those low-decay-probability protons and electrons. On the other hand, the probability of an antielectron or positron being annihilated is essentially one, and hence our electric circuits are devoid of positrons.

At the macroscopic level, virtually *every* quantum mechanical probability approaches either zero or one. Thus the quantum mechanical probability that beer will tunnel through its surrounding glass "potential barrier" pitcher onto the table top is almost identically zero, while the probability of finding a planet in its "Bohr" orbit around the sun is almost identically one. Moreover, while the behavior of individual entities at the atomic level can be described only in terms of probabilities, the overall statistical behavior of aggregates of such entities can be described to a high degree of exactness and sophistication.

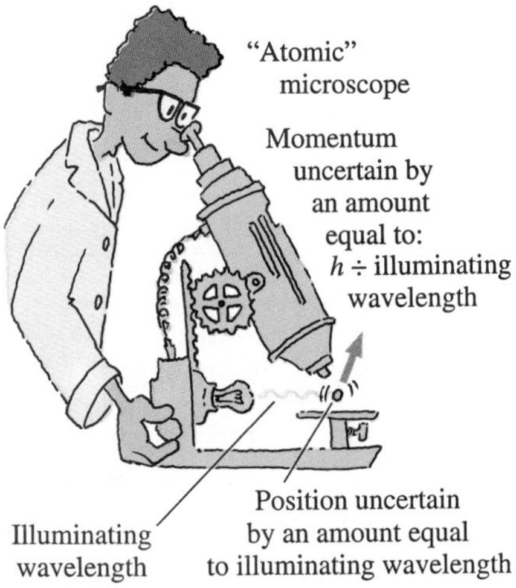

The radioactive decay of a single nucleus can be described only in terms of a probable elapsed time before the decay occurs or the probability that the decay will occur over a specified time interval. If we measure the length of time it takes before a single nucleus decays and then repeat that measurement on another single nucleus, we will find in general that the decay times are very much different. If, however, we consider a large aggregate of such nuclei, then we can determine to a very high degree of precision the time it will take for half of the nuclei to decay or, conversely, the percentages of decayed and undecayed nuclei that will be observed after some specified time. Moreover, if we repeat our measurements on a different aggregate of nuclei, we will obtain identical results. It is then this exactness and predictability of the behavior of large aggregates that manifests itself in the deterministic behavior observed for nature at the macroscopic or "large atomic aggregate" level of our perception.

The indeterminate nature of the universe also has an impact on our ability to make measurements of that universe. The nature of this effect is summarized in the concept developed by Heisenberg, which is called the *uncertainty principle*. The uncertainty principle states in essence that there is a fundamental intrinsic limit on the accuracy to which we can make simultaneous measurements in the microscopic world. This limit arises from the fact that in any such measurement there is an interaction between the observer and the object that is to be measured (Figure 13.18). It is these interactions which, in turn, inject an inherent uncertainty into the measurement process. Thus no matter how well designed and accurate

"Atomic" microscope

Momentum uncertain by an amount equal to: $h \div$ illuminating wavelength

Position uncertain by an amount equal to illuminating wavelength

Illuminating wavelength

FIGURE 13.18 · · · · · · · · · ·

The Uncertainty Principle If an attempt is made to precisely locate the position of the particle, the momentum from the light used to illuminate the particle will impart a momentum to the particle, thereby creating an uncertainty in the particle's momentum. Thus there is an intrinsic limit to the accuracy to which positions and momenta can be simultaneously measured. This concept is called the uncertainty principle.

a piece of scientific equipment may be, according to quantum mechanics, it can measure nature no more accurately than the inherent constraints placed on it by the uncertainty principle. What a blow to the experimental scientist!

To verify this, let us consider the most common mathematical statement of the uncertainty principle:

$$\left(\begin{array}{c}\text{uncertainty}\\\text{in position}\end{array}\right)\left(\begin{array}{c}\text{uncertainty}\\\text{in momentum}\end{array}\right) \geq \hbar \; .$$

Let us further assume we have a particle we wish to examine under an idealized atomic microscope. If we wish to locate the position of the particle, one way to do it would be to shine light on the particle, thereby illuminating its position. To reveal the position of the particles as accurately as possible we should employ light at the smallest possible wavelengths. Unfortunately for the observer, small-wavelength light has large momentum, which is imparted to the particle being observed. The harder we try to locate the particle's position with ever smaller wavelengths of light, the more momentum we impart to the particle. In effect, what we gain in the particle's position accuracy is lost in its momentum accuracy, all in accordance with the uncertainty principle.

If we envision the same exercise using the moon as our particle and the light from the sun to illuminate its position, we find that the position and momentum can be measured simultaneously to a very high degree of accuracy. It is thus tempting to regard such measurements as a violation of the uncertainty principle, but let us consider the numbers of involved. The most accurate possible determination of the moon's position, if we are most generous, has an uncertainty of about 10^{-3} m, and its momentum an uncertainty of about 10^{12} kg-m/sec. Even if we could determine the mass of the moon to an accuracy of one kilogram or one part in 10^{23}, we would still be stuck with an uncertainty in the lunar momentum of about 10^{-3} kg-m/sec. Thus the product of the position and momentum uncertainties, even under the wildest imaginable accuracies, is only 10^{-6}, or some 28 powers of 10 more than what would be needed to violate the uncertainty principle at the macroscopic level! We thus conclude that for all intents and purposes, the photon-moon interactions do not cause us measurement problems at the everyday level, a result that is in keeping with what we observe.

With a bit of mathematical manipulation the uncertainty principle may be stated in the following form:

$$\begin{array}{c}\text{uncertainty in total}\\\text{energy of a system}\end{array} \times \begin{array}{c}\text{intervals of time over}\\\text{which system is observed}\end{array} \geq \hbar \; .$$

This is an amazing result, for it says that even a quantity as scientifically sacrosanct as the total energy of a system suffers from a fundamental limit of uncertainty. It has also led to a revolutionary view of how forces in nature seem to operate. As a result of the uncertainty principle, for example, it is possible for a proton to "fluctuate" between itself and two component particles, the neutron and a π-meson, over extremely short periods of time. The lifetimes of the component particles are so short that we cannot observe them directly and hence refer to them as "virtual" particles. Under the proper circumstances, however, these virtual particles can be exchanged from one proton to another during their lifetimes. As a result of such back and forth collisions between virtual particles and their "parent" particles, there is an exchange of momentum at the atomic level

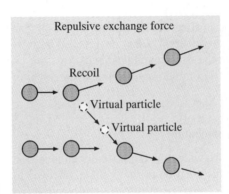
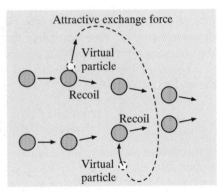

FIGURE 13.19 ··········
Exchange Forces As a result of back and forth collisions between virtual particles and their ''parent'' particles, an exchange of momentum occurs at the atomic level over a finite time interval that manifests itself as an exchange force.

(Figure 13.19). Recall that a change of momentum over a time interval constitutes a force. Thus forces at the atomic level are viewed quantum mechanically as the exchange of momentum between the entities involved via these virtual ''exchange'' particles. Each of the four basic interactions in nature is believed to operate in this fashion and the exchange particles have been identified for three of the interactions: strong and weak nuclear and electromagnetic. Only the gravitational interaction and its hypothesized exchange particle, the ''graviton,'' have thus far defied this view of forces.

Various individuals often tend to assign a significance to the uncertainty principle that far transcends the world of the atom. Many social scientists delight in claiming that, in light of the uncertainty principle, the only difference between measurement in the social sciences and the physical sciences is in the magnitude of the respective ''$\hbar$'s.'' Others are in general disconcerted by the demise of a comforting, securely deterministic physical world. At the everyday level, of course, the universe is, for all intents and purposes exact, deterministic, and causal. It is only when our perception descends many orders of magnitude into the realm of the atom that we find a portrait of nature that is quantized, probabilistic, and defiant of exact measurement. It is a seemingly alien view of an alien world. Quantum mechanics, however, continues to survive and even flourish as a physical theory because, in keeping with the scientific method, its predictions and consequences continue to be impressively verified by our observational data. In short, quantum mechanics continues to be the best theoretical game in town.

WORKED EXAMPLE

In a certain experiment, the position of an electron can be measured to within 2.0×10^{-9} m. Find the uncertainty in the momentum of the electron.

SOLUTION: From the uncertainty principle, we have that

$$\begin{pmatrix}\text{uncertainty}\\ \text{in position}\end{pmatrix} \begin{pmatrix}\text{uncertainty}\\ \text{in momentum}\end{pmatrix} \geq \hbar \; .$$

Continued

> ## WORKED EXAMPLE
>
> *Continued*
>
> Since the uncertainty in the position of the electron is 2.0×10^{-9} m and $\hbar = 1.05 \times 10^{-34}$ joule-sec, the uncertainty in the electron's momentum is
>
> $$(2.0 \times 10^{-9}) \left(\begin{array}{c} \text{uncertainty in} \\ \text{momentum} \end{array} \right) \geq 1.05 \times 10^{-34}$$
>
> or
>
> $$\text{uncertainty in momentum} \geq 0.5 \times 10^{-25} \text{ kg-m/sec}$$
> $$\geq 5 \times 10^{-26} \text{ kg-m/sec .}$$
>
> Thus, the momentum of the electron cannot be determined with an uncertainty any less than 5×10^{-26} kg-m/sec.

1. Discuss the basic concepts in quantum mechanics. How do these ideas compare with our perceptions of the everyday world?

2. What is a wave function? Are there any analogues to wave functions in the everyday world?

3. Describe what is meant by the term "wave-particle quality"?

4. What is the uncertainty principle? Of what importance is the uncertainty principle?

5. Discuss Einstein's comment that "God does not play dice with the universe."

1. Suppose that a particle's position can be measured to an uncertainty of 5×10^{-10} m. What is the uncertainty in the momentum of the particle?

2. If a particle has an uncertainty of 5×10^{-10} kg m/sec, to what accuracy can its position be measured?

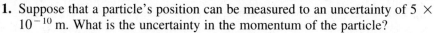

Chapter Review Problems

1. Suppose the nucleus of a hypothetical atom were the size of the sun, whose diameter is equal to 1.4×10^6 km. On this same scale, how large would the entire atom be? Look up the distances between the sun and the planets and comment on the claim that atoms are like "tiny solar systems."

2. A photon has a frequency of 5×10^{14} Hz. Find the energy and momentum of the photon.

3. Yellow light has a wavelength of about 5.8×10^{-7} m. Find the energy and momentum of a yellow light photon.

4. Find the wavelength, energy, and momentum associated with the 60-Hz frequency of household electrical current. Comment on how difficult it might be to direct such radiation.

5. At the atomic level, physicists often use a unit of energy called the electron volt, where one electron volt is equal to 1.6×10^{-19} joules. If a photon has an energy of one electron volt, what is its frequency and wavelength?

6. The first absorptive line in the hydrogen-Balmer series has a wavelength of 6.56×10^{-7} m. What is the energy associated with the electronic transition that gives rise to this line?

D ——————————— 3.5 eV

C ——————————— 3.0 eV

B ——————————— 2.0 eV

A ——————————— 0.0 eV

7. An atom has the set of energy levels shown at the left. If an electron is in level A, what are the possible energies in electrons volts (eV) that the electron can absorb in order to make a transition to a higher energy level?

8. A 1-kg object is moving at a speed of 0.1 m/sec. Find the de Broglie wavelength for this object. Is this a detectable wavelength? Explain.

9. What momentum would an electron possess if it had a de Broglie wavelength equal to the wavelength of green light or 5.5×10^{-7} m?

10. If the mass of the electron is 9.1×10^{-31} kg, at what speed must the electron in the previous problem be moving in order to have a "green" de Broglie wavelength?

11. Find the de Broglie wavelength of an electron having an energy of one electron volt or 1.6×10^{-19} joules.

12. Assuming that a neutron is spherically shaped and that the volume of a sphere is equal to $\frac{4}{3}\pi \, (radius)^3$, find the "classical" density of a neutron if its radius is 1.0×10^{-15} m and its mass is 1.7×10^{-27} kg.

13. If an alpha particle strikes an oxygen-16 nucleus, what substance will result, and what will its atomic weight be? (*Hint:* Refer to the periodic table in Appendix C.)

14. The isotope carbon-14 emits a beta particle in the form of an electron. What substance will result and what will its atomic weight be?

15. The isotope neon-23 emits a beta particle. What element will result from this reaction and what will its atomic weight be?

16. The isotope radon-222 decays into polonium with a half-life of 3.8 days. How long will it take before a sample of 100% radon-222 will decay into 25% radon and 75% polonium?

17. It is noted that after 30 seconds, a pure sample of phosphorus-34 has decayed to $\frac{1}{64}$ phosphorus-34 and $\frac{63}{64}$ sulphur-34. What is the half-life of the phosphorus-34 isotope and what particle is emitted in the decay process?

18. The isotope sodium-24 decays into magnesium-24 with a half-life of about 15 hours. How much sodium-24 will be left from an originally pure sample after 2.5 days?

19. An archaeologist exploring a prehistoric human cave dwelling finds fragments of clothing that have a carbon-14 to carbon-12 ratio of 1.0×10^{-13}. If the carbon-14 to carbon-12 ratio in living creatures is 16.0×10^{-13} and the half-life of the carbon-14 isotope is 5700 years, how old is the clothing? (*Hint:* Assume that the carbon-14 began to decay with the death of the plant or animal used to make the clothing.)

20. The isotope potassium-40 decays into the isotope argon-40 with a half-life of about 1.3 billion years. When a certain moon rock was formed, it had all potassium-40 and no argon-40 in its composition. It is now found to be made up of ⅞ argon-40 and ⅛ potassium-40. What is the age of this moon rock?

21. If a particle has an uncertainty in its momentum of 10^{-6} kg-m/sec, to what accuracy can we measure its position?

22. Suppose we can measure the position of a particle to an uncertainty of 1.0×10^{-10} m. How much uncertainty is there in the measurement of this particle's momentum?

23. An attempt to measure the position and momentum of a certain atom simultaneously results in an observed uncertainty of 1.0×10^{-8} m in the atom's position and our uncertainty of 4×10^{-26} kg-m/sec in the atom's momentum. What is the value of $\hbar$ for these data?

24. Show that if the product of the uncertainty of momentum and position is greater than or equal to h, then so is the product of the uncertainties in the particle's energy and in the interval of time over which the particle is observed.

1. How do atoms resemble the solar system? How are they different? Could the solar system be an atom in someone else's universe? Explain.

2. Describe what Rutherford would have observed in his alpha particle scattering experiment if Thompson's "plum pudding" model for the atom had been correct.

3. Suppose most of the alpha particles in Rutherford's experiment didn't pass through the gold foil. What, if anything, would Rutherford been able to conclude about atomic structure from such a result?

4. Why do you suppose Rutherford used gold foil as opposed to some other type of metallic foil such as tin or aluminum?

5. Which of the following combinations are possible to observe in a given spectrum? Explain your answers.
    a. continuous spectrum only
    b. absorption lines only
    c. emission lines only
    d. continuous spectrum with absorption lines
    e. continuous spectrum with emission lines
    f. absorption lines and emission lines
    g. continuous spectrum with absorption and emission lines.

6. Describe how you might develop the analysis of spectral lines into a "potentially powerful tool for chemical analysis."

7. Why do you suppose that ionizing an atom will cause its pattern of spectral lines to change?

8. Why do alpha and beta ''rays'' follow curved paths in Figure 13.10, while the gamma rays do not?

9. How do you suppose a beam of neutrons would behave if passed through the apparatus shown in Figure 13.10?

10. The helium isotope helium-3, consisting of one neutron and two protons, is found to be stable. Why is this unusual? Can you offer an explanation for the stability of the helium-3 isotope?

11. Discuss the possibility that nuclear interactions other than the strong and weak nuclear interactions might exist. Speculate on the properties of such interactions.

12. The most common form of fusion occurs when hydrogen is converted into helium. Speculate on other possible fusion reactions in nature.

13. Describe what it might be like to take a voyage into Bohr's ''water drop'' nucleus.

14. Suppose that stable ''superheavy'' elements exist that have atomic numbers around 114, 126, and 164. Speculate on what properties such substances might possess.

15. Is it practical to make gold by the transmutation of elements? Explain.

16. Why do you suppose that radioactive processes are expressed in terms of half-lives?

17. How would you go about experimentally detecting the presence of neutrinos?

18. It has been suggested that the entities at the atomic level be named ''wavicles.'' Discuss the pros and cons of such terminology.

19. Speculate on the possibilities that the quarks themselves may be composed of ''sub-quark'' particles.

20. Discuss how you might go about detecting a particle, such as a quark, which has a charge of $1/3$ or $2/3$ of the presently accepted fundamental unit of electric charge.

21. Do the planets' orbits about the sun have quantized energy states? If so, why aren't we able to detect them?

22. From what you have read in this chapter, would you expect a ''quantum'' magnetic field to exist? Explain.

23. Compare and contrast the statistical descriptions of human behavior encountered in the social sciences with the ideas of quantum theory.

24. Explain what is meant by the statement in the text that the only difference between physics and the social sciences is the size of h in these respective fields.

25. Does the quantum theory satisfy the requirements of the correspondence principle? Explain.

26. Discuss the implications that the uncertainty principle has for the scientific method.

Planck's Constant

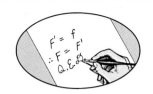

Planck's constant h permeates the microscopic world of the atom and quantum mechanics. It appears, among other places, in the mathematical expression for the blackbody curve and is fundamentally tied to what are currently believed to be the basic building blocks or quanta of photon energy, angular momentum, and magnetic field. It even shows up in its modified form of $\hbar$ in Heisenberg's famed uncertainty principle. Because of its importance in our theoretical descriptions of atomic and nuclear phenomena, an evaluation of Planck's constant in terms of our macroscopic standards of measurement becomes of crucial importance if we are to have a better grasp of the events occurring at the frontiers of the microscopic world.

One apparatus by which Planck's constant can be measured is that used to investigate the properties of the photoelectric effect (Figure 12.31). If light shines on the negatively charged plate of a phototube (Figure 13.20), a given photon will impart an energy to some given electron that is equal to $h \times$ frequency of photon. When this energy is absorbed by the electron, the energy must first be used in an attempt to surmount an energy barrier at the surface of the metal plate, which is called the *work function*. If the energy imparted by the photon is less than the work function, the electron cannot leave the plate surface. If the energy absorbed by the electron is greater than or equal to that of the work function, then the electron leaves the plate surface with a kinetic energy that is equal to the energy left over from ''getting past'' the work function of the plate. For any electron that escapes the surface of the metal, we have

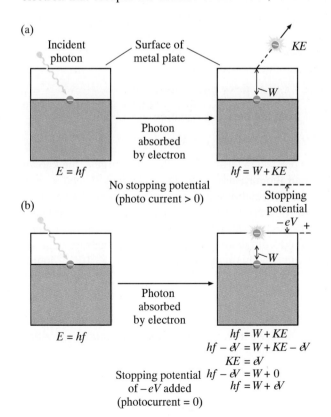

FIGURE 13.20 ··········

Energy Transfer in the Photoelectric Effect   (a) When an incident photon strikes an electron with a sufficient amount of energy hf, the electron surmounts the work function energy W and acquires a kinetic energy KE. (b) If a stopping potential energy eV is now applied to the electron such that $-eV = $ KE, then no photocurrent flows and $hf = W + eV$.

$$\frac{\text{energy of}}{\text{photon}} = \frac{\text{work function of}}{\text{metal plate}} + \frac{\text{kinetic energy}}{\text{of electron}} .$$

The electrons can, however, be "stopped" from arriving at the anode by adjusting the electric potential between the plates. The value of the plate potential at which the electrons are just prevented from reaching the surface of the plate is called the *stopping potential*. For an electron, the electrical potential energy associated with the stopping potential is equal to the electron charge $e \times$ stopping potential. Thus, if the electron is prevented from leaving the surface of the plate by some value V of the stopping potential, we can write that

$$eV = \frac{\text{kinetic energy}}{\text{of electron}} .$$

If we let f equal the frequency of the photon, we have that the energy of the photon is

$$\frac{\text{energy of}}{\text{photon}} = hf .$$

Substituting these two equations into our earlier statement and denoting the work function by W, we have

$$hf = W + eV .$$

Suppose now we shine light having a frequency f onto the plate and measure the stopping potential V needed to prevent the flow of photocurrent from the plate surface. For that result, we have.

$$hf_1 = W + eV_1 .$$

If we repeat the experiment with light of a different frequency f_2, we would find that it will take a different value V_2 of the stopping potential to prevent the flow of photocurrent from the plate, and we could write

$$hf_2 = W + eV_2 .$$

If we subtract this equation from the previous equation, we obtain

$$h(f_1 - f_2) = e(V_1 - V_2) .$$

Thus, Planck's constant is expressed in terms of the electric charge e, the known light frequencies f_1 and f_2, and the measured stopping potentials V_1 and V_2 and can readily be calculated from this relationship. In practice, many different stopping potentials are measured for many different frequencies and the results are plotted on a graph of f versus V. The slope of the resulting straight line is then equal to h/e and h is determined from that slope value. Precise measurements of Planck's constant yield a value for this quantity of 6.626176×10^{-34} joule-seconds.

QUESTIONS

1. From the final expression for h, what are the units of h? What other quantities in physics, if any, have the same units?

2. What is the final expression for h if wavelengths of light are used instead of frequencies?

3. Can you think of another way you might go about measuring the value of h?

1. Light having a frequency of 7.3×10^{14} Hz shines on the plates of a phototube. A stopping potential of 2.5 volts is required to prevent the flow of photocurrent across the plate gap. When the frequency of incident light is adjusted to 4.5×10^{14} Hz, the stopping voltage changes to 1.3 volts. If the charge on the electron is 1.6×10^{-19} C, what is the value for Planck's constant for these data?

PHYSICS FORMULATION 13.2

"Absolute" Atomic Masses

The system of atomic masses which has been developed to describe the chemical elements and their isotopes is a scale which is relative in nature. Thus, we can say that the element carbon has an atomic mass of 12 and helium has an atomic mass of 4, we can conclude that carbon atoms are three times more massive than helium atoms, but unless we can somehow tie such atomic masses into a standard of mass such as the gram or kilogram, the system is of limited value. In short, it would be highly desirable to determine the value of an atomic mass unit in kilograms.

One of the more ingenious devices by which such measurements can be accomplished is illustrated in Figure 13.21 and is referred to as a *mass spectrometer*. The mass spectrometer consists of an ion source in which the atoms or molecules of interest are ionized through high-energy electromagnetic radiation or electron bombardment. Having been ionized, the now-charged atoms are shot into the "velocity selector," an evacuated chamber in which an electric field E_1 and a magnetic field B_1 have been set up perpendicular to each other. As the charged particle enters the velocity selector at some velocity v, it is subject to an electrostatic force from the electric field which is given by

$$\frac{\text{electrostatic}}{\text{force}} = \text{charge} \times E_1$$

and an electromagnetic force given by

$$\frac{\text{electromagnetic}}{\text{force}} = \text{charge} \times v \times B_1 .$$

Because the electric and magnetic fields are perpendicular to each other, these forces act in opposite directions. If a given charged particle is to make it to the opening into the measuring chamber, it must move in a straight line across the velocity selector chamber in order to do so. The particle, therefore, can experience no net deflections from the magnetic and electric fields. For such "selected" particles, we have

$$\frac{\text{electromagnetic}}{\text{force}} = \frac{\text{electrostatic}}{\text{force}} .$$

Substituting the respective expressions from the above fields

$$\text{charge} \times v \times B = \text{charge} \times E .$$

FIGURE 13.21 ···········

The Mass Spectrometer Ions having a known electric charge, usually e, are shot into the velocity selector chamber where the electric field E and the magnetic field B are adjusted so that the ion beam is directed along a straight line through the entrance aperture into the measuring chamber. On entering the measuring chamber, the beam is deflected by the magnetic field B_2 and strikes the detector at a measured distance $2R$ from the entrance aperture. The atomic weight in kilograms of the ions in the beam can then be determined from the values of the ion charge, the measured value of $2R$, and the adjusted values E_1, B_1, and B_2 of the electric and magnetic fields.

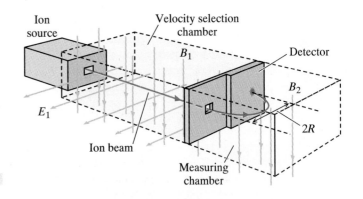

If we solve this equation for the "selected" particle velocity v, we have

$$v = \frac{E_1}{B_1} .$$

Thus, the particles so selected by the known values of E_1 and B_1 now enter the measuring chamber, where they are subjected to a second magnetic field of strength B_2, but no electric field. The charged particle responds to its new environment by deflecting into a curved path, eventually striking a detector such as a photographic plate, fluorescent screen, etc., at a measurable location. The centripetal force F_c required to produce such a curved path is given in general by

$$F_c = m \times \frac{v^2}{R} ,$$

where R is the radius of curvature of the path and m is the mass of the particle. The centripetal force is provided by the electromagnetic force on the charged particle or

$$F_c = \frac{\text{electromagnetic}}{\text{force}} = \text{charge} \times v \times B_2 .$$

Equating the two expressions for F_c yields

$$m \frac{v^2}{R} = \text{charge} \times v \times B_2 ,$$

and solving for the mass of the particle, we obtain

$$m = \frac{\text{charge} \times R \times B_2}{v} .$$

If we are dealing with particles that have one unit of electrical charge e, such as a beam of protons, electrons, or singly ionized atoms and molecules, then the charge $= e$ and

$$m = \frac{e \times R \times B_2}{v} .$$

The radius of curvature R can be found by noting where the particles strike the detector relative to the opening into the measuring chamber. Thus, all of the quantities on the right side of this equation are either known or have been measured, and the mass of the particle can then be readily calculated.

High-precision measurements conducted in this fashion have yielded a value of $1.6726485 \times 10^{-27}$ kg for the mass of the proton and a value of 9.109534×10^{-31} kg for the mass of the electron.

QUESTIONS

1. Could a mass spectrometer be used to measure the mass of a neutron? If not, how might a neutron mass be measured?

2. Discuss how you might use a mass spectrometer to conduct chemical analyses of various substances.

3. Suppose you had a sample of an element that contained a mixture of the isotopes of that element. How could you use a mass spectrometer to separate and isolate the various isotopes in such a sample?

PROBLEMS

1. A stream of protons passes through the velocity selector of a certain mass spectrometer in which the electric field is set at 2×10^5 N/C. To what value must the magnetic field be adjusted so that protons having a velocity of 1.0×10^5 m/sec make it into the measuring chamber?

2. Protons entering the measuring chamber of a mass spectrometer at a velocity of 4.0×10^6 m/sec are observed to strike a detector at a distance 0.4 m from the entrance aperture. If the magnetic field in the measuring chamber is 0.2 Tesla and the unit electric charge is 1.6×10^{-19} C, what is the mass of the proton for these data? Assume that the protons are moving in a circular path and that the distance between the entrance aperture and the impact point on the detector is equal to the diameter of that path.

14 PHYSICS AND
MODERN TECHNOLOGY

*O*ver the past four centuries, the scientific method has allowed us to gain considerable insight into the workings of the physical world. The progression of that insight has been accompanied by the historical appearance of a wondrous and awesome technology as human beings have sought to employ these empirically determined principles of nature in the solution of numerous useful and practical problems. Such efforts have resulted, especially during the twentieth century, in a myriad of technological marvels and achievements that have affected virtually every aspect of our daily lives. Much of the science fiction technology of a generation ago has become the ''science fact'' technology of the present. Moreover, this ''technological acceleration'' has occurred on such a short time scale that a number of social scientists have expressed serious concerns over the possible effects such a rapid rate of profound technological change might ultimately have on the human psyche.

Physics is, of course, one of the scientific cornerstones of this vast technology. The basic principles of physics form the underpinnings of a great many of the inventions and devices that, in just a few short decades, have become basic to the functioning of modern civilization. In the preceding chapters, we have, from time to time, examined a few of these devices such as the hydraulic lift, the telescope, and the electric motor, each of which relies almost exclusively on a single concept of physics, such as pressure, refraction, or electromagnetism. Having been introduced to a wide spectrum of the concepts and principles of physics, we are now ready to examine the innovative and imaginative ways in which these concepts and principles have been *combined* in the creation and development of some of the ''high-tech'' devices used in contemporary society.

14.1 Modern Communications

No aspect of our technological advances has had more impact on civilization than those achieved in the area of long-range communication. Messages and news that barely a century and a half ago took days or even weeks to wend their way around the globe can now be almost instantaneously delivered worldwide, courtesy of modern communications technology. The overall process by which sights and sounds are delivered across oceans and continents involves a most ingenious blending of a variety of physical principles.

The first step in this communication process is the conversion of sounds and visual images into electronic signal currents. For sound waves this is accomplished by the use of a microphone, which contains a membrane that is free to vibrate. In one type of microphone, the vibrating membrane is attached to a cylindrical-shaped magnet, which has been inserted into a coil of wire as shown in Figure 14.1(a). As the sound waves strike the microphone membrane, the membrane begins to vibrate. The attached magnet also begins to oscillate in phase with these vibrations. As it does so, it induces an electric current in the coil, which changes in tandem with the magnet's oscillations. Thus the variations that occur in the sound vibrations picked up by the microphone membrane are converted into corresponding variations in the microphone's output current.

One can obtain a similar result by placing the vibrating membrane in contact with a layer of a granulated substance such as carbon as shown in Figure 14.1(b). As the membrane vibrates, the granules experience a variable compression, which

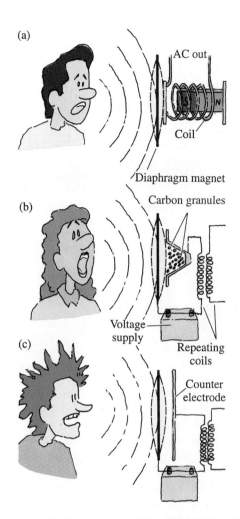

(a)

AC out

Coil

Diaphragm magnet

(b)

Carbon granules

Voltage supply

Repeating coils

(c)

Counter electrode

FIGURE 14.1 ············

Microphone Designs   In design (a) the diaphragm is attached to a movable magnet, which induces the electrical current signal in output wire. In design (b) the variable electrical resistance in the carbon granules induces the output signal, and in design (c) the variable capacitance produces the desired fluctuations in the output signal current.

in turn produces a corresponding variability in the electrical resistance of the layer. If a constant electric potential is applied across the layer, the current flow across the layer will vary inversely with the resistance of the layer, thereby converting the sound variations into corresponding variations in the output electrical current.

An even more sensitive sound conversion system can be achieved by hooking the vibrating membrane up as one side of a parallel plate capacitor as shown in Figure 14.1(c). As the membrane vibrates, the distance between the plates and, hence, the capacitance changes in tandem with the membrane vibrations. If a constant electric potential is applied across the capacitor, the size of the electric charge contained on the capacitor plates will vary along with the plate separation. In response to the variation in the capacitance, the electric charge flow effectively produces an electric current whose variations correspond to the membrane's vibrations.

The conversion of visual images into their corresponding electronic signal currents is a somewhat more involved process. One begins by first focusing the given image onto a signal plate whose front face is composed of an array of several hundred thousand globules of photoresponsive materials such as cesium/cesium oxide, and silver/silver oxide colloids. When light from the image strikes

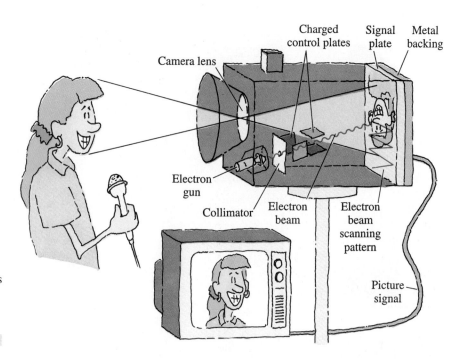

Camera lens

Charged
control plates

Signal
plate

Metal
backing

Electron
gun

Collimator

Electron
beam

Electron
beam
scanning
pattern

Picture
signal

FIGURE 14.2 ·········

The Electronic Camera Tube   As the beam of electrons scans the various levels of brightness in the image on the signal plate, a variable picture signal is produced as an output.

the signal plate, the array of globules acquires a positive photoelectric charge whose distribution in size and location directly corresponds to the distribution of the light intensity of the incident image. A beam of electrons is then directed toward the signal plate from a hot electron-emitting filament called an *electron gun,* which has been connected to the signal plate as shown in Figure 14.2. When the electron beam hits a charged globule, the globule discharges an electrical impulse into the circuit that is proportional to the amount of charge on the globule and, hence, the light intensity of the incident image at that point on the signal plate.

As the electron beam is emitted toward the signal plate, it is made to pass through two sets of parallel plates, called deflecting or control plates, on which an electric potential has been placed. The plate sets are perpendicular to one another so that the emerging beam of electrons can be directed toward any location on the surface of the signal plates by adjusting the electric potentials on each set of control plates. To obtain electrical impulses from each part of the signal plate, the electron beam is moved row by row downward across the signal plate. This process is called scanning and an entire signal plate is usually scanned over time scales of the order of $\frac{1}{60}$ of a second. Each complete scan of the signal plate by the electron beam produces a series of output electrical impulses called the picture signal.

For the transmission of black-and-white picture signals, one such scanning circuit is sufficient, but if we seek to transmit color picture signals, then the incident image must first be broken up into a set of three images, one red, one blue, and one green, by means of an array of mirrors and color filters. Each of these images is then projected onto its own signal plate and the images are scanned separately but simultaneously to produce a set of three outgoing picture or video signals corresponding to the images recorded in the additive primary colors of red, blue, and green.

PHYSICS FACET

14.1 The Cathode-Ray Oscilloscope

One of the ''must'' items in any late-night sci-fi movie laboratory is a seemingly mysterious device called an oscilloscope. Oscilloscopes generally come in an oblong box with a circular viewing screen, and when they are turned on, the screen displays a variety of geometric curves that flicker, flash, pulsate, and even change shape. In reality, the oscilloscope is a highly useful tool in electronics technology and is, in principle, not all that mysterious.

The cathode-ray oscilloscope consists of a vacuum tube, which has an electronic gun at one end, two pairs of deflector plates near the center, and a fluorescent screen at the other end. When an alternating voltage is applied to the horizontal or x plates, the electron beam emanating from the oscilloscope's electron gun is deflected back and forth and produces a glowing horizontal line along the x axis of the fluorescent display screen. When a similar alternating voltage is applied across the vertical or y plates, the electron beam is deflected up and down and produces a glowing vertical line along the y axis of the fluorescent display screen. If we apply a ''sawtooth'' voltage, which increases linearly in time and then suddenly drops to zero at regular intervals to the x plates of the oscilloscope, the electron beam will appear to ''sweep'' along the x axis of the display screen and then suddenly return back to its starting point for the next ''tooth'' of voltage. If the frequency of the sweep is high enough, the residual brightness on the display screen will give the appearance of a single horizontal line. If we now introduce an ''unknown'' signal to the vertical or y plates and match the frequency of the x axis sweep to that of the signal to be studied, the electron beam will repeatedly trace out a seemingly stationary portrait of that signal.

The oscilloscope can thus be used to analyze the amplitudes, frequencies, and waveforms of variable electrical potentials whose frequencies range from small fractions of a hertz to thousands of megahertz. Oscilloscopes can also accurately measure time intervals between electrical pulses in a signal current which are less than 10^{-7} seconds in length. In short, any piece of information that can be converted into electronic signals can be displayed on an oscilloscope screen for analysis. Small wonder then that this instrument is to be found in virtually every research laboratory, electronics shop, and radio and television repair store in the world.

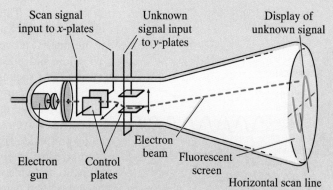

··· The Cathode Ray Oscilloscope A known input scan signal is placed across the x-control plates and the unknown signal is placed across the y-control plates. By matching the frequency of the x-plate scan to that of the unknown signal, a two-dimensional portrait of the unknown signal is created on the fluorescent screen.

The actual transmission of these electronic audio and video output signal currents over long distances is accomplished by using a high-frequency electromagnetic carrier wave. The pure carrier wave is generated from a transmitting circuit in which a source of alternating voltage having a predetermined or ''assigned'' frequency is applied to the primary of a transformer as shown in Figure 14.3. The transformer secondary is connected to a long piece of conducting wire, which serves as the antenna. The alternating current produced across the

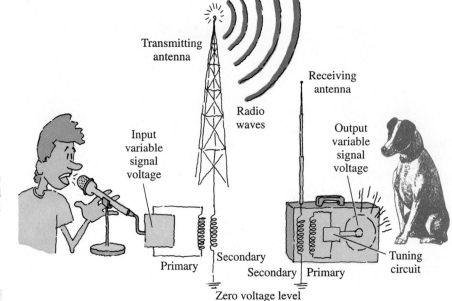

FIGURE 14.3 ··········
Radio and Television Transmission and Reception Sounds and pictures are converted into electromagnetic sound and picture signals, which are transmitted and received via antennas. The received sound and picture signals are then converted back into the sounds and pictures initially converted and transmitted.

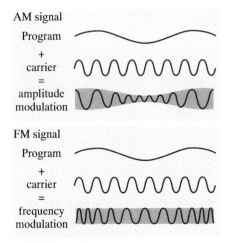

FIGURE 14.4 ··········
Modulation If a given signal is transmitted by modifying the amplitude of an electromagnetic carrier wave, the process is called amplitude modulation or AM. If a signal wave is transmitted by modifying the frequency of an electromagnetic wave carrier, the process is called frequency modulation or FM.

primary of the transformer by the input voltage induces electric charge oscillations in the antenna. The resulting charge flow up and down the antenna generates an electromagnetic wave having the same frequency as that of the charge oscillations in the antenna as well as that of the initial input voltage to the transformer primary. The audio and video signals to be transmitted on the carrier wave are first amplified by means of vacuum tube triodes or transistors and then blended or "mixed" in with the input current to the transformer primary. This blending can be accomplished by using the audio and video signals to modify either the amplitude of the carrier wave oscillations or their frequency (Figure 14.4). Transmissions of audio and video signals by carrier waves whose amplitudes have been modified is referred to as *amplitude modulation* or *AM* transmission, while transmission of audio and video signals by carrier waves whose oscillation frequencies have been modified is referred to as *frequency modulation* or *FM* transmissions. The audio and video signals are thus radiated in all directions at the speed of light from the transmitting antenna in the form of AM or FM electromagnetic waves.

To allow these waves to be broadcast over as large an area of the earth as possible, transmitting antennas are constructed as tall as possible and are often situated atop high buildings and mountains so that their transmissions can "peek" around more of the earth's curvatures [Figure 14.5(a)]. The transmission process is also aided and abetted by a layer of charged particles in the earth's atmosphere called the ionosphere. Located in a band extending from about 80 to 700 km above the earth, the ionospheric layer reflects outgoing low-frequency radio waves back toward the earth, thereby allowing a given radio transmitter to literally bounce radio signals around the globe as they alternately reflect off the earth's surface and ionosphere.

Because the electromagnetic properties of the ionospheric plasma are constantly changing with the times of the day or night, the flux level of charged particles incident from the sun, etc., the ability of the ionosphere to reflect certain frequencies of radio waves can vary quite dramatically. Thus there are times

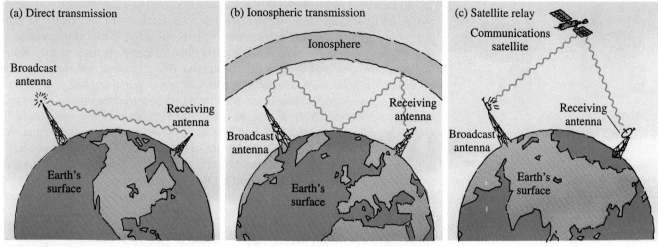

FIGURE 14.5 ··········
Long Distance Radio and Television Transmission (a) Direct; (b) reflection; (c) relay.

when we may be able to receive clearly the broadcasts of a radio station hundreds of kilometers away, while the broadcast reception of a much closer station operating at a different frequency will be far worse.

The ionosphere does not reflect the higher frequency radio waves, and worldwide transmissions of these waves are made possible by using communications satellites. These satellites are placed in orbit above the earth and serve as relay stations in which a signal from one location on the earth is picked up and retransmitted to a second location [Figure 14.5(c)]. This process is enhanced considerably by the use of "satellite dishes," which serve to collect and focus the waves as they are transmitted and received.

Whether the transmitted waves are bounced off the ionosphere, relayed by satellites, or simply travel on a direct line from the broadcast antenna, the task of "picking up" these waves for a receiving device at a distant location falls to the receiving antenna. Like its transmitter counterpart, a receiving antenna is in essence a long piece of conducting wire. Incident electromagnetic waves induce oscillations in the electric charges residing in the receiving antenna, which in turn induce a variable current flow in the receiver circuit primary as shown in Figure 14.3.

A given receiver antenna can be bombarded from afar by a multitude of electromagnetic carrier waves from a multitude of transmission sources. Hence, it is necessary to somehow sift out the waves we wish to receive from this carrier wave backdrop. This sorting process is performed by hooking the receiver transformer primary to a tuning circuit. The electronic components of the tuning circuit are set up in such a way that the currents induced in the receiver transformer primary by the charge oscillations in the receiving antenna will pass through the tuning circuit only if their frequency is equal to the resonant or tuning frequency of the circuit. All other frequencies are blocked or filtered out by the tuning circuit. If the tuning circuit is properly designed, the value of this tuning frequency can be easily adjusted so that we may "tune in" on whatever frequency we wish to have pass through the tuning circuit.

Input signal

FIGURE 14.6 · · · · · · · · · ·
The Speaker Incoming signal currents induce the magnet to vibrate. Since the magnet is attached to a diaphragm, it too vibrates, producing the sound head coming from the speaker.

The current that emerges from the tuning circuit is a far weaker version of the mixed signal and carrier wave current that was initially fed into the transmitter primary; therefore, it must be amplified to a useful level. If this amplification is properly performed, the emergent current will be virtually identical to that which was originally fed into the broadcast antenna. This amplified current must now go through a process of demodulation in which the signal current is separated from the carrier current. The signal current in turn is separated into its audio and visual components. The audio signal is fed into a speaker circuit consisting of a coil wrapped around a magnet, which is attached to a movable membrane. As the variable audio signal current flows through the coil, the coil produces a variable magnetic field. The magnet at the center of the coil responds to this current by oscillating in tandem with the magnetic field variations it experiences from the coil. As a result, the attached movable membrane also vibrates, thus reproducing the initially broadcast sound (Figure 14.6).

The video signal is meanwhile fed into a picture tube, which, like its camera tube counterpart, consists of an electron gun, grid, and control plates (Figure 14.7). In the case of the picture tube, however, the signal plate of the video camera is replaced by a fluorescent screen having a chemical coating of phosphors, substances such as phosphorous or zinc sulfide, which will exhibit fluorescence when impacted by a beam of high-speed electrons. The video signal fed into the picture tube alters the voltages on the grid and control plates in such a way that the electron beam scans the fluorescent screen in exactly the same way as that which scanned the signal plate.

As the electron beam strikes a given phosphor, the phosphor will briefly glow with an intensity that is proportional to the intensity of the incident electron beam. The resulting scan thus produces a screen of glowing phosphors having various degrees of brightness. Because the glow of the phosphor will tend to linger both as an image on the screen as well as a residual image on the retina of the human eye, the image produced on the screen can be made to appear to

FIGURE 14.7 · · · · · · · · · ·
The Electronic Picture Tube   The incoming electronic picture signals are converted into visual images by reversing the processes of the electronic camera tube.

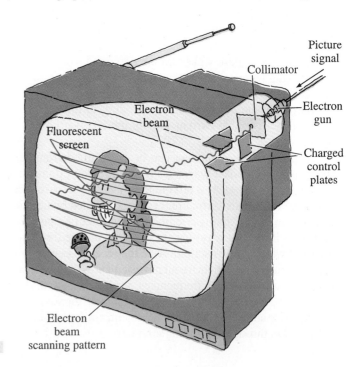

(a)

(b)

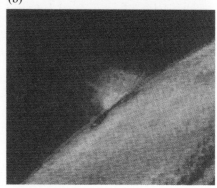

(d)

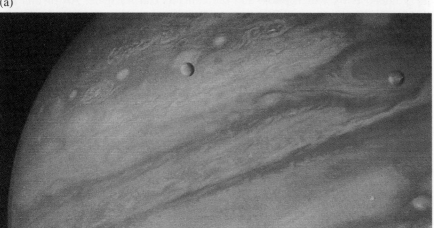

(c)

FIGURE 14.8 ··········

Interplanetary Telemetry   Telemetry signals transmitted from distant space probes have given scientists close-up views of (a) the planet Saturn, (b) the surface of Mars, (c) a volcanic eruption on the Jovian satellite Io, and (d) of Jupiter itself.

be present continuously if each successive scan is done quickly enough, usually of the order of $1/60$ of a second or so. Color video signals are reproduced by three screens, each of which fluoresces in one of the three additive primary colors of red, blue, and green. The incoming video signal current corresponding to the red image is fed into the red image producing picture tube, and so on. The three images are then blended by mirrors in a full-color final image to be enjoyed by the viewer.

One of the more spectacular and ''far out'' applications of modern communications technology has come in the area of telemetry and remote sensing. Data-collecting devices have been put aboard satellites and spacecraft and then placed in orbit hundreds of kilometers above the earth, or in some cases, hurled millions of kilometers across interplanetary space to rendezvous with the distant worlds of our solar system. While on their assigned missions, these instrument packages transmit their findings and observations back to earth for interpretation and analysis using long-range communication techniques that are virtually identical to those just discussed (Figure 14.8).

The major advantage of exploring space in this fashion is that we need not worry about having life support systems for human astronauts aboard the spacecraft. The instrument packages that are sent into space may end up being cooked in the hellishly hostile Venusian surface environment, exposed to the high-energy interplanetary emanations from the Sun, or simply leave the solar system forever. Whatever happens they are regarded in the end as being expendable entities.

The information sent back from these distant and not so distant satellites and space probes has revealed a universe that is at once exciting, beautiful, and complex. One orbiting unmanned astronomical observatory, called IRAS (Infrared Astronomical Satellite), for example, peered at the universe with infrared detectors from above the glowing, distorting, and absorbing gaseous layers of the earth's atmosphere and discovered thousands of infrared sources in the sky, the analysis of which will keep astronomers engaged for decades.

Unmanned satellites orbiting above the earth also routinely perform a variety of more practical services as well. Weather satellites transmit daily pictures of terrestrial weather conditions, which have led to more accurate weather forecasting; communication satellites relay long-wavelength telecommunications from one place on the earth to the other; and navigational satellites provide satellite guideposts by which pilots and sea captains can navigate. All of this and much more has been opened up to us because we can communicate with instruments that operate from afar without the direct presence of a human being.

1. Compare and contrast AM and FM electromagnetic transmissions.
2. Describe the principles of physics involved in the transmission and reception of a color television program.
3. Explain why a satellite relay system is necessary in order to broadcast television programs around the world, while radio waves can be broadcast worldwide without such a satellite.
4. How is a television camera different from a television receiver? How are they the same?
5. Summarize the various types of energy conversion that occur in the transmission and the reception of a radio or television program.

14.2 Photoduplication

The words "carbon copy" are often used to denote someone or something that is an exact duplicate. The term, of course, hearkens back to an era not that many years ago when the duplication of a given document was accomplished through the use of thin sheets of carbon-coated paper. When sandwiched between sheets of regular paper, any indention made on the top or original document would be left as a black carbon imprint of that same indention on the carbon copy. Such a process is, by modern standards, both tedious and inefficient.

For better or for worse, copies of any document can now be created with incredible speed and accuracy through the use of a process called photoduplication

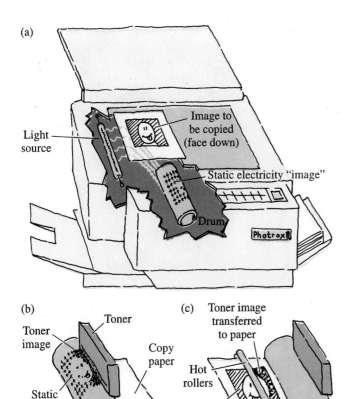

(a)

Light source

Image to be copied (face down)

Static electricity "image"

Drum

Photrox

(b)

Toner image

Toner

Copy paper

Static electricity "image"

Drum

(c) Toner image transferred to paper

Hot rollers

Final copy

Drum

FIGURE 14.9 ··········

The Photocopying Machine   (a) The image to be copied is focused onto a charged cylindrical drum having a thin photoconducting surface. Electric charge is retained in the dark areas of the image and dissipated in the lighter areas. (b) This static electricity image is then rotated past a carbon toner, which is attracted to the charge patterns on the drum. (c) The resulting toner pattern is transferred to a sheet of copy paper where it is baked onto the paper by hot rollers.

or photocopying. A schematic of a typical photocopying device is shown in Figure 14.9. The principle element in the photocopier is a large cylindrical drum that has been coated with a layer of a photoconducting substance. A uniform layer of electric charge is placed on the drum by rotating it against a positively charged strip called a corona. When the document to be copied is illuminated, its image is reflected by means of mirrors onto the drum. In the bright areas of the image, the light increases the photoconductor's electrical conductivity and the electric charges dissipate. Meanwhile, the dark areas of the image do not produce such changes and the static charges on the drum in those locations remain.

The image of the document is now in the form of a "picture" pattern of static electric charges on the surface of the drum. This static charge pattern is then rotated past a reservoir of black carbon powder, called toner, which has been given a negative electrostatic charge. The positively charged image pattern draws the negatively charged toner particles on the drum to create an image on the drum surface which is now outlined in carbon toner. As the drum rotates further, it comes in contact with a piece of copy paper that has been given an electric charge of its own. As the paper and drum make contact, the carbon toner is electrostatically transferred to the copy paper with the same image as it had on the drum surface. The toner-paper combination is finally passed between a pair of heated rollers that "bake" the carbon toner image onto the copy paper, thus producing the final copy of the original image. As the drum rotates further,

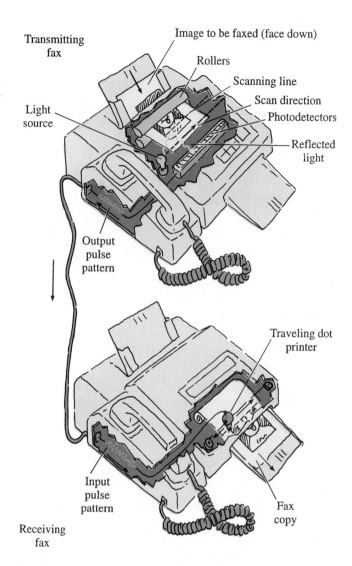

Transmitting fax

Image to be faxed (face down)

Rollers

Scanning line

Scan direction

Photodetectors

Light source

Reflected light

Output pulse pattern

Traveling dot printer

Input pulse pattern

Fax copy

Receiving fax

FIGURE 14.10 ·········

The Fax Machine As the beam of light scans a given image, the light reflected from the image is converted into a pattern of electric pulses. This pulse pattern is transmitted via phone to the receiving machine, which converts the incoming pulse pattern back into the original image.

any leftover toner and electric charges are removed, and the drum is ready for the next image to be copied. Photocopies made in such a fashion can be produced quickly and efficiently, and very often the copy is indistinguishable from the original.

In one important variant of the photocopying process, documents, photographs, texts, etc., can be transmitted over long distances by means of a facsimile or fax machine (Figure 14.10). When a fax is to be sent, the sending and receiving machines must first be connected via telephone so that they can adjust themselves to send and receive information at the same rate. The "sending" fax machine then scans the document to be sent with a beam of light and converts the varying reflected light intensities into a series of electronic pulses. This pattern of pulses is sent via telephone to the receiving fax machine, which receives the pattern of pulses and recreates the original image using a traveling dot printer, which prints the image as an array of tiny dots in accordance with the pulse pattern it receives.

14.2 Physics, Technology, and World History

Almost from the onset of the Scientific Revolution, governments discovered that a nation's political power is very much dependent on that nation's scientific and technological prowess. Thus, the Western European imperialism of the eighteenth and nineteenth centuries was due in no small part to the use of technology as an instrument for subjugation. Among the technological "have" nations, science and technology have often played crucial roles in the course of human history. In the summer of 1940, for example, with the German war machine at the throat of Western Civilization, only 23 miles of English Channel separated humanity from what Winston Churchill was to refer to as ". . . a new Dark Age made more sinister and protracted by the lights of perverted science. . . ."

Thus began the Battle of Britain in which the German airforce, the Luftwaffe, sought to gain air superiority in the skies over England as a necessary prelude to a German invasion. But the German Luftwaffe was turned away not only by the courage and heroism of the pilots of the Royal Air Force, but also by a new British technological device not yet perfected by the Germans in 1940—radar. With radar, the pilots of the RAF were spared the burden of doing standing patrols without adequate pilots or planes. In fact, they could wait on the ground until the German aircraft were actually about to attack. Then, with the use of radar, the fighters of the RAF were directed toward the approaching German formations, knowing the speed, altitude, and numbers of their adversaries. Military historians almost universally agree that without radar, the Battle of Britain almost certainly would have been won by the Germans.

World War II also brought with it the development of the first nuclear weapons, a state of affairs that was very much accelerated by virtue of the fact that at the onset of war in 1939, the German efforts to construct such devices seemed to be well ahead of those of the Allies. The chilling prospect of an Adolph Hitler armed with nuclear bombs prompted no less a personage than Albert Einstein to write his famed letter to President Roosevelt in the autumn of 1939, pleading that the United States give top technological priority to the development of such weapons of its own.

Fortunately, the German effort came to naught, but sadly the Nuclear Age was nevertheless thrust on us in the summer of 1945, and along with it all of the attendant adjustments in the foreign policies of the nations now forced to live in a nuclear world.

In more recent times, when the reforms of the Nobel Peace Prize-winning President Mikhail Gorbachev of the Soviet Union were threatened by a coup attempt in August 1991 by the hard-line Communist elements in the Soviet government, it was the technology of communication that allowed information to flow freely in and out of Moscow via television, short-wave radio, and fax transmissions. In the end, it was the technology of communication that was a key factor in the ultimate failure of that attempt. As one young anticoup Soviet journalist put it, "The media were our tanks and our armored vehicles."

With such events as these in the past, we can only speculate on the impact science and technology might have on future history.

••• **British Radar Antenna in 1940** The so-called CH or Chain Home radar network covering the approaches to England was a key factor in the British victory over the Germans in the Battle of Britain in the summer of 1940.

1. Describe how a photocopy is made.

2. What are some of the practical uses of a photocopying machine? What are some of the disadvantages?

3. Compare and contrast a fax machine with a photocopying machine.

4. Discuss some of the principles of physics employed in a photocopying machine and a fax machine.

5. Why are fax machines so important in contemporary society?

14.3 Computers

Over the past several decades perhaps the most astounding ''grand entrance'' onto the technological scene belongs to the electronic computer. Because of its ability to perform an array of tasks as diverse as solving complex theoretical problems in science and engineering, setting up next term's class schedule, or billing you for all of those long-distance phone calls made to home or a distant lover, the computer has taken on a position of considerable influence and importance in our society. So much so that many have become concerned over the rise in the level of computerized mischief such as invasion of privacy, embezzlement, espionage, and other forms of ''computer crime.''

Sociological considerations aside, however, the operation of a computer is a most imaginative blend of the principles of physics, mathematics, and logic. To begin with, all of the numbers, letters, and other symbols employed in our language are ''translated'' or coded for the computer in the form of a series of electronic current pulses. In one type of computer, the digital computer, information is coded in the form of the binary number system in which there are only two digits, 0 and 1. The two digits of the binary number system are most convenient in computer technology because they can be readily represented electronically by + and − voltage and magnetically by N and S magnetic poles. To express a letter or number in the binary code, the electronic pulse representations of 0 and 1 are grouped together in one or more sets of four-digit arrays called bits. For example, the number 13 in binary notation would be 1101, the letter D would be 0001 0011, etc.

Associated with this binary numbering scheme is a binary system of mathematical logic known as Boolean algebra, which postulates that only two constants exist, 0 and 1. The advantage of using this seemingly peculiar form of algebra is that all three of the basic operations defined in Boolean algebra can be simulated by the use of electronic circuits called logic gates or logical functions. The fundamental logic function in digital circuits are referred to as AND, OR, and NOT gates (Figure 14.11) and can be created on a highly miniaturized scale through the use of junction diodes and transistors (Physics Formulation 14.2). Moreover, these basic logic circuits can be assembled into integrated circuit combinations that permit the performance of ever more tedious and complex tasks.

To perform these functions, however, the computer must also have the ability to store electronic pulse patterns for future use or recall. One early technique by

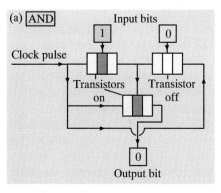

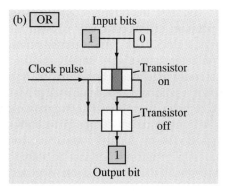

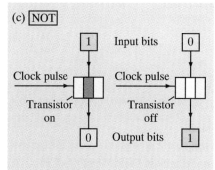

FIGURE 14.11
The Basic Logic Circuits of a Computer By suitable use of switches and electronics, the three basic logic gates of (a) AND, (b) OR, and (c) NOT can be created.

which this was done was through the use of *magnetic cores,* tiny rings made of ferrite or some other readily magnetized material. If an insulated wire is passed through the center of the core, the core can be magnetized in one of two polarities, depending on the direction of current through the wire. Thus the core can be used to store data in binary form. A "0" electronic pulse magnetizes it in one polarity direction and a "1" electronic pulse magnetizes it in the other.

To obtain a large memory, the cores are arranged into a magnetic core plane array as shown in Figure 14.12. In this figure, each core center is mounted at the intersection point of two insulated magnetizing wires. A given core at a given location in the array can then be magnetized by sending one-half of the magnetizing current pulse through each of the wires intersecting at that particular core center. The information stored on the magnetized cores can be retrieved by running a sense wire through all of the cores in the array. When the information from a given core is to be retrieved, a current pulse having the opposite polarity from that which produced the initial core magnetization is sent down the appropriate pair of magnetizing wires. As the core demagnetizes, a pulse of current is induced in the sense wire and appears as an output signal for further use or analysis.

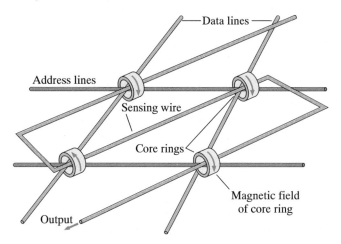

FIGURE 14.12
The Magnetic Memory Core of a Computer Information is stored at a given core ring by magnetizing that ring with current pulses sent down the appropriate pair of data and address lines. The information is retrieved by sending demagnetizing current pulses down the same pair of lines. As the core ring demagnetizes, it induces a pulse of current in the sense wire. This pulse then appears as an output electronic signal.

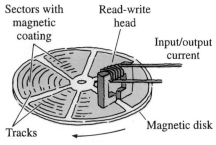

FIGURE 14.13 ···········

The Magnetic Disk Memory of a Computer As the magnetic disk rotates, input current flow patterns passing through the electromagnetic coil in the read-write head induce changing magnetic fields, which align the magnetic domains present in a track of the magnetic disk into a corresponding pattern of small magnetic fields or bands on the disk.

Because the timing sequence of such tasks is crucially important, the entire operation of the computer is controlled by a current pulse generator called a digital clock. The current pulses produced by the digital clock and fed into the computer have a very small period ($\sim 10^{-6}$ sec) whose precise value is very carefully regulated. By having the response of the computer's logic circuits set to these computer clock pulses, it is possible for a computer to perform a sequence of complex operations at extremely high speeds and in the proper order.

In more recent computer technology, the magnetic memory core has been superseded by highly compacted computer memory systems in which vast amounts of information can be stored and accessed. One such system employs a magnetic disk, which consists of a disk coated with a thin layer of magnetic material. This magnetic coating is divided into tracks, which are circles concentric to the center of the disk, and sections, which are pie-shaped segments radiating outward from the center of the disk (Figure 14.13).

Information is stored onto and retrieved from the disk by means of a read-write head placed in close proximity to the disk. This read-write head is little more than a small iron-core electromagnet whose coil terminals serve as a source of input and output voltage. Information is stored onto the disk by alternating the direction of the current flow through the coil in the read-write head. As the current flows through the read-write head coil, it induces a magnetic field in the core of the coil. The resulting field lines in turn align the magnetic domains in the disk's magnetic coating. When the current is reversed, the domains are aligned in the opposite direction. As the disk is rotated past the read-write electromagnet, the pattern of current reversals can thus be stored in the form of a series of north-south magnetic domain alignments on the magnetic disk (Figure 14.14). The information so stored can be retrieved from the disk by rotating it past the read-

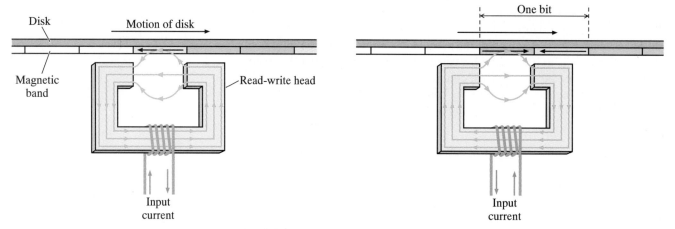

FIGURE 14.14 ···········

Writing onto a Magnetic Disk As current passes through the coil in the read-write head, it induces a magnetic field, which magnetizes a small area of the surface of the disk called a band. Each bit is registered on the disk as two tiny bands of magnetic fields. Each new individual bit is delineated by a reversal in the magnetization of the previous band. If the band following the first bit has the same magnetic field direction, a 0 has been stored. A reversal of the magnetic field direction in the second band denotes a 1.

write head. As the pattern of magnetic field alignments in a given disk track rotates past the read-write head, it induces a pattern of electric currents in the read-write head coil that is exactly the same as the current pattern that was initially "written" onto the disk.

Incredible advances in the art of miniaturization in recent years have led to the development of computer memory "chips" in which large amounts of information can be stored in almost fantastically tiny volumes of space. One form of computer memory chip is the random-access memory chip or RAM chip. A typical RAM chip is extremely small, often having dimensions of less than a millimeter, but nonetheless containing thousands of storage cells, each of which can store one bit of binary code. As in the case of the magnetic memory core, each storage cell in a RAM chip is connected into a two-dimensional network of address and data wires or lines.

Instead of a magnetic ring, however, the RAM chip storage cell makes use of the transistor-capacitor combination shown in Figure 14.15. To store a given bit of information, an electrical pulse is first sent to all of the cells along a single address line. This pulse has the effect of switching on all of the transistors along that line. The bit to be stored is now sent along a data line in the form of a binary code pattern of "on" and "off" electrical pulses. Each "on" pulse causes a small electrical charge to build up and be stored on the plates of the capacitor in one of the storage cells, while each "off" pulse leaves the capacitor uncharged. When the data are to be read, each charged cell releases its electrical charge in the form of an electrical pulse down the data line, while the uncharged cells release no such pulses. Thus the initial pattern of on-off electrical pulses can be retrieved. The storage cells in a RAM chip are usually arranged in groups of eight so that one byte of information can be stored in each such array.

RAM chips have the disadvantage that they can store and retrieve data only when an electric current is flowing. A second type of computer memory chip, called a read-only memory or ROM chip, stores information when there is no electric current present. Once again, the basic network of address and data lines is present, but now each cell in a ROM chip consists of the arrangement of solid-state diodes shown in Figure 14.16. For the cell that stores a 1 bit, the intersection of the address and data wires is connected by a diode while a cell that stores a 0 bit does not have such a connection. This memory is then "read" by sending a pulse down an address line. The pulse diverts through the system of diodes and emerges in the data lines as a pattern of 0 and 1 bits. As the name implies, the ROM is a permanent memory chip, which can be read from but not written to.

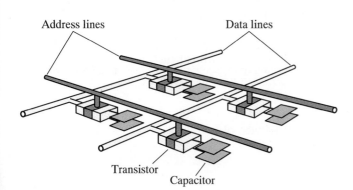

Address lines   Data lines

Transistor
Capacitor

FIGURE 14.15 ···········

The Random-Access Memory (RAM) Chip   Each cell in the address line and data line array consists of a transistor (see Physics Facet 14.2) and capacitor. Binary 1's are stored as electrical charges on the capacitors, while binary 0's are not.

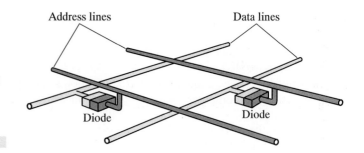

Address lines   Data lines

Diode   Diode

FIGURE 14.16 ···········

The Read-Only Memory (ROM) Chip   Built into the ROM memory are binary 1 cells in which address lines and data lines are connected by a diode and binary 0 cells, which do not have such connections.

The development of such miniaturized solid-state circuitry for use in computers has generated what has become in effect a "computer revolution." The computer chips in our digital watches tell us when we should go to class and when a horribly boring lecture should end. The computer chips in our microwave ovens tell us when dinner is ready, and those in our automobiles inform us of possible car trouble on the way. Computerized cameras tell us when to shoot a photograph. On our way to and from classes, computers control whether an intersection signal light will be red, yellow, or green. On a more grandiose scale, complex physical systems, such as biochemical reactions, which heretofore totally defied analysis by scientists and engineers can now be attacked through the use of these marvelous machines. In fact the performance level of some of the "supercomputers" has become so impressive as to give rise to an area of knowledge called artificial intelligence. The literature of science fiction, of course, abounds with stories of computer *coups d'état* of one sort of other, to which some computer scientists whimsically respond by saying that computers which "do what I say rather than what I mean" are too stupid to take over the earth, while others make the equally whimsical claim that perhaps they are too smart to want to take over!

WORKED EXAMPLE

A certain RAM chip has a total of one million cells. How many bits of binary code can such a chip store? How many bytes of information?

SOLUTION: Each cell in the chip can store one bit of information, therefore, the chip can store a total of one million bits of binary code. Since eight bits of binary code are required to specify one byte or character of information, the total memory capacity of this chip would be

$$\text{total memory} = \frac{1{,}000{,}000 \text{ bits}}{8 \text{ bits/byte}}$$

$$= 125{,}000 \text{ bytes}$$

or

125 kilobytes of memory .

1. Why do computers employ a binary code?
2. Describe the AND, OR, and NOT digital circuits.
3. Compare and contrast a magnetic core with a magnetic disk.
4. Summarize the various ways in which information can be stored in a computer.
5. Give several examples in which computers have been integrated into some other device or machine such as an automobile, etc.

1. If a RAM chip is in the shape of a square 1 millimeter on a side and has a memory consisting of 10,000 cells, find the average area occupied by each cell.
2. If a certain ROM chip has a memory of one megabyte, how many cells does this chip possess? How many bits of binary code can this chip store?

14.4 Lasers

Since the early 1960s one of the more familiar technological acronyms that has taken up residence in the language is the term "laser." Based on the phrase "*l*ight *a*mplification by *s*timulated *e*mission of *r*adiation," the laser operates essentially as advertised by this acronymic expression.

As we have seen in an earlier chapter, the light incident on a given medium can cause the atoms in the medium to be excited to higher energy states. Ordinarily, these excited atoms will randomly reemit the energy they absorbed from the incident light and return to their original energy states on time scales of the order of 10^{-9} second. Many atoms, however, exhibit excited energy states whose lifetimes are of the order of 10^{-3} seconds, or several orders of magnitude longer than typical excited state lifetimes. Such excited states are referred to as *metastable states,* and in order for a laser to function, its active medium must always contain a substance having at least one of these metastable energy states.

One of the earliest types of lasers employed a rod made of ruby as its active medium. Ruby is an impure form of aluminum oxide in which some of the aluminum atoms in the crystal have been replaced by chromium atoms having a metastable state. If the ruby rod is exposed to a burst of green light having a wavelength of 5.5×10^{-7} m, then the chromium atoms in the ruby crystal will be excited to a higher energy state as shown in Figure 14.17. Once in the higher state, the electrons can decay back into the initial ground state, but also have a high possibility of dropping down into an intermediate metastable state. If the bursts of green light are sufficiently intense and frequent, it is possible to create a "population inversion" in which more of the chromium atoms are in their metastable state than in their ground state.

This excitation process using light as the exciting agent is called optical pumping and is made possible because of the relativity long times that atoms can spend in the metastable state. The population inversion occurs because it is

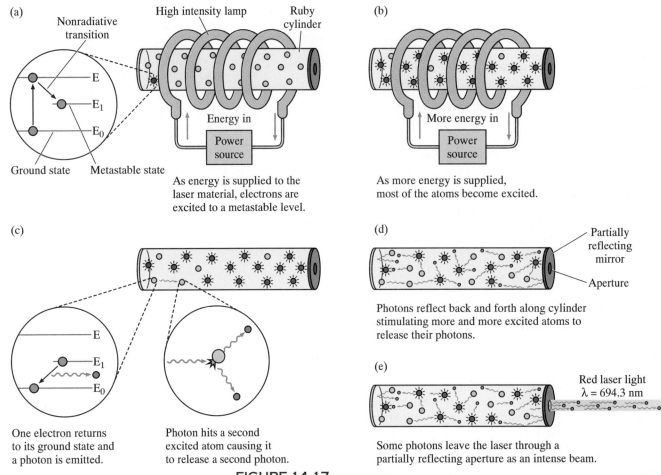

FIGURE 14.17
The Operation of a Laser

now possible to pump atoms into their metastable states faster than they can spontaneously decay out of them. At some point, however, a few of the chromium atoms in the metastable state begin to spontaneously decay back into the ground state, and as they do so, they emit photons having wavelengths in the deep red region of about 6.9×10^{-7} meters. As these red photons move through the ruby rod, they resonate with other chromium atoms in the metastable state and stimulate these atoms to also decay to the ground state, thereby emitting more red photons. Most of this light escapes randomly through the sides of the rod, but the light that travels along the principal axis of the rod is contained by mirrors placed at each end of the rod. One of the mirrors is totally reflective, while the other is partially transmissive so as to allow some of the red photons to escape at that end of the rod. As these red photons bounce back and forth from one end of the rod to the other they stimulate a cascade of red photons, many of which emerge at the partially reflecting end of the rod as a laser light beam.

The ruby laser is an example of a pulsed laser in which the laser light is created in short single bursts. In other types of lasers, such as helium-neon and

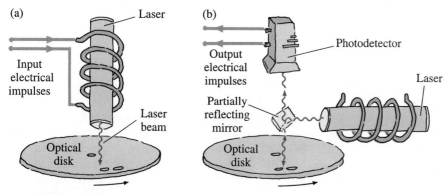

(c)

FIGURE 14.18 ··········

Optical Memory   (a) Information can be stored on an optical disk when a pattern of electrical input pulses causes a laser to turn on and off with the same pattern. When the laser is on, it etches an elongated pit into a highly reflecting rotating disk. When the laser is off, no such pit is produced. The result is a pattern of pits on the disk that represents the information "written" on the disk by the laser. (b) Information can be read from an optical disk by directing a laser beam at the disk as it rotates. If no pit is present under the laser beam, it is reflected into a photodetector where it produces an electric current. If a pit passes under the beam, no reflection occurs to the photodetector, and no electric current is produced. The emerging pattern of electric current pulses reproduces the initial input pattern of electric current pulses "written" onto the disk. (c) A microscopic view of the laser-etched pits on an optical disk.

carbon dioxide lasers, the optical pumping can be done on a continuing basis, thus producing a continuous beam of laser light.

The beam of light that emerges from a laser is highly intense, collimated, monochromatic, and coherent, all of which are most useful properties for both science and technology alike. As a result, lasers and laser technology have found a wide variety of applications over the past three decades. For example, laser beams can be used in various types of medical surgery. In one form of eye surgery, which corrects a condition called a detached retina, a laser beam is aimed right through the lens of the eye to fuse the retina properly to the rest of the eye. Although the surgeon's laser beam has an intensity sufficient to perform the desired fusion of tissue, its monochromatic visible region wavelength allows it to pass through the lens of the human eye without harming it. Thus armed, the eye surgeon can perform the operation without having to make an incision.

Lasers can also be employed in data storage and retrieval. In an optical analogue to the magnetic disk, binary coded information is first fed into a laser circuit in the form of a pattern of electrical pulses. Each "1" pulse turns the laser on while a "0" or no pulse input turns the laser off. As the laser blinks out its binary coded information, its beam etches a series of pits along a track of a rotating thin "optical" disk of metal, much as the read-write head creates a pattern of magnetic alignments along a track on a magnetic disk. When the information is to be retrieved, a laser beam is directed at the rotating disk by means of a semitransparent mirror as shown in Figure 14.18(b). When the beam strikes an unpitted part of the disk, it is reflected back to the mirror where a part of the beam continues in a straight line path to a photodetector. When struck by

FIGURE 14.19 ··········

The Compact Disk The compact disk is simply an optical disk onto which recorded music has been stored. The reproductive quality and fidelity achieved from a compact disk is generally regarded as being superior to that obtained from magnetic tape or vinyl disks.

the beam, the photodetector produces an electric current. As the disk rotates to a pit, the laser beam is not reflected back to the photodetector and no electric current flows. In this way, the initial pattern of electrical pulses that produced the etched pattern of pits on the optical disk can be retrieved. One of the most familiar examples of such techniques is the compact disk on which music can be recorded and played back with an extremely high degree of quality and fidelity (Figure 14.19).

Perhaps the most visually striking application of laser optics is the creation of three-dimensional images called *holograms*. A given holographic image or hologram is produced by the optical arrangement shown schematically in Figure 14.20. A laser beam is split into two parts by a partially reflecting mirror. One part of the beam illuminates the object to be recorded while the other is spread out over the photographic plate as a reference beam. Because the laser beam has all of its corresponding wave crests and troughs precisely aligned on top of each other, the film records not only the intensity of the light reflected from the object, as in an ordinary photograph, but also includes every other piece of information contained in the light reflected from the object, including its wavelengths, amplitudes, and phases. A developed holographic ''negative'' or hologram will show only a pattern of tiny interference fringes, which are not unlike a fingerprint in appearance. When the hologram is illuminated at the same angle as the initial reference beam, it will produce a multiplicity of two-dimensional images. As your eyes view the hologram (Figure 14.21), each sees a different two-dimensional image, just as they would when viewing a real three-dimensional object. As such, your brain interprets this pair of two-dimensional images as a solid three-dimensional object.

One very familiar example of holography can be found in the so-called bar code scanner at your local supermarket. Each item in stock has a ''bar code''

FIGURE 14.20 ··········

Holography A hologram of a given object is created by first dividing a laser beam into an ''object'' beam, which illuminates the object, and a reference beam. The reference beam and some of the light reflected by the object from the object beam are recombined and recorded on a photographic plate to produce a hologram of the object. If the hologram is illuminated at the same angle as the initial reference beam, a binocular view of the hologram will seem to have a solid or three-dimensional appearance (see Figure 14.21).

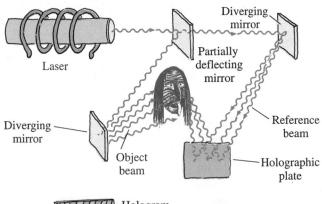

FIGURE 14.21 ··········

Viewing a Hologram If light shines on a hologram at the same angle as the reference beam initially used to make the hologram, the reflected beam is broken up by the pattern on the hologram into a continuum of object images. When two such images are viewed with binocular vision, the observer perceives a three-dimensional image of the object.

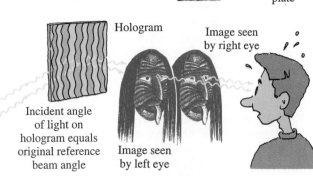

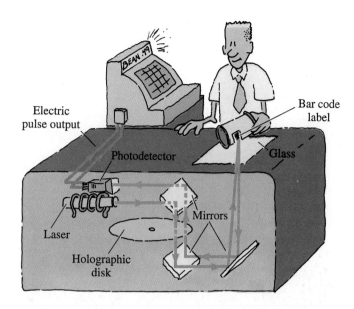

FIGURE 14.22 ⋯⋯⋯⋯

The Bar Code Scanner A laser beam is directed through a holographic disk, which consists of a set of three-dimensional images of bar codes and serves to focus the laser beam onto the surface of the item being scanned. As the code is scanned, it produces a variation in the intensity of the reflected beam. A photodetector converts this pattern of light intensities into a corresponding pattern of electrical pulses, which is in turn fed into the store computer to give a read-out on the item's description and price.

label, which is printed with a pattern of vertical black and white bars of variable widths denoting the pertinent information concerning a given item. The bar code scanner (Figure 14.22) sends a laser beam through a rotating holographic disk containing three-dimensional images of bar codes. This disk serves to focus the laser beam onto the label even though the label might not be flat up against the scanner window or indeed, as for the label on a cylindrical can, not flat at all. As the bar code label for the given item is moved across the laser beam, the reflected beam varies in intensity according to the pattern of the black and white bars in the item code. The reflected beam is then fed into a photodetector where the variations in the intensity of the reflected laser beam are converted into corresponding variations of electrical current. This pattern of electric current variations is then fed into the store's computer which, among other things, rings up the description and price of the item on the customer's receipt.

The laser thus provides us with yet another example of a device that appeared on the technological scene, quickly became assimilated into both everyday society as well as the technology, and finally became an enabling agent by which progress and advances are made in seemingly unrelated areas of human endeavor. So our technology continues its headlong acceleration into the future.

1. Summarize the physical principles involved in the operation of a laser.
2. Describe some practical uses for lasers.
3. What is a metastable state in an atom? Why are such states important in the operation of a laser?
4. Discuss ways in which atoms can be ''pumped'' into higher energy states.
5. Briefly describe the characteristics of a laser beam.

REVIEW
14.4
QUESTIONS

THOUGHT QUESTIONS

1. Is it possible for AM and FM signals to be broadcast at the same time on the same carrier wave? Explain.

2. Can a microphone be used as a speaker and vice versa? Explain.

3. Discuss how you think a tape recorder might work.

4. Science fiction stories sometimes describe teleportation devices that can transport solid objects over large distances. Describe how such a device might work and the technical problems that one might encounter in attempting to build one.

5. How might you go about producing color photocopies?

6. Can photocopies be made of three-dimensional objects? Explain.

7. Would superconducting material be of any use to a person who designs computers? Explain.

8. If you had access to a supercomputer, what sorts of problems would you try to have it solve (aside from next week's homework assignment!)?

9. What sorts of devices would employ computers that had memories stored on

    a. magnetic disks;   **b.** microchips;   **c.** optical disks.

10. Compare and contrast the operation of a computer with that of the human brain.

11. Discuss the extent to which computers have become an integral part of our society. Do you think that this is a positive thing?

12. Discuss the possibility that a ''supercomputer'' might one day take over the world.

13. Are there any limits to the storage capacity of a RAM chip having a given size? Explain.

14. Describe the difficulties that might arise were one to attempt to construct an X-ray or gamma-ray laser.

15. Can you think of an example of a laser-type process that occurs naturally in the physical world. Describe such a process.

16. Can lasers exist that make use of absorption lines rather than emission lines? Explain.

17. Can laser beams ever be seen? If so, under what circumstances?

18. What color(s) might the beam from a hydrogen laser be? (*Hint:* See Chapter 13.)

19. Since laser beams are of such high intensity, does this mean that lasers give off more energy than they absorb? Explain.

20. Can you think of any practical applications for holograms?

21. Describe examples of how advances in one segment of our technology can lead to advances in other segments.

22. Discuss a science fiction story from years past in which a fictional device is described that has now become a technological reality.

23. Pick out a device such as a fax machine that is not mentioned in the text and describe the physical principles by which the device operates.

24. Describe the relationship between science and technology in contemporary society.

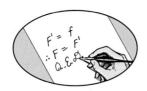

The Electron Microscope

Recall that at the microscopic level of nature every particle has an associated de Broglie wavelength, which is given by

$$\text{wavelength} \ = \ \frac{\text{Planck's constant}}{\text{momentum}} \ .$$

In particular, an electron emerging from an electron gun at a constant nonrelativistic velocity will have an associated momentum equal to the mass of the electron × the electron velocity and an associated de Broglie wavelength equal to

$$\text{electron wavelength} \ = \ \frac{\text{Planck's constant}}{\text{mass} \ \times \ \text{velocity}} \ .$$

If the electrons in a beam emerging from an electron gun are moving at a typical velocity of 5×10^6 m/sec, then the associated wavelength of the electrons in the beam will be

$$\text{electron wavelength} \ = \ \frac{6.63 \ \times \ 10^{-34} \ \text{J/sec}}{9.1 \ \times \ 10^{-31} \ \text{kg} \ \times \ 5 \ \times \ 10^6 \ \text{m/sec}}$$
$$= \ 1.5 \ \times \ 10^{-10} \ \text{m} \ ,$$

a value of the order of a thousand times smaller than the smallest wavelengths of visible light. Thus a beam of electrons emerging from an electron gun cannot only be thought of as a beam of charged particles, but also as a beam of very small wavelength electron waves. Since the degree of diffraction or blurring that occurs in the formation of an image by an optical instrument is reduced proportionally to the reduction in the size of the wavelength, it is theoretically possible to obtain images a thousand times more detailed than those formed at visual wavelengths—if we can somehow focus the electron waves into decipherable images. This is the basic premise of an amazing instrument called the electron microscope.

One version of the electron microscope, called the *transmission electron microscope,* is shown in Figure 14.23 and is essentially an electronic analogue to the optical microscope described in Chapter 12. Electrons from an electron gun are formed into a parallel beam, which is then directed through the sample to be viewed. The emerging electron beam is then focused onto a fluorescent screen for direct viewing or onto a photographic plate for analysis at a later time. The focusing of the electron beam is done by passing the beam through the center of a donut-shaped magnetic lens, an electromagnet whose magnetic field serves to deflect the electrons into a beam pattern. This beam pattern can be converging, diverging, or parallel depending on the size and direction of the current flow through the coil of the electromagnet.

Although the image resolution attained by transmission electron microscopes is often more than a thousand times that attained by optical microscopes, the ability of the transmission electron microscope to produce a good depth of field (proper focus for all parts of a three-dimensional image), is very limited. The thickness of samples prepared for use in transmission electron microscopes usually has to be less than one ten-millionth of a meter thick! The preparation of uncontaminated samples this thin, however, is not performed without considerable difficulty.

Transmission Electron Microscope

Scanning Electron Microscope

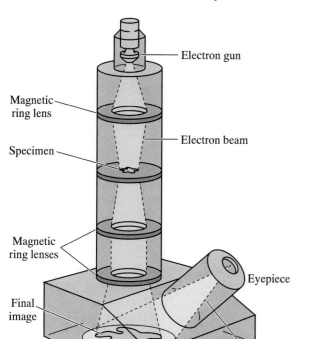
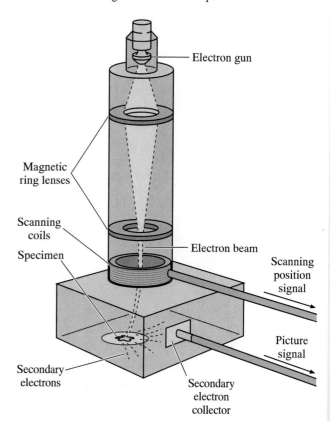

FIGURE 14.23 ··········
The Electron Microscope   Schematic diagrams of a transmission electron microscope (left) and a scanning electron microscope (right).

These difficulties are overcome to some degree by the *scanning electron microscope* in which a given sample is systematically scanned by an electron beam that has been sharply focused onto the sample. As the electrons from this focused primary beam strike the sample, they liberate secondary electrons from the sample itself. The flow of secondary electrons emanating from the surface of the sample varies with the composition and topography of the sample and, as the scan proceeds, the flow of secondary electrons constitutes a variable signal current, which is collected, amplified, and fed into a cathode-ray tube. The cathode-ray tube functions almost exactly the same way as a television tube, in that an incoming signal current is converted into a variable beam of electrons, which produces a variable intensity image on a fluorescent screen. In this case, however, the sweep of the electron beam across the fluorescent screen is identical to the sweep of the primary electron beam as it scans across the sample, thereby generating an image of the sample as it is being scanned.

The scanning electron microscope can thus be employed to analyze the thicker samples that cannot be readily observed using transmission electron microscopes. Unfortunately, the resolution of detail achieved by a scanning electron microscope

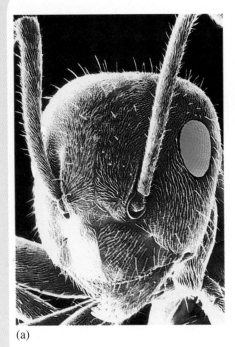

(a)

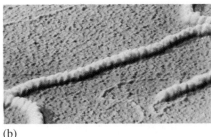

(b)

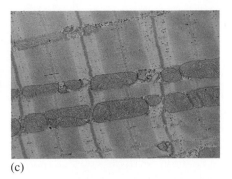

(c)

FIGURE 14.24 ··········

Electron Microscope Photography   Scanning electron microscope photograph of (a) the head of a garden ant (false colored). Transmission electron microscope photographs of (b) the molecular structure of a sample of DNA-protein complex and (c) cardiac muscle tissue.

is only about one-tenth as great as that of transmission electron microscopes, but it's still hundreds of times better than that attained by optical microscopes.

The increased resolution with which the electron microscope permits us to view the microcosmic world has been of inestimable value to many areas of scientific endeavor such as chemistry and solid-state physics, but nowhere has the impact been greater than in the realm of the biological and medical sciences (Figure 14.24). For example, disease-carrying viruses, unresolvable in optical microscopes, can be readily detected with electron microscopes. These high-resolution instruments have also yielded a bonanza of important information about the structure and operation of living cells that simply would not have been otherwise obtained. How many human lives have been saved or extended as a result of the knowledge provided by these remarkable instruments!

QUESTIONS

1. Compare and contrast an electron microscope with the optical microscope described in Chapter 12.

2. Summarize the characteristics of the transmission electron microscope and those of the scanning electron microscope.

3. What are some of the uses that can be made of (a) a transmission electron microscope and (b) a scanning electron microscope.

PROBLEMS

1. Find the de Broglie wavelength of an electron moving at 0.01 times the speed of light.

2. If the wavelength of yellow light is about 5.8×10^{-7} m, at what speeds should electrons in the beam of an electron microscope be moving in order to produce a resolution that is 1000 times better than that possible with yellow light illumination?

3. If the electron speeds in an electron microscope beam are 0.008 that of the speed of light, how does the resolution of this electron microscope compare with that for an optical microscope that uses red light illumination at 6.6×10^{-7} m?

Solid-State Electronics

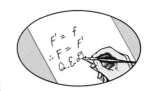

Recall that semiconducting materials in general are substances whose electrical conductivity can be altered by the addition of tiny amounts of impurities to their crystalline structure. In particular, let us consider the Group IV elements, which are located in the column headed by the element carbon in the periodic table. Two of the elements in this group, silicon and germanium, have four valence electrons and four electron vacancies in their outer shells. In forming pure crystals, the atoms of such elements will share each other's valence electrons as covalent bonds. As a result, all of the electrons in a crystal of pure silicon or germanium are tightly bound within the framework of the crystal lattice, and these substances will not readily conduct electric current.

Suppose, however, we subject a crystal of pure silicon to a process called *doping* in which atoms of a second element called the *dopant* are introduced into the crystal as an impurity. In the course of such doping, some of the silicon atoms are replaced at their lattice sites by atoms of the dopant. If the dopant element has five valence electrons, as in the case of elements such as arsenic, antimony, and phosphorus, then at each lattice site occupied by a dopant atom, one electron will be left over after covalent bonding occurs. If an electric potential is applied across the mixture, these "extra" electrons are free to move through the crystal and the system will now conduct an electric current [Figure 14.25(a)].

Because the current-carrying entities in such systems are the negatively charged electrons, a semiconductor so generated is referred to as a negative or *n-type* semiconductor. On the other hand, if the dopant atom has only three valence electrons, such as for the elements aluminum, boron, and gallium, then there will not be enough electrons to form four covalent bonds and an electron vacancy or "hole" will be left over at a given lattice site. A neighboring electron can move into such a hole to complete the covalent bond, but in doing so it must vacate its original position, thereby creating a new hole at a new location in the crystal. Thus if an electric potential is applied to this type of mixture, the electron holes, like the electrons in an *n*-type semiconductor, have the ability to move within the crystal and the system has the ability to conduct electric current [Figure 14.25(b)]. In this case, however, the current-carrying entity is in effect a

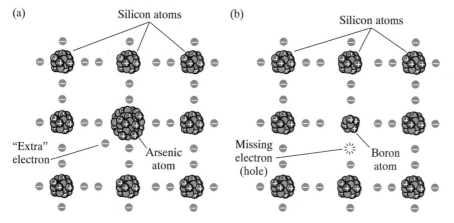

(a) Silicon atoms (b) Silicon atoms

"Extra" electron   Arsenic atom   Missing electron (hole)   Boron atom

FIGURE 14.25 ··········

The Atomic Structure for *n*- and *p*-type Semiconductors   (a) For the *n*-type semiconductor, the extra negatively charged arsenic electron is free to move about the solid when an electric potential is applied. (b) For the *p*-type semiconductor, the extra positively charged electron hole created by a lack of boron electrons can also be transported about the solid by means of an electric potential.

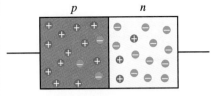

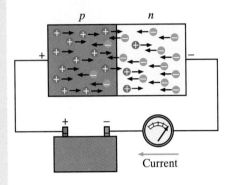

Current

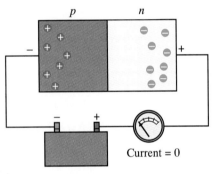

Current = 0

FIGURE 14.26 ···········

The *p-n* Junction When a positive potential is applied to the *p* side, a current will flow as shown. If a negative potential is applied to the *p* side, current does not flow.

missing negative charge or the equivalent of a positive charge. Such semiconductors are thus referred to as positive or *p-type* semiconductor.

If a *p*-type semiconductor is attached to an *n*-type semiconductor, the interface boundary is called a *p-n junction*. If we connect the *p*-type side of the junction to a positive electric potential and the *n*-type side to a negative electric potential as shown in Figure 14.26, then the presence of such an electric potential across the *p-n* junction will cause the electrons in the *n*-type semiconductor to migrate away from the negative terminal and toward the junction boundary. At the same time the holes in the *p*-type semiconductor will migrate away from the positive terminal and toward the junction boundary. At the boundary, free electrons can now fill up vacant holes approaching from the other side of the junction. As the electron jumps across the *p-n* boundary to fill the hole, it is replaced at the negative terminal by an electron that in effect has been pulled out of the *p*-type semiconductor at the positive terminal. The result is a net flow of electric current across the *p-n* junction from the *n*-type semiconductor to the *p*-type semiconductor. If the terminals are reversed so that the positive electric potential is connected to the *n*-type semiconductor and the negative electric potential is connected to the *p*-type semiconductor, we would find that the electrons in the *n*-type semiconductor migrate toward the positive terminal and away from the junction boundary.

In a similar fashion, the holes in the *p*-type semiconductors migrate toward the negative terminal and away from the junction boundary. The net result is the formation of a *depletion region* along both sides of the *p-n* junction across which no electric current is able to flow. The *p-n* junction thus permits the flow of electric current across itself in the direction of the *p*-type semiconductor, but will not allow electric current to flow in the reverse direction toward the *n*-type semiconductor. Such a device is referred to as a semiconductor diode or a junction diode.

If we interface *n*-type and *p*-type semiconductors into a three-layer sandwiched array as shown in Figure 14.27, the resulting system is called a *transistor*. The three semiconductor layers of a transistor are referred to as the *emitter,* the *base,* and the *collector,* where the base is always the center layer of the system. Transistors that have an *n*-type semiconductor as the base and *p*-type semiconductors for the emitter and collector are referred to as *pnp transistors*. Those that have a *p*-type semiconductor as the base and *n*-type semiconductors for the emitter and collector are referred to as *npn transistors*.

To understand how a transistor functions, consider the *npn* transistor shown in Figure 14.28 in which a voltage has been placed across the transistor at the emitter and collector terminals. Such a voltage will, depending on the polarity of the potential, produce a depletion zone at one or the other of the two *p-n* junctions. hence, electric current cannot flow in either direction across the transistor terminals.

FIGURE 14.27 ···········

The Transistor (a) If the emitter and collector are *n*-type semiconductors and the base is a *p*-type semiconductor, then the resulting transistor is said to be an *npn* transistor. (b) If the emitter and collector are *p*-type semiconductors and the base is an *n*-type semiconductor, the resulting transistor is said to be a *pnp* transistor.

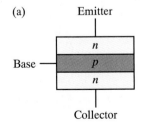

(a)

Emitter

Base —

| n |
| p |
| n |

Collector

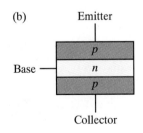

(b)

Emitter

Base —

| p |
| n |
| p |

Collector

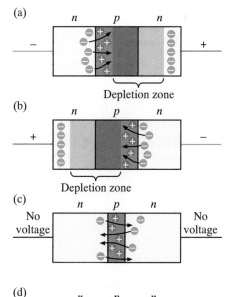

(a)

n   p   n

Depletion zone

(b)

n   p   n

Depletion zone

(c)

No   n   p   n   No
voltage   voltage

(d)

n   p   n

+ Bias

(e)

n   p   n

– Bias

FIGURE 14.28 · · · · · · · · ·

The Operation of a Transistor   If the base layer of a transistor is too thick as in (a) and (b), the depletion zone will be too large in width and no current will flow. If the base layer is very thin as in (c), then it is possible for charges to migrate across the depletion zone, thus enabling a current flow to be controlled by an appropriate bias potential as shown in (d) and (e).

If however, the base layer is very thin and is weakly doped, then it is possible for charge carriers to drift across the base layer. The rate at which this flow of charge carry occurs can be dramatically controlled by placing a *bias voltage* on the base layer. For example, if the base layer has a negative electric potential, the drift of the negative charge carriers will be repulsed away from the base while a positive electric potential will enhance the drift rate of the negative charge carriers toward the base layer. By varying the electrical potential on the base layer of the transistor, the current flow between the emitter layer and the collector layer can be enhanced, reduced, or even shut off. Thus the transistor can function as either or both a current switch or electronic amplifier.

Solid-state electronic circuit elements such as the transistor and junction diode possess considerable advantages. The solid-state elements are durable, compact, reliable, power efficient, and require no warm-up time. Most importantly, solid-state elements can be readily and cheaply incorporated into *integrated* circuits in which all of the component elements of a given electronic circuit, transistors, capacitors, etc., which are required for the circuit to perform its desired function are properly arranged and mounted on a small, single chip of semiconducting material, which is usually silicon. Amazingly, it is currently possible to produce integrated circuits that are so miniaturized that the density of electronic components on a given semiconductor chip is more than 500,000 elements per square centimeter! It is little wonder that solid-state electronic elements have become the backbone of our modern electronics technology.

QUESTIONS

1. Compare and contrast a *p*-type semiconductor and an *n*-type semiconductor.
2. Briefly discuss how junction diodes and transistors operate.
3. Discuss some possible chemical element combinations that could be used in a *p-n* junction.
4. Is it possible to have liquid or gaseous *p-n* junctions? Explain.
5. Draw the analogous diagram to Figure 14.27 for a *pnp* transistor.

$E=mc^2$

In 1894 the American Nobel Prize winning physicist Albert Michelson wrote that ''The more important fundamental laws and facts of physical science have all been discovered . . . the possibility of their ever being supplemented in consequence of new discoveries is exceedingly remote.'' Statements such as this were, of course, born of a naive confidence on the part of nineteenth century science that bordered on arrogance. We have already seen in the last chapter how the ''classical'' physics of the two centuries after Newton began to experience serious difficulties in its attempts to account for phenomena at the atomic level, and in the end, had to give way to the theory of quantum mechanics. The difficulties encountered by classical physics, however, were not confined solely to the realm of the atom. Other effects and phenomena that could not be explained by classical physics became evident by the turn of the twentieth century and inspired the development of the second of the twin theoretical pillars on which modern physics rests, that of relativity theory.

15.1 Macroscopic Failures of Classical Theory

In addition to the difficulties encountered at the microscopic level of atoms and nuclei, classical physics was unable to ''save the phenomenon'' in a number of instances at the macroscopic level as well. Literally and figuratively, perhaps the most glaring example of failed classical theory came in its dealings with that most familiar and important of celestial objects, the Sun. The basic problem was alluded to in an earlier chapter, when we described the explanation offered by the Greek Anaxagoras for solar energy production. In Anaxagoras's view, the Sun was a white-hot sphere of stone. The basic scientific objection to this idea hinged on the fact that such a heated stone would cool off far too quickly to provide a long-term source of heat and light.

While Anaxagoras's explanation is now obviously naive, similar difficulties befell alternate proposals for the Sun's energy production. Astronomers have determined that the Sun radiates energy at a rate of about 4×10^{26} watts. Moreover, the fossil evidence in Earth's strata indicates that the Sun has been shining at or very near this rate for at least several hundred million years. The dilemma created for classical physics by these numbers is simply that no type of energy known in the nineteenth century, including chemical energy, rotational energy, and gravitational potential energy could be stored in the Sun in amounts sufficiently large for the Sun to shine at its present rate for the required length of time. For example, if the Sun generated its energy by means of a chemical reaction that released 10^8 joules of energy for each kilogram of material consumed, the Sun would have a total of 10^7 joules/kg $\times$ 2×10^{30} kg or 2×10^{37} joules of energy available. If it now radiates that energy at a rate of 4×10^{26} joules/sec, the Sun can last for a total of $2 \times 10^{37}/4 \times 10^{26}$ or 5×10^{10} sec, a lifetime of about 1700 years. This, of course, doesn't even cover the span of human history on the planet.

The longest lifetime that could be generated for the Sun from classical energy sources was proposed late in the nineteenth century by the English physicist James Jeans, who theorized that the Sun was slowly shrinking, and in the process was converting gravitational potential energy into electromagnetic energy.

Unfortunately, in such a scenario, the radius of the Sun would have had to be equal to the radius of Earth's orbit only 20 million years ago, a time span that would not even reach one-third of the way back to the age of the dinosaurs in our geological history. Clearly a source of energy far different from anything classical physics could offer was powering the Sun and the distant stars.

Orbiting the Sun at a mean distance of some 58 million kilometers is the tiny, airless, and rugged planet Mercury. The planet itself is not particularly noteworthy for our purposes except for the fact that its motion about the Sun exhibits some peculiar characteristics. As Mercury orbits the Sun, the line joining the near and far points of its orbit or *line of apsides* changes its orientation in space. Most of this apsidal motion can be attributed to small-scale gravitational tugs from other planets, most notably Venus and Earth. However, a small fraction of this apsidal motion—about $1/100$ of a degree per century of it—cannot be attributed to gravitational forces from known objects in the solar system.

Fresh from a magnificent mathematical triumph in which he correctly co-predicted the existence and location of the planet Neptune with John Adams of England, the French mathematician Urbain Leverrier hypothesized in 1850 that the anomalous motions of Mercury were due to an as-yet undiscovered planet between Mercury and the Sun, which was even prematurely named Vulcan (but not, of course, the Vulcan Mr. Spock hails from!). Unfortunately, the Newtonian mechanics that served Leverrier so very well in his predictive discovery of Neptune, deserted him in the case of Vulcan. Despite a number of carefully conducted searches, astronomers could not find Vulcan and were forced to conclude that the planet simply did not exist. But if Vulcan did not exist, what then was the cause of the anomalous behavior in Mercury's orbital motion? Once more classical physics was unable to provide an acceptable answer.

Back on Earth, meanwhile, in the last years of the nineteenth century, Albert Michelson and American chemist Edward Morley designed an experimental apparatus that allowed them to measure the speed of light to a degree of precision not previously attained. In the course of their work, they made the remarkable experimental discovery that the measured value of the speed of light in a vacuum is always the same, regardless of the relative motion between the source of the light and the observer. This observation is the result of an experiment often referred to as the Michelson-Morley experiment, and it is, of course, totally at odds with the principles of Newtonian mechanics.

Consider, for example, Figure 15.1 in which a spaceship is approaching an asteroid with a landing light on at a velocity equal to $c/2$ or half the speed of light. Newtonian mechanics would predict that the observer on the asteroid would measure the velocity of the approaching wave front to be $c + c/2$ or $3c/2$. The

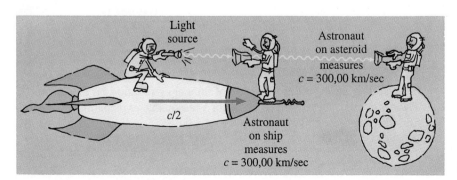

Light source

Astronaut on asteroid measures $c = 300,00$ km/sec

$c/2$

Astronaut on ship measures $c = 300,00$ km/sec

FIGURE 15.1 ·············
The Constancy of the Speed of Light Even though the spaceship is approaching the asteroid at half the speed of light, both astronauts will measure the same value of 300,000 km/sec for c.

observer on the spaceship, on the other hand, measures the velocity of the light waves to be *c*. The empirical result is that both observers measure a value exactly equal to *c*! Also implied by this result is the deduction that no entity in nature has the ability to exceed the speed of light. To date, no form of energy or particle has been found that violates this "ultimate" speed limit. This fact, as we shall see, suggests that the familiar three dimensions do not provide an adequate coordinate description of events and objects in space.

In light of these and other contradictions and discrepancies between observation and theory in the macroscopic world, it was plain to a careful observer at the turn of the twentieth century that a new and fundamentally different view of the physical world was sorely needed. That view was to be provided by a young Bavarian named Albert Einstein who, as we shall see in the next section, published a series of papers from 1905 to 1916 in which he set forth a new world view called relativity theory.

WORKED EXAMPLE

Diesel fuel gives off 4.4×10^7 joules of energy for each kilogram of fuel burned. Assuming the Sun is a gigantic 2×10^{30} kg ball of diesel fuel, how long can the Sun shine at its present rate of 4×10^{26} watts? Is this a reasonable value? Explain.

SOLUTION: Since the Sun has a mass of 2×10^{30} kg, and each kilogram is capable of liberating 4.4×10^7 joules of energy, the total energy available to the "diesel-ball" Sun is

$$\text{total available energy} = 2 \times 10^{30} \text{ kg} \times 4.4 \times 10^7 \text{ J/kg}$$
$$= 8.8 \times 10^{37} \text{ J} .$$

The Sun, however, is radiating energy at a rate of 4×10^{26} watts, or 4×10^{26} J/sec. The lifetime of the "diesel-ball" Sun is thus

$$\text{lifetime} = \frac{\text{total available energy}}{\text{rate energy is radiated}}$$
$$= \frac{8.8 \times 10^{37} \text{ J}}{4 \times 10^{26} \text{ J/sec}}$$
$$= 2.2 \times 10^{11} \text{ sec} .$$

There are 3.15×10^7 seconds in one year, so

$$\text{lifetime} = 2.2 \times 10^{11} \text{ sec} \times \frac{1 \text{ year}}{3.15 \times 10^{7 \text{ sec}}}$$
$$= 0.70 \times 10^4 \text{ yrs} = 7 \times 10^3 \text{ yrs} .$$

The "diesel-ball" Sun can thus shine for a total of 7000 years, a time interval that barely covers the span of human civilization on the earth. Therefore, the "diesel-ball" Sun is not a reasonable explanation for the Sun's energy production.

1. Why can't the Sun be shining from chemical energy?
2. What are some other possible mechanisms by which the Sun could produce its energy?
3. Discuss the theoretical difficulties that the orbital motion of Mercury presented to nineteenth century astronomers.
4. Describe two of the characteristics of the speed of light that are in disagreement with the principles of Newtonian mechanics.
5. Summarize some of the phenomena in nature that cannot be explained by Newtonian physics.

1. A certain solid fuel compound gives off 1.0×10^{11} joules of energy for each kilogram burned. Assuming the Sun was composed entirely of this compound when it was formed, how long can the Sun shine at its present rate if its energy is produced by the burning of this compound?
2. If the Sun were initially rotating at its "breakup" rotational velocity, it would have approximately 7×10^{40} joules of energy available to radiate as heat and light. How long can the Sun last on this supply of energy if it shines at its present rate of 4×10^{26} watts? Do you think this is a reasonable lifetime for the Sun?

15.2 Special Relativity

The development of relativity theory was, in essence, a two-step process. In 1905 Albert Einstein described the behavior of quantities measured in "inertial" reference frames or reference frames moving at constant velocities relative to one another. This theory is known as the *special theory of relativity*. Special relativity is based on two fundamental ideas. The first is often called the *principle of relativity* and can be stated as follows:

> All laws of nature have exactly the same form in any reference frame that is moving at a constant velocity.

In effect, the principle of relativity tells us that the idea of the universality of nature's laws applies not only to one's time and location in the universe but also to any reference frame moving at a constant velocity. It is a reasonable assumption to make, and is certainly a necessary one if we are to avoid the chaotic view that each little reference frame in each little corner of the universe is governed by a different set of natural laws.

The second basic postulate stated by Einstein in his theory of special relativity was the assumption that:

> The measured value for the speed of light in a vacuum will be the same for any observer in any reference frame moving at a constant velocity.

15.1 Albert Einstein (1879–1955)

Perhaps no single human being is more representative of the embodiment of the science of the twentieth century than Albert Einstein. Born of Jewish heritage in the German province of Bavaria in 1879, Einstein's early years were remarkable by their seeming lack of academic and scientific promise. At one point, he dropped out of secondary school and finally managed to graduate from the University of Zurich in Switzerland in 1900. Because of his lackluster performance at Zurich, he was unable to obtain an academic position and instead settled for a position at the Swiss patent office in Bern.

The fortuitous combination of large amounts of free time and the quiet isolation afforded by this position provided Einstein with the opportunity to develop a number of radically new scientific concepts in an atmosphere devoid of academic distractions and paperwork. In 1905 Einstein published his theory of special relativity and a scant 11 years later his more comprehensive theory of general relativity. Unlike the development of quantum mechanics, which was the result of a sort of "corporate" effort on the part of a number of different scientists, relativity theory is the creation of Einstein alone. The ongoing empirical verification of the wondrous effects and phenomena predicted by that theory many decades ago has built an enduring monument to him which is anything but lackluster.

••• **Albert Einstein (1879–1955)**

Interestingly, in his original formulation of special relativity, Einstein advanced the idea of the constancy of the speed of light as a postulate. Historical evidence indicates that he was almost surely unaware of the results of the Michelson-Morley experiment when he published his first paper on special relativity in 1905, since he makes no mention of them at that time. It is thus a fine tribute to Einstein's insights that what he intuitively perceived as theoretical truth turned out to be an experimental truth as well.

Let us examine the consequences of these two ideas. We have already described the dilemma created by two moving observers as a result of the constancy of the speed of light (Figure 15.1). In a similar vein, suppose we have a pair of charges moving in opposite directions at speeds of $v/2$, as shown in Figure 15.2. To an "outside" at-rest observer such as the reader, two forces appear to be acting on the negative charge: the coulomb electrostatic force and the electro-

PHYSICS FACET

Continued

Ironies abound in Einstein's career. Despite his famed "God does not play dice with the universe" lament against the ideas and concepts of quantum theory, Einstein won the Nobel Prize in 1921, not for his theories of special and general relativity, but for his work on the photoelectric effect in which he demonstrated that light behaves as if it were indeed composed of the dreaded quanta.

By Einstein's own admission, the "worst mistake of my career" came when he incorporated the infamous cosmological constant into his general relativity theory in order to explain a universe that seemed to be static in the face of the prediction by his own general relativity theory that the universe must be expanding. Even as Einstein was proceeding with this exercise in "mathematical prestidigitation," the American astronomer Vesto Slipher was obtaining the first glimmerings that the universe was indeed expanding, an observational fact that another American astronomer Edwin Hubble would firmly nail down a generation later.

Because Einstein came to personify twentieth century science more than any other individual, the stereotypical scientist is more often than not endowed with a variety of "Einsteinian" qualities, including rumpled clothes, unruly hair, and the like. The detached, socially insensitive dimension of the scientific stereotype, however, is cruelly ironic in the case of Einstein. Although his causes were not always popular, his social voice was nonetheless in evidence, whether it was a pacifist voice in 1914 Germany, a voice forced to leave its native homeland forever with the Nazi ascendancy in 1933, or a pro-disarmament and anti-McCarthy voice in the America of the early 1950s. On the eve of America's entry into World War II, Einstein urged the development of the atomic bomb, lest Western civilization be caught naked before a similarly armed Adolf Hitler. Yet, this same man would personally apologize in tears to the visiting Japanese Nobel Prize winning physicist Hideki Yukawa when that weapon ultimately came to be used against a far less sinister and dangerous adversary. Hardly the type of scientist one sees in the late-night sci-fi flicks!

magnetic force produced by the motions of the two charges. If we were situated on the negative charge, however, the positive charge would appear to be moving toward us at a velocity v. The coulomb force between the charges would still be observed, but there would be no electromagnetic force because in this frame the velocity of the negative charge would be zero. Therefore, it would appear that for electromagnetic phenomena the principle of relativity does not hold up because two observers situated in two frames of reference moving at constant velocities appear to observe two different force laws.

In both cases, the dilemma is created by the fact that we are simultaneously requiring that the principle of relativity and the constancy of the speed of light be satisfied along with the idea that the fundamental measurable quantities of nature, time, length, and mass are also invariant with constant velocity reference frames. In the Newtonian view of nature, units such as the meter, kilogram, and

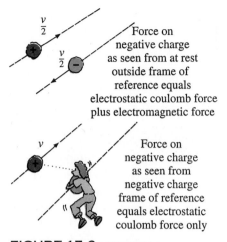

FIGURE 15.2 ··········
The Relativity of Electric and Magnetic Fields To the reader, the total force between the moving charges consists of a coulomb electrostatic force and a current–produced electromagnetic force [see Figure 11.7(b)]. An observer situated on one of the charges sees only a coulomb electrostatic force.

second are assigned an absoluteness and constancy that permits their use in a wide variety of situations. Thus, the seconds ticked off by a watch on board a speeding spaceship are, in the Newtonian view, the same as those ticked off by an identical watch on Earth. But, as our two dilemmas suggest, we can't have it both ways.

Einstein essentially opted to advance the very radical idea that the values we measure for quantities such as time, length, and mass are not invariant with the motions of reference frames, but, quite the contrary, are intimately dependent on such motions. Thus, the time kept by a clock moving with respect to a stationary observer is perceived by that observer to run more slowly than an identical clock stationary at the observer's side (Physics Formulation 15.1). Lengths of moving objects are perceived as being shorter than the lengths of identical stationary objects, and so on. Such an approach quickly resolves the two dilemmas proposed above. For example, in the case of the spaceship bearing down on the asteroid, the reason both observers measure the same value for the speed of light is due to the fact that the measured quantities by which that velocity is determined, length and time, are perceived differently by the two observers. But the difference is such that a constant value of c is maintained for the light's distance traveled/ elapsed time ratio. In a similar fashion, the "relative" nature of length and time measurements serves also to resolve the seeming contradictions generated by electric charges in motion.

At first glance, the idea that our perception of time, length, and mass depends on our frame of reference may seem very distant from our "everyday" experiences, but it really isn't. For example, our perception of the passage of time may vary considerably depending on what we are experiencing or, if you like, what our "frame of reference" happens to be. Time always seems to pass far too quickly when a student is taking a difficult exam and far too slowly during a tooth extraction.

The basic principles of special relativity have led to a number of other interesting and important results. In one extraordinary result, Einstein was able to demonstrate that mass and energy are in effect equivalent quantities, which are related to one another by what is perhaps the most widely known equation in the history of science:

$$E = mc^2 ,$$

or the energy E locked up in a mass m is equal to the product of the mass m and the square of the speed of light c. Mass thus can be thought of as a sort of "freeze-dried" form of energy. Because of the large size of the square of the speed of light, a rather small amount of mass is capable of becoming a tremendous amount of energy. It is precisely by making such a conversion that the Sun and distant stars are able to generate their large amounts of energy over geological time scales. It is also the development in recent decades of such mass-to-energy conversion techniques that has, for better or for worse, placed virtually limitless power in the hands of human beings.

As we have already indicated, the constancy of the speed of light in all reference frames moving at constant velocity implies that the speed of light also represents an "ultimate" speed limit in the universe. Since light travels some 7.5 times around the world in one second, any light-propagated event will be transmitted virtually instantaneously around the globe. As a result, a terrestrial event such as a volcanic eruption, appears to occur at virtually the same time for everyone on the earth. The situation in outer space, however, is a far different matter. Even moving at a rate of 300,000 km/sec, it takes sunlight over eight minutes to make the 150 million kilometer journey from the Sun to Earth and over five hours to reach Pluto, the last known outpost of the solar system. On leaving the solar system, it takes sunlight over four years to reach the nearest star and some 30,000 years to reach the center of the Milky Way galaxy about which our Sun revolves.

So vast are the distances to celestial objects that astronomers often measure them in units of the *light year.* One light year is defined as the distance light travels in a year or about 9×10^{12} km. Such vast distances combined with the idea of an ultimate speed limit raise some interesting points regarding our perception of ''space'' in the universe. Suppose, for example, that a certain star is located 100 light years from the earth. Since by definition 100 light years is the distance light travels in 100 years, the light by which we see that star left the star 100 years ago. Were the star to explode, the light from that event would take 100 years to reach us. We are thus left with the eerie possibility that some of the stars we now observe may in fact no longer exist!

As a result of such considerations, Einstein concluded that our perception of the time at which a given event occurs is intimately tied into our location in the universe (Figure 15.3). In short, there is no ''absolute'' or ''standard'' time that exists throughout the universe. Einstein reasoned that the addition of a fourth dimension, time, to create a *space-time* geometry of the universe thus provides a more complete description of that universe. The above star, for example, may be located at a fixed location in spatial coordinates throughout its life cycle from initial formation to final explosion, but depending on the location of a given observer in space-time, the star may exist, may not yet exist, or may have ceased to exist for that observer. If we were to take a journey in a starship at relativistic speeds, when we returned to Earth, we would find that although we have returned to the same spatial position, i.e., Earth, we have returned to an Earth and its inhabitants that have aged more rapidly than we. In other words, the earth would now be in a different time coordinate of space-time.

One of the noteworthy aspects of time in our space-time universe is the fact that it always flows in a forward or ''positive'' direction. Thus, regardless of how we took our space voyage, on our return, Earth would always be farther along in time than we. In this sense, we can always travel forward in time, but never backward. Suppose, for example, that we journeyed toward our exploding star and intercepted the radiation from that explosion at a distance of 50 light years (Figure 15.4). Suppose further that, on seeing the event, we determined that Earth's inhabitants must be warned. If we turned our ship about and tried to return to Earth in time to warn of the imminent blast, we would find that the best speed we could manage is that of light from the blast itself. As a result, we would arrive at the earth behind the light from the blast and, hence, would fail in our effort. We learned of the event 50 years before the earth's population but had no way of moving backward in time to warn them.

This observed forward flow of time in turn gives rise to the validity of an important concept in science called *causality,* which simply states that any effect cannot precede its cause. Thus, it is impossible, for example, to travel backward in time and have a fatal automobile accident with oneself. So as much as we would love to be able to go back in time and retake that flunked final exam or relive that ecstatic weekend with the person of our dreams, causality says we can't, except in our memories.

Interestingly, relativistic calculations allow for the possibility of an ''inverse'' universe in which time runs backward, and the speed of light represents a lower bound to the speeds in this other universe. Particles that would exist in such a

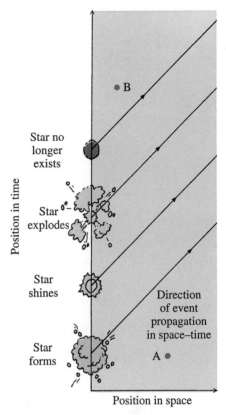

Position in time

Star no longer exists

Star explodes

Star shines

Star forms

B

Direction of event propagation in space–time

A

Position in space

FIGURE 15.3

Space-Time Because of the finite velocity of light, and the vastness of outer space, our perception of events depends not only on our spatial position, but also on our position in time. In this case, an observer situated at position A in space-time does not know that the star has yet formed, while an observer situated at position B in space-time has seen the star go through its entire life cycle.

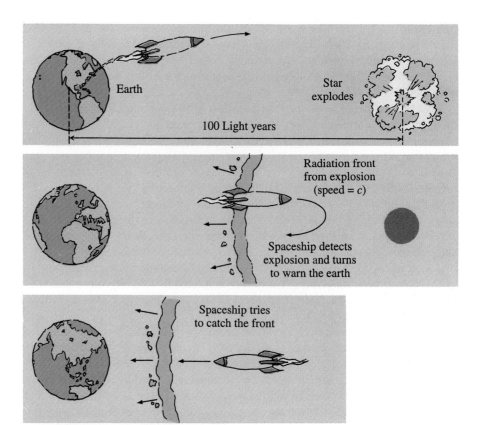

FIGURE 15.4 ··········

A Race Against Space-Time   Even
though the spaceship detects the radiation
front coming from an exploding star
some 50 years before it arrives at the
earth, the spaceship cannot move any
faster than *c,* and therefore cannot
overtake the radiation front in order to
warn the earth in time.

universe have even been named ''tachyons,'' but to date there is no experimental
evidence for the actual existence of tachyons or the universe that they presumably
populate. At this juncture, we may also speculate about whether or not our universe
can be even better described by the addition of other dimensions whose nature
has yet to be ascertained.

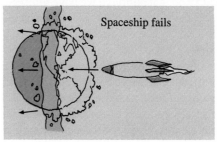

WORKED EXAMPLE

In a certain nuclear reaction, the total mass of the material before the
reaction was found to be 236.053 g, and 236.022 g afterward. How much
energy was released in this reaction?

Continued

WORKED EXAMPLE

Continued

SOLUTION: The mass has been reduced by an amount equal to 236.053 − 236.022 or 0.031 g, which is 3.1×10^{-5} kg. If we assume that this "missing mass" has been converted into its equivalent energy, then we have

$$E = mc^2$$

and for this case,

$$E = 3.1 \times 10^{-5} \times (3 \times 10^8)^2 \text{ J}$$
$$= 3.1 \times 10^{-5} \times 9 \times 10^{16} \text{ J}$$
$$= 2.79 \times 10^{12} \text{ J}$$

Thus, 2.79×10^{12} joules of energy was liberated in this reaction.

---

1. What are the basic assumptions made in special relativity?
2. Compare and contrast the Newtonian and relativistic views of time, length, and mass.
3. How are mass and energy related to each other in relativity theory? How does this relationship compare with that between mass and energy in Newtonian theory?
4. Why is there no "absolute" or "standard" time in the universe?
5. Discuss the concept of causality. Give an example of causality in your everyday life.

REVIEW
15.2
QUESTIONS

---

1. The mass of the earth is about 6×10^{24} kg. How much energy would be produced if the earth's mass could somehow be converted into its equivalent energy?
2. How much mass would be required to produce a total of 1000 joules of energy if the mass were completely converted into its equivalent energy?

REVIEW
15.2
PROBLEMS

15.3 Experimental Verification of Special Relativity Theory

Perhaps the most amazing aspect of relativity theory is that virtually all of its marvelous predictions have thus far been verified empirically.

The earth's atmosphere, for example, is bombarded by high-energy atomic particles from outer space, which are called cosmic "rays." Composed primarily of high-speed hydrogen and helium nuclei (protons and alpha particles), these particles crash into the upper layers of the earth's atmosphere, fragmenting the nuclei of the atmospheric atoms. As a result of this fragmentation process, a type of particle called a μ-meson (see Chapter 13) is produced at an altitude of about 10,000 meters above the earth's surface and travels at a rate of $0.999c$.

These particles can also be produced in laboratories here on the earth. In the at-rest earth-based frame of reference, the average lifetime of a μ-meson is about 2×10^{-6} seconds. The distance an average μ-meson can thus travel through the earth's atmosphere before decay is $0.999 \times 3 \times 10^8$ m/sec $\times 2 \times 10^{-6}$ sec or about 6×10^2 meters. If the at-rest lifetime were correct, we should not be able to observe any μ-mesons at the surface of the earth. Since we do observe these particles at the earth's surface, something is wrong. The difficulty is resolved if we recalculate the lifetime of the μ-meson taking into account the fact that it is in a frame of reference that is moving toward an observer on the surface of the earth at a rate of $0.999c$. Because of that relative motion, we find that the earth-based observer now sees a longer lifetime for the μ-meson that is equal to 5×10^{-5} sec rather than the 2×10^{-6} sec observed for the at-rest μ-mesons.

As a result, the average μ-meson moving at $0.999c$ now appears to the earth-based observer to travel $0.999 \times 3 \times 10^8$ m/sec $\times 5 \times 10^{-5}$ sec or about 15,000 meters before decaying. The average μ-meson can thus strike the surface of the earth before it decays, in accordance with observed results (Figure 15.5). This same result must also be observed by someone who is moving along with the μ-meson. Since that observer is at rest relative to the average μ-meson, he or she must measure a value of 2×10^{-6} sec for the lifetime of that μ-meson, or the same as the at-rest lifetime measured by the observer on the earth's surface. Because of the motion of the observer "on board" the μ-meson relative to the

FIGURE 15.5 ···········
Time Dilation and Length Contraction in Relativity Theory   In the left frame, μ-mesons, as seen from the earth's surface, have lifetimes that are time-dilated, thereby permitting them to strike the earth's surface. In the right frame, the earth's atmosphere as seen from the incoming μ-mesons has a contracted thickness, which permits them to strike the earth's surface in an at-rest lifetime for the μ-meson. In both cases, μ-mesons strike the earth's surface, a result which is in agreement with observation.

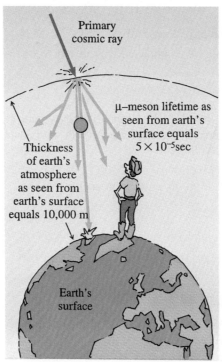
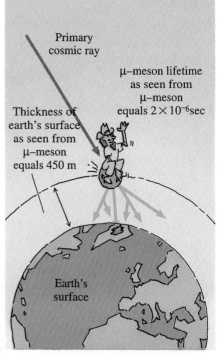

earth's atmosphere, however, the depth of the atmosphere appears to that observer to be about 450 meters. Since the average μ-meson can travel about 600 meters before decaying in this frame of reference, it too will strike the earth's surface. Thus, while the thickness of the earth's atmosphere and the decay time for the average μ-meson are perceived differently by the two observers, both observe the same empirically verifiable result that the average μ-meson strikes the earth's surface. Such could not be the case if the time dilations and length contractions predicted by relativity theory did not occur.

As indicated earlier, special relativity also predicts that the mass of a particle moving past an observer will be perceived as having a larger mass than an identical particle at rest relative to the observer. This prediction of special relativity has been verified through the use of nuclear particle accelerators of various types.

One example of such an accelerator, called a cyclotron, consists of a pair of hollow, dee-shaped plates, one of which is electrically charged positively and the other negatively. Both dee plates are then evacuated and placed in a uniform magnetic field (Figure 15.6). When a charged particle is placed in the center of the cyclotron, it will move toward the dee plate having the opposite electrical charge. As it does so, the magnetic field will exert a force on the moving charge and cause it to move in a curved path. As the charged particle approaches the oppositely charged dee plate, the cyclotron suddenly reverses the electrical charge on the dee plate, so that the particle is now attracted to the other dee plate.

By alternating the charge of the dee plates at just the right frequency, electrons and other charged particles can be accelerated to enormous velocities. In the course of such experiments, it has been found that as the particle accelerates to

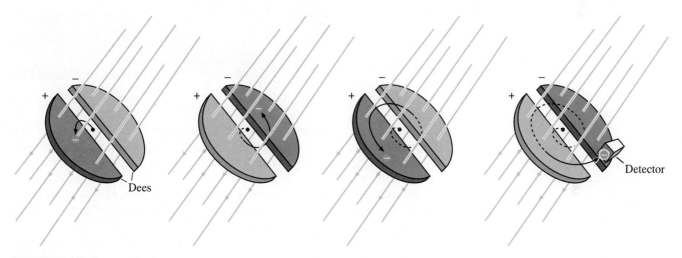

FIGURE 15.6 ··········
The Operation of a Cyclotron   When an electrically charged particle such as an electron or proton is introduced into the center of the cyclotron, it is pulled toward the oppositely charged dee plate. As the particle moves toward the dee plate, it is deflected into a curved path by the presence of a magnetic field. As the particle nears the dee plate, the electrical charge polarity is reversed. The particle is now attracted to the second dee plate and the process is repeated. By alternately changing the charge polarity at just the right times, the particle can be accelerated to enormous velocities.

FIGURE 15.7 ··········

A Nuclear Explosion The most frightening experimental verification of the principles of special relativity is the conversion of mass into its equivalent energy via a nuclear detonation.

relativistic speeds, the timing of the dee plate charge reversal is thrown off by an increase in the observed masses of the accelerated particle. If the relativistic mass increase is allowed for in setting the timing of the dee plate charge reversal, the cyclotron works perfectly. As powerful and sophisticated as these particle accelerators can be, however, none of them has ever been able to accelerate a particle to speeds equal to the speed of light. Thus, the speed of light continues to be the ultimate speed limit in the universe, both theoretically and empirically.

A wide variety of nuclear interactions exists in which the equivalency of mass and energy has been empirically verified. High-energy gamma rays, for example, can be made to interact in such a way that particles such as electrons and positrons can be produced. The most familiar and perhaps the most terrifying manifestation of relativistic mass-energy equivalence are the fission and fusion reactions in nuclear weapons in which vast amounts of energy are released by converting mass into its equivalent energy (Figure 15.7). In all cases, whether mass is converted into energy or vice versa, the interaction always occurs in accordance with $E = mc^2$.

1. How do cosmic rays demonstrate time dilation in special relativity?
2. Describe an experiment in which the perceived mass increase for a moving particle predicted by special relativity is observed to occur in the laboratory.
3. Explain how a cyclotron operates.
4. How can a cyclotron or other particle accelerator be used to demonstrate that the speed of light is the ultimate speed limit of the universe?
5. In what way(s) do nuclear reactions demonstrate the validity of special relativity theory?

15.4 General Relativity

The theory of special relativity, as we have seen, deals with the consequences of observations made from inertial or constant velocity frames of reference. Scarcely a decade after publishing his theory of special relativity, Einstein considered the consequences of observations made from reference frames that were accelerating rather than moving at constant velocity. The result of those efforts was a new view of the phenomenon of gravity, which we now call the theory of general relativity.

The basic principle underlying general relativity is the *principle of equivalence,* which in essence states that there is no way to observationally distinguish an accelerating reference frame from one that is inertial, but in a constant gravitational field. To illustrate this concept, consider the astronauts shown inside a sealed spaceship, which is at rest on the surface of the earth (Figure 15.8). The astronauts experience the "tug" of gravity and can perform a number of experiments with falling bodies that establish the existence of a gravitational acceleration. Suppose now the astronauts nap for a time during which the spaceship is transported to deep space, far from any gravitating celestial objects, and then accelerated in the spaceship's "upward" direction at exactly 9.8 m/sec², the acceleration of gravity at the earth's surface.

Upon awakening, the astronauts would again feel the "downward" tug just as we experience a downward tug when an elevator accelerates upward. Since the upward acceleration is 9.8 m/sec² the downward "pull" experienced by the astronauts would be exactly that which was experienced when the ship was on the earth's surface. Any object that is released inside of the spaceship cabin would appear to an observer outside of the ship to float or remain motionless with respect to the at-rest outside reference frame. The "floor" of the cabin, however, would accelerate toward the floating object and as a result, the astronauts inside the cabin would perceive the object as "falling" to the floor in a direction opposite the ship's 9.8 m/sec² upward acceleration.

Such a result would be the same for *any* object released inside the ship's cabin. Thus, if two objects of unequal mass were simultaneously released inside the cabin, they would appear to fall toward the cabin floor at exactly the same

The Principle of Equivalence   According to the principle of equivalence, the astronauts cannot tell the difference between a reference frame that is accelerating (left) and an at-rest frame located in a gravitational field (right).

rate, just as they do here on the earth. In short, there is no experiment that the astronauts could devise that would permit them to tell whether they were in a sealed spaceship resting on the earth's surface or in a sealed spaceship accelerating in deep space at a rate equal to the 9.8 m/sec^2 acceleration of gravity at the earth's surface.

Surely there must be some way for the astronauts to decide. Suppose that in deep space a flash of light moving horizontally were to enter the cabin just as the accelerating spaceship passed by as shown in Figure 15.9. In the time required for the light flash to move in a straight line across the cabin, the ship also would accelerate upward. As a result, the astronauts would observe the light flash to "drop" or "deflect" downward from a horizontal straight-line path. A similar effect would be observed if a particle moving in a horizontal straight-line path were to enter the cabin.

Suppose now that the ship were returned to earth and the particle and light flash entered the cabin of the spaceship moving horizontally in a straight line. The particle would be tugged on by the earth's gravitational field as it moved across the cabin, once more would deflect downward, and the equivalence principle would be satisfied. The light flash, however, has no mass and therefore should not be deflected as it crosses the earthbound spaceship cabin. Surely this is an experiment by which the equivalence principle could be violated. At this juncture, Einstein offered several profound insights.

First of all, he concluded that the equivalence principle holds not only for phenomena involving gravitating masses, but for *any* phenomenon in nature.

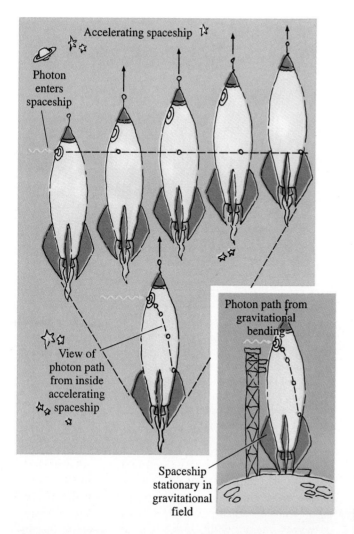

Accelerating spaceship

Photon
enters
spaceship

View of
photon path
from inside
accelerating
spaceship

Photon path from
gravitational
bending

Spaceship
stationary in
gravitational
field

FIGURE 15.9

Equivalence and the Gravitational Bending of Light If a beam of light enters an accelerating spaceship, it appears to deflect toward the bottom of the spaceship. If the spaceship is stationary in a gravitational field, the principle of equivalence requires that the light beam be somehow deflected by the gravitational field.

Thus while the light flash had no gravitating mass, it could nevertheless be deflected somehow by the earth's uniform gravitational field. This, in turn, was made possible by the fact that the presence of mass in space-time causes a "warping" or distortion of the space-time in the vicinity of that mass. This is an amazing view of nature that has far-reaching theoretical and experimental consequences. Basically Einstein portrays space-time as a gigantic topographical map whose hills and valleys are determined by the presence or lack of gravitating masses. These space-time hills and valleys in turn determine the way a given mass, such as the planet Mercury, moves in space-time. Thus in the Einsteinian view of the universe, a symbiotic relationship exists between space-time and mass. Mass determines the topography of space-time and the topography of space-time determines the behavior of mass. Such distortion effects are generally not all that large, and they are virtually nonmeasurable for the space-time in the vicinity of the earth's gravitation field. Einstein, however, argued that space-time in the vicinity of larger, more massive objects such as the sun is sufficiently large to permit the detection of such light bending—if it exists.

From special relativity we can also see that if our spaceship is accelerated in deep space, on-board clocks would appear to run more slowly than the clock

15.2 Gravity Waves

One of the more unusual consequences of the theory of general relativity is the possible existence of what Einstein referred to as *gravity waves*. As we have seen, space-time is distorted in the vicinity of a massive object and is undistorted if no mass is present. Suppose now that a massive object exhibits some sort of vibrational motion, such as mutual orbital motion with a second massive object. As the object moves about in its orbit, the distortions of space-time produced by its mass move along with it. Thus the vibrational motion of the massive object induces a vibrational distortion of space-time in the vicinity of the orbit of the object, as shown in the figure.

Analogous to Maxwell's work a half-century earlier, Einstein was able to demonstrate that, like electromagnetic waves, these vibrational distortions of space-time or gravity waves would not remain confined to the vicinity of the oscillating mass but would instead radiate outward from the mass at the speed of light. Gravity waves would thus propagate through space as ''ripples'' in space-time. In theory, as these ripples pass through space-time they should be able to induce oscillations in a given mass or array of masses, just as electromagnetic waves can cause electrical charges to oscillate.

Unfortunately, any potential source massive enough to produce detectable gravity waves must be celestial in nature and thus very far away from the earth. This in turn means that even if a vibrating mass produces gravity waves of enormous power, by the time those waves reach the earth, their intensity will have been considerably reduced. For example, one of the potentially more powerful sources of gravity waves is a binary star called UV Leonis. Theoretically, this system radiates gravity waves having a power

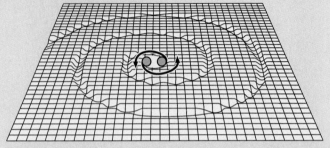

∙∙∙ Gravity Waves If a mass oscillates in space-time, such as in a closed orbital motion, general relativity theory predicts that the resulting disturbances in space-time will propagate outward in all directions at the speed of light, just as electromagnetic waves will propagate outward from an oscillating electric charge. In this three-dimensional analogue illustration, the gravity waves show up as distortions in the ''straight-line'' fabric of the space-time grid.

of about 6×10^{23} watts. By the time these gravity waves have traveled the 220 light year distance from UV Leonis to the earth, however, the intensity of these waves is only 10^{-10} watts per square *kilometer!* Moreover, once the gravity waves interact with masses here on the earth, that interaction is about 10^{36} times weaker than electromagnetic wave interactions. To add to the woes of the experimental physicist attempting to detect these tiny gravitating mass oscillations are the myriad of everyday shakes, jiggles, and bumps that tend to screen off any vibrational effects that might be due to gravity waves. The unambiguous detection of gravity waves thus represents a most difficult experimental problem that has not yet been solved.

of an at-rest observer due to the velocity associated with that acceleration (Figure 15.10). If the ship returned to the earth's surface and was now at rest with respect to our outside observer, the on-board and at-rest clocks would run at identical rates. The astronaut on board the spaceship however would be able to detect the difference between the rate at which the clock ran when it was accelerated in outer space and the rate at which it runs on the earth's surface. Einstein resolved this seeming violation of the equivalence principle by predicting

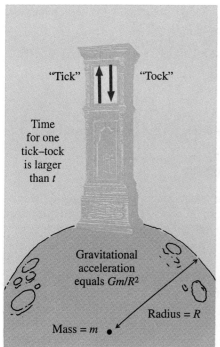

FIGURE 15.10 ·············

Time Dilation in a Gravitational Field If a clock is placed in a gravitational acceleration of GM/R^2, it runs more slowly by a ratio $t'/t = 1/\sqrt{1 - 2GM/Rc^2}$.

that a clock placed in a gravitational field would run more slowly compared to a clock in an at-rest reference frame away from the gravitation. Thus, the rate at which the astronaut's clock runs stays the same for the astronaut regardless of whether he or she is accelerating in deep space or is at rest in the earth's gravitational field. This leads to the remarkable prediction that the rate at which a clock runs is dependent on the intensity of the gravitational field.

In a similar fashion Einstein argued that as light leaves a gravitating mass its wavelength should become slightly larger in value, and its frequency correspondingly smaller. This effect, called the *gravitational red-shift* (Figure 15.11), is too small to be readily observed in the vicinity of the earth, but should be detectable in the light from celestial objects having high-intensity gravitational fields.

As he had done with his theory of special relativity, Einstein used his theory of general relativity to predict a number of amazing and seemingly fantastic effects and, once again, as we shall see in the next section, his predictions proved to be correct.

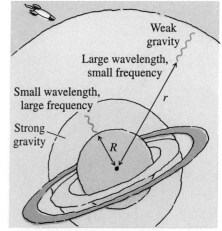

FIGURE 15.11 ·············

The Gravitational Red-Shift When a photon moves from a strong gravitational field to a region of zero gravitational field, its frequency is decreased and its wavelength increased due to effects predicted by the theory of general relativity.

1. Discuss the basic assumptions made in general relativity.

2. Compare and contrast the theory of special relativity with the theory of general relativity.

3. What is the principle of equivalence? Why is it important in general relativity?

REVIEW 15.4 QUESTIONS

4. Describe some phenomena in nature which are predicted by general relativity.

5. How is gravity represented in the following?

a. Newtonian theory   **b.** special relativity   **c.** general relativity.

15.5 Experimental Verification of General Relativity Theory

The intrinsic weakness of gravitational interactions in nature has made the empirical verification of Einstein's theory of general relativity one of the most challenging problems in all of experimental science. Investigators have nevertheless managed to perform a number of ingenious observational tests of general relativity.

Einstein himself proposed that the gravitational bending of starlight might be observed in the vicinity of the Sun at the time of a total solar eclipse, when the Sun's light is sufficiently dim so as to permit the detection of background stars (Figure 15.12). In 1919, the first such observations were performed during a solar eclipse when astronomers carefully measured the positions of stars in the vicinity of the eclipsed Sun and then the position of these same stars six months later when the Sun was no longer in that region of the sky. The results of these and similar observations made during subsequent solar eclipses indicate that gravitational bending of starlight does indeed exist. Unfortunately, the measurements have not been accurate enough to determine precisely the angle of deflection for a light ray passing the edge of the Sun's disk.

In recent years even more dramatic qualitative evidence for gravitational bending of starlight has been found with the discovery of the "gravitational lens" effect in astronomy. In this effect, the light from a very remote object in space such as a galaxy or quasar is distorted as it passes a very massive nearby galaxy (Figure 15.13). The distortion of space-time in the vicinity of the Sun also provides

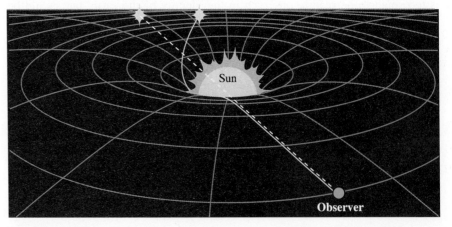

FIGURE 15.12 ··········
Gravitational Bending of Starlight In this three-dimensional analogue to Einstein's view of space-time warps, when light from a distant star passes close to the Sun, the slight warp in the fabric of space-time in the vicinity of the Sun generated by the sun's gravitating mass produces the deflection of starlight as shown.

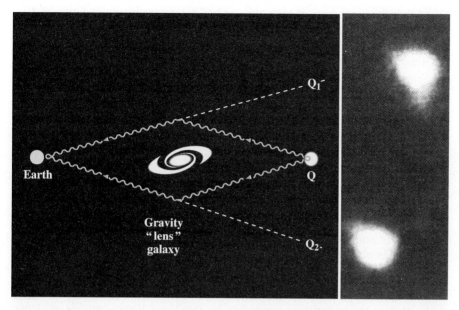

FIGURE 15.13 ·········
The Gravitational Lens Effect As light from the distant quasar Q passes the closer "lensing" galaxy, the quasar's light is deflected by the warp in space-time in the vicinity of the galaxy. As seen from the earth, the quasar now appears as the "twin" quasar images Q_1 and Q_2. Sometimes gravitational lensing, as this phenomenon is called, can produce luminous arcs of light as well.

an explanation for the unexplained anomalies in the orbital motion of the planet Mercury. In the two-dimensional analogue shown in Figure 15.14, we can see that as the planet moves through its perihelion point or near point to the Sun, it encounters a slight distortion in space-time due to the Sun's mass. As a result, it leaves the sun's vicinity along a slightly different orbital path. The successful explanation of this peculiarity in Mercury's orbital motion stands as one of the major triumphs of the theory of general relativity. The verification of the existence of "warps" in space-time leads to the possible existence of a number of phenomena such as gravity waves and black holes, which seem to be almost straight out of the realm of science fiction (see Chapter 16).

Other experimental tests of general relativity have been performed using pairs of highly accurate atomic clocks, one of which is flown at an altitude of several kilometers, while the identical companion clock is left behind to keep time in the more intense ground-level gravitational field. In each case, it was

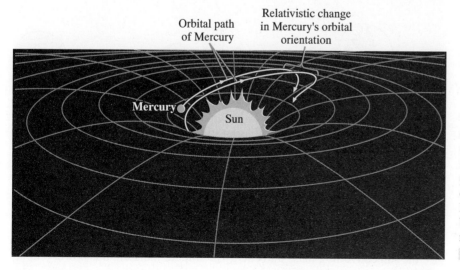

FIGURE 15.14 ·········
The Perihelion Advance in the Orbit of the Planet Mercury The same space-time warp in the vicinity of the sun shown in Figure 15.12 also causes the perihelion point in Mercury's orbit to change its position. This perturbation in Mercury's motion can thus be explained by general relativity without having to resort to the Newtonian concept of force.

found that the clock kept at high altitude had run faster by about 1 part in 10^7, which was in accordance with the result predicted by general relativity for two such clocks situated in different regions of the earth's gravitational field.

Empirical verification of the existence of the gravitational red-shift predicted by general relativity has also been obtained. In one of the more impressive experiments done at Harvard University or anywhere else, Robert Pond and Glen Rebka were able to measure the gravitational red-shift of gamma rays moving through a 23-meter vertical shaft to an accuracy of about 2 parts in 10^{15}! Measurements of such minute effects are not only empirically impressive but also serve to emphasize that the relativistic marvels described in this chapter become important only when we are dealing with the very great distances, high speeds, and enormous gravitating masses we find at the "large-scale" end of the universe.

REVIEW 15.5 QUESTIONS

1. What are some of the problems in attempting to verify experimentally the principles of general relativity?
2. Explain how an eclipse of the sun can be used to verify one of the predictions made by general relativity.
3. How does the experiment conducted by Pond and Rebka provide evidence for the theory of general relativity?
4. What is the gravitational red-shift?
5. What is a "gravitational lens"?

Chapter Review Problems

1. An alien spaceship A and a starship B are moving relative to the planet P. If c is the speed of light, how fast does a beam of light appear to be moving if it is observed:
   a. from A coming from B?
   b. from A coming from P?
   c. from B coming from P?

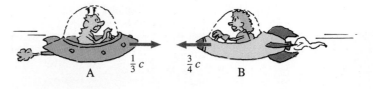

$\frac{1}{3}c$ A $\frac{3}{4}c$ B   P

2. How much energy will be produced if one kilogram of matter is converted into its equivalent energy?
3. How much mass would be required to produce one joule of energy if the mass is completely converted into its equivalent energy?
4. How much energy will be produced if one cubic centimeter of water were converted into its equivalent energy?
5. The sun radiates energy at a rate of about 4×10^{26} watts. At what rate is the sun converting mass into energy in order to maintain this power output?

6. Show that the units of mc^2 are units of energy.

7. When four protons are fused into a helium nucleus, the atomic mass of each of the protons is 1.0078 and that of the helium nucleus is 4.0026. If the mass of one atomic mass unit is 1.6606×10^{-27} kg, how much energy is released in each such reaction?

8. An electron has a mass of 6.1×10^{-31} kg. How much energy would be required to create an electron? To what frequency photon will such an energy correspond?

9. How much mass would be necessary to create a blue photon with a wavelength of 4.7×10^{-7} m if the mass were converted into its equivalent energy?

10. Find the wavelength and frequency of a photon created from the conversion of a particle with a mass of 5.0×10^{-30} kg into its equivalent energy.

11. A neutron having a mass of 1.67×10^{-27} kg is converted into its equivalent energy in the form of a single photon. Find the frequency and the wavelength of such a photon.

12. What is the *total* energy associated with an electron moving at a speed of 30,000 m/sec if the rest mass of the electron is 9.1×10^{-30} kg?

13. One electron volt is equal to 1.6×10^{-19} J. What mass is needed for the production of one electron volt of energy?

14. When one kilogram of coal is burned, it liberates about 1.0×10^7 joules of energy. If the Sun were a big lump of burning coal having a mass of 2×10^{30} kg, how long could the sun continue to shine at its present rate of 4.0×10^{26} watts? Is this a reasonable value for the lifetime of our sun? Explain.

15. The total amount of magnetic energy available to the Sun is about 7.2×10^{29} joules. How long could the Sun continue to shine at its present rate of 4×10^{26} watts if it were converting magnetic energy into heat and light? Is this a reasonable explanation for the Sun's energy production? Explain.

16. If the Sun were to convert all of its mass into its equivalent energy, how long would a 2×10^{30} kg mass Sun shine at a rate of 4.0×10^{26} watts?

17. The kinetic energy of the earth in its orbit is about 5.4×10^{32} joules. What is the equivalent mass for such an energy? How does this compare with the mass of the earth (6.0×10^{24} kg)?

18. An electron and a positron annihilate each other. If the masses of these two particles are each 9.11×10^{-27} kg, how much energy is released? If the energy is emitted in the form of a photon, find the frequency and wavelength of the photon.

19. If light travels at a rate of 300,000 km/sec, find the distance in km to a star 25 light years away.

20. Suppose that you were in a universe in which the interstellar medium was occupied by a substance with an index of refraction of 1.5. Find the number of kilometers in a light year for such a universe.

21. If a fortune were counted to the same accuracy as the Pond-Rebka experiment (2 parts out of 10^{15}), how large would the fortune be in dollars if the uncertainty is one penny?

22. If a time interval were measured to the same accuracy as the Pond-Rebka experiment (2 parts out of 10^{15}), how long would a time interval be in years if it were uncertain by one second?

23. If the distance could be measured to the same accuracy as the Pond-Rebka experiment (2 parts out of 10^{15}), how large would a total distance be in kilometers if it were uncertain by one centimeter?

24. At what magnitude and direction should the spaceship to the left be accelerated so as to give the crew the sensation of being back on the surface of the earth where the acceleration of gravity is 9.8 m/sec^2?

THOUGHT QUESTIONS

1. Suppose a form of radiation or some other entity were discovered that could move faster than the speed of light. What impact would such a discovery have on relativity theory?

2. Discuss the possibility that some of the stars we now see might no longer exist for an observer situated somewhere else in interstellar space.

3. Are there any counterparts to relativity theory in other branches of science? Explain.

4. Suppose that light traveled at an infinite velocity. What effect would this have on the concept of space-time?

5. Can you think of nonrelativistic effects that might cause Mercury's apsidal motion? If so, how might you observationally attempt to verify such effects?

6. Give an example in everyday life in which our perception of each of the following quantities depend on what we are experiencing:
   a. time   **b.** length   **c.** mass.

7. Compare and contrast the view of mass as seen from
   a. Newtonian theory
   b. special relativity
   c. general relativity.

8. Can Newtonian theory account for the bending of light in a gravitational field? If so, what sorts of assumptions have to be made about the nature of light?

9. Describe the relativistic effects that could be perceived for
   a. energy   **b.** momentum   **c.** density.

10. How might you distinguish between a gravitational red-shift and a red-shift produced by the Doppler effect?

11. What impact, if any, do you think relativity theory has had on the nonscientific world in the twentieth century?

12. Discuss the idea of "simultaneous" events. What effect does the finite velocity of light have on our perception of simultaneous events?

13. Suppose that a scientist of today made comments similar to those quoted for Michelson at the start of this chapter. How would you regard them? Explain.

14. Why do you suppose that Einstein developed his theory of general relativity after his theory of special relativity?

15. Do special and general relativity satisfy the correspondence principle? Explain.

16. Describe how you might observationally detect the existence of tachyons.

17. Discuss the possibility that our universe may have a fifth dimension. What sort of dimension might it be?

18. What do we mean when we say that time ''flows forward.'' Can time ever flow backward or stand still? Explain.

19. Do you think that relativity theory is consistent with quantum theory? Explain.

20. Discuss some of the effects and phenomena that are predicted by relativity theory, but have not yet been observed.

21. Describe some of the difficulties involved in trying to verify observationally the predictions of relativity theory.

22. Speculate on the possibility of creating a ''time machine'' in which a person could go ''forward'' or ''backward'' in time.

23. Why do you suppose atomic clocks are used in experiments involving tests of general relativity?

24. Why is the concept of space-time used in relativity theory?

25. What are gravity waves? How do they compare with electromagnetic waves?

26. Discuss the relationship between the principle of relativity and the principle of universality.

PHYSICS FORMULATION 15.1
Relativistic Contractions and Dilations

One of the seemingly mysterious consequences of special relativity theory is the claim that the values observed for the fundamental quantities of physics, length, time, and mass depend on the relative motion between the observer and the entity being observed.

To understand why this is so and to gain an idea of the actual magnitude of such effects, let us consider the observer in Figure 15.15 who is watching two clocks, one of which is at rest relative to the observer and one moving at a velocity v as shown. We will also assume that the clocks' ticks and tocks are in the form of a pulse of light that is emitted from a source, strikes a mirror for the "tick," and returns to the source for the "tock."

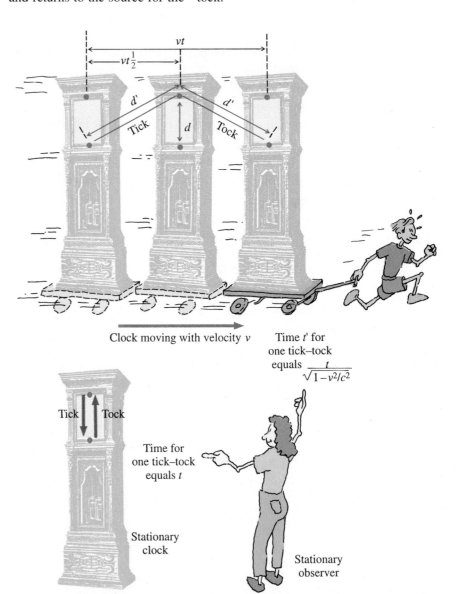

Clock moving with velocity v

Time t' for one tick–tock equals $\dfrac{t}{\sqrt{1 - v^2/c^2}}$

Tick   Tock

Time for one tick–tock equals t

Stationary clock

Stationary observer

FIGURE 15.15 · · · · · · · · · · ·
Time Dilation in Special Relativity   The "tick-tock" of the flashes from a light clock in motion at a velocity v relative to the stationary observer appears to be a longer interval of time when compared to the same "tick-tock" measured for an identical clock at rest with respect to the observer. The ratio of the two time intervals t'/t is equal to $1/\sqrt{1 - v^2/c^2}$.

For the stationary clock, the time t between successive ''tocks'' is simply the time t it takes for the flash to travel the distance d to the mirror at the speed of light and the return is simply $2d/c$. On the other hand, if the clock is moving past the observer at a velocity v, the time t' it takes the flash to travel the distance d' and then return is equal to $2d'/c$. Since the clock moves a horizontal distance vt' over that same time interval, from the Pythagorean theorem for right triangles, we have

$$(d')^2 = \left(\frac{vt'}{2}\right)^2 + (d)^2 = \left(\frac{ct'}{2}\right)^2 .$$

Now $d = ct/2$, and we may write

$$\left(\frac{vt'}{2}\right)^2 + \left(\frac{ct}{2}\right)^2 = \left(\frac{ct'}{2}\right)^2 .$$

If we solve this equation for t', we have

$$t' = \frac{t}{\sqrt{1 - v^2/c^2}} .$$

Since v can never exceed the value of c, the quantity $\sqrt{1 - v^2/c^2}$ ranges between 0 and 1. The interval between successive tocks that is observed for the moving clock will always be larger than the value observed for the stationary clock. This effect, called ''time dilation,'' tells us that clocks moving relative to an observer will be perceived to run more slowly by that observer than those at rest to that observer. An observer moving along with the clock, of course, sees no time dilation effect in that reference frame.

The quantity $\sqrt{1 - v^2/c^2}$ appears throughout the algebra of special relativity. An observer moving past a meterstick, for example, will perceive that the length of the meterstick is shorter by a factor of $\sqrt{1 - v^2/c^2}$ when compared to the length of a meterstick moving along with the observer, and the mass m of a moving particle will be increased by a factor of $1/\sqrt{1 - v^2/c^2}$ over its ''rest'' mass or mass observed at zero relative velocity.

This quantity in a certain sense is a measure of the degree to which special relativistic effects are present in a given situation. A relativistic time dilation of 1%, for example, would require that a clock be moving at a rate of about $0.14c$. This corresponds to a speed of some 42,000 km/sec or one circumference of the earth per second! We thus can see that the fascinating relativistic effects described in this chapter begin to make themselves manifest only at comparatively large speeds.

For accelerating reference frames, a similar dependence of observed quantities on the state of relative motion, in this case the relative acceleration, is observed. Suppose a clock is experiencing a uniform acceleration a. The principle of equivalence tells us that this acceleration can be replaced by an equivalent gravitational acceleration arising from a gravitating mass M at a distance r from the clock. Such an acceleration would have a value of

$$a = \frac{GM}{r^2} .$$

Also, had the clock fallen under the influence of such an acceleration from an infinitely large distance to the distance r, it would, from the results in Physics Formulation 6.2, have acquired a velocity v given by

$$\frac{1}{2} mv^2 = \frac{GMm}{r} ,$$

where m is the mass of the clock. Solving this equation for v^2 yields

$$v^2 = \frac{2GM}{r} .$$

We have, from our earlier analysis, that the time t' observed for a moving clock is related to the time t for the at rest clock by

$$t' = \frac{t}{\sqrt{1 - v^2/c^2}} .$$

Substituting the previous expression for v^2 into this expression yields the following:

$$t' = \frac{t}{\sqrt{1 - (GM/rc^2)}} .$$

This equation permits us to make calculations involving general relativistic effects in the vicinity of gravitating masses in space, such as the earth or sun.

Once more, however, the effects are quite small under "everyday" conditions. A clock running in the earth's gravitational field at the earth's surface would run 1.0000002 times slower than the same clock operating in space far away from any gravitating objects.

QUESTIONS

1. Indicate how, if at all, the constancy of the speed of light enters into the derivation of the relationship between t and t'.

2. How would you observationally test the validity of one of the amazing claims (your choice) made in this formulation?

3. Discuss the impact that the relativistic effects described in this formulation might have on the standards of length, time, and mass.

PROBLEMS

1. At what fraction of the speed of light must an object be moving in order that the quantity $1/\sqrt{1 - v^2/c^2} = 10.0$?

2. At what speed must an observer pass a meterstick so that it appears to be only 50 cm long?

3. If a clock in the earth's surface gravitational field runs 1.0000002 times slower than the same clock in free space, how much time will elapse in what reference frame before a difference of one second is observed between the two clocks?

16 ASTROPHYSICS

The sky above us has long been a source of great fascination for human beings. The overpowering brilliance of the Sun, the dark markings and changing phases of the Moon, and the plodding movement of the planets across the backdrop of the seemingly "fixed" stars are but a few of the phenomena that beckon us to study this "final frontier." Along with the fascination, however, comes great frustration as well. Despite the fact that on some nights the moon and stars seem so close that you can reach out and touch them, the reality is that these denizens of the sky, with only a precious few exceptions, are not at all "touchable" to humans. Instead, they lie at distances so remote as to challenge not only our ability to obtain information about them, but to even comprehend the vastness of the distances themselves. Despite this fundamental limitation, we have nonetheless been able to ingeniously piece together a portrait of a cosmic realm which, like its microscopic counterpart world of the atom, is filled with a wondrous array of entities and phenomena whose behavior and physical characteristics are the province of the branch of physics known as astrophysics.

16.1 Astrophysical Techniques

The astronomer or astrophysicist who seeks the scientific secrets of outer space is faced with the basic difficulty that their objects of interest are incredibly remote, and, therefore, distinctly inaccessible for controlled experimentation. In a laboratory setting here on Earth, for example, we can readily and accurately measure a number of physical characteristics of a given object, including its mass, temperature, and chemical composition. If, however, we wish to measure those same characteristics for a celestial object such as the Sun or Moon, we find that they are too big, too far away, and in the case of the Sun, much too hot to lend themselves to direct measurements in a terrestrial laboratory. The astrophysicist must, therefore, perform his or her experiments on these objects *in situ,* or in their natural states. Space age efforts to bring the terrestrial laboratory to the object via space probes have produced a number of spectacular contacts and encounters with many of the worlds of our solar system, but have left the remainder of the cosmos essentially untouched. Thus, the astrophysicist necessarily continues to observe and deduce from afar.

The seeming inaccessibility of the heavens was overcome in part by the invention of the telescope early in the seventeenth century. Armed with telescopes having ever-increasing image quality and light-collecting ability, we can, in a very real sense, bring celestial objects into terrestrial laboratories, by collecting and analyzing the light that we receive from them. At this point, we make a most vital appeal to the principle of universality. If the way nature operates is independent of time and location here on earth, then it is reasonable to apply this very same principle to the whole of outer space. Gravitation operates among celestial objects in precisely the same way it does here on Earth. The composition and behavior of matter, down to the most elusive neutrino, is precisely the same in outer space as it is on Earth. The properties of electromagnetic radiation produced in outer space are precisely the same as those of electromagnetic radiation produced on Earth. In short, all of the physical universe is regarded as being consistent, or, in a variation of an old astrological refrain, "As below, so above." Having made this fundamental assumption, which has yet to be contradicted by any of our observations and measurements, we are now in a position to interpret objects and phenomena observed from afar in terms of the principles by which objects and phenomena are observed to behave here on the earth.

For example, if we observe a binary star system and find that the stars orbit each other in elliptical orbits (Figure 16.1) rather than in orbits shaped like squares or triangles, we assume that the stars are moving in one of the orbital paths predicted for two objects moving in each other's gravitational force, just as the Moon moves in an elliptical path under the influence of Earth's gravitational force. By interpreting the motions of the stars about each other in terms of a simple gravitating two-body system, astronomers can deduce some of the important physical characteristics about the stars, most notably their masses. Recall from Chapter 5 that objects moving about each other in gravitationally produced circular or elliptical orbits must orbit in such a way as to satisfy the *harmonic law:*

$$\text{mass } 1 + \text{mass } 2 = \frac{4\pi^2}{G} \frac{(\text{mean orbital distance})^3}{(\text{time for one orbit})^2}.$$

For the Sun, the mass of any of the planets is several orders of magnitude less than that of the Sun and for this case, mass 2 can be thought of as being essentially zero. If one measures the mean orbital distance and the time for one orbit for a given planet, the Sun's mass can then be readily calculated by using these measured values in the harmonic law. For binary star systems in which neither of the masses can be neglected, we must somehow obtain a second piece of information concerning the masses of the system, usually in the form of a known or measurable mass ratio mass 1/mass 2, before the individual masses of the stars in such systems can be calculated.

If the light from a star is broken up into an array of its component wavelengths by means of a device called a spectrograph, the resulting spectrum exhibits the

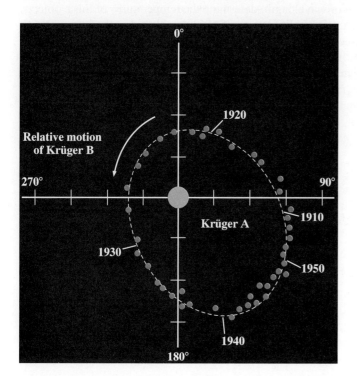

FIGURE 16.1 ··········
The Orbit of the Binary Star Krüger 60   The position of Krüger 60 B relative to that of Krüger 60 A has been plotted for several decades' worth of observations. The resulting elliptical orbit is precisely what we would expect for two objects moving in each other's gravitational force.

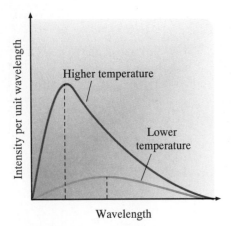

FIGURE 16.2 ···········
**Wien's Law and the Stefan-Boltzmann
Law** As the temperature of an
incandescent light source increases, the
wavelength of maximum intensity
decreases, and the total area under the
curve or total surface intensity at all
wavelengths increases as the fourth
power of the temperature.

same characteristics of an incandescent source shining through a relatively cool
layer of gas. The distribution of intensity as a function of wavelength has the
same hump-shaped appearance as that from a terrestrially radiating blackbody,
and the dark lines of missing intensity have the same wavelengths of spectral
lines of known substances here on the earth.

We know from observations made on radiating incandescent light sources
in terrestrial laboratories that the wavelength at which the maximum value of
the observed intensity occurs or *wavelength of maximum intensity* is related to
the temperature of the object by an equation called *Wien's law,* which states:

$$\frac{\text{wavelength of}}{\text{maximum intensity}} \times \frac{\text{Kelvin}}{\text{temperature}} = \frac{\text{constant}}{\text{value}} \; .$$

Interestingly, the constant value is the same for *any* incandescent light source
and has a value of 0.00290 meter-K. In addition, if an incandescent source has
a given energy intensity at its surface, then the surface intensity of radiant energy
emitted at all wavelengths and the surface temperature are related by a second
equation, called the *Stefan-Boltzmann law,* which states:

$$\text{surface energy intensity} = \text{constant} \times \left(\frac{\text{Kelvin}}{\text{temperature}}\right)^4,$$

where the constant in this equation is known as the Stefan-Boltzmann constant
and has a value of 5.67×10^{-8} watts/meter$^2 \cdot$ K^4. Thus, an increase in the
temperature of an incandescent light source will produce a marked increase in
the total intensity of an object's radiant energy and also downward displacement
in the position of the wavelength of maximum intensity (Figure 16.2). By assuming that the radiant energy of any incandescent celestial object is operating
under these same laws, one can, by measuring the distribution of observed intensity as a function of wavelength, determine the temperatures of these objects.

The principle of universality is also of considerable aid in the interpretation
of the Fraunhofer lines that appear in the spectra of the Sun and stars (Figure 16.3).
Since the wavelengths of these lines correspond almost exactly to those produced
by terrestrial elements in terrestrial laboratories, the obvious conclusion is that
the atoms that give rise to these lines in remote celestial objects are identical to
those found here on the earth. In short, the atoms out of which stars and stardust

FIGURE 16.3 ···········
The Spectrum of the Star Vega The continuous distribution
of intensity with wavelength is that of an incandescent
blackbody here on Earth. The wavelength ''dips'' or absorption
lines in the spectrum correspond in wavelength to those
observed for hydrogen and other elements here on the earth.

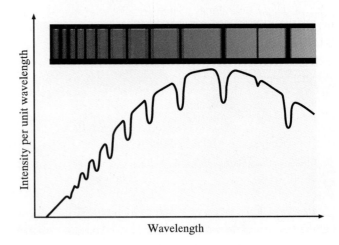

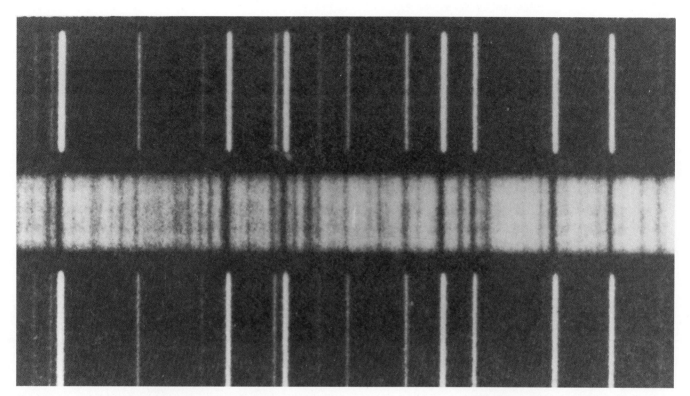

FIGURE 16.4 ··········
Celestial Radial Velocities Because of
the Doppler effect, the spectral lines of
this star (center strip) are displaced
relative to the at-rest lines of the
reference or comparison spectra (top and
bottom strips).

are composed are equivalent to those out of which the Earth and earthdust are
composed. This fact permits us to conduct fairly accurate chemical analyses of
incandescent celestial objects (see Physics Facet 16.1).

Precise measurement of the wavelengths of the lines contained in stellar
spectra reveals that, while the patterns of the lines present are identical to those
formed for elements on Earth (hydrogen Balmer pattern, etc.), small-scale differences exist between the values of the "laboratory" wavelengths measured
here on Earth for these lines and "observed" wavelengths measured in the
spectrum of the star (Figure 16.4).

These small-scale differences between observed and laboratory wavelengths
are interpreted by astrophysicists as being due to that component of the object's
relative motion, which is directed along the line of sight to the object, either
toward or away from the observer. This component of the object's relative motion
is called the *radial velocity* of the object. It produces a change in the observed
wavelength as a result of the Doppler effect (see Physics Formulation 9.1) occurring for light waves. The relationship between this *Doppler shift* in wavelength
and the object's observed radial velocity is as follows:

$$\text{radial velocity} = \left(\frac{\text{observed wavelength} - \text{laboratory wavelength}}{\text{laboratory wavelength}} \right) \times \text{speed of light},$$

where the radial velocity is a velocity of approach toward the observer if the
value of observed wavelength minus the laboratory wavelength is less than zero,

16.1 Astrochemistry

One of the most powerful techniques of astrophysics is that which permits the chemical analyses of objects thousands of light years away. This marvelous feat is accomplished by analyzing the absorption and emission lines found in the spectra of stars and then comparing these stellar wavelengths to those of substances found on Earth. Perhaps the most impressive vindication of such procedures came in the last third of the last century. In 1868 the French and English astronomers Pierre Janssen and Normal Lockyer discovered absorption lines in the solar spectrum that could not be identified with any earthly counterparts. Lockyer became convinced that the unidentifiable lines were from a new element, which he called ''helium,'' a Greek word meaning ''element of the sun.'' This bold assertion was confirmed nearly 30 years later when in 1895 the Scottish chemist William Ramsay was able to isolate this same element here on Earth. The terrestrial discovery of helium was thus a most dramatic affirmation of the principle of universality.

Detailed astrophysical studies of the elements and their cosmic abundances have led to a reasonably accurate picture of the composition of the universe called the *cosmic abundance scale*. By mass, all but about 1.8% of the matter in the universe is either hydrogen or helium. All of the elements heavier than helium must share that last 1.8%. The sharing is not done evenly, however, and some elements such as carbon, oxygen, and neon are more abundant than their neighbors in the periodic table such as fluorine, phosphorus, and chlorine. A plot of the atomic number of an element versus its relative abundance in the universe, shown in the figure, has the appearance of a jagged sawtooth-like plot superimposed on an exponential decay curve.

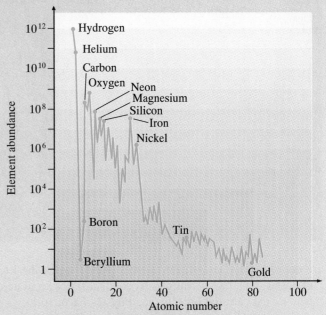

··· The Cosmic Abundance Curve Nearly all of the features of this plot of atomic number versus element abundance can be explained by stellar nucleosynthesis processes.

In an attempt to explain both the origins of the elements as well as the features of the cosmic abundance curve, astrophysicists have developed the theory of *nucleosynthesis* in which the elements and their respective abundances are the result of nuclear reactions occurring at the centers of highly evolved stars. To aid in developing and verifying

and is a velocity of recession if this difference is greater than zero. Any relative motion between Earth and the object will produce such a Doppler shift. This fact has enabled us to demonstrate spectroscopically the orbital motion of Earth as well as many binary stars, pulsation and rotation in stars, and the expansion of the universe in general.

In Chapter 12, we learned that the observed intensity of a light source at a given distance is defined as the amount of power per unit area received from that source at that distance. The relationship between the intrinsic power of the source,

Continued

such a theory, terrestrial physicists have measured countless nuclear cross sections or probabilities that a given nuclear reaction will occur. The end products of highly probable reactions will become the most abundant elements, while those less probable will result in the formation of less abundant elements. Thus the capture of a helium nucleus of alpha particle by a carbon nucleus in the reaction

carbon-12 + helium-4 → oxygen-16 ,

which produces the element oxygen, is a highly probable reaction, while the capture of a hydrogen nucleus or proton by a carbon nucleus,

carbon-12 + hydrogen-1 → nitrogen-13 ,

which yields the element nitrogen is a far less probable reaction than the previous one for oxygen. Therefore, we expect nitrogen to be less abundant than oxygen, which it is, Earth's atmosphere notwithstanding.

The theory of nucleosynthesis invokes only a few basic nuclear reactions in order to explain cosmic abundance. One such reaction is heavy element fusion in which elements heavier than hydrogen are fused into ever-heavier elements. For example, helium can be fused into carbon by the following reaction:

helium-4 + helium-4 + helium-4 → carbon-12 ,

and carbon can be fused into the element magnesium by

carbon-12 + carbon-12 → magnesium-24 .

Helium nuclei readily participate in nuclear reactions to form heavier elements in reactions such as that described earlier. Neutrons and protons can also be captured by nuclei in reactions such as the neutron capture, which is believed to produce the element antimony out of tin:

tin-119 + neutron → antimony-120 + beta particle ,

and the proton capture process, which is believed to produce the element xenon out of iodine as follows:

iodine-123 + proton → xenon-124 .

All of the elements in the periodic table are thought to have originated from one or more of these types of nuclear reactions. The glaring exceptions to this claim are the light elements lithium, beryllium, and boron. All suffer from the problem that any nuclear conditions that would create these elements would also almost instantly destroy them. To a certain degree, their presence in the universe can be accounted for by the Big Bang theory, but such are the difficulties with these elements that astrophysicists have referred to the unknown process by which they are created and preserved as the "x-process."

Theoretical predictions by nuclear physicists have indicated that stable nuclei of elements may exist at atomic numbers well beyond the current frontier of 104. An interesting possibility is that such "superheavy" elements, like their less massive brethren, may also have been created in the deep interiors of advanced age stars, and like helium, await their discovery among the myriad of absorption lines existent in the spectra of these stars.

---

the distance to the source, and the observed intensity of the source is as follows (see Physic Formulation 12.1)

$$\frac{\text{intensity}}{\text{of source}} = \frac{\text{power of source}}{4\pi \times (\text{distance to source})^2} .$$

For the astrophysicist, any optical light source bright enough to be detected in a telescope is bright enough to have its observed light intensity measured. To be sure, these observed intensities are most often quite small by terrestrial standards.

The faintest celestial objects detectable in our largest telescopes, for example, have observed intensities of about 5×10^{-19} watts/m², the intensity that would be observed emanating from a candle burning at a distance of 120,000 kilometers! Once the intensity of a given celestial light source is measured, the preceding equation can be employed in one of two ways. For a star that is relatively close to the Sun, its distance can be measured by a triangulation process in which the object's projected position relative to the more distant background stars is measured from opposite ends of a diameter of the earth's orbit (Figure 16.5). The distance between Earth and the Sun is accurately known to be slightly less than 150 million kilometers and the *parallax angle* of the star can be measured from the amount of displacement it appears to exhibit against the background stars when observed from the two different positions in Earth's orbit. From our discussion in Physics Formulation 4.1, if an angle, in this case the parallax angle, subtends a known length, in this case the earth's orbital radius, then the distance between the apex of the angle (the star's position) and the known length (Earth's position) can be readily calculated. Once the distance is obtained to a star whose observed intensity has been measured, the power of the source can be calculated by

$$\text{power of} \atop \text{source} = {\text{intensity} \atop \text{of source}} \times 4\pi \times \left(\text{distance to} \atop \text{source}\right)^2 .$$

Unfortunately, there are only a few hundred stars whose distances can be measured accurately by the method of trigonometric parallax or triangulation. To deal with celestial objects at even greater distances, we must once more resort to the principle of universality to gain the information we seek. Accordingly, we

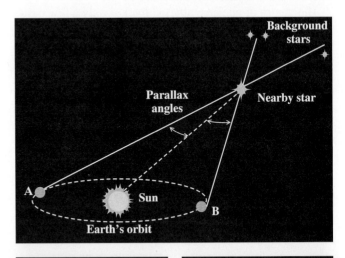

FIGURE 16.5 ··········

The Method of Trigonometric Parallax   As Earth orbits around the Sun, the apparent position of a nearby star relative to the more distant background stars will change. This effect is called the parallax effect, and half of the total amount of the parallactic shift is defined as the star's parallax angle.

View from A

View from B

assume that the laws of physics and chemistry that have worked to create a given type of celestial object are the same as those that have worked to create that same type of object at any other location in space. As a result, *all* of the physical characteristics of both objects are assumed to be virtually the same. In particular, the intrinsic power output of every object in a given class of celestial objects is assumed to be the same as any other object possessing those observed characteristics. Thus, if a distant star has all of the observed physical characteristics of the Sun, such as temperature, composition, etc., then by invoking the principle of universality, we can assume that the power output of the distant star is the same as that of the Sun. If the intrinsic power of the distant source is known and the observed intensity has been measured, then the relationship between observed intensity, intrinsic power, and distance can be written as follows

$$\text{distance to source} = \sqrt{\frac{\text{power of source}}{4\pi \times \text{intensity of source}}} \; .$$

Thus, by making alternating use of these last two equations, astronomers and astrophysicists can essentially "boot strap" distances and power outputs for celestial objects from the relatively close confines of the Solar System out to the very edges of the observable universe.

Such are but a few of the principles and techniques by which we have skillfully and painstakingly woven the tiny thread of light received from distant cosmic light sources into a tapestry of the universe that far transcends in vastness, beauty, and complexity anything that could have been envisioned by human beings even as recently as a century ago.

WORKED EXAMPLE

A star has a wavelength of maximum intensity of 2.5×10^{-7} m. What is the temperature of this star?

SOLUTION: From Wien's law, we have

$$\begin{matrix} \text{wavelength of} \\ \text{maximum intensity} \end{matrix} \times \text{temperature} = 0.003 \text{ m-K} \; .$$

Solving for the temperature yields in general that

$$\text{temperature} = 0.003/\text{wavelength of maximum intensity} \; .$$

Since the wavelength of maximum intensity is 2.5×10^{-7} m, we have that

$$\begin{aligned} \text{temperature} &= .003 \text{ mK}/(2.5 \times 10^{-7} \text{ m}) \\ &= 3.0 \times 10^{-3} \text{ K}/(2.5 \times 10^{-7}) \\ &= 1.2 \times 10^{4} \text{ K} = 12{,}000 \text{ K} \; . \end{aligned}$$

Thus the temperature of this star is 12,000 K.

REVIEW 16.1 QUESTIONS

1. Describe the difficulties human beings encounter in the study of celestial objects.
2. How is the principle of universality employed by the astrophysicist?
3. What is Wein's law?
4. Explain what is meant by a radial velocity. How are radial velocities measured?
5. Discuss the relationship between the power of an incandescent source, the intensity of the source, and the distance to the source.

REVIEW 16.1 PROBLEMS

1. A spectral line of hydrogen that normally occurs in the laboratory at 6.563×10^{-7} m is observed in the spectrum of a star to be at 6.565×10^{-7} m. Find the magnitude and the direction of the radial velocity of this star.
2. A certain star suddenly increases its temperature by a factor of 3. Compare the surface intensity and wavelength of maximum intensity for the star before and after the increase. Assume that the radius of the star does not change during this process.

16.2 The Astrophysics of the Sun

By far the most dominating and important of the celestial objects is the Sun. Its presence in the sky turns night into day, and without its energy, no life could exist on the earth. In recognition of this crucial role in our daily lives, the Sun has very often in the span of human history been revered as a deity by cultures around the world. Modern science still acknowledges the Sun's preeminent importance for us, but only in the context of the Sun as a physical object displaying physical phenomena that obey physical laws.

To gain some insight regarding the Sun, it is first useful to obtain a few of its overall physical characteristics. By employing a variety of techniques (see Chapter Review Problem 1), it has been determined that the mean Earth-Sun distance is just under 150 million kilometers or 1.5×10^{11} meters. The Sun's angular radius is about 0.0092 radians. Recalling from Physics Formulation 4.1 that

$$\text{linear radius} = \text{angular radius} \times \text{distance} ,$$

we find that the Sun's linear radius is

$$\text{Sun's linear radius} = 0.0092 \times 1.5 \times 10^{11} \text{ meters}$$
$$= 1.38 \times 10^{9} \text{ meters}$$
$$= 6.9 \times 10^{8} \text{ meters} .$$

In other words, the linear size of the Sun is about one hundred times that of Earth.

Measurement of the intensity of the Sun's radiant energy according to wavelength yields the plot shown in Figure 16.6. Overall, the curve resembles that of

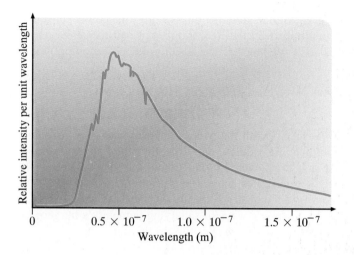

y-axis: Relative intensity per unit wavelength

x-axis: Wavelength (m)

0   0.5 × 10⁻⁷   1.0 × 10⁻⁷   1.5 × 10⁻⁷

FIGURE 16.6 • • • • • • • • • •

The Intensity Distribution of the Sun's Energy The plot of wavelength versus intensity per unit wavelength for the Sun's output of radiant energy has the overall properties of blackbody radiation, but with wavelengths missing due to electronic absorptions.

a blackbody with a wavelength of maximum energy roughly equal to 4.8×10^{-7} m. From Wien's law, this corresponds to a temperature of about 5800 K at the Sun's visible surface layers or *photosphere*. This plot also shows large- and small-scale departures from an idealized blackbody curve that are in the form of thousands of absorption lines (Chapter 13, Figure 13.3). Using the strengths and wavelengths of these lines, astrophysicists have been able to determine that the Sun is composed primarily of hydrogen (78.5% by weight), helium (19.7% by weight), and 1.8% by weight of 65 other chemical elements which have been identified in the solar spectrum.

If we carefully measure Earth's orbital motion about the Sun, we find that it takes 3.16×10^7 seconds to make one complete orbit about the Sun. Substituting this value along with Earth's mean orbital distance into the harmonic law and neglecting the mass of the Earth, we obtain

$$\text{mass of Sun} = \frac{4\pi^2}{G} \frac{(\text{mean orbital distance})^3}{(\text{time for one orbit})^2}$$

$$= \frac{4(3.14)^2}{6.67 \times 10^{-11}} \frac{(1.5 \times 10^{11})^3}{(3.1 \times 10^7)^2} \text{ kg}$$

$$= 2.0 \times 10^{30} \text{ kg} .$$

This mass is more than 330,000 times that of Earth. We were, thus, indeed justified in neglecting Earth's mass in our calculation!

For the scientist, the Sun's energy production holds the greatest interest and fascination. The Sun's energy sustains life on the earth. The Sun's energy reflects off each of the planets, satellites, and meteoroids in the solar system, thus, permitting us to view these objects. As viewed from Earth, the intensity of the Sun's radiant energy is referred to as the *solar constant* and has a measured value of 1360 watts/meter².

If we imagine that the Sun is a light source surrounded by a sphere whose radius is equal to that of the Earth's orbit, then, as we have seen, the surface area of this sphere is equal to $4\pi \times (\text{distance})^2$ or $4 \times \pi \times (1.5 \times 10^{11})^2$ m² $= 28.3 \times 10^{22}$ m². If we assume that the Sun is emitting its radiant energy into all directions of space in exactly the same way, then each square meter on this sphere receives the same 1360 watts of power that we observe for the solar constant.

The total power output of the Sun in all directions is simply the power per square meter, or 1360 watts, multiplied by the total number of square meters on the surface of our sphere (28.3×10^{22} meters^2). The sun's total power output is thus given by

$$\text{Sun's total power} = 1360 \times 28.3 \times 10^{22} \text{ watts}$$
$$= 3.84 \times 10^{26} \text{ watts} .$$

To place this value in perspective, all the world's industry currently consumes energy at a rate of about 6×10^{20} joules/year. The Sun's total power output of 4×10^{26} watts corresponds to an energy production of 4×10^{26} joules *in one second.* The Sun, therefore, generates about one million times more energy in one second than all the world's industry currently consumes in an entire year! It is little wonder that those seeking ways of satisfying humanity's burgeoning energy demands have long cast covetous eyes toward the sun!

Perhaps the most fundamental astrophysical question we can ask of the Sun is how such a vast amount of solar power is achieved. The question has long been asked by those searching for practical explanations of this phenomenon. It will be recalled, for example, that the Greek philosopher Anaxagoras claimed that the sun radiated its energy because it was a white-hot stone, an idea that was challenged by Aristotle on the grounds that a white-hot stone would quickly cool, and the Sun displayed no such tendency. The Anaxagoras-Aristotle exchange, however, was to become the central theme of the efforts to deduce the mechanism by which the Sun produces its energy.

Fossil evidence from the rock strata in the earth's crust betrays an existence of life forms dating back at least one billion years and perhaps as long ago as 2.5 billion years. Moreover, creatures such as the horseshoe crab, skate, and nautilus have existed virtually unchanged on Earth for hundreds of millions of years. Picturesquely referred to by palentologists as "living fossils," these creatures offer mute testimony to the fact that for them to survive these eons, the Sun's current level of power production could not possibly have varied during those same eons by more than a few percent from its currently observed level.

To illustrate the limitations such an observation places on the possible mechanisms by which the Sun produces its energy, let us assume an explanation more reasonable than Anaxagoras's white-hot stone model or the "burning Sun" models from Chapter 15. Let us assume that somehow the Sun is converting its rotational energy into radiant energy. From our discussion in Chapters 4 and 6, the total rotational energy of the sun is given by

$$\begin{array}{c}\text{rotational energy}\\\text{of the Sun}\end{array} = \frac{1}{2} \times \left(\begin{array}{c}\text{Sun's moment}\\\text{of inertia}\end{array}\right) \times \left(\begin{array}{c}\text{Sun's angular}\\\text{velocity}\end{array}\right)^2 .$$

Observations of the Sun's rotation rate tell us that the Sun's present angular velocity is about 2.8×10^{-6} radians/second. If we regard the Sun as a uniform sphere, then its moment of inertia is $\frac{2}{5} \times$ mass $\times$ (radius)^2. From measured values for the Sun's mass and radius, we have that the moment of inertia for the Sun is $\frac{2}{5} \times 2.0 \times 10^{30} \times (6.9 \times 10^8)^2$ kg-m^2 or 3.8×10^{47} kg-m^2. The total rotational energy available for the Sun to shine is then

$$\begin{array}{c}\text{rotational energy}\\\text{of the Sun}\end{array} = \frac{1}{2} \times 3.8 \times 10^{47} \times (2.8 \times 10^{-6})^2 \text{ joules}$$

$$= 1.5 \times 10^{36} \text{ joules} .$$

Therefore, the total amount of energy available for the Sun to convert into radiant energy is 1.5×10^{36} joules. The Sun, however, is using this energy at a rate of 3.84×10^{26} watts or 3.84×10^{26} joules/second. The total time the Sun can shine in this fashion is simply the total available energy divided by the rate at which this energy is radiated away, or

$$\frac{\text{``lifetime''}}{\text{of Sun}} = \frac{\text{total available energy}}{\text{power output}}$$

$$= \frac{1.5 \times 10^{36} \text{ joules}}{3.84 \times 10^{26} \text{ joules/sec}}$$

$$= 3.9 \times 10^{9} \text{ seconds or } 124 \text{ years!}$$

Thus, even if the sun could somehow convert all of its rotational energy into radiant energy, it would barely last a century. Similar analyses of other forms of energy-generating mechanisms such as conversion of magnetic energy and gravitational potential energy into radiant energy suffer the same fate as those of rotational energy and the white-hot stone: The Sun manifestly cannot possibly shine long enough. Such was the state of affairs at the turn of the twentieth century when Einstein demonstrated in his theory of special relativity that mass and its equivalent energy were related by the famed $E = mc^2$ equation. At long last, there was a potential reservoir of energy large enough to power the Sun throughout the Earth's geological history (see Chapter Review Problem 10).

We can pinpoint the actual nuclear process by which matter in the Sun is converted into energy by recalling from spectroscopic abundance analysis that the Sun's composition by weight is 78.5% hydrogen and 19.7% helium. The two most abundant substances in the Sun are hydrogen and helium, the reactant and product elements, respectively, of the thermonuclear fusion of hydrogen into helium. From all of these observations, astrophysicists can conclude, almost by a process of elimination, that the energy produced by the Sun is the result of protons or hydrogen atom nuclei colliding with each other at enormous velocities and fusing together into helium nuclei. As the fusion occurs, some of the mass of the four reacting protons is converted into its equivalent energy. For every 1000 kilograms of hydrogen mass that initially react in a fusion process, 993 kilograms end up as helium, and the remaining 7-kg mass is converted into its equivalent energy of $7 \times (3 \times 10^8)^2 = 63 \times 10^{16}$ joules.

If nuclear collisions are going to occur at this level of violence, however, a temperature of some 15 million kelvins and a pressure of 340 *billion* atmospheres is needed. For inhabitants of Earth, such temperatures and pressures are almost unimaginable, but at the center of the Sun, where the self-gravitating weight of 2×10^{30} kg of gas layered 7×10^8 meters deep comes crushing in from all directions, such extreme conditions of temperature and pressure can and do occur. This sort of environment obviously renders any direct observation of the Sun's interior totally impossible, since any instrument probe would, under these conditions, be instantly vaporized. We therefore must once more rely on the principle of universality to gain information about an otherwise inaccessible region of astrophysical interest. Using this concept, the astrophysicist can set forth the laws and conditions that the gas on the inside of the Sun must satisfy in order to be consistent with the observed behavior of gases on Earth. These laws and conditions are written in the form of a set of mathematical equations that describes the physical conditions at any location in the Sun's interior. For example, since

TABLE 16.1   A Model Solar Interior

Distance from Center	Temperature	Pressure	Density
0 meters	15.5×10^6 K	3.4×10^{11} atm	1.6×10^5 kg/m³
7×10^7	13.0×10^6	1.5×10^{11}	9×10^4
14×10^7	9.5×10^6	5.2×10^{10}	4×10^4
21×10^7	6.7×10^6	1.2×10^{10}	1.3×10^4
28×10^7	4.8×10^6	2.3×10^9	4×10^3
35×10^7	3.4×10^6	4.7×10^8	1×10^3
42×10^7	2.2×10^6	1.0×10^8	4×10^2
49×10^7	1.2×10^6	1.2×10^7	80
56×10^7	0.7×10^6	1.5×10^6	18
63×10^7	0.3×10^6	8.7×10^4	2
70×10^7	0.006×10^6	0.14	3×10^{-4}

the Sun is neither expanding nor contracting in its size, we may conclude that all of the net forces operating in the solar interior are in balance, a condition referred to as a state of *hydrostatic equilibrium.*

Using the measured values of the Sun's mass, radius, chemical composition, and total power output in conjunction with this set of equations, astrophysicists can, making use of high-speed computers, solve these equations and produce what is called a *model interior* in which the physical conditions existing inside of the Sun are tabulated for various distances from the Sun's center (Table 16.1).

These types of calculations indicate that all of the fusion processes occur at or very near the Sun's center. As each photon of energy is produced, it travels only a centimeter or so before it is absorbed and randomly reemitted by one of the closely packed atoms at the Sun's center. As a result, a given photon must make a laborious "random walk" journey from the center of the Sun out to the less dense outer layers (Figure 16.7). This journey lasts on the average of about

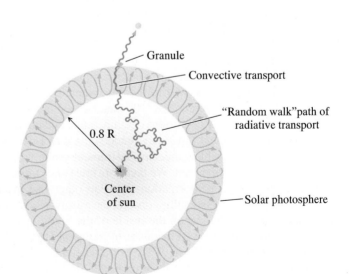

Granule

Convective transport

"Random walk" path of radiative transport

0.8 R

Center of sun

Solar photosphere

FIGURE 16.7

Energy Transport in the Sun   Photons created at the center of the Sun are absorbed and reemitted in a "random walk" path until they reach about four-fifths of the solar radius where the energy is then transported to the outer layers of the Sun via large-scale convection currents.

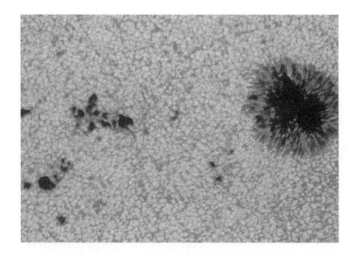

FIGURE 16.8 ············

The Solar Granulation in the Vicinity of Sunspots The tops of the solar convective cells appear as this mottled pattern in the solar photosphere. The bright areas are regions of hot upwelling gas, while the dark lanes are regions of descending cooler gas.

ten million years. Once the photon travels about 80 percent of the distance to the Sun's outer layers, it is quickly transported the remaining distance by means of enormous convective cells, which carry the energy to the base of the outer layers of the sun in about a month or so.

At this point, one can obtain at least some observational verification of the correctness of this picture of solar energy transport. High-resolution photographs of the Sun's photosphere reveal the existence of a mottled pattern of light and dark regions, referred to collectively as the *solar granulation.* Observations of the individual granules indicate that they are the tops of gigantic convective cells roughly 1000 kilometers in diameter. The bright regions are areas where the hot upward moving gas "tops out" in the Sun's photosphere, cools off, and falls back into the solar interior along the surrounding lanes of the darker and cooler regions of descending gas (Figure 16.8).

Once the energy produced at the Sun's center reaches the photosphere, it begins to interact with the hot gases of the Sun's atmospheric layers in a wide variety of complicated ways (Figure 16.9). The oldest known and most familiar of these phenomena are the *sunspots,* which are irregular-shaped dark regions in

FIGURE 16.9 ············

Some Aspects of Solar Activity (a) A solar flare and eruptive prominence; (b) sunspots, faculae, and a solar prominence; and (c) coronal hot and cold spots.

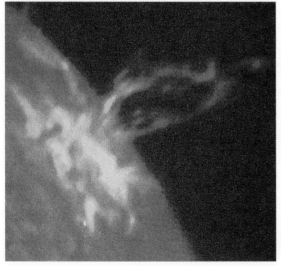

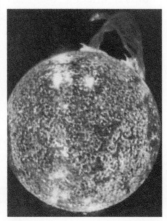

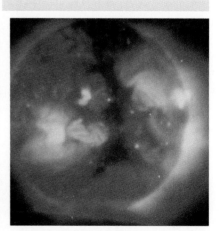

(a)

(b)

(c)

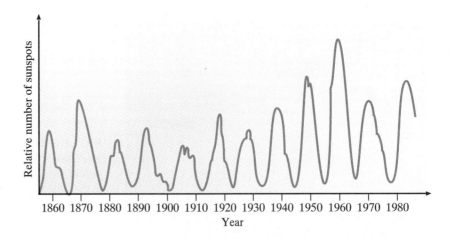

FIGURE 16.10 ··········

The Sunspot Cycle The number of spots observed on the Sun's disk varies with an 11-year periodicity. Other forms of solar activity such as flares, etc., vary in the same way.

the solar atmosphere that are cooler than the surrounding photospheric regions by some 1500 K. The reduced temperatures of the sunspots cause them to appear darker than the higher temperature regions surrounding them. They are believed to be caused by an inhibition of energy flow into these regions by the presence of strong, localized magnetic fields.

Faculae and *plages* also occur in regions of the photosphere that also have highly localized magnetic fields. In this instance, however, the presence of the magnetic field seems to superheat the region to higher temperatures than the surrounding photospheric regions, and these features have the appearance of bright, granulated clouds. From time to time, huge jets of gas will, in a matter of minutes or hours, shoot thousands of kilometers into space from the Sun's photosphere. These jets, called *prominences,* are often twisted and bent by magnetic fields and can take on a variety of ever-changing shapes. Prominences that are not at the edge of the Sun project onto the solar disk as dark, wispy features called *filaments.*

Flares are perhaps the most impressive manifestation of solar activity phenomena. Sometimes a highly localized area of the Sun will suddenly brighten up over a period of only a few minutes. As it does so, a wide variety of emissions, including X rays, bursts of radio waves, and high-velocity atomic particles, are hurled into the interplanetary medium. To make life even more interesting for the astrophysicist, these phenomena and events seem to come and go over an 11-year cycle that is popularly referred to as the sunspot cycle (Figure 16.10), but is in effect a solar activity cycle, in which not only the number of sunspots but the observed occurrences of flares, plages, and prominences also increase and decrease with the same 11-year cycle. These and other solar atmospheric phenomena are at best only partly explained by modern astrophysics.

WORKED EXAMPLE

How far from the Sun would Earth have to be in order for Earth to have a solar constant of 50 watts/meter^2?

Continued

WORKED EXAMPLE

Continued

SOLUTION: For an incandescent light source we have in general that

$$\frac{\text{distance}}{\text{to source}} = \sqrt{\text{power}/4\pi \times \text{intensity of source}} \ .$$

If the new solar constant is 50 watts/meter^2, then we have

$$\frac{\text{distance}}{\text{to source}} = \sqrt{3.84 \times 10^{26}/4 \times 3.14 \times 50}$$

$$= \sqrt{0.0061 \times 10^{26}}$$

$$= \sqrt{61 \times 10^{22}}$$

$$= 7.8 \times 10^{11} \text{ meters} \ .$$

The sun would have to be at a distance of 7.8×10^{11} meters in order to exhibit a solar constant of 50 watts/meter^2. This is a little over five times the sun's present distance.

1. How does the astrophysicist determine the radius, mass, temperature, and power output of the Sun?
2. Why do we believe that the sun shines as a result of nuclear fusion processes?
3. What is a model solar or stellar interior? Why is it necessary for astrophysicists to make use of them?
4. Describe the journey of a photon of energy from the time it is formed until it reaches the Sun's outer layers.
5. Summarize the various phenomena that occur in the Sun's photospheric layers.

REVIEW 16.2 QUESTIONS

1. If life on Earth is estimated to date back about 2.5 billion years, how much energy would the Sun have radiated into space in that time if it has maintained a constant power output throughout this time interval?
2. If the solar constant were measured on Earth to be 2000 watts/m^2, what would the Sun's total power output be?

REVIEW 16.2 PROBLEMS

16.3 Physical Characteristics of the Stars

From the initial realization that Earth orbits about the Sun, astrophysicists simultaneously surmised that the stars necessarily had to be situated at enormous distances from the Sun, or otherwise the nearest of them would exhibit relatively large parallax angles as Earth proceeded in its orbital motion. For more than two centuries after the invention of the telescope, astronomers' efforts to measure the parallax angles of the distant stars came to naught. When the first parallax angles were finally measured in the middle of the last century, they confirmed the long suspected fact that the stars were orders of magnitude more distant than the outermost planets. Thus, the nearest star, called Alpha Centauri, is found to be located at a distance of some 4×10^{16} meters from the Sun, or about 6000 times further away than the last known planetary outpost of the Solar System. On this same scale, if the Earth-Sun distance were 1 meter, the distance between Earth and Alpha Centauri would be 270 kilometers!

Because stellar distances are so large, and the parallax angles measured to obtain them so small, astronomers employ a unit of distance called the *parsec,* which is defined as the distance to a star having a parallax angle of one arcsecond or $\frac{1}{3600}$ of a degree. Using these units, the relationship between the distance to a star in parsecs and its parallax angle in arcseconds is simply

$$\text{distance (in parsecs)} = \frac{1}{\text{parallax angle (in arcseconds)}}.$$

With suitable conversion calculations, we can show that one parsec is equal to about 3.1×10^{16} meters and one arcsecond is about 4.8×10^{-6} radians. Astronomers also employ a unit of distance called the *light year,* which was defined in the last chapter as the distance a photon of light can travel in one year moving at a rate of 300,000 km/sec. One light year is thus equal to about 9.5×10^{15} m or about 0.3 parsec.

Despite the fact that the stars are at such large distances, about 5000 of them are bright enough to be seen with the naked eye. Clearly, for stars to appear as bright as they do at the distances at which they are located, the stellar energy outputs required are comparable to that of the Sun. For example, the observed intensity of the light coming from the star Alpha Centauri is approximately 3.0×10^{-8} watts/m^2. From the measurements of the parallax angle of Alpha Centauri, the distance to this star is about 1.32 parsecs or about 4.1×10^{16} meters. If we calculate the total power output of Alpha Centauri into all directions of space in a fashion similar to our previous calculation for the Sun, we obtain

$$\begin{aligned} \text{power of} \atop \text{Alpha Centauri} &= (3.0 \times 10^{-8}) \times 4\pi \times (4.1 \times 10^{16})^2 \text{ watts} \\ &= 6.34 \times 10^{26} \text{ watts} \end{aligned}$$

or about 1.6 times that of the Sun's power output. Similar calculations for other stars reveal a rather large range of values for stellar power production. Stars exist that produce energy at a rate over one million times that of the Sun, while others have a power output that is barely a millionth that of the Sun. Thus, the Sun's power output is about average, not remarkably large nor small as stars go.

Because the Sun seems to be a typical star in terms of its energy production as well as its other physical characteristics, astrophysicists most often measure stellar properties in terms of ''solar'' units. Thus, the Sun's power output of 3.84×10^{26} watts is defined as one solar power unit or one *solar luminosity.* A star having a power output of seven solar luminosities would be radiating a total power of $7 \times 3.84 \times 10^{26}$ watts $= 26.9 \times 10^{26}$ watts. Similar solar units are defined for the mass, radius, and photospheric temperature. In this system, one solar mass unit is defined as 2.0×10^{30} kg, one solar radius unit is equal to 6.9×10^8 meters, and one solar photospheric temperature unit is equal to 5800 K.

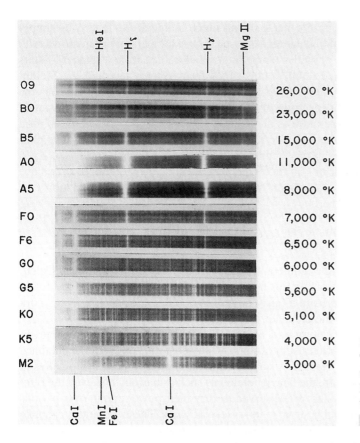

FIGURE 16.11 ···········

The Temperature Sequence of Stellar Spectra As the temperature of a star changes, the appearance of its spectral line pattern can alter dramatically as these astronomical spectral types O–M illustrate.

Stellar masses are determined from observing the orbital motions of binary stars and making use of the harmonic law. Results of such determinations indicate that stars range in mass from about 0.1 solar mass unit to nearly 100 solar masses. Temperatures of stars can be obtained from intensity spectra and by employing either Wien's law or the Stefan-Boltzmann law. Early in this century, it was discovered that the patterns of absorption lines in stars of differing temperatures can be arranged into the sequence of spectral types as shown in Figure 16.11 because as the temperature of a given gaseous mixture changes, the energy levels occupied by the electrons of the atoms in the gas will change as well. Thus, an electron in a given energy level will drop down or *thermally deexcite* to a lower level at a lower temperature or *thermally excite* up to a higher level at a higher temperature (Figure 16.12). As a result, the wavelength pattern of spectral lines that atoms in a given gas can absorb is controlled in large measure by the temperature of the gas.

FIGURE 16.12 ···········

An Electronic View of the Stellar Temperature Sequence As the temperature of a star changes, so do the energy states of the electrons of the atoms in its atmosphere. Thus, a line capable of absorbing a given wavelength photon at one temperature will not be able to absorb it at a different temperature, thus giving rise to the changes in the pattern of absorption lines exhibited by the star.

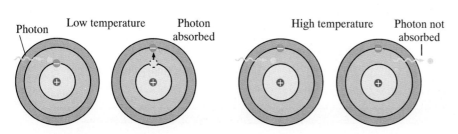

We can make use of this fact to obtain stellar temperatures readily by simply comparing the spectral line patterns of a star whose temperature is unknown with the spectral sequence of stellar temperatures determined from "standard" stars having well-determined temperature values from Wien's law and the Stefan-Boltzmann law. The standard star line spectrum that most closely matches the patterns of lines in the spectrum of the star of unknown temperature is assumed to have the same temperature as the unknown star. This process is referred to in astronomy as the method of *spectral classification*. Almost all stars have photospheric temperatures in the range of 2,000 K to 50,000 K.

If the photospheric or "surface" temperature and power output of a star have been determined, one can obtain the radius of that star from the relation (see Physics Formulation 16.1):

$$\text{radius of star} \atop \text{(in solar radii)} = \sqrt{\frac{\text{power of star}}{(\text{temperature of star})^4}},$$

where the power and the temperature are both expressed in solar units. Calculations of stellar radii done in this fashion have produced the amazing result that stars can range in size from a fairly large metropolitan area such as New York or Los Angeles up to the diameter of the entire planetary system!

Additional investigation and observations have uncovered even more diversity in physical characteristics of stars. While most stars have similar chemical compositions, certain classes of stars have remarkable abundance anomalies in elements such as carbon, the heavier elements such as barium, and even the rare-earth elements. Some stars are observed to vary their power output over periods of time ranging from an hour or so up to several years. These variations can take the form of pulsations in some stars or sudden outbursts of gargantuan proportions in others. Some stars have photospheric densities so low as to be almost indistinguishable from the interstellar medium, while others have densities in the trillions of kg/m³. Some stars rotate so rapidly that they are barely able to hold themselves together gravitationally, while others rotate so slowly that their spinning is barely perceptible.

These characteristics can be combined in a variety of ways to produce a variety of types of stars. Thus, while stars in an overall way possess the properties of an incandescent ball of gas, they simultaneously retain a great deal of individuality as well. In short, by the 1920s, astronomers and astrophysicists had uncovered a plethora of different types of stars, which leads immediately to the basic question of how the plethora came to be.

WORKED EXAMPLE

The bright star Antares has a temperature of one-half that of the sun and a power output that is 10,000 times that of the sun. What is the radius of Antares in solar radii?

SOLUTION: In general, we have that

$$\text{radius of star} \atop \text{(in solar radii)} = \sqrt{\frac{\text{power output (solar units)}}{\text{temperature (solar units)}^4}}.$$

Continued

> ## WORKED EXAMPLE
>
> *Continued*
>
> For Antares the power output in solar units is equal to 10,000 and the temperature in solar units is equal to one-half. Thus, we have that
>
> $$\text{radius of Antares} = \sqrt{10,000/(1/2)^4}$$
>
> $$= \sqrt{10,000/(1/16)}$$
>
> $$= 400 \text{ solar radii .}$$
>
> Thus the radius of the star Antares, which is one of a class of stars called supergiants, has a radius comparable to the orbital radius of the planet Mars!

1. How are the distances to stars obtained? How do these distances compare with the distances to the planets?

2. Explain how stars having the same chemical composition but different temperatures can exhibit differences in their spectral line patterns.

3. What physical laws are employed in the determination of stellar temperatures?

4. Discuss the information that can be obtained about a star from an analysis of the spectrum of its light.

5. Briefly describe how the following are obtained for stars:
   a. power output   **b.** temperature   **c.** radius   **d.** mass.

REVIEW 16.3 QUESTIONS

1. The bright star Vega exhibits a parallax angle of 0.123 arcseconds. Find the distance to Vega in
   a. parsecs   **b.** light years   **c.** meters.

2. The bright star Aldebaran has a temperature of about three-quarters that of the sun and a power output 100 times that of the sun. What is the radius of Aldebaran compared to that of the sun?

REVIEW 16.3 PROBLEMS

16.4 Stellar Life Cycles

In many ways there is a striking resemblance between the astrophysical problem of accounting for all of the "species" of stars found in space and the biological problem of accounting for all of the species of life found on Earth. In 1859, Charles Darwin shook the Victorian era by making the claim in his book *On the Origin of Species* that the variety of species of plants and animals observed on Earth were the product of responses of life forms to a variety of evolutionary "forces," such as climate and food supply over millions of years of time. Among other things, Darwin convincingly demonstrated the fact that the earth's eco-system is not now nor had it ever been a static, "immutable" entity, bur rather one that is ever-changing. Those changes, however, most often occur over time scales that dwarf a single human being's lifetime or indeed the entire span of human existence on the planet. In attempting to explain the origin of stellar "species," the astrophysicist too assumes that the time scale for virtually all cosmic change is comparable to and in fact exceeds those of Earth's geological and paleontological history. Like the paleontologist or archaeologist who care-fully sifts through the sand and soil of a key site, seeking out bits and pieces of the mosaic of the earth's living past, so also does the astrophysicist scan the heavens for revealing bits and pieces of a cosmic past that are often little more than the wispy remains of a long-ago exploded star.

Scientists, however, have ascertained that the development of species as envisioned by Darwin does not apply to stars. Clusters of stars formed at the same time in the same part of space and subject to the same "cosmic evolutionary forces" should, in the Darwinian view, result in all of the stars in the cluster becoming stars of the same species. One need only observe the variety of colors and power output exhibited by the member stars of a given cluster to demonstrate the error of such a proposition. Even though the Darwinian scenario of evolution is not that of the stars, the life scientist can still offer aid and comfort to the astrophysicist in this matter.

Associated with every life form on the earth is a life cycle in which a given creature is born, grows up, reproduces itself, and dies. During the course of such life cycles, the physical characteristics of a given creature can change drastically. Butterflies, for example, lay eggs from which caterpillars hatch. After chomping on selected plants and flowers for a few weeks or so, the caterpillar spins itself into a silken cocoon from which the adult butterfly emerges to mate and ultimately lay more eggs, thus beginning the cycle over.

We now believe that this concept of life cycles can be used to account for the existence of stellar species. In this view, there are actually only a few basic types of stars, but as each goes through a life cycle having a variety of stages with a variety of physical characteristics, the result is the seeming plethora of different stellar species. Unfortunately, unlike the entomologist who can carefully watch and monitor the life cycle of a butterfly as it unfolds over weeks or months, the astrophysicist is confronted with stellar life cycles that run into billions of years, compared to which a human life span is but a wink of an eye. Despite such difficulties, astrophysicists have employed a blend of the principle of uni-versality, direct measurements and observations, and computer models and sim-ulations to piece together a portrait of a star's life cycle that is not only consistent with our knowledge of the behavior of the physical world, but also is able to account for a large fraction of the stellar species we have observed and cataloged.

Astrophysicists are virtually certain that stars have their beginnings in vast clouds of gas and dust such as those that line the band of light in the sky we call the Milky Way. On such gas cloud in the constellation of Orion is over 1200 light years in diameter (Figure 16.13). From our discussion of thermodynamics, however, we know that there is a tendency toward disorder in the physical world that can only be overcome in a given system with an input of energy. Gas and dust clouds in the interstellar medium are in a very high state of disorder and are hence not at all about to coalesce into highly ordered spheres of gas on their

FIGURE 16.13
The Great Nebula of Orion This gigantic cloud of gas, dust, and newly formed stars is over 1200 light years in diameter.

(a)

(b)

FIGURE 16.14

Triggering Mechanisms in Star Formation   (a) A collision of hot and cold gas and dust clouds; (b) a shock front from a supernova explosion.

own. Accordingly, astrophysicists have deduced the existence of several processes at work in the interstellar medium, called *triggering mechanisms,* by which star formation is induced or triggered in these otherwise stagnant gas and dust clouds (Figure 16.14). Each of these processes is thought to produce a high-velocity shock wave, which fragments the given gas or dust cloud into compacted segments called protostars. Such shock fronts can be produced by collisions of high and low temperature gas and dust clouds, stellar explosions, interstellar winds produced by the formation of large mass stars, and even from violent processes occurring at the center of the Milky Way.

Once the protostar is formed, the remainder of the star's life can be thought of as a duel between self-gravitation, which seeks to collapse the star into an ever-smaller size, and the star's power production, which seeks to blow the star apart. At first, the protostar has little or no power production with which to resist the contraction induced by the self-gravitation of the mass within the protostar. As a result, the protostar continues to shrink in size from an object roughly one to two light years across down to an object roughly the size of the Sun. As the protostar continues its contraction, its density increases to the point where it becomes opaque to visible radiant energy and thus appears as a dark irregular-shaped object called a *globule* silhouetted against the backdrop of more distant stars and glowing gas (Figure 16.15).

As the contraction proceeds, the temperature and pressure at the center of the protostar continue to increase until they are sufficient to initiate thermonuclear fusion reactions at the center of the protostar. The energy release halts the

FIGURE 16.15

Globules   When a star first forms into a protostar, it becomes opaque to visible light and appears as a dark, irregular-shaped globule against the background sky glow from stars and fluorescing gas.

contractions of the protostar, but not without an attendant amount of flare and prominence activity, which in effect propel the protostar's outer layers back into the interstellar medium. It is this material that is believed to accrete eventually into planets and satellites. At the onset of nuclear fusion reactions at the center of the protostar, the gas ball takes on an equilibrium configuration in which forces associated with the outward flow of power are balanced by the inward pull of the protostar's self-gravitation. The protostar is then a full-fledged star.

The physical characteristics taken on by stars in this equilibrium stage are almost totally controlled by the star's mass. A star formed out of a 20 solar mass globule will attain a power output 4300 times that of the Sun, a radius five times that of the Sun, and a surface temperature three times that of the Sun. On the other hand, a star that forms out of only one-fifth of a solar mass of material will shine with only 0.007 of the power output of the Sun, will be only one-third as large, and possesses a temperature about half that of the Sun. Thus, there is an entire series of possible equilibrium configurations, depending on the mass of the cloud out of which the stars are formed. Astrophysicists refer to this set of equilibrium configurations as the *main sequence,* and a star that is in this equilibrium phase of its life cycle is said to be in its *main sequence stage* (Figure 16.16).

As the star continues to shine in its main sequence stage, the helium atoms formed from the hydrogen fusion sink to the center of the star and begin to form

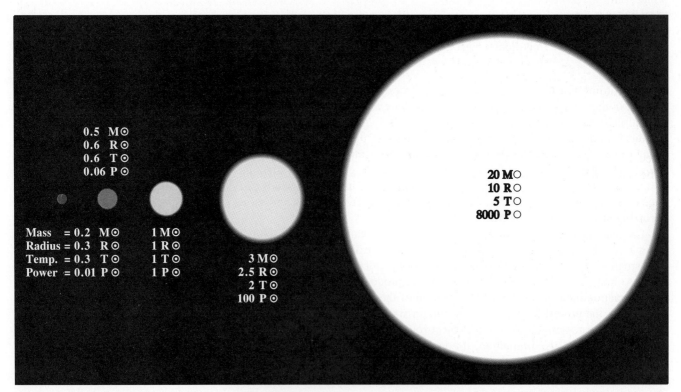

FIGURE 16.16 ··········

The Main Sequence Stage As a star forms, it takes on one of a set of stable configurations of physical characteristics, which are dependent on the total mass of the star.

a helium core. In the more massive main-sequence stars with larger power output, this process occurs more rapidly than in the lower mass main-sequence stars. Eventually the size of the helium core, which is not producing any energy of its own, grows to the point where it begins to contract gravitationally in size. As it does so, the power produced at the core of the star increases due to a release of gravitational potential energy as the core shrinks to a configuration of smaller gravitational potential. The rate of hydrogen fusion also increases due to an increase of the core temperature. As a result of the increase of the power output of the star, its outer layers expand in an attempt to settle into a new equilibrium configuration. The increase in energy, however, cannot heat this new stellar volume as efficiently and the star actually has a cooler photospheric temperature. Such a star having an increased power output, very much increased radius, and cooler surface temperature is said now to be in its *red giant stage*. The time that a star spends in its main sequence stage before the changes at the core turn it into a red giant depends on how massive the star was in its main sequence stage. Computer models tell us that more massive stars evolve much more rapidly than less massive stars. Thus, the core changes that produce the red giant phase of a star's life occur within about 20 million years for a 10 solar mass star, but will take as long as two trillion years for a star with one-fifth of the Sun's mass.

The ultimate fate of a given star once it reaches the red giant stage once more depends almost entirely on the star's mass (Figure 16.17). The lower mass stars gently eject their outermost layers to form a shell of gas that can be as large as a light year across. Left behind are the hot core regions, which produce enough high-energy photons to cause the surrounding shell of gas to fluoresce. The resulting object has the appearance of a dimly glowing planetary disk when viewed through a telescope and is referred to as a planetary nebula (Figure 16.18). Eventually, the gas shell dissipates into the interstellar medium and abandoned core cools and contracts into a small compact object called a white dwarf in which one solar mass of material has been gravitationally compacted into an object the size of the earth, with the resulting density equal to about 2×10^9 kg/m^3. The compacting is finally stopped by the counter pressure of the quantum-mechanical crowding effects of the electrons of the atoms in the white dwarf. The star then locks into the final stable configuration of mass and radius in which it will remain until finally cooling into a dark, cold stellar hulk called a black dwarf. This is the ultimate fate of our own Sun in another five to six billion years.

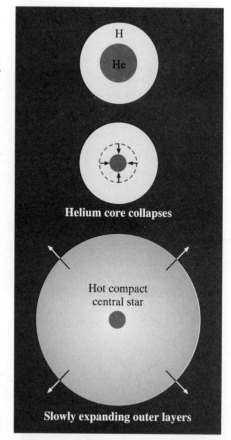

FIGURE 16.17 ···········
The Post–Red Giant Stages of a Lower Mass Star The outer layers of the star are gently pushed into the interstellar medium, leaving behind an Earth-sized compact core region.

FIGURE 16.18 ···········
A Planetary Nebula The outer layers of this lower mass star have detached, leaving behind a small, hot central star that will eventually cool progressively into a white dwarf and finally a black dwarf.

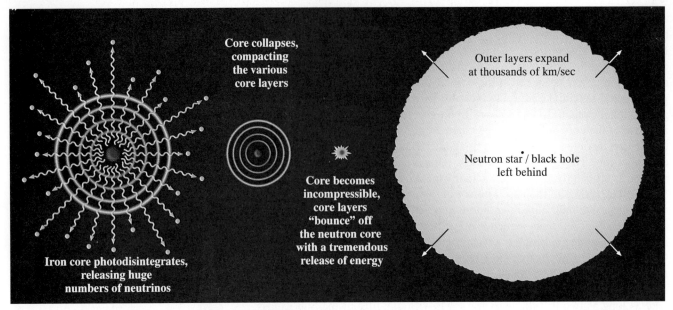

Core collapses, compacting the various core layers

Iron core photodisintegrates, releasing huge numbers of neutrinos

Core becomes incompressible, core layers "bounce" off the neutron core with a tremendous release of energy

Outer layers expand at thousands of km/sec

Neutron star / black hole left behind

FIGURE 16.19 ··········
The Post Red Giant Stages of a Higher Mass Star After creating a series of fusion layers of elements up to iron in atomic weight, the central core collapses to neutron densities and the outer layers fall into the core and then ''bounce'' off, creating a huge explosion that implodes the cores and sends the outer layers off into space at thousands of kilometers per second.

The fate of more massive stars is not nearly as gentle. Stars having masses larger than about 10 solar masses or so have sufficient mass to crush and thermonuclearly fuse helium atoms at their cores into a series of shells of progressively heavier elements, helium into carbon, carbon into neon, neon into oxygen, oxygen into silicon, and finally silicon into iron (Figure 16.19). When iron nuclei are produced at the cores of such stars, the iron core regions collapse, and in roughly one-tenth of a second the core temperature rises to the almost unimaginable values of 5 to 10 billion Kelvins. The radiant energy from an incandescent gas at such a temperature consists of extraordinarily high-energy, small-wavelength gamma-ray photons that shatter the iron nuclei into alpha particle fragments by means of a process called *photodisintegration*. In the next tenth of a second, the protons and electrons in the erstwhile iron of the star core are crushed into neutrons in a process called *neutronization*. This latter process is accompanied by a huge burst of neutrinos into the interstellar medium.

When the central core density reaches that at which protons and neutrons are packed into atomic nuclei densities (about 4×10^{17} kg/m^3), the core suddenly becomes very incompressible and rigid. As the outer layers of the star rush toward the now incompressible core of the star at speeds in excess of one-tenth that of light, they create huge temperatures and pressures, which in effect cause them to ''bounce'' off the core. This effect, which is called the *nuclear bounce effect* or *core bounce,* occurs in a matter of milliseconds and is accompanied by a blast of incredible proportions, which astrophysicists have called a supernova explosion (Figure 16.20). Supernova detonations are of such magnitude that for a few days they can shine with a power output ten billion times that of the sun!

FIGURE 16.20 ·············
A Supernova in a Distant Galaxy   The power output of this exploding star was roughly ten billion times that of the sun.

The outer layers of the supernova, which bounce off the core, are returned to the interstellar medium in the form of gaseous supernova remnants. The core left behind is now a sphere of closely packed neutrons barely 30 kilometers in diameter. If the mass of the core is less than about three solar mass units, then the quantum-mechanical crowding effects of the neutrons create a counterpressure that prevents the core from any further gravitational contraction. The star thus takes on a stable configuration and becomes a *neutron star.*

Although all of the details have yet to be worked out, it is believed that such compacted neutron stars rotate very rapidly, as much as 30 times per second, and also possess very highly compacted and therefore very high intensity magnetic fields. As the neutron star rapidly rotates, charged particles at the surface of the star are accelerated by the magnetic field along the field lines. These accelerated particles emit beams of electromagnetic radiation from the magnetic polar regions of the neutron star. As the star rotates, these twin beams of electromagnetic waves sweep through space much as the beam from a lighthouse or airport beacon sweeps through the darkness. These beams of electromagnetic cones emanating from rapidly rotating neutron stars manifest themselves to us as the flashing *pulsars,* in which radio and visual electromagnetic waves have been observed to wink on and off with periods measured in fractions of a second. Perhaps the most famous pulsar is that buried in the Crab Nebula, an object known to be the gaseous remnant of a supernova explosion that was observed to occur in 1054 A.D. (Figure 16.21).

FIGURE 16.21 ···········
A Supernova Remnant   This object, called the Crab Nebula, is the remnant of a star that exploded in 1054 A.D. Nearly a thousand years later the object consists of a flashing neutron star, called a pulsar, embedded in the star's blown-off outer layers.

(a)   (b)

Eventually, as the rotational energy of the neutron star is converted into electromagnetic radiation, the neutron star slows its rotation, and the light from its "beacon" is forever darkened. It then joins its larger brothers and sisters, the black dwarfs, as a cold, lightless, burned out stellar corpse.

If the neutron core left behind in a supernova explosion exceeds three solar mass units, then the degenerate gas pressure from the quantum-mechanical crowding effects of the packed neutrons can no longer balance the self-gravitational effects, and the core collapses further into an object the size of a typical college campus. At a radius of less than two kilometers and a mass in excess of three solar mass units or about 6×10^{30} kg, the densities in these objects now soar to values in excess of 1.0×10^{21} kg/m^3. Moreover, the velocity of escape from the surfaces of such objects exceeds the speed of light, and since no energy or mass entity has been observed to exceed this speed, any particle or photon caught in the gravitational clutches of such an object cannot escape. The object in effect becomes a sink into which masses and energy can fall, but from which they cannot escape. In short, the object disappears from the observable universe. Such objects, therefore, are appropriately referred to as *black holes,* and are perhaps the most intriguing of all the conjurings of modern astrophysics (Physics Formulation 16.2).

Black holes present us with a number of interesting challenges and important questions. Absolutely essential to our investigations of any celestial object is our ability to detect its presence via electromagnetic radiation. By their very nature, black holes do not permit such a direct revelation of themselves. Instead, we must seek out these elusive entities by searching for whatever observable effects they might have on their surroundings. Proceeding along these lines, we have discovered a number of objects and systems whose behavior strongly suggests the presence of a black hole.

The most studied of these is an object in the summertime constellation of Cygnus the Swan, which is called Cygnus X-1. Cygnus X-1 is a binary system consisting of a hot, very large, blue "supergiant" star, which is gravitationally paired with an unseen source of X rays. The model that seems to best fit the measurements and observations made on this system is illustrated in Figure 16.22. It is believed that the blue star, like others of its type, emits a large-scale stream of particles called a stellar wind. Part of this wind is intercepted by an orbiting black hole, whose crushing gravitational field compacts the incoming gases such that their temperatures rise to values in excess of 2 million Kelvins. At these temperatures, the gas becomes a plasma.

FIGURE 16.22 ···········

Cygnus X-1   At the stellar level, Cygnus X-1 is probably the best example of a black hole in action. Stellar "winds" of material from a large, hot star are gravitationally captured and compacted into a hot plasma by the orbiting black hole. As the charged particles spiral in toward the black hole, they radiate the X rays we see here on Earth.

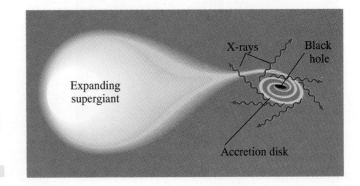

As the now-charged atoms in the plasma spiral toward the black hole, they are in effect charges oscillating back and forth and, as such, produce the X rays we see on Earth. In a certain sense, the X rays coming from Cygnus X-1 can be thought of as the last cry for help by the atoms in the plasma just before they cascade over the black hole's event horizon and into oblivion. The ultimate fate of the energy and matter that makes such a plunge is a topic of considerable speculation in current astrophysics, as is the fate of the black holes. We know that the end of a star's life sees vast amounts of stellar material returned to the gas and dust clouds of the Milky Way, and a lesser amount locked forever into cold, dark, and compact spheroids of mass. The stellar material that has collapsed into black holes could possibly end up in either or neither of these scenarios.

WORKED EXAMPLE

A typical white dwarf star has a mass equal to about 2×10^{30} kg and a radius of about 8×10^6 m. Find the density of this white dwarf. Assume that the white dwarf is spherical in shape and therefore has a volume of $\frac{4}{3}\pi(\text{radius})^3$.

SOLUTION: In general, the density of a substance is given by

$$\text{density} = \frac{\text{mass}}{\text{volume}}.$$

For a spherical white dwarf star,

$$\text{density} = \frac{\text{mass}}{\frac{4}{3}\pi(\text{radius})^3}.$$

Substituting the values for the mass and radius of the white dwarf, we obtain

$$\text{density} = \frac{2 \times 10^{30} \text{ kg}}{\frac{4}{3}\pi \times (8 \times 10^6 \text{m})^3}$$

$$= \frac{2 \times 10^{30}}{4.2 \times 512 \times 10^{18}} \text{ kg/m}^3$$

$$= \frac{2 \times 10^{30}}{2150 \times 10^{18}} \text{ kg/m}^3$$

$$= \frac{9.3 \times 10^{-4} \times 10^{30}}{10^{18}} \text{ kg/m}^3$$

$$= 9.3 \times 10^8 \text{ kg/m}^3 .$$

By contrast, the density of water is 1000 kg/m^3 and that of gold is 1.9×10^4 kg/m^3.

1. Compare and contrast the theory of stellar life cycles with the biological theory of evolution.
2. How do stars form?
3. What is meant by the main sequence stage?
4. Describe the various forms of stellar "death."
5. Discuss the differences in the life cycle of a star having a large mass with one having a small mass.

1. Verify by direct calculation the statement in the text that a main sequence star having one-half the temperature of the sun and a radius one-third that of the sun will shine with a power output of the sun.
2. If the density of a certain gas cloud is 5×10^{-20} kg/m^3, how much volume of this gas cloud will be required to form a star having a mass equal to the sun's mass of 2×10^{30} kg? If the volume were in the shape of a cube, how large would each side be?

16.5 Cosmology

If one could stand off at a distance from Earth that is farther than the distances to the gas, dust, and star clouds of the Milky Way, the distances to the exterior galaxies, and even the distances to the most remote quasars, one presumably could obtain an overview of the entirety of all that we see in the physical universe. While we are not physically able to make such a journey, we can nonetheless deduce at least in part what such an overview would hold in store for us. The astrophysical attempt to get such a "big picture" is referred to as *cosmology*.

In the course of our measurements and observations of celestial objects, we have found a number of general characteristics of the universe that provide us with some clues as to what the overview of the universe is currently like, and, most interestingly, what that overview has been like in the past and what it might be like in the future. The most striking observation about the universe is that it seems to be expanding in all directions. If we peer outside of our own Milky Way and look at other similar aggregates of stars, called galaxies, we find that the spectral line patterns of each and every one of them is displaced toward longer wavelengths than their terrestrial laboratory wavelengths would dictate. Such a displacement can be explained only in terms of a Doppler shift arising from a velocity of recession. Moreover, the more distant galaxies have red shifts, which indicate larger velocities of recession. In other words, the more distant the galaxy, the faster it is receding from us (Figure 16.23).

If we observe elemental abundances, we find that certain elements and isotopes, most notably the isotope deuterium, ought not to exist at their observed abundance if they were created in some sort of stellar nucleosynthesis process. As quickly as a deuterium isotope is created, for example, it should immediately be transmutated into helium under the same physical conditions that created it.

PHYSICS FACET

16.2 A Universe of Galaxies

Thinkers from as early as the time of the Greek Democritus thought of the Milky Way as a vast aggregate of stars, but it wasn't until the first telescopic observations of Galileo that such a view could be directly verified. In the first years of the twentieth century, the perceived universe was neatly packaged into an oblate-spheroid-shaped Milky Way roughly 30,000 light years × 6,000 light years in size, which was referred to as the Kapteyn universe, named after the turn of the century Dutch astronomer who helped develop this view.

The Kapteyn universe was able to account for the locations of nearly all of the observed interstellar celestial objects, including stars, dust clouds, and glowing nebulae by the simple, verifiable view that they all belonged to the Milky Way "universe." One could see, for example, that stars tended to congregate along the lanes of the Milky Way and thus had to be a part of this system. Two classes of objects, however, did not fit into the portrait.

The first was a class of objects called globular clusters, which are huge balls of tens or even hundreds of thousands of stars that seemed to have no particular statistical preference as to whether they were close to or away from the plane of the Milky Way. The second class consisted of objects that seemed to be like the gaseous nebulae of the Milky Way, but were remarkable in that they possessed definite geometric shapes, most notably ellipses and spirals. These objects were even more unusual than the globular clusters in that they seemed to "avoid" the plane of the Milky Way. The further the angular distance from the plane of the Milky Way, the higher the number of such objects that could be seen.

The keys to unlocking these mysteries came in 1908, courtesy of an astronomer at the Harvard Observatory named Henrietta Leavitt. Leavitt found that for a certain class of pulsating variable stars, called *cepheid variables,* a mathematical relationship existed between the period of light variation and the intrinsic mean power output of these stars. In short, to obtain the mean power output of a given cepheid, one needed only to measure its period of light variation. By determining the observed mean intensity of

the light from a given cepheid variable and combining this result with its mean power output, its distance could be determined.

In 1916, another Harvard astronomer, Harlow Shapley, discovered cepheid variables in the globular clusters and, by means of the same procedure used by Leavitt, was able to deduce that the globular clusters formed a roughly spherical array about a center some 30,000 light years from the sun. Moreover, this sphere was about 100,000 light years across. Since Shapley's work, astronomers have uncovered a Milky Way that is a flat, spiral-shaped system consisting of 200 billion younger stars encased within a spherical halo of globular star clusters and other galactically ancient objects and stars.

Less than a decade later, Edwin Hubble, an astronomer at Mt. Wilson Observatory in Southern California, found cepheid variables embedded in the spiral and elliptically shaped "nebulae" off the Milky Way plane. This time, however, measurements of the distances to the cepheids

••• **The Hercules Cluster of Galaxies**   Almost every image on this photograph is a galaxy comparable in size and stellar content to the Milky Way.

PHYSICS FACET

Continued

and hence to the system in which they were located reached into the millions of light years, well outside even Harlow Shapley's newly enlarged Milky Way. Hubble demonstrated that these objects were in fact other Milky Ways, "island universes," which astronomers now call galaxies.

Obscuration by the gas and dust in our own Milky Way prevents us from seeing many of these objects close to the Milky Way plane. But away from that obscuration plane, astronomers have literally found a universe of galaxies stretched in all directions as far as the 12 billion light years

that our telescopes can see. Each of these galaxies is a system of stars as vast and complex as our own Milky Way Galaxy. There are spiral and elliptically shaped galaxies as well as those with no regular shapes at all. There are galaxies that appear to be exploding on scales measured in thousands of light years, and those that may have their centers occupied by vast black holes with masses a billion or more times larger than the Sun and capable of consuming the entire galaxy in which they are located. All of these discoveries occurred in the span of a single human lifetime!

CLUSTER NEBULA IN	DISTANCE IN LIGHT-YEARS	RED-SHIFTS
VIRGO	78,000,000	H+K → 1,200 KM/SEC
URSA MAJOR	1,000,000,000	15,000 KM/SEC
CORONA BOREALIS	1,400,000,000	22,000 KM/SEC
BOOTES	2,500,000,000	39,000 KM/SEC
HYDRA	3,960,000,000	61,000 KM/SEC

FIGURE 16.23 ·········
The Hubble Law In the 1920s, American astronomer Edwin Hubble discovered that more distant galaxies were receding faster than those close to the earth. Since the light from the more distant objects is "older" light, the earlier recessional speeds of the universe were larger than they were in subsequent time. The expansion rate of the universe is thus slowing in time.

Nonetheless, deuterium, and fairly large amounts of it, exist in nature. Thus, some other mechanism for element production must exist besides that found to occur at the centers of stars.

A scan of the heavens also reveals that a background of radiant energy exists that permeates the interstellar medium, much as the sky glow from the lights of a big city permeates the atmosphere above that city. If a blackbody were to be placed in this background radiation field, it would heat up or cool off to an equilibrium temperature of about 3 K. This all-pervasive radiation field is thus often referred to as the 3-K cosmic microwave background, and seems to be a fundamental characteristic of the universe as a whole (Figure 16.24).

Finally, if one carefully measures the distances and directions of the millions of galaxies in space, one finds that the galaxies are not uniformly distributed, but instead possess a lumpy, filamentary structure throughout space (Figure 16.25). To explain these and other observed characteristics of the universe as a whole, astrophysicists have developed an overview of the universe that is at differing levels both simple and complex. It is a view that is popularly referred to as the *Big Bang*.

Put very simply, the Big Bang theory states that about 15 to 20 billion years ago, an event occurred, the Big Bang, in which the universe, as we observe it currently, was set in motion. Combining observations of the universe as we presently observe it with the laws of nature as we presently observe them, astrophysicists have pieced together a fascinating scenario of what the Big Bang must have been like at the beginning of its existence in time past as well as its ultimate fate in future time.

Any discussion of the Big Bang theory should be prefaced with a basic philosophical acknowledgment that science currently has no way of describing the universe prior to the Big Bang, nor can it even describe the Big Bang in the opening flash of its existence. In our discussion of quantum mechanics, we learned that there is a fundamental limit to our ability to perform accurate measurements on the physical world. This limit is expressed in the form of the uncertainty principle. If we extrapolate our presently expanding universe backward in relativistic space-time, we come to a point at which the universe has a diameter so small that the *entire universe* lies within this fundamental uncertainty zone and, therefore, outside of our ability to describe it. To be sure, this diameter, referred to by cosmologists as the *Planck length* is an exceedingly small 4.1×10^{-35} meters, or about 10^{-20} times smaller than the diameter of a proton. The light travel time across this "cutoff" diameter is equal to

$$\text{time} = \frac{\text{distance}}{\text{velocity}}$$

$$= \frac{4.1 \times 10^{-35} \text{ meters}}{3.0 \times 10^{8} \text{ meters/sec}}$$

$$= 1.4 \times 10^{-43} \text{ seconds} .$$

Thus, during the first 1.4×10^{-43} sec of the existence of the Big Bang, the matter and energy contained in our entire universe was so packed as to render our physics powerless to analyze it. This cutoff time of 1.4×10^{-43} sec, where the Big Bang makes the transition from a physically indescribable to a physically describable state is called the *Planck time*. We can gain a rough idea of how long ago this occurred by measuring the rates at which galaxies are moving away

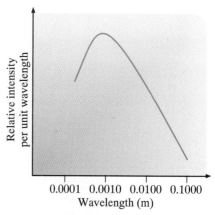

FIGURE 16.24
The 3-K Cosmic Microwave Background Energy "left over" from the Big Bang permeates the present universe in the form of a background radiation, which corresponds to that which would be emitted by a blackbody at a temperature of 3 K.

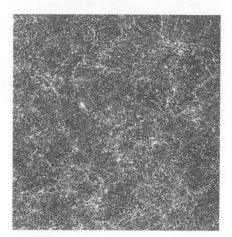

FIGURE 16.25
Large-Scale Structure of the Universe If one plots the locations of hundreds of thousands of galaxies whose distances are known, one obtains this whispy, filamentary structure.

16.3 Antimatter

The experimental confirmation of the existence of the positron and other antiparticles predicted by Dirac in the late 1920s raises the fascinating possibility that "anti-atoms" can exist whose nuclei are made up of anti-neutrons and negatively charged anti-protons, which are in turn surrounded by positively charged anti-electrons or positrons. Such *antimatter* would obviously have a hard time of it in our matter-dominated corner of the universe, since *any* antimatter contact with *any* matter could result in the mutual annihilation of both the matter and antimatter, with the total mass being completely converted into its equivalent relativistic energy.

Dirac's work in particular demands that in the course of a Big Bang *both* particles and antiparticles ought to be produced in equal numbers. What we in fact observe, at least in our local neighborhood of the universe, is an environment that is almost completely dominated by matter particles. This imbalance is referred to by cosmologists as a *particle asymmetry* in which somehow the Big Bang occurred in such a way that the creation of matter was "preferred" over that of antimatter. One possible "preference" mechanism can be developed by assuming that in the opening stages of the Big Bang expansion the universe went through a cosmological phase transition not unlike a solid melting into a liquid or a liquid vaporizing into a gas. In the course of this phase transition, the various fundamental interactions of nature "precipitate out" at different times. Thus, for a brief instant, the strong nuclear interaction was present in the universe in the absence of the weak nuclear interaction. The result, so the argument goes, was a preference for the formation of particles over antiparticles.

Others speculate that the "missing" antimatter could well exist in the form of antigalaxies. Here in the vast and remote isolations provided by intergalactic space, it is quite possible that antimatter galaxies and their antimatter constituent gases, dust, and stars could have formed in an earlier era of the universe and have continued to exist into the present without fear of annihilation. Unfortunately, the only way that matter can be distinguished from antimatter and, hence, galaxies distinguished from antigalaxies is through the short-range nuclear interactions each exhibits, and to make such determination from extragalactic distances is simply not within the scope of contemporary experimental science.

In more recent years, cosmologists have also speculated on the existence of "parallel universes," which have been invoked to explain, among other things, the ultimate fate of energy and matter that disappear beyond the event horizons of the black holes of our own universe. Along similar lines, one of the suggestions for a parallel universe is an antimatter universe in which the formation process favored the creation of antimatter particles over matter particles. Such a universe would be virtually identical to our own, except it would be made entirely of antimatter. Antimatter students would take antimatter exams from antimatter instructors. Obviously, any relationships between ourselves and the inhabitants of such an antimatter universe would have to be strictly platonic!

from us as well as their present distances. An estimate for the age of the Big Bang and hence for the universe itself can be had by calculating the time it takes a galaxy moving at its observed velocity of recession to traverse its presently observed distance. Thus, the giant elliptical galaxy M87 in the constellation of Virgo has a distance of 13×10^6 parsecs or about 40×10^{22} meters. It is also moving away from us at a rate of 1050 km/sec or 1.05×10^6 m/sec. Hence, the time it took M87 to traverse its 40×10^{22} meter distance moving at its present speed of 1.05×10^6 m/sec is

$$\text{time} = \frac{\text{distance}}{\text{velocity}}$$

$$= \frac{40 \times 10^{22} \text{ meters}}{1.05 \times 10^6 \text{ meters/sec}}$$

$$= 38 \times 10^{16} \text{ seconds},$$

which corresponds to about 12 billion years. As one might expect, such assumptions are too simplistic. We know, for example, that the expansion rate of the universe is slowing because of the self-gravitation of the mass that lies within it. Thus, the velocities of recession of the galaxies whose light left much earlier in the cosmological history of the universe, say, several billion years ago, are found to be higher than those such as M87 whose light left there much more recently. Accordingly, astrophysicists estimate that the universe is probably 15 to 20 billion years old.

Although we have no way of scientifically describing the opening instant of the Big Bang, we can describe the physical properties of the universe after the Planck time and to some degree recreate those conditions in high-energy particle accelerators here on Earth. The universe in the infinitesimal time intervals after the Planck time of the Big Bang was a universe of pure energy locked up in an almost unimaginably high-temperature, rapidly expanding space-time. Astrophysicists have estimated that the temperature of this "radiation-dominated" universe cooled from 10^{32} K to 10^{13} K in the first second of its existence, and it is much of this energy that we now see spread throughout the present volume of the universe as the 3-K cosmic microwave background.

Amidst this cauldron of energy where atoms meet galaxies and quarks meet quasars, the four basic interactions in nature—gravitational, electromagnetic, strong nuclear, and weak nuclear—were indistinguishable from each other because the virtual particles invoked to explain these interactions all behave like photons at such high temperatures. As the temperature of the universe dropped, the "photonic" properties of these virtual particles changed into those we presently observe. First gravitons "precipitated out" of this "photonic fireball" at about 10^{32} K at the instant of Planck time. The strong nuclear interactions produced gluons at 10^{27} K at 10^{-35} sec, followed by the weakons of weak nuclear interactions at 10^{15} K and 10^{-12} sec, thus leaving the photons as the sole "keepers" of electromagnetic interactions. The next second colliding gamma-ray photons were transformed into subatomic particles having relativistically equivalent masses, a process that has been observed on a far less grandiose scale in particle accelerators and is referred to as *pair production.*

In the first second after the Big Bang, the temperature of the rapidly expanding universe cooled off to the point where many of the particles, particularly protons and electrons, were "left behind." The conditions that had both created and annihilated them were no longer present and so those particles that happened to exist when the conditions changed could no longer be destroyed, nor could new particles be created. Although the gamma rays could no longer participate in a large amount of pair production, their level of intensity was nonetheless able to prevent the "left behind" protons and neutrons from combining with one another into heavier nuclei.

This situation changed when the universe was about three minutes old and had cooled off to a temperature of less than one billion K. Under these conditions, protons and neutrons could collide and ''stick'' together as deuterium nuclei without fear of being immediately broken part by the ever-present gamma rays. Once deuterium nuclei could form, they could then participate in nuclear reactions with each other as well as with protons and neutrons to form helium nuclei and, to a lesser extent, those of the light elements, lithium, barium, and boron. As the expansion and cooling of the universe rapidly continued the conditions necessary for such reactions were expanded and cooled out of existence, leaving behind these nuclei ''frozen'' in the abundances we observe for them today.

Finally, after a million years or so, the continuing expansion of the universe reached a point where the temperature had dropped to 3000 K. By now, the photons no longer possessed sufficient energy to even keep electrons away from hydrogen and helium nuclei and there were wholesale combinations of electrons and nuclei into atoms. Hydrogen and helium are both virtually transparent to the photons in a 3000-K radiation field, so the universe suddenly became one of atoms and photons that could now move freely among those newly formed atoms. Astrophysics often refer to this event as the *era of decoupling* or recombination. Upon the formation of atoms, the next 1 to 2 billion years saw the gravitational collapse of those atoms into the wispy, bubbly, filamentary structure of galaxies we observe today. It is out of this structure that stars have been cycling and recycling through their lives over the past 13 to 18 billion years.

The final fate of this vast expense of gas, dust, stars, and galaxies is very much dependent on the total mass that exists within it. Too little mass and the self-gravity of the galaxies cannot slow the expansion enough to prevent the universe from expanding forever. Too much mass and, just as an upward thrown ball is gravitationally yanked back to the earth, the self-gravity will stop the expansion, causing the universe to collapse onto itself in an event cosmologists have called the Big Crunch. Unfortunately, we can only place a lower limit to the total mass that might be present in the universe, and, therefore, can make no definite statements concerning what may be in store for us billions of years down our cosmological time line.

WORKED EXAMPLE

Suppose the universe was created in a Big Bang and set in motion 6000 years ago. A galaxy that is now found to be 2 million light years distant had to have moved at what speed in order to be located in its presently observed distance? Is this a reasonable result? Explain.

SOLUTION: To be located at its present position, the galaxy would have had to move at a rate of

$$\text{velocity} = \frac{\text{distance}}{\text{time}}$$

$$= \frac{2 \times 10^6 \text{ light years}}{6000 \text{ years}}.$$

Since one light year $= 9.5 \times 10^{15}$ m and 1 year $= 3.16 \times 10^7$ sec, we have

Continued

WORKED EXAMPLE

Continued

$$\text{velocity} = \frac{2 \times 10^6 \text{ light years} \times 9.5 \times 10^{15} \text{ m/light year}}{6000 \text{ years} \times 3.16 \times 10^7 \text{ sec/year}}$$

$$= \frac{19 \times 10^{21}}{1.9 \times 10^4 \times 10^7} \text{ m/sec}$$

$$= 10 \times 10^{21-11}$$

$$= 10 \times 10^{10} \text{ m/sec}$$

$$= 1.0 \times 10^{11} \text{ m/sec} .$$

Since the speed of light is 3×10^8 m/sec, this velocity is over 300 times that of the "ultimate" speed limit in the universe. Therefore, this is *not* a reasonable result.

---

1. Describe some of the overall properties that have been observed for the universe as a whole.

2. What is the Big Bang theory? How does the Big Bang theory account for the overall observed properties of the universe?

3. What is meant by Planck time? What is its significance in cosmology?

4. Describe the various stages of the Big Bang universe that are believed to have occurred between the Planck time and the present.

5. Discuss how the universe might possibly evolve in the future.

REVIEW 16.5 QUESTIONS

---

1. Calculate the wavelength of maximum intensity for the 3-K background radiation. Where in the electromagnetic spectrum is this peak value located?

2. A cluster of galaxies is observed to be receding from the earth at a velocity of 5×10^5 m/sec. If the cluster of galaxies is known to be 8×10^6 parsecs distant and 1 parsec is equal to 3.1×10^{16} m, estimate the age of the universe from these data.

REVIEW 16.5 PROBLEMS

Chapter Review Problems

1. A radar pulse is directed toward the sun. After a total of 975 seconds, the return echo is received. If the radar waves travel at the speed of light, find the Sun's distance.

2. Verify by direct calculation that the distance light travels in a year is about 9.46×10^{12} km, assuming that light travels 300,000 km/sec.

3. If the sun's temperature is 5800 K, find its wavelength of maximum intensity.

4. If the temperature at the core of a high mass star is 5 billion K, find the wavelength of maximum intensity for such an incandescent source. How does your calculated value for the wavelength compare with the diameter of an atomic nucleus (about 10^{-14} m)?

5. At what wavelength would we expect to see the maximum intensity from 3-K cosmic background radiation? Does your result explain why it is formally referred to as the 3-K cosmic microwave background?

6. A certain star has a temperature of 3000 K. Find the surface intensity of the radiation leaving this star.

7. A star is found to exhibit a parallax angle of 0.04 arcseconds. Find the distance to this star in

   a. parsecs   **b.** meters.

8. Suppose that the Sun had an observed angular diameter of 0.005 radians, but was still located 1.5×10^{11} meters away. How large would the Sun's linear diameter be?

9. An astronaut notes that the Sun's apparent angular diameter is 0.015 radians. How far is the astronaut from the Sun?

10. Assuming that the Sun has a mass of 2.0×10^{30} kg and a power output of 4.0×10^{26} watts, find:

    a. the total amount of relativistic energy available in the Sun's mass

    b. the "lifetime" of the Sun if the Sun radiates this energy at its present rate of 4.0×10^{26} watts.

11. How many kilograms of matter per second is the sun converting into its equivalent energy if it is shining with a power output of 4.0×10^{26} watts?

12. The distance between the plant Mercury and the Sun is about 0.4 that between Earth and the Sun. What is the value of the solar constant for the planet Mercury?

13. A spacecraft operating on solar cells must have a sunlight intensity of at least 1.0 watt/meter^2 in order to function properly. How far from the Sun can this spacecraft go before it will not be able to operate?

14. A solar-type star, i.e., one with the same power output as the Sun, is observed in a star cluster and found to have an observed intensity of 4.0×10^{-15} watts/meter^2. What is the distance to this star cluster?

15. A certain star has an observed parallax angle of 0.008 arcseconds. Find the distance to this star in parsecs. If this star were moved to a distance of 500 parsecs, what parallax angle would it now exhibit?

16. A star known to be 4.0×10^{18} meters from the earth has an observed intensity of 2.0×10^{-12} watts/meter^2. What is the power output for this star?

17. If a star has a radius of 3.45×10^9 meters, what is the radius of this star in solar radii?

18. If a star has a mass of 12 solar mass units, what is the mass of this star in kilograms?

19. The sum of the masses of the stars in a binary star system is found to be 10 solar mass units. Find the masses of the individual stars in this system if:

    a. the masses of both stars are the same

    b. one of the stars is a solar-type star

    c. the mass ratio is equal to 4.0.

20. The companion to the bright star Sirius has a temperature of one-half that of the Sun and a power output that is 1.0×10^{-4} that of the Sun. What is the radius of this star in solar radii?

21. A brown dwarf star has a radius and mass about one-tenth that of the Sun, and a temperature one-sixth that of the Sun. Find the power output of such a star in terms of that of the Sun.

22. A star at 4000 K suddenly increases its power output by a factor of 16. If the star's radius remains fixed during the outburst, what is the new temperature of the star?

23. Suppose that the measurements of the cosmic background radiation indicated that the wavelength of maximum intensity was located at one centimeter. Find the temperature of this background radiation.

24. The galaxy Messier 51 has a radial velocity of about 400,000 m/sec. If this galaxy is 3.9×10^6 parsecs distant, and if 1 parsec = 3.1×10^{16} m, estimate the age of the universe from the time of the Big Bang.

THOUGHT QUESTIONS

1. Describe how the various space missions to the Moon and other planets have verified the principle of universality.

2. Astrophysics has sometimes been referred to as an "observational" science rather than an "experimental" science. Discuss this claim.

3. It is known that Earth's oceans are at least two billion years old. What restrictions, if any, does this fact place on the Sun's power output throughout Earth's geological history?

4. Do you think that nuclear *fission* processes could play a significant role in the Sun's energy production? Explain.

5. From the plot shown in Figure 16.6, how is the Sun's intensity distribution similar to that of a blackbody? How is it different?

6. Besides hydrostatic equilibrium, can you think of any other conditions that could be invoked to describe the state of the gases inside of the Sun? Explain.

7. The eighteenth century English astronomer Sir William Herschel believed that the sun had a cool interior that was inhabited. How would you respond to such a claim?

8. Can you think of any practical applications that might arise out of astrophysical studies of the Sun and stars?

9. If the distance to a star could not be obtained, which, if any, of its physical characteristics could still be deduced by astrophysicists? Which could not? Explain.

10. Why can't the stars have always been exactly the way we see them now?

11. Discuss how you would obtain mean densities of stars. How do you think mean stellar densities would compare with water?

12. Can you think of possible triggering mechanisms for star formation that weren't discussed in the text? Describe how they might work.

13. From the theory of stellar life cycles, do you judge planetary systems to be numerous or rare among the stars. Explain.

14. How does Wien's law provide us with an explanation for the observed colors of stars? On the basis of this explanation, is a red star relatively hot or cool compared to a blue star?

15. Discuss how astrophysicists would use stellar models to theoretically "evolve" a star.

16. Compare and contrast the life cycle of a star with that of a typical life form here on the earth.

17. Discuss the scientific questions raised by the existence of black holes.

18. Astronomers believe that the very youngest stars have relatively high abundances of elements heavier than hydrogen and helium, while the very oldest stars have a low abundance of elements heavier than hydrogen and helium. How might the theory of stellar life cycles account for such an effect?

19. How might particle accelerators help us in understanding the element building processes occurring inside of stars?

20. Astronomers have observed stars that seem to be small, compact stars made up primarily of helium. Can you explain such objects in terms of the theory of stellar life cycles?

21. Why would the exceedingly close crowding of electrons in white dwarf stars and neutrons in neutron stars produce a quantum mechanical "counter-pressure," which resists further crowding in these objects?

22. Can there be a field of experimental cosmology? Explain.

23. What can high-energy particle accelerators tell us, if anything, about the early conditions of the Big Bang?

24. In what ways might the opening infinitesimal time intervals of the Big Bang be similar to those of a Big Crunch?

25. How would the observed radial velocities of the most distant galaxies compare with those at a closer distance if:
    a. the expansion rate of the universe has been constant throughout cosmological time
    b. the expansion rate of the universe has been increasing throughout cosmological time.

26. Discuss the possibility that "alternate" or "parallel" universes exist.

Stellar Radii

When we measured the linear diameter of the Sun, the procedure was relatively straightforward. The distance and angular diameter of the Sun were measured and then used to calculate the linear diameter of the Sun. Unfortunately, the vastness of stellar distances renders direct measurement of stellar angular diameters extraordinarily difficult for nearby stars and impossible for the more remote ones. Some direct measurements of stellar radii have been made by observing binary stars that eclipse each other's light and through the use of interferometric devices, but such determinations have to date been limited to a comparatively small number of stars. As a result, the astrophysicist must resort to more indirect techniques to obtain stellar sizes.

If a star has a certain power output P_*, then at the star's photospheric layers or "surface" that total power is equal to the product of the intensity of the radiant energy at the surface and the total area of the star's surface, or

$$P_* = \text{surface intensity} \times \text{surface area} .$$

The surface intensity is given by the Stefan-Boltzmann law or

$$\text{surface intensity} = \sigma T_*^4 ,$$

where σ is the Stefan-Boltzmann constant and T_* is the star's photospheric or surface temperature. If the star is in the shape of a sphere having a radius R_*, then the total surface area of the star is given by

$$\text{surface area} = 4\pi R_*^2 .$$

Substituting these last two equations into the expression for P_*, we obtain

$$P_* = (\sigma T_*^4)(4\pi R_*^2) .$$

If we write this equation for the Sun, we obtain

$$P_\odot = (\sigma T_\odot^4)(4\pi R_\odot^2) ,$$

where $P_\odot$, $T_\odot$, and $R_\odot$ are the power output of the Sun, the temperature of the Sun, and the radius of the Sun, respectively. Dividing the last two equations by each other yields

$$\frac{P_*}{P_\odot} = \frac{\sigma T_*^4 \; 4\pi R_*^2}{\sigma T_\odot^4 \; 4\pi R_\odot^2} ,$$

which becomes

$$\frac{P_*}{P_\odot} = \left(\frac{T_*}{T_\odot}\right)^4 \left(\frac{R_*}{R_\odot}\right)^2 .$$

Solving this equation for $R_*/R_\odot$ yields

$$\frac{R_*}{R_\odot} = \sqrt{\frac{P_*/P_\odot}{(T_*/T_\odot)^4}} .$$

If we measure R_* and P_* in solar units, this equation becomes

$$\frac{R_*}{\text{(solar units)}} = \sqrt{\frac{P_* \text{ (solar units)}}{[T_* \text{ (solar units)}]^4}} .$$

For example, if a star has a power output 1600 times that of the Sun and a temperature twice that of the Sun, the radius of such a star would be

$$R_* = \sqrt{\frac{1600}{(2)^4}} = \sqrt{100}$$

or

$$R_* = 10 .$$

The star would have a linear size 10 times that of the sun.

QUESTIONS

1. Describe how astronomers might use an eclipsing binary star system to obtain the radii of the stars in that system.
2. What assumptions have been made to obtain the equations in this formulation?
3. Why do you suppose it is important for the astrophysicist to obtain the linear sizes of stars?

PROBLEMS

1. The companion to the bright star Procyon has a temperature about 1.5 times that of the Sun and a power output of $\frac{1}{5,500}$ that of the Sun. What is the radius of this object in solar radii and in meters? How does this compare in size to the diameter of Earth?
2. When a star the size of the Sun evolves into a red giant star, its radius increases by a factor of about 200 and its power output increases by a factor of 2500. What is the temperature of such a red giant in terms of the Sun's temperature?
3. A certain star initially has a power output that is seven times that of the Sun. If the star's temperature suddenly tripled and its radius doubled, what would the power output of the star be after the outburst in solar power units?

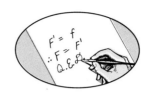

PHYSICS FORMULATION 16.2

Black Holes

We learned in Chapter 5 that for an object to escape the gravitational pull of a second object having a mass M and a radius R, it must acquire an escape velocity v_e, which is given by

$$v_e = \sqrt{\frac{2GM}{R}} \; .$$

From this expression, we can see that as the value of the object's radius decreases in size, the velocity required to escape from the object increases. The ultimate escape velocity, however, is that ultimate of all velocities, the speed of light. If we let $v_e = c$, then

$$c = \sqrt{\frac{2GM}{R}} \; .$$

Squaring both sides of this equation, we obtain

$$c^2 = \frac{2GM}{R}$$

and solving for R gives us

$$R = \frac{2GM}{c^2} \; .$$

Thus, if a certain mass M could somehow be compacted into a sphere having a radius equal to $2GM/c^2$, no form of particle or radian energy could move fast enough to escape it. The radius at which this happens is called the *Schwartzschild radius* and the object itself is referred to as a black hole because of the shape of the relativistic space-time fabric that exists in its vicinity (Figure 16.26).

In theory, black holes can come in a variety of sizes (see Table 16.2), but in reality, any mechanism capable of crushing an object's mass to a size less than its Schwartzschild radius is generally a mechanism that requires considerable mass as well. Thus, we expect to see black holes form out of the final evolutionary stages of large mass stars and, on a larger scale, at the centers of galaxies.

Black holes create some interesting questions in modern physics. If the energy and matter falling into a black hole are forever trapped, where do they go? What is the nature of matter crushed to densities larger than that of the nucleon? Do such objects change and evolve as they are showered with particles and energy? It is here that we arrive at a "triple point" in human thinking in which the slightest change can carry us in and out of the realms of science fact, science speculation, and science fiction.

Considerable problems arise when attempts are made to verify observationally the actual existence of such objects because, by their very nature, the light by which we would be able to see them is forever trapped. Black holes, however, can still gravitationally affect their surroundings and it is this fact that enables the astrophysicist to detect their presence. The system known as Cygnus X-1 is thought to be a binary system whose component is a black hole in which six solar masses of material have been compacted into an object 17 km across. The black hole makes its presence felt by trapping and compacting the gaseous stellar

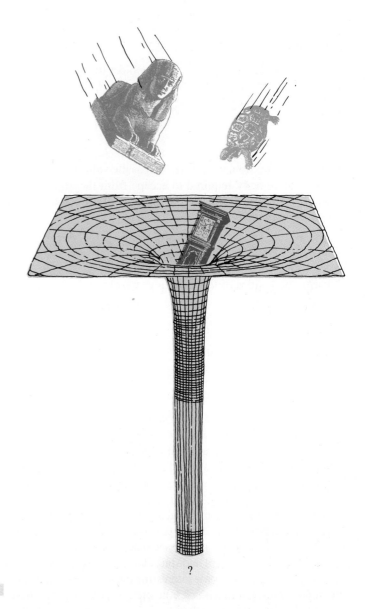

FIGURE 16.26 ··········

A Black Hole In this three-dimensional rendition of four-dimensional space-time, the fabric of space-time is warped by the black hole's gravitating mass in such a way as to form a ''hole'' into which all incident energy and matter will fall.

wind emanations from its supergiant stellar neighbor into a charged plasma, which is then accelerated toward the Schwartzschild radius of the black hole. So accelerated, the charged plasma radiates X rays until it reaches the Schwartzschild radius, at which time it plunges past the event horizon of the black hole and disappears from our observable universe.

Centers of galaxies are now believed to harbor similar objects, although on a far larger scale. Here the mass of the black hole is believed to be on the order of one billion solar masses and its radius is about 5.9×10^{14} m or some 4000 times the Earth-Sun distance. Light could transverse the diameter of such an object in 21 days if it were able to get through. Like Cygnus X-1, we would expect to see such a black hole exert observable effects on its surroundings, and we do. The nuclei of galaxies are among the most astrophysically active regions

TABLE 16.2 Some Properties of Black Holes

Mass	Radius	Density
100 kg (human being)	1.5×10^{-25} meter	7.1×10^{75} kg/m³
6×10^{24} kg (earth mass)	9 mm	2.9×10^{34}
2×10^{30} kg (sun solar mass)	2.9 km	2.0×10^{19}
4×10^{41} kg (Milky Way mass galaxy)	5.9×10^{14} m (0.06 light years)	4.6×10^{-4}

known. At the far edges of the observable universe, galaxies are so distant that they are detectable only by the black hole/matter interactions at their centers, which manifest themselves to astrophysicists as quasars, exploding galaxies, and a wide range of other exotic extragalactic phenomena.

The mean density, $\bar{\rho}$, of a black hole can be calculated by assuming a spherical shape for the black hole in which case the volume $V = \dfrac{4}{3}\pi R^3$,

$$\bar{\rho} = \frac{M}{V} = \frac{M}{\dfrac{4}{3}\pi R^3}$$

$$= \frac{3M}{4\pi R^3}.$$

Calculation of $\bar{\rho}$ for black holes having different masses (Table 16.2) reveals the fascinating fact that as the black hole becomes more massive, the density of material required to produce it decreases markedly. Thus, a black hole made up of one solar mass of material compacted down to a Schartzschild radius of 2900 meters will have a density of about 2×10^{19} kg/m³ or about a hundred times that of neutrons. The gigantic black holes believed to exist at the centers of galaxies, however, require only a density one ten-millionth that of water to sustain themselves. All of this leads to a most intriguing possibility. The mean density of a black hole having a radius equal to that of the entire observable universe is about 10^{-26} kg/m³. At present, the observed density of the observable matter in the universe is about 3×10^{-28} kg/m³. However, to explain how clusters of galaxies and many individual galaxies hold themselves together, astrophysicists must assume that large amounts of dark, nonluminous mass exist in these objects. Thus, the true average density of the universe may be higher than 10^{-26} kg/m³ in which case all of the observable universe is in reality the inside of a huge black hole. What a marvelous prank of nature that would be!

QUESTIONS

1. From the density results in Table 16.2, comment on why we do not seem to observe black holes less massive than about three solar masses.

2. Do black holes in our universe suggest the existence of "parallel" or "alternate" universes? Explain.
3. Discuss the possible relationship between black holes and the principle of the conservation of energy.

PROBLEMS

1. Calculate the mass of a black hole having a radius equal to that of the solar system (6.0×10^{12} m).
2. Derive an expression for the mean density of a black hole that involves only the radius of the black hole.
3. Find the radius and mass of a black hole which has
   a. the same density as water (1000 kg/m^3)
   b. the same density as a neutron (4×10^{17} kg/m^3).

17 PHYSICS, POWER, AND THE ENVIRONMENT

*O*ne of the premier hallmarks of the technological revolution of the past century and a half has been the development of a diversity of devices that can perform a wider, more sophisticated range of tasks than ever before at a rate faster than ever before. As varied as these devices are in their purpose and performance, however, a fundamental thread is common to all: Each and every one requires that energy somehow be delivered to it in order to operate.

17.1 Energy and Power in the Modern World

Energy has long been a basic and unifying concept within the scientific framework. Over the past 150 years or so, it has also come to occupy a position of similar importance in our technology as well. Moreover, as human technological advances have swept forward at breathtaking speed, the number and variety of energy-consuming devices have jumped dramatically. With these increases comes a demand not only for greater amounts of energy but also for an increased rate at which that energy is delivered, i.e., increased power production (see Figure 17.1).

In the course of producing and delivering energy at ever-increasing rates, however, a number of factors related to those processes have made themselves manifest. Throughout this text, we have discussed a variety of ways in which energy appears in the physical world. Unfortunately, not all of the energy forms we've studied can be directly used to operate our televisions and refrigerators. The thermal energy generated by a burning lump of coal, for example, cannot be used directly to operate a washing machine. To be ''useful'' to the washing machine, the thermal energy emitted by the burning piece of coal must somehow be converted into electrical energy. Such energy conversion processes will, in varying degrees, create pollution effects, which are often highly undesirable. As it gives off its thermal energy, the burning lump of coal will also exude polluting gases in the form of smoke, which in turn can interact with the surrounding environment in a harmful way. Thus the conversion of chemical energy from gasoline into the mechanical energy used to run our cars, buses, and airplanes has resulted in the toxic ''brown clouds'' of air pollution that all too often hang over the major cities of the world (Figure 17.2).

Similar environmental impacts can result in the course of removing energy-producing materials such as coal and oil from the earth's crust and transporting them to more convenient locations for conversion into usable power. The ravages to the earth's surface resulting from the strip mining of coal and of oil spills at sea are but two examples of such polluting effects. Sadly, the mining, production, and energy conversion processes are sometimes accompanied by a direct negative impact on human beings, even to the extent of causing death. Human lives, for example, have been greatly shortened by the effects of black lung disease contracted by mining coal, and have been lost outright by accidents as diverse as collapsing offshore oil rigs, mine explosions and cave-ins, and the Chernobyl nuclear plant meltdown.

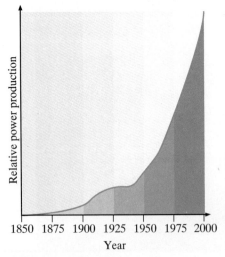

FIGURE 17.1 ••••••••••

Power Production in Recent History   The demand for energy has, like the world's population, increased exponentially over the past 150 years.

FIGURE 17.2
Pollution in Contemporary Society

In the course of using the energy we produce, we find that yet another form of pollution makes its appearance. The second law of thermodynamics tells us that in any energy conversion process by which useful energy is produced, a certain amount of "wasted" energy, usually thermal energy, is produced as well. Moreover, the principle of the conservation of energy further requires that as we employ the useful energy, it is not depleted, but rather must somehow end up somewhere in the environment. Once again, this "used" energy almost always ends up as thermal energy. The result of having these two sources of thermal energy end up in the environment is an effect referred to as thermal pollution, a pollution source that makes its impact by raising environmental temperatures.

The environment in which all of this occurs is a 12-mile-thick layer about the earth called the *biosphere*. Extending from the highest mountains on the earth's surface to its deepest oceanic trenches, the biosphere contains all of the known life-forms on earth and indeed is currently the only region in all the universe where we can state with absolute certainty that life-forms exist. An increasing human population requiring increasing levels of energy and power production with their attendant increasing levels of environmental pollution have, in the present century, begun to threaten seriously the delicate balance of physical conditions within the biosphere that have made life and the evolution of life possible on earth for hundreds of millions of years.

In the present century we have been faced with more and more difficult decisions regarding the environmental impact of increased energy and power production versus the sociological and political impact of not increasing such production. Unfortunately, all too often, a serious lack of quantitative data exists on which to base our energy-versus-environment decisions. To what extent can the rain forests of the Amazon basin be cleared for agricultural purposes without seriously impacting the world's climate? How much carbon dioxide can the world's industry eject into the earth's atmosphere without triggering a disastrous greenhouse effect? How environmentally safe is nuclear power production? Dozens of such questions relating to energy and the environment can be asked and the answers to them debated vigorously. We cannot fully explore these complex questions in depth here, but we can provide some insight into the role the principles of physics play in what is certainly one of the most fundamental issues we will face over the next century.

1. Why has there been a demand on the part of humanity for ever-increasing amounts of energy?

2. Discuss an example of a negative impact on the environment that arises from:
    a. mining energy sources
    b. transporting energy from one location to another
    c. converting energy from one form to another.

3. Why is thermal pollution a consequence of *any* energy use?

4. What is the biosphere? Research the evidence for and against the existence of other biospheres in the universe.

5. What are some of the physical characteristics of the earth's biosphere that make it suitable for life-forms? Do these conditions exist on any of the other planets or satellites of the solar system? How do we know?

17.2 Electrical Energy

Far and away the most useful, convenient, and important form of energy in our modern civilization is electrical energy. Electrical energy provides the power by which our light bulbs, television sets, and computers operate. Even the cars, planes, trains, and buses that transport millions of us around town and around the globe using mechanical energy require electrical energy in order to function properly.

The most direct way to deliver electrical energy is via a battery in which the electric charges within the battery have been separated in some fashion, thereby creating an electric potential between the battery terminals. Whenever an electrical device is connected to the battery terminals, electric current flows from one terminal, through the device, and on to the other terminal. Batteries are a convenient source of electricity in that they can be carried along to any desired location. The electrical energy needed in the operation of an automobile, for example, is provided by the electrical potential energy stored in a battery that is carried along with the car as it moves. Similarly, battery-operated toys, radios, wristwatches, etc., have become so commonplace that one of the more familiar manufacturer's warnings in our society is "batteries not included."

If, however, we wish to deliver electrical power on a much larger scale, such as that needed to satisfy the power requirements of a major city, then the use of such battery-delivered electrical energy would be woefully inadequate. These difficulties are surmounted by the use of alternating electrical potential. In Chapter 11, we saw that it is possible to convert rotational mechanical energy into alternating electrical potential energy by rotating a coil of conducting wire in the presence of a magnetic field. Electrical generators that operate in this fashion can be constructed on a scale sufficiently large as to be able to supply the amounts of electrical power necessary to service large population centers. In addition, we have also seen how it is possible to transmit alternating electrical potential efficiently over very large distances through the use of step-up and step-down transformers. A schematic of the process is illustrated in Figure 11.21. Rotational energy is used to produce an alternating high-current, low-voltage electrical potential by means of an electric generator. A step-up transformer converts the electrical energy into a high-voltage, low-current mode, which permits it to be transmitted over large distances with minimal power losses.

At the destination sites, the electrical energy is changed back into a low-voltage, high-current mode by step-down transformers, which deliver alternating electrical potential to individual outlets as 120-volt and/or 220-volt output. If the distribution site is sprawled out over a large area, the step-down process may occur in two voltage jumps. The first is a step-down of the electrical power to an intermediate current and voltage, which is delivered in that mode to distribution transformers located throughout the area to be serviced. These distribution transformers then provide the final step-down of the electric power for individual use.

Electrical power can be produced in great quantities and in several convenient ways. It can also be supplied over large distances, and is virtually the only form of power universally employable by the variety of machines and devices used in our society. Unfortunately, the impressive manifestations of electrical power that occur in nature, such as the lightning bolt, are at present inaccessible for our use. Instead, we must obtain our electrical power by conversion from other, more readily available energy forms. This can be accomplished in one of two basic ways. We can employ the given energy source to somehow separate electrical charges, thereby setting up an electric potential capable of delivering electrical energy; or we can use the given energy source to create mechanical rotational kinetic energy in the rotating shaft of an electric generator, which in turn produces an alternating electrical potential output.

1. Why do we make a major effort to convert other energy forms into electrical energy?

2. What are the advantages and disadvantages of a battery as a source of electrical energy?

3. How is electrical energy supplied to a big city?

4. Discuss two basic ways in which electrical energy can be obtained from other energy sources.

5. List 10 devices used in our modern world that run on electrical energy.

17.3 Geophysical Energy

Several sources of energy exist in the biosphere that can be used to create electrical power in a fairly direct fashion. Among these energy sources is flowing water, wind, and geothermal activity. Flowing water and wind have long been used by human beings to perform a variety of tasks. As water flows from a higher to a lower altitude, its gravitational potential energy is converted into kinetic energy. If this kinetic energy-laden water is channeled so as to strike the blades of a water wheel as shown in Figure 17.3(a), the water will impart a rotational motion and, hence, a rotational kinetic energy to the wheel. In earlier eras of human

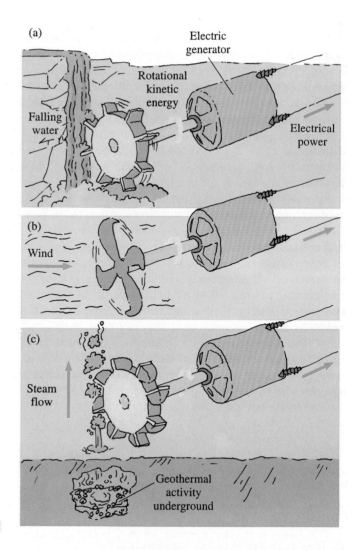

FIGURE 17.3 ··········

Geophysical Energy With suitable devices, various forms of geophysical energy including (a) falling water, (b) wind, and (c) geothermal activity can be converted first into rotational kinetic energy and then into electric power.

history, this rotational energy was used to perform tasks such as grinding grain into flour, but now serves to turn the shafts of electric generators to produce what is referred to as hydroelectric power. Roughly 20% of all electric power produced in the world today is estimated to be hydroelectric.

As the earth moves through space, the sun and moon exert a differential gravitational pull on its water, which is called a tidal force (see Physics Facet 5.2). As a result of this tidal force, the level of water along the seacoasts is raised and lowered at regular intervals. As the water rushes in toward shore at high tide and then back out to sea at low tide, this tidal water flow can be used to drive electrical generators in much the same fashion as the cascading water described earlier. The use of tides to generate electrical power is particularly effective at locations on the earth such as the Bay of Fundy in Eastern Canada, the Bering Strait in Alaska, etc., where the difference in the water levels at high and low tides can be as much as 10 meters or more.

The windmill is another device used to convert the kinetic energy of a fluid flow into the rotational energy necessary to drive an electric generator. In this instance, the wind strikes the blades of the windmill in such a way as to cause them to rotate [Figure 17.3(b)]. Because the density of air is only about a thousandth that of water, the kinetic energy associated with a given wind flow will be far less than that of a corresponding flow of water, and hence the ability of a windmill to generate large amounts of electrical power is somewhat limited.

In addition to the energy sources present on the earth's surface, other geophysical energy sources are buried deep within the earth. If we could somehow make Jules Verne's famed "journey to the center of the earth," we would find that the earth's interior is a hot place indeed. The temperature at the earth's core, for example, is estimated to be some 4500°C. This temperature decreases outward to the more habitable values of the biosphere, but there are a number of regions around the globe where portions of the earth's crust have temperatures of several hundred degrees Celsius at depths of only a few kilometers. The result is an area of geothermal activity in which heat from the earth's interior is transported to the surface in the form of hot springs, steam vents, geysers, "paint pots," and other geothermal phenomena.

Perhaps the most famous of the geothermal areas are the geyser basins in Yellowstone National Park, but geothermal regions also exist in other locations such as California, Iceland, Mexico, and the North Island of New Zealand. The existence of high-temperature reservoir of thermal energy so close to the earth's surface has made it possible for geothermal power plants to "tap" this thermal energy for conversion into electrical power. Very simply, a well shaft is drilled into the high-pressure underground steam domes associated with these regions and as the high-pressure steam vents up the well shaft at large velocities, its kinetic energy can be used once again to impart rotational kinetic energy to the shaft of an electric generator.

Geophysical energy sources have several attractive features. First of all, its use is virtually pollution free and its environmental impact comparatively low. Whether it's low enough, however, is a matter that is often hotly debated almost every time a proposal is made for the construction of a new energy conversion facility. The construction of a tidal or hydroelectric power dam does have an environmental impact on the immediate area, in particular on the aquatic life of the area. Therefore, as much consideration as possible should be given toward deciding the degree to which an impact of this sort can be tolerated or perhaps even circumvented. Adjustments can be made, such as adding "fish ladders" to dams built in the Pacific Northwest for the benefit of spawning salmon.

Geophysical energy forms are also generally *renewable,* i.e., once used they will be replaced in some fashion by natural processes. Thus, the water that drives the turbines of a hydroelectric power plant will flow into a lake or ocean, be evaporated into the atmosphere by the sun's rays, transported by the movements in the earth's atmosphere to a higher altitude, condense as rain falling to the earth, and finally run downhill to once more pass through the turbines of the hydroelectric power plant. Similar regenerative cycles exist for wind, tidal, and geothermal energy sources as well. Thus the geophysical energy supplies can never be exhausted. Unfortunately, what *can* be exhausted are the number of available sites at which such energy can be readily captured. The number of geothermal basins located around the globe is highly restricted, as are the possible sites for tidal dams. Sites suitable for the production of hydroelectric power and wind power are far more plentiful, but are still not adequate to satisfy current worldwide demands for energy and power. To satisfy that demand, we must seek out energy sources elsewhere, some of which we discuss in the following sections.

1. Describe how flowing water can be used to produce electrical energy.
2. How is electrical energy produced from the wind?
3. What is a geothermal area? Where are such areas to be found on the earth?
4. Explain what is meant by a renewable energy form.
5. Can you think of any other sources of geophysical energy not mentioned in the text?

17.4 Chemical Energy

Without question the most plentiful and important source of energy in the world at the present time is the energy released from chemical reactions, in particular, those reactions that are combustive in nature. Every electronic configuration about an atom or molecule has associated with it a given amount of electrical potential energy. To change that configuration, energy must either be put into the system, thereby increasing the electrical potential of the configuration, or given off by the system as it lowers its total electrical potential energy.

We have already described how the electronic configuration of a given substance can be changed through the absorption and emission of energy in the form of photons. Similar changes in electronic configurations can also be produced in chemical reactions as well. Whenever two or more substances chemically interact, the configuration of valence electrons that characterizes the reactants is changed into a different configuration of valence electrons for the products produced in the reaction. If the electron configuration of the products has less electronic potential energy than that for the reactants, then the difference is given off as "liberated" energy and the reaction is said to be *exothermic*. On the other hand, if the potential energy of the electronic configuration of the products is larger than that for the reactants, then the energy difference had to have been absorbed from an exterior source. This type of reaction is said to be *endothermic*. Like the absorption and emission of photons, a given chemical reaction can absorb or emit energy depending on the circumstances. For example, the gases hydrogen (H_2) and oxygen (O_2) exothermally react with one another according to the following reaction:

$$2H_2 + O_2 \rightarrow 2H_2O + energy \ .$$

If 4 grams of hydrogen gas react with 32 grams of oxygen gas in this fashion, 36 grams of water are produced with about 485,000 joules of energy. The liberation of the 485,000 joules of energy arises from the fact that the total electrical potential energy of the electron configurations of the individual gases is larger than those of the water vapor molecules that are formed (Figure 17.4). If we wish

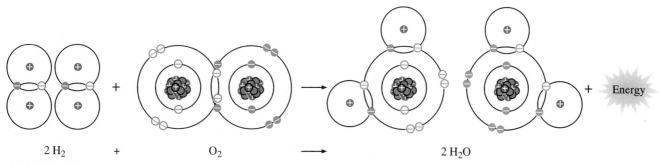

$$2\,H_2 \qquad + \qquad O_2 \qquad \longrightarrow \qquad 2\,H_2O$$

FIGURE 17.4 ···········
Chemical Energy   When a chemical reaction occurs, very often the atoms shift from a configuration of higher electronic potential energy to one of lower electronic potential energy. When this occurs, chemical energy is released in the reaction.

to somehow reverse this reaction, we would have to put the 485,000-joule energy difference back into the system as follows:

$$2H_2O \;+\; energy \rightarrow 2H_2 \;+\; O_2 \;.$$

If we wish to make use of chemical energy, ideally we would like to find naturally occurring substances whose electronic configurations have a comparatively large amount of associated electrical potential energy that can be liberated readily through chemical reactions with other substances. The most important of such materials is a class of substances called fossil fuels. As the name suggests, fossil fuels are the fossilized remains of plant and animal life from earlier eras in the earth's geological history. When these life-forms died eons ago, their remains were subjected to a variety of geological effects, most notably heat and high pressure. The result was that these organic remains were converted into substances such as coal, oil, and natural gas whose molecules are characterized by high-potential-energy electronic configurations.

The most common technique for liberating the chemical energy stored in fossil fuels is by oxidizing them according to the following "generic" chemical reaction:

$$fossil\ fuel \;+\; O_2 \rightarrow H_2O \;+\; CO_2 \;+\; energy \;.$$

In writing this chemical reaction we have assumed that the fossil fuel is in the form of a pure hydrocarbon, i.e., a material composed *only* of atoms of hydrogen and carbon such as methane gas (CH_4). We have also assumed that a sufficient amount of oxygen is available to burn the fuel cleanly so that the only products of the reaction are water (H_2O) and carbon dioxide (CO_2). The energy liberated form this type of chemical reaction will not of itself run our electrical devices, but it can be used to power the operation of gasoline and steam engines, whose output of rotational kinetic energy serves to turn the shafts of our electric generators. In a similar fashion, the energy locked up in the organic fuels of more recent vintage, such as wood and peat, can also be used to produce electrical power.

The fossil fuels, however, presently represent the most important source of our current energy production. Unfortunately, coal, oil, and natural gas are not renewable energy sources. The processes by which these fuels were created in the earth's crust occurred over time periods spanning millions of years. The processes by which they will be used up will occur over a time scale of at most three or four centuries. Moreover, the idealized fuel burn equation presented earlier assumed that a pure hydrocarbon fuel burned in a "pure" fashion. In reality, fossil fuels are neither completely pure nor do they burn completely clean. Most often fossil fuels contain contaminants such as sulfur that, once liberated, participate in a variety of secondary chemical reactions that can cause considerable harm to the environment in the form of "acid" rains, "killer" fogs, and the like. Hydrocarbons that are not completely oxidized in a fuel burn produce other hydrocarbons as reaction products. With the help of ultraviolet radiation from the sun, these products combine in a variety of photochemical reactions to produce the smog for which many of our world's great cities have sadly become infamous. Even the carbon dioxide produced by a clean hydrocarbon enjoying a clean burn is capable of environmental mischief.

A number of climatologists are concerned that the ever-increasing levels of carbon dioxide in the earth's atmosphere (an increase of 10% in the last 50 years alone) will create a "greenhouse effect," in which the carbon dioxide present in our atmosphere will trap the energy from the sun in a fashion similar to the way the glass of a greenhouse traps solar energy. The immediate result of such an effect will be the increase of the mean temperature of the biosphere. Should such a scenario actually be played out, climatologists fear that the resulting rise in the earth's mean surface temperatures might be sufficient to cause the earth's polar ice caps to melt, thereby causing an increase of as much as 10 meters in the water level of the earth's oceans. A quick glance at a topographical map of the world will quickly reveal that an increase in water level of this sort would wreak considerable havoc with a great many of the world's population centers.

A second, more extreme fear is that a "runaway" greenhouse effect might occur in which the temperature increase would drive the carbon dioxide dissolved in the earth's oceans back into the atmosphere where it would cause more solar energy to be trapped, leading to yet higher temperatures that drive yet more dissolved carbon dioxide into the atmosphere, and so on. The incredibly hostile conditions known to exist currently on the surface of the planet Venus were almost certainly created at least in part by just such a process.

The earth's atmosphere, oceans, and climate are, however, complicated entities, and some climatologists take a far less pessimistic view of such matters. For example, in one alternate scenario, the increase in the earth's mean temperature results in an increase in the rate at which water evaporates from the oceans. That increase then creates a more pronounced cloud cover around the earth. The clouds so produced have the effect of reflecting a higher fraction of the sun's energy back into space before it has the opportunity to be trapped in the earth's atmospheric "greenhouse." The net result is thus a reduction in the earth's mean temperature. Pessimism, however, lurks here as well, for there are those scientists who envision this cooling as the beginning stage of a downward temperature cycle via which we will climatically spiral back into an Ice Age! Long before any of this is likely to happen, however, the supply of the world's fossil fuels will have already begun to run out, and in the absence of an alternate energy source that can match the low cost, convenience, and magnitude of the power production afforded us by the fossil fuels over the past two centuries, the planet's population faces a distinct drop in its future standard of living.

REVIEW 17.4 QUESTIONS

1. What is meant by an exothermic chemical reaction? An endothermic reaction? Look up an example of each.

2. Why will a lump of coal burn but not a lump of glass?

3. Compare and contrast organic fossil fuels such as coal and oil with the more contemporary organic fuels such as wood and peat.

4. Discuss the advantages and disadvantages associated with the use of chemical energy.

5. Explain how the greenhouse effect operates. Are there any examples in nature of a greenhouse effect?

17.5 Nuclear Energy

Early on the morning of July 16, 1945, the desert landscape outside Alamogordo, New Mexico, was lit up briefly and brilliantly by a fireball that was to dramatically alter the course of human history. That detonation at the Trinity test site released in one split second the explosive power of some 20,000 tons of TNT and represented the first technologically initiated, large-scale liberation of a new form of energy that was both awesome and terrifying in its implications for humanity. For better or worse, not only had Einstein's assurances 40 years earlier of the existence of a near infinitude of energy locked up in the mass of the universe been dramatically affirmed, but this new energy source was now, for the first time in history, available to human beings. That new energy source was, of course, nuclear energy, and in the years since that early morning moment on the New Mexico desert nearly a half-century ago, the suggestion that nuclear energy represents the best possible solution to the world's present and future power needs has become a most controversial one indeed.

Recall from Chapter 13 that two basic methods are currently available to human technology by which nuclear energy can be extracted from its equivalent mass. The first of these is the process called *nuclear fission,* in which the nuclei of atoms are broken apart by incident particles. In the process, mass is converted into its equivalent energy. For most substances, such fission processes cannot be sustained past the first particle-nucleus collision and the reaction ceases at that point. However, a certain class of materials, called fissionable materials, exhibits the distinctive characteristic that when their nuclei are struck by a certain kind of incident particle, the resulting nuclear fission produces two or more of those same nucleus-splitting particles. These "second-generation" particles in turn strike other nuclei and the process continues as a chain reaction.

One commonly used fissionable material is the uranium-235 isotope. When a uranium-235 nucleus is struck by a neutron, it breaks up into two major fission fragments, one of which has an atomic weight of about 140 and the other of about 95. The atomic weights and the atomic numbers (numbers of protons) of the nuclei of the major fission products can and do vary, but the atomic numbers of the two fragments must always add up to the 92 protons present in the initial

17.1 Nuclear War: The Ultimate Catastrophe

The famed British writer H. G. Wells once said that the human history of the twentieth century would be a race between education and catastrophe. Even a visionary of the stature of an H. G. Wells, however, could not possibly have imagined the nature or the magnitude of the catastrophe of which he spoke.

On August 6, 1945, scarcely three weeks after the Trinity nuclear test, a second nuclear device was detonated. This time the explosion site was not in the New Mexico desert but over the Japanese city of Hiroshima. The world had its first brief look at nuclear war.

Since then, nuclear devices have become far more powerful than the 20-kiloton weapon that leveled Hiroshima and the means to deliver them far more sophisticated than the lumbering B-29 that carried that weapon to its fateful destination. A single typical thermonuclear warhead in a modern nuclear arsenal carries more explosive power than all of the shells, bombs, and bullets fired or detonated throughout all of the battles of World War II. There are roughly 50,000 such warheads poised around the world ready to be delivered within minutes to any spot on earth.

We can only speculate, for example, via computer simulations, on what it would be like for the world to endure a "nuclear exchange" in which all or a large fraction of these 50,000 thermonuclear warheads were to be delivered and detonated all at once. Strangely enough, the opening detonations in such an exchange would probably not claim a single human life. Instead they would be high-altitude bursts designed to produce an "emp" (*electromagnetic pulse*), which is a surge of electromagnetic energy that has the effect of rendering any piece of unshielded electronic equipment virtually useless. Thus the first casualty of a nuclear war would be virtually all of twentieth century technology. Military targets and population centers would very

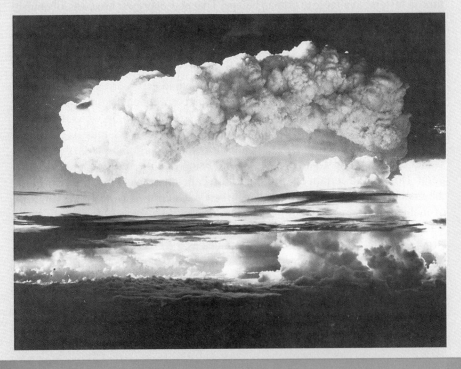

··· **A Hydrogen Fusion Explosion** The first terrestrial large-scale release of thermonuclear fusion energy occurred with this blast on the Central Pacific islet of Elugelab in the Eniwetok Atoll on November 1, 1952. The detonation produced a fireball 5 kilometers in diameter and completely vaporized Elugelab, leaving in its place in the atoll reef a crater over 3 kilometers wide and 1 kilometer deep.

Continued

quickly be next as hundreds of followup warheads would rain down on their assigned blast sites, with each warhead capable of producing a fireball 5 kilometers in diameter and virtually complete destruction over a circular area 30 kilometers in diameter.

Associated with each and every blast would be an enormous radioactive cloud that would expand in size to dozens of miles in diameter and billow all the way into the earth's stratosphere. Much of the contaminated material in such clouds would drop back to the earth as radioactive fallout, thereby contaminating large tracts of the earth's surface that may have been unaffected by the initial blast. A sufficiently large number of detonations can produce so much dust in the earth's atmosphere that the amount of sunlight reaching the earth's surface would be significantly reduced. A most severe cooling effect called ''nuclear winter'' would be experienced at the earth's surface. It has also been speculated that such a nuclear exchange could cause great damage to a protective layer of the earth's atmosphere called the ozone layer, which presently screens the surface of the earth from the higher energy ultraviolet rays of the sun. With little or no ozone layer, the nuclear winter would be followed by an ''ultraviolet summer'' in which an unprotected terrestrial surface would now be raked by high-energy solar ultraviolet photons.

To make a long story short, such a nuclear exchange would be the ultimate in environmental disaster and could well mark the end of the ability of the biosphere to sustain life of any kind. Will such a catastrophe befall us? Wells never tells us who finally wins the race.

uranium-235 nucleus. Thus, one way in which a uranium-235 nucleus can be split is into a barium-137 nucleus, which has 56 protons, and a krypton-83 nucleus, which has 36 protons. The nuclei of major fission fragments are quite often very unstable and will undergo additional radioactive decay into more stable nuclear isotopes. We will return later to the implications these fission fragments and their secondary radioactive decay reactions hold for nuclear power production.

The fission of the uranium-235 nucleus is also accompanied by the conversion of some of its mass into approximately 3.2×10^{-11} joules of energy. By itself, such an amount is hardly the stuff of which whole cities are powered, but the fission of the uranium-235 nucleus produces yet another product—at least two and occasionally as many as five additional free neutrons. These neutrons can now strike additional uranium-235 nuclei, causing them to split in the fashion described above. The overall result of this continued cycle of neutron production and nucleus splitting is a very rapid buildup of the number of free neutrons present, the number of major fission fragments, and, most importantly, the total amount of energy released [Figure 17.5(a)]. If nothing is done to stop or slow down the chain reaction process, the energy release quickly takes on the explosive dimensions of the Trinity site detonation: An energy of the order of 10^{14} joules is released in a small fraction of a second. This order of magnitude of energy *is* capable of servicing a city for a year or so, provided we can control the rate at which it is released.

The key factor in the buildup of energy in a fission reaction is the rate at which the free neutrons are produced. If that rate could somehow be held at a

(a)

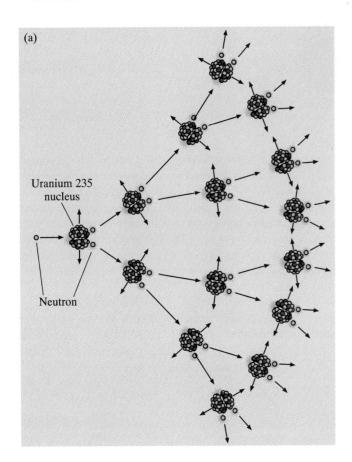

Uranium 235 nucleus

Neutron

(b)

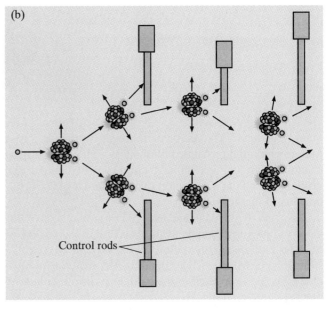

Control rods

FIGURE 17.5 ··········

Nuclear Fission (a) In an uncontrolled fission process, the reactions continue to build in number until an explosive release of energy results. (b) By placing neutron-absorbing control rods into the site of the reactions, the number of free neutrons and, hence, the rate of energy release can now be controlled.

fixed value, then the corresponding rate of energy generated from the fission processes could also be kept at a constant value. In pursuing this concept further, one finds that a number of substances exist whose nuclei have the ability to absorb free neutrons without experiencing any further nuclear reactions themselves. Thus the control of the rate at which energy is produced can in theory be accomplished by somehow making use of these neutron-absorbing materials. A schematic of the neutron controlling process is shown in Figure 17.5(b). If a neutron-absorbing substance such as cadmium is placed into the region where the uranium fission is occurring, then some of the free neutrons will be absorbed before they can strike other uranium nuclei. The degree to which the neutron absorbers intrude into the region where the uranium fission is taking place controls the degree to which the uranium fissions and their associated energy release are produced.

On the other hand, the fission process can be aided considerably by introducing a moderating substance such as graphite, which serves to slow down the free neutrons as they are created, thereby increasing the probability that they will strike other uranium nuclei. A controlled nuclear fission reactor thus consists of a core made up of fuel elements of fissionable material embedded in a neutron moderating medium. Neutron-absorbing control or damping rods can be inserted into this array to provide a means of regulating the flux of neutrons in the core (Figure 17.6). The equilibrium rate at which the nuclear fission energy is created is then controlled by the relative presence of the fissionable fuel, moderating medium, and control rods.

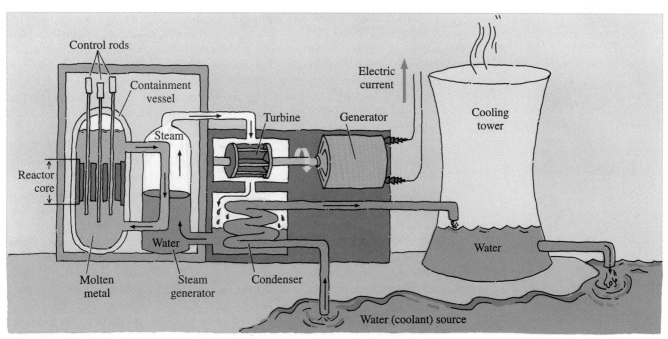

FIGURE 17.6 ··········

A Nuclear Reactor The release of thermal energy in a controlled fission reaction is used to heat water and produce steam. The steam then is used to drive the turbine of an electrical generator to produce electrical power.

The energy released in such a controlled nuclear fission reactor is given off in the form of thermal energy and short-wavelength photons such as gamma rays. This energy must be converted into more useful energy forms. Conversion is accomplished by immersing the reactor into a vat of liquid as shown in Figure 17.6. The energy from the reactor heats the liquid in the closed vat to temperatures of about 600°C. This hot liquid is then used to produce steam from water in a heating vat. The steam in turn drives a turbine, producing rotational kinetic energy, which can be used to generate electricity in the familiar fashion, or mechanical energy to perform such tasks as turning the propellers on a "nuclear" ship. The process can be made cyclic by transporting the steam through a steam-condensing unit and returning the resulting water back to the heating vat.

The production of energy from nuclear fission reactors has an important advantage in that the process is more efficient and pollution free than a corresponding level of energy production gleaned from fossil fuels. Most importantly, however, nuclear fission processes represent a potentially inexhaustible supply of energy. At first glance, this is not very obvious. The uranium-235 isotope is not very abundant in the earth's crust and if we used the fission of this isotope to produce our total power needs, we would run out of the world's supply of uranium-235 in a few short decades. The intriguing possibility exists, however, that the world's supply of fissionable material can be expanded through a process called nuclear "breeding."

In essence, nuclear breeding is a process by which fissionable material can be created in nuclear reactors from relatively abundant isotopes. The most common nuclear breeding process is initiated when a uranium-238 isotope absorbs a neutron to become uranium-239. Uranium-239 decays to the isotope neptunium-239 by a beta decay with a half-life of about 24 minutes. The neptunium-239 isotope then undergoes a second beta decay, this time with a half-life of 2.4 days, and becomes the isotope plutonium-239. Plutonium is also an unstable nucleus and eventually decays by emitting an alpha particle with a very long half-life of 24,000 years. The importance of this process lies in the fact that the plutonium-239 nucleus is, like that of uranium-235, capable of sustaining a nuclear fission chain reaction.

A similar nuclear breeding process results in the conversion of the fairly common isotope thorium-232 into the long-lived and fissionable isotope uranium-233. An ordinary nuclear fission reactor can be converted into a breeder nuclear reactor simply by placing a shell of either uranium-238 or thorium-232 about the reactor core. The reactor continues to generate its energy as before, but now produces an additional supply of fissionable material. By expanding the supply of fissionable material in this way, the breeder reactor can not only provide the nuclear fuel for its own future operation but for the future operation of additional breeder reactors as well.

Although it possesses a number of important advantages, nuclear power production unfortunately also suffers from disadvantages of considerable proportions. One of the products resulting from nuclear fission reactions, for example, is the major fission fragments described earlier. Many of these ''nuclear waste'' fragments are highly radioactive isotopes and are immediately dangerous; others are capable of being subtly absorbed into the environment where they can cause long-term problems. The isotope strontium-90, for example, is a radioactive secondary product formed from the 2.6-minute half-life decay of the fission fragment rubidium-90. Strontium is in the same chemical family as calcium. Thus, like calcium, it is a constituent of the bones and teeth of animals and humans. Since the physiological chemistry of animals and humans generally cannot distinguish isotopes of the same element, the radioactive isotope is biochemically treated the same way as the nonradioactive isotope of the same element. Of course, the two isotopes are not the same and the ingestion of the radioactive isotope can result in the alteration and disruption of cell tissue, an effect that can lead to the formation of mutant cancerous cells, genetic damage, etc.

In addition to the major fission fragments, the chemistry of the nuclear fuel plutonium makes it a highly dangerous substance in its own right. The question of how and if one can safely deal with nuclear waste materials is of crucial importance in any future expansions of nuclear power generation. Finally, the ever-present specter of nuclear fuel or a nuclear reactor gone awry still exists. If a shipment of coal or oil falls into the hands of a terrorist group or the reckless leader of a country, there would be little if any cause for concern. That would certainly not be the case, however, if a shipment of fissionable material sufficient in size to build one or more nuclear weapons were to fall into similar hands. The worst instance to date of a nuclear reactor gone awry occurred in April 1986 when the Chernobyl nuclear reactor in the Soviet Union malfunctioned, with the result that a considerable amount of radioactive material was released into the surrounding region and dozens of lives were lost. The total long-term environmental impact of this tragic nuclear accident may not be known for decades.

An alternate, less dangerous method of nuclear energy production is possible through the use of the nuclear fusion process. Such energy is liberated when the nuclei of hydrogen atoms are fused at very high temperatures and densities into helium nuclei. As we have already noted, nuclear fusion is the basic process by which the sun and stars are able to generate their energy on such a large scale. Although we have had the capability since 1952 to reproduce fusion processes in an uncontrolled way in the form of thermonuclear detonations, we have not yet been able to produce such an energy release in a controlled way. The principle difficulty thus far in getting a controlled nuclear fusion reaction to work lies in our inability to contain the hydrogen nuclei long enough at temperatures and densities high enough to trigger the onset of nuclear fusion and its attendant liberation of substantial amounts of energy. Unfortunately, despite a great deal of effort on its behalf, it is highly doubtful that a controlled fusion reactor will be operational before the end of the present century.

Our nuclear energy options are thus restricted, at least for the time being, to nuclear fission, and the ongoing controversy over the desirability of producing such power on a large scale is the classic confrontation between an ever-increasing demand for more power production and the dangers associated with the methods of such production.

1. Describe what is meant by nuclear energy. How is such energy produced?
2. Compare and contrast nuclear fission and nuclear fusion. Which process do you suppose would have the least impact on the environment? Explain.
3. Discuss how nuclear energy, once produced, is then converted into electrical energy.
4. Why do you suppose that nuclear energy is so controversial politically?
5. What is a breeder reactor? Why is such a device important in the production of nuclear power?

17.6 Solar Energy

Human beings have long recognized the sun as a giver and sustainer of life, and while we no longer worship this hot ball of gas as a deity, modern science nonetheless acknowledges that without it, life on earth would be impossible. The key to such predominance lies in the energy that is produced by this important celestial entity. The sun is now and has been a veritable bonanza of energy production. The sun's energy warms our planet to the temperature levels that permit our existence, provides the "photo" part of the photosynthesis reactions carried out in the plant life of our world, and has, with few exceptions, been responsible for providing the reservoirs of energy, whether wind, flowing water, coal, or oil, to which we currently have access. Even in the realm of nuclear power, the sun has solved the controlled fusion containment problem deep in its core by simply self-gravitating, and in its photospheric flaring processes, it tantalizes us with a possible solution to our own fusion containment problem. For all of this, however, it is still raw sunlight that may or may not provide the answers to some of the energy-related dilemmas described earlier in this chapter.

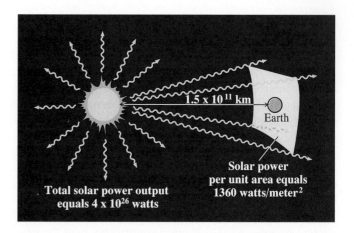

FIGURE 17.7 ··········

Solar Energy The sun produces some 4×10^{26} watts of power. At a distance of 150 million kilometers, the earth-sun distance, the intensity of this power is 1360 watts per square meter.

In terms of sheer magnitude, no other energy source can come close to that represented by sunlight. A square meter located at a distance from the sun equal to the earth's orbital radius of 150 million kilometers receives about 1360 watts of power from the sun, which corresponds to an energy rate of 1360 joules per second (Figure 17.7). Such a power level, of itself, is not exactly awesome. Fourteen 100-watt light bulbs will light up your dorm room or apartment handsomely, but not an entire town or city. Two aspects of that 1360 watts, however, make solar energy a power source worthy of careful consideration.

First, the sun will keep radiating that 1360 watts of power on our square meter at an undiminished rate second after second, year after year, for the next few billion years or so. Moreover, as we have already seen in Chapter 15, the sun generates about a million times more energy in one second than all of the world's industry consumes in an entire year! Solar energy clearly represents the ultimate in inexhaustibility. While that is true, a difficulty arises in that the earth intercepts only a very small fraction of this impressive energy output. Moreover, the earth's atmosphere reflects, scatters, absorbs, and otherwise prevents more than 70% of the energy that is intercepted by the earth from directly reaching the earth's surface. From an environmental viewpoint, it's just as well, for if we were to intercept a significantly larger amount of direct solar energy, we would have undoubtedly been long since cooked and boiled off the planet. From an energetic viewpoint, however, it means that here on the earth's surface we don't have direct access to the bulk of the energy that the sun radiates into space. Instead, we must be content, for the time being, with somehow making use of the 200 watts/square meter worth of the sun's power in the visible region to which we do have direct access.

Most of us in our childhood days at one time or another livened up an otherwise boring hot summer day by exploring the fun that can be had from enhancing the effect of the sun's energy with Aunt Bessie's reading glass. Of course, we were not so much interested in a possible solution to the world's energy problems as we were in the incineration of ant hills or zapping the kid down the block with a dose of focused solar energy, but the basic optical principles underlying the success of the reading glass in concentrating the sun's energy in fact provide us with what is perhaps the simplest means of converting the sun's direct energy into more useful forms. A schematic of one version of the process is shown in Figure 17.8.

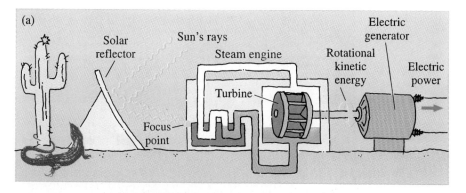

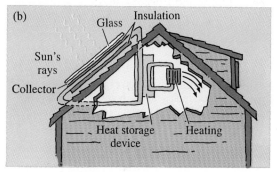

FIGURE 17.8 ··········
Solar Power Utilization (a) Reflectors can be used to focus the sun's incoming energy to create steam to run a steam engine. The output rotational kinetic energy of the steam engine is then used to produce electrical power via an electrical generator. (b) Solar heating can be accomplished by means of collectors that store the incoming solar energy.

Rays incident from the sun are collected and brought to a focal point by a set of large reflectors, much as starlight is collected and focused by a large reflecting telescope mirror. The amount of solar power per unit area collected at the focal point is sufficient to boil water in quantities adequate to run a steam engine. The steam engine's output of rotational kinetic energy then drives an electric generator for the final conversion into electrical energy.

Along similar lines, it is also possible to accumulate solar energy by means of a solar collector in which a layer of material capable of efficiently absorbing the sun's energy is interspersed with an array of water carrying circulation tubes. As the water circulates through the absorbing layer, it is heated up and carries the absorbed solar energy away from the collector in the form of thermal energy. This hot water can now be stored in a hot water reservoir for later use as kitchen or bathroom hot water, or to heat homes and offices.

Another important technique by which solar energy can be converted to more useful forms is through the use of photovoltaic or "solar" cells. Solar cells use the sun's energy to create an electric charge flow, thus generating electricity directly without the "middleman" energy conversions. One such type of solar cell consists of a *p-n* junction in which the *p*-type semiconducting layer is made thin enough so as to be transparent to the passage of sunlight (Figure 17.9). When the cell is placed in sunlight with the thin *p*-type semiconducting layer facing the sun, the incoming solar photons pass through the *p*-type layer and strike the crystal lattice of the *n*-type layer. As the photons strike the *n*-type layer they knock electrons in the *n*-type layer out of their bound states and across the junction boundary into the *p*-type layer, where it quickly fills one of the *p*-type layer's electron holes.

The net effect is the creation of a charge separation between the negatively charged free electron, which has been knocked across the *p-n* boundary to fill

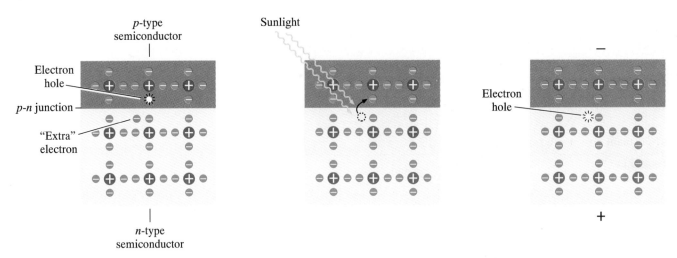

FIGURE 17.9 ···········
The Solar Cell   If a cell having a *p-n* junction with a thin *p*-layer is exposed to sunlight, the electrons will jump across the *p-n* boundary to the holes in the *p*-layer, thus creating a charge separation and with it an electric potential capable of running electrical devices.

an electron hole, and the positively charged electron hole, which has been left behind in the *n*-type layer. The charge separation thus produced creates an electric potential between the two charges, and if the incoming sunlight can create such charge separations on a scale large enough, the resulting electrical potential that is built up between the *p* and *n* layers is capable of running electrical devices. Solar cells of this type are already used to some degree as small-scale power sources for pocket calculators, etc. They have also served as an important source of power for the operation of satellites and space probes.

If we try to extend the successes of solar cells to a larger scale, however, we find that although solar energy itself is free, producing the solar cells necessary to convert it into electrical energy is not. The high cost currently associated with the production of solar cells prevents this type of large-scale solar energy production from being cost competitive with other contemporary forms of energy production. In addition there are geophysical and meteorological restrictions associated with the large-scale use of solar energy. As the earth spins on its axis, every location on the earth spends on the average one-half of its time facing away from the sun and hence is unavailable for the reception of solar energy. Weather conditions can obviously play a crucial role in terms of the availability of solar energy. The availability of solar energy is very high at those locations, such as the desert areas of the world, where the percentage of clear, sunny days during the year is very high, but it is very low where the percentage of such days is correspondingly low. In recent years some questions have been raised over the possible environmental impact of the large-scale use of some of the materials such as gallium arsenide and indium phosphide, which are employed in the manufacture of solar cells. Nevertheless solar energy continues to rank along with controlled nuclear fusion as one of the most tantalizing of the possible sources for clean and limitless energy.

1. Discuss the various ways that the sun supplies energy to the earth.

2. Why is direct sunlight such a potentially important sources of energy for human technology?

3. Describe the various ways in which solar energy can be converted into more useful forms of energy.

4. Explain how a solar cell operates.

5. Summarize the advantages and disadvantages of solar energy use.

17.7 Physics and the Future

If we were to examine the story lines of a reasonably large sample of recent-vintage fictional forays into future history, we would find that the overviews of the future of humanity would take on what the statisticians refer to as a bimodal distribution. One group of storytellers would tend to optimistically cluster about the view of an ever-advancing, ever-achieving human technology with all of the attendant blessings of such a world. The second group could perhaps best be described as purveyors of pessimism regarding the ultimate fate of human beings. The enthralling aspect of this dichotomy of viewpoints is that sometime over the next 100 years, one or the other of them will have basically come true. The jackpot question, of course, is which view will it be?

The astounding advance of science and technology since the Scientific Revolution has brought with it a set of most critical challenges, one of which involves the nuclear processes described earlier. When the first test of a thermonuclear fusion device or "hydrogen bomb" vaporized the entire island of Elugelab in the Eniwetok Atoll in the autumn of 1952, human beings had their first access to unlimited destructive power. In the years that have elapsed since, the destructive capability of the thermonuclear arsenals of the world powers has reached a level of unspeakable horror. In addition to the nuclear arsenals of the superpowers, a number of smaller nations around the world either now have or soon will possess smaller scale nuclear arsenals of their own, thereby enhancing the possibility of a late twentieth century version of the "Shot at Sarajevo." * Unless these nuclear arsenals are somehow reduced or eliminated, humanity faces the distinct possibility of going out with a bang rather than a whimper.

It is also entirely possible that humanity could well go out with a whimper rather than a bang. Another spin-off product of twentieth century technology, particularly in the health and medical areas, has been a rapid increase in the number of people living on the planet. From the first appearance of human beings on the earth to the year 1650, the human population increased to a total of 500 million people. Two hundred years later in 1850, the population doubled to one billion, and then to two billion by 1930. By 1975 the world's population stood

---

*The start of World War I was triggered when Archduke Ferdinand of Austria-Hungary was assassinated in the city of Sarajevo in modern Yugoslavia on June 18, 1914.

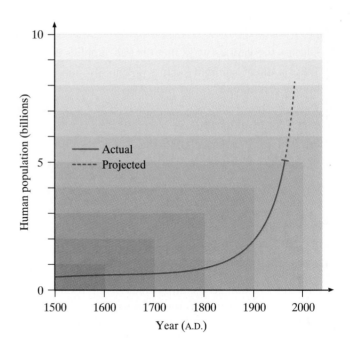

FIGURE 17.10 ··········

World Population Growth   The world's population is currently increasing in an exponential fashion. At the present time, one human being is born on our planet every three seconds.

at four billion people and is projected to be eight billion by the year 2005. This type of life-form growth behavior is well known in the biological sciences and is referred to as exponential growth. Bacteria cultures, for example, can exhibit such growth behavior.

Notice the similarity between the population growth curve in Figure 17.10 and the energy consumption curve in Figure 17.1, because it points out one of the major impacts of this kind of population growth. Throughout this chapter we have described the environmental pressures that result from attempting to increase our energy production significantly. The larger the production increase, the more environmental pressure. One can argue that should our flirtations with achieving ready access to a clean, inexhaustible energy supply be rewarded, the environmental problems would be taken care of as well. Unfortunately, human beings place all sorts of other pressures on the environment besides a demand for more energy. Whenever any life-form places pressure on the environment, a quantity results from that relationship that is called the *carrying capacity* (Figure 17.11). Carrying capacity can be defined as the total number of life-forms that can be supported by the environment in question without destroying the resources of the environment. If a life-form enjoys a population boom that rapidly carries it past the carrying capacity of the environment, a population crash must inevitably occur in which the bulk of the life-forms are destroyed. The laws of biological science are blunt and unforgiving in this matter: Exceed the carrying capacity of the environment and grimly suffer through a "die-back" population crash.

There are alternate situations in which the population growth slows down as the total population approaches the carrying capacity of the environment. This type of growth is referred to as *logistic growth* and represents a far less drastic scenario than a population crash. This asymptotic approach of a population level to a constant value if often called a condition of zero population growth.

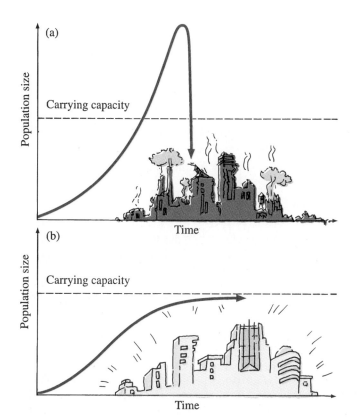

FIGURE 17.11 ··········

Carrying Capacity   In any biosystem a given life-form population can approach the environment's carrying capacity in one of two ways. In scenario (a), the population overshoots the carrying capacity, in which case the population is catastrophically reduced below the carrying capacity. In scenario (b), the population growth rate can gradually level off as the carrying capacity is approached.

Unfortunately, the world's population growth rate presently appears headed for a population crash. It is estimated that the carrying capacity of the earth's biosphere is somewhere around eight billion people. Thus if our current population trends continue into the next century, humanity is in danger of a population overshoot, with the obligatory consequences.

None of this is necessarily inevitable. The same human intellect and ingenuity that have forged the theories, principles, ideas, and devices described in this text are also capable of successfully addressing the problems we collectively and universally face as a species. We do in fact have the power to cast our lot with the optimists, if we only choose to do so.

Through our science we have thus forged a picture of a universe that is vast, complex, and exciting. We may not, however, have even begun to envision it all. Words as appropriate today as they were two and a half centuries ago are those of Sir Isaac Newton:

> I do not know what I may appear to the world, but to myself I seem to have been like a boy playing on the sea-shore, and diverting myself now and then finding a smoother pebble or a prettier shell than ordinary, while the great ocean of truth lay all undiscovered before me.

One can only speculate on what pebbles, what shells, and what oceans lie ahead for us and for our physics.

THOUGHT QUESTIONS

1. Describe a form of energy not mentioned in this chapter and how you would convert it into electrical energy.

2. Which of the energy sources described in this chapter are *not* due either directly or indirectly to the sun?

3. Why do you suppose that:
   a. it is so hot at the center of the earth?
   b. high-temperature regions exist relatively close to the earth's surface?

4. Does a breeder reactor violate the principle of the conservation of energy? Explain.

5. Find the location of the nearest power substation and distribution transformer in your neighborhood. Compare and contrast the two.

6. Give an example of an appliance or other entertainment or work-saving device that does *not* make use of electrical energy in some way.

7. How might you try to capture and store the electrical energy contained in the discharge of a lightning bolt? What problems might you encounter in such an endeavor?

8. Can you think of a way in which the energy contained in an ocean wave might be converted into useful energy? Explain how this might be done.

9. How would you convert the energy of a volcano into useful forms? What problems do you envision in such a process?

10. It has been suggested that we might find fossil fuels in abundance on the other planets of the solar system. Comment on this idea.

11. Why do you suppose that it is desirable to have a warship such as an aircraft carrier or submarine powered by a nuclear reactor?

12. Why do you suppose that solar cells are used so extensively aboard satellites and spacecraft?

13. Give some examples of devices you use that are powered by solar cells.

14. One method proposed for the large-scale generation of solar energy involves orbiting a huge array of solar cells about the earth and sending the energy gathered by these cells back to earth in the form of an electromagnetic microwave beam. Discuss the advantages and disadvantages of such a scheme.

15. List some of the regions on the earth where the use of solar energy would not be practical because of the high percentage of cloudy days.

16. Can you think of an energy source here on the earth that is *not* directly or indirectly related to the sun?

17. Would moonlight be a useful source of energy? Explain.

18. Compare and contrast the mathematical pattern associated with the world's population growth and that for the development of a nuclear fission reaction.

19. It has been suggested that fuels such as wood, peat, and dry animal dung could be used to supply our energy needs when coal and oil run out. Comment on this possibility.

20. Discuss the relationship between the world's population level and the world's energy production.

21. Of all the energy sources mentioned in this chapter, identify those that are:
   a. renewable   **b.** nonrenewable.

22. Can you think of any renewable energy source not mentioned in the text?

Glossary

aberration of starlight Effect in which the apparent position of a star in the sky is displaced due to the interaction between the velocity of light and the earth's orbital velocity. (page 351)

abscissa The horizontal or x-axis on a graph. (page 30)

absolute zero Temperature at which no further thermal energy can be extracted from a given object; absolute zero ($-273.15°$ Celsius) is the lowest temperature possible. (pages 222, 234)

acceleration The time rate of change of velocity in a specified direction. (page 52)

action = reaction Principle in physics in which a force (action) exerted on an object will cause the object to exert an equal and opposite force (reaction) on the agent exerting the force; Newton's Third Law of Motion. (page 60)

action potential Shift of electric potential in a neuron caused by stimulant chemicals such as hormones or neurotransmitters. (page 125)

additive color mixing Creation of a given color by adding the additive primary colors of blue, green, and red to each other in the proper proportions of intensity. (page 363)

additive primary colors Basic colors used in additive color mixing processes; the additive colors are blue, green, and red. (page 363)

alpha particle A nuclear particle consisting of two protons and two neutrons; a helium atom nucleus. (page 384)

alternating voltage Voltage whose sign and magnitude changes with time. (page 326)

amorphous solid Solid that exhibits no definite repeating pattern in the arrangement of its atoms. (page 192)

ampere The amount of electric current flowing in each of two long, parallel wires that will produce a force of 2×10^{-7} newtons per meter of wire length when the wires are separated by a distance of exactly one meter; a current flow roughly equal to 6.25×10^{18} unit electric charges per second. (pages 294, 315)

amplitude The maximum displacement distance of a wave motion from its average or equilibrium position. (page 257)

amplitude modulation (AM) Process in which a carrier wave amplitude is modified by a signal wave. (page 428)

angle of incidence Angle between an incoming light ray and the line normal to the surface at the ray's point of contact with the surface. (page 356)

angle of reflection Angle between a reflected light ray and the line normal to the surface at the ray's point of contact with the surface. (page 356)

angular acceleration The change in an object's angular velocity per unit time. (page 93)

angular displacement The change in an object's angular orientation. (page 93)

angular momentum The "quantity" of rotational motion in a given object which is equal to the product of mass, velocity, and orbital radius for a revolving object and to the product of the angular velocity and the moment of inertia for a rotating object. (page 101)

angular velocity The change in an object's angular displacement per unit time. (page 93)

antimatter Atoms made up of positively charged anti-electrons (positrons), negatively charged antiprotons, and antineutrons. (page 516)

antiparticle A particle whose characteristics are such that when it collides with its particle counterpart, both particle and antiparticle will be totally annihilated with their masses converted into their equivalent relativistic energy. (page 403)

antinode Point in a standing wave that experiences the maximum displacement from the equilibrium position of the wave motion. (page 260)

apparent brightness The amount of power per unit area received from a source of electromagnetic energy. (page 347)

Archimedes' Principle Statement that a fluid will exert a buoyant force on any body floating on or immersed in it which is equal to the weight of the fluid displaced by the floating or immersed body. (pages 203–204)

arcsecond An angle equal to 1/3600 of a degree. (page 500)

atmosphere Layer of gases surrounding a planet, satellite, or star; the gas pressure of the earth's atmosphere at sea level or 100,000 N/m². (pages 203, 290)

atom The smallest division of a given chemical element that retains all of the properties of that element. (page 187)

atomic clock A type of clock that keeps time from the highly regular oscillations of electrons between the energy levels of a given atom such as cesium. (page 389)

atomic mass The mass of an atom or molecule measured in atomic mass units. (page 420)

atomic mass unit In physics, one twelfth of the mass of an atom of the most common isotope of the element carbon; in chemistry, one sixteenth of the mean-mass of an oxygen atom; the approximate mass of a hydrogen atom or about 1.67×10^{-27} kg. (page 420)

atomic number The total number of protons in a given element or isotope. (page 187)

aurora The light radiated by atoms and molecules in the earth's upper atmosphere that have been struck by high energy particles coming from the sun. (page 320)

autotransformer A type of transformer in which both the primary and secondary terminals are connected to a single coil. (page 341)

axis of rotation An imaginary line through a given object about which that object appears to rotate. (page 89)

barycenter The point between two mutually revolving objects for which the product of the mass and the distance to the barycenter is equal for both objects. (page 134)

base (electronics) The layer between the emitter and the collector in a transistor; the layer on which the bias voltage in a transistor is placed. (page 452)

beat A periodic variation in sound intensity that occurs when two sound waves having different frequencies constructively interfere; a beat wave. (page 269)

Bernoulli's Principle The statement that the quantity $P + \frac{1}{2}\rho v^2 + \rho gh$ has the same value for a given fluid at any position within the fluid. (pages 205–206)

beta decay The process in which a neutron decays into a proton and emits a high-velocity electron from the nucleus; beta decay can also occur when a proton decays into a neutron and emits a high-velocity positron. (page 403)

beta particle A high velocity electron or positron that is emitted from the nucleus of an atom during beta decay. (page 396)

bias voltage The voltage placed on the base layer of a transistor that controls the rate at which charge carriers flow between the emitter layer and the collector layer of the transistor. (page 453)

Big Crunch A hypothesized end to the universe in which all of the material in the universe ultimately collapses in on itself. (page 518)

Big Bang An event thought to have occurred on the order of 15 billion years ago in which the universe as we now

observe it was set in motion. (page 515)

binary number system A number system based entirely on the two numbers 0 and 1. (page 436)

biosphere The region on the earth extending from the deepest part of the oceans to the highest mountain peaks which is inhabited by life forms. (page 531)

Biot-Savart Law The statement that the magnetic field strength from a current-carrying wire varies directly with the current in the wire and inversely with the perpendicular distance to the wire. (page 337)

Boolean algebra A system of mathematical logic that postulates that only two constants exist, 0 and 1. (page 436)

black hole An object whose gravity is so strong that its escape velocity exceeds the speed of light. (page 510)

British thermal unit (BTU) The amount of thermal energy required to heat one pound of water one degree Fahrenheit. (page 223)

boiling point The temperature at which a given substance is changed from its liquid state into its gaseous state under a pressure of one atmosphere. (page 229)

bow wave The V-shaped or cone-shaped wave created by a source of wave motion moving through a fluid more rapidly than the waves it creates; a shock wave. (page 262)

Boyle's Law The statement that the product of the pressure and the volume for a gas at constant temperature has a constant value. (pages 231, 251)

breeder nuclear reactor A nuclear reactor in which fissionable material is produced in addition to nuclear energy. (page 544)

bulk deformation A deformation in which the volume of an object is altered by the exertion of outside forces; volume stress. (page 211)

bulk modulus The ratio of the volume stress to the corresponding change of volume per unit volume for a given object or substance. (page 212)

buoyant force The upward force exerted by a fluid on an immersed or floating object in reaction to the force exerted by the object to initially displace the fluid. (page 203)

calorie The amount of thermal energy required to raise the temperature of one gram of water by one degree Celsius; the calorie which is employed in food and dieting is equal to one kilocalorie or 1000 calories. (page 223)

capacitance The ratio of the total stored charge to the voltage for two oppositely charged conducting plates that are separated by insulating material; a measure of the ability of a capacitor to store electric charges. (page 439)

capacitor A device, usually in the form of two parallel conducting plates separated by a layer of insulating material, that is used to store electric charges. (page 439)

capillary action (capillarity) A phenomenon in which the interplay of adhesive and cohesive forces in a liquid with the liquid's surface tension can cause the liquid in a tube immersed in a reservoir to rise or fall below the level of the liquid in the reservoir. (page 196)

capture orbit An orbit in which one object will neither crash into the object about which it is orbiting nor escape from it; a circular or elliptical orbit. (page 135)

Carnot engine An idealized heat engine in which all of the input thermal energy originates from a single hot reservoir at a single temperature and all of the engine's unused or rejected energy emerges at a single lower temperature cold reservoir. (page 240)

carrier wave A transmitted electromagnetic wave whose amplitude or frequency has been slightly modified or modulated by input audio and/or visual signals in order to convey information over long distances. (page 427)

carrying capacity The total number of life forms that can be supported by a given environment without destroying the life-sustaining resources of that environment. (page 550)

causality The concept that any cause cannot be preceded in time by its effect. (page 463)

Celsius temperature scale A temperature scale in which 0°C is the temperature of an ice/water mixture and 100°C is the temperature of a water/steam mixture. (page 222)

center of gravity The point at which all of the mass of a given object seems to be concentrated. (pages 107,129)

centrifugal force An apparent "away from the center" force observed in a rotating reference frame. (page 90)

centrifuge A device whose rotational effects can separate mixtures of particles of various sizes, fluids of differing densities, etc. into individual layers. (page 90)

centripetal acceleration A "center-seeking" acceleration that produces a change in an object's direction of motion. (page 118)

centripetal force A "center-seeking" force that produces a change in an object's direction of motion; the product of an object's mass and its centripetal acceleration. (page 89)

cepheid variables A type of pulsating star whose luminosity is mathematically related to its period of pulsation or light variation through a period-luminosity law. (page 513)

chain reaction A reaction in which some of the products

of the reaction are the same as some of its initial reactants, thereby allowing the reaction to become self-sustaining. (page 401)

charging by contact A process of creating a net electrical charge on an object by touching the given object with a second, electrically charged object. (page 285)

charging by friction A process of creating a net electric charge on an object by rubbing that object with another. (page 284)

charging by induction A process of creating a net electric charge on an object by first bringing an electrically charged second object close to the given object and then bleeding the repelled charges from the object by means of a third "touching" object. (page 285)

Charles' Law The statement that for a gas at constant pressure, the ratio of the volume of the gas to its temperature has a constant value. (page 232)

coil A spring-like set of current loops used to create an enhanced and uniform magnetic field within the array. (page 327)

collector The layer of a transistor toward which electric charge carriers flow from the emitter. (page 452)

color The human perception of the variety of electromagnetic waves between wavelengths of 4×10^{-7} m and 7.5×10^{-7} m. (page 362)

color mixing A process in which any color is produced by adding various proportions of the additive primary colors of blue, green, and red, or by combining various proportions of the subtractive primary colors of cyan, expanded yellow, and magenta. (page 363)

commutator rings A set of rings in an electric generator or an electric motor that permits voltage to be transferred into (motor) or out of (generator) the device. (page 325)

conduction A mechanism of energy transport in a substance in which thermal energy is transported by propagating an increased agitation of atoms and molecules throughout the given substance. (page 294)

conductor A substance that readily transports electric charges or thermal energy. (page 294)

conservative force A force in which the total amount of work done against the force to move from one position to another does not depend on the path taken between the two positions. (page 168)

constructive interference A phenomenon in which the crests of a given wave pattern are coincident with those of a second wave pattern, resulting in a wave effect that is enhanced or reinforced over the effects of the individual waves by themselves. (page 260)

control rod A neutron absorbing rod placed in a nuclear

reactor that controls the number of free neutrons present in the reactor, thereby controlling the rate at which nuclear fission occurs; a damping rod. (page 542)

convection A mechanism of energy transport in which thermal energy is transported from one region to another by fluid motions. (page 237)

core bounce An event in the life cycle of a massive star in which the star's outer layers collapse and then recoil from a small, imcompressible core with an attendant release of a considerable amount of energy; the nuclear bounce effect. (page 508)

Coriolis effect The apparent deflection of the straight line motion of an object when viewed in a rotating frame of reference. (page 90)

correspondence principle The statement that any new or revised scientific theory must not only account for the discrepancy between experimental results and the predictions of the old theory, but also must be successful where the old theory was successful. (page 20)

cosmic abundance scale The table of the relative abundances of all of the chemical elements and isotopes found in the universe. (page 488)

cosmic rays Small, high-speed atomic particles, usually protons, which exist in space. (pages 405, 465)

coulomb A unit of electric charge corresponding to the total electric charge residing on either 6.25×10^{18} electrons or 6.25×10^{18} protons. (page 315)

Coulomb's Law The statement that the electrostatic force between two electric charges is directly proportional to the product of the two charges and inversely proportional to the square of the distance between the charges. (page 287)

Coulomb's Law of Magnetic Forces The statement that the force between two magnets is directly proportional to the product of the magnetic pole strengths and inversely proportional to the square of the distance between the magnetic poles. (page 302)

covalent bond A chemical bond in which two atoms share their valence electrons. (page 191)

crusades A series of military expeditions mounted by Western Europe over the years 1096–1270 for the purpose of capturing the ''Holy Lands'' of the Near East for Christianity. (page 11)

crystalline solid A solid in which the component atoms are arranged in a regular and repeating three-dimensional pattern. (page 192)

current balance A device that measures the force between two current-carrying wires by balancing that force with known weights acting over a known lever arm. (page 336)

curvilinear motion Motion that is directed along some sort of curved or circular path as opposed to a straight-line path. (page 88)

damping rod A neutron-absorbing rod placed in a nuclear reactor that controls the number of free neutrons present in the reactor, thereby controlling the rate at which nuclear fission occurs; a control rod. (page 542)

Dark Ages The historical period in Western European history extending from the Fall of Rome in 476 A.D. to about 1000 A.D. (page 10)

de Broglie wavelength The wavelength associated with a particle such as an electron which is equal to the ratio of the Planck's constant to the momentum of the particle. (page 392)

decay probability The probability that a given nuclear decay process will occur for a given atom in a given time interval. (page 399)

decibel A measure of relative sound intensities that corresponds to a ratio of about 1.26 to 1 in actual power intensity. (page 276)

decibel scale A scale of sound intensities in which a difference of one bel corresponds to a tenfold increase or decrease in the power intensity of the second, and whose zero point is a power intensity of 1.0×10^{-12} watts/m^2. (page 276)

density The mass per unit volume of a given substance. (page 192)

depletion region The region formed at the boundary of a *p-n* semiconductor junction as a result of an applied voltage in which no electric current is able to flow. (page 452)

destructive interference A phenomenon in which the crests of a given wave pattern are coincident with the troughs of a second wave pattern, resulting in the destruction of the wave pattern at the point of coincidence. (page 260)

deuterium An isotope of hydrogen having one proton and one electron in its nucleus. (page 512)

deuteron The nucleus of a deuterium atom. (page 398)

diatonic scale A scale of musical notes each note of which has a characteristic frequency or pitch. (page 263)

differential rotation A type of rotational motion in which different regions of an object rotate at different angular velocities. (page 99)

diffraction A phenomenon in which light passing through a narrow opening or past a solid edge creates a pattern of light and dark areas due to constructive and destructive interference. (page 367)

diffuse reflection A type of reflection in which different parts of a rough surface will reflect various portions of an incident wavefront in different angles. (page 356)

diode A device that permits electric current to flow in one direction in a given circuit and prevents it from flowing in the opposite direction. (page 439)

direct proportion A mathematical relationship between two quantities A and B which can be expressed in the form A = constant $\times$ B or B = constant $\times$ A; a direct variation. (page 124)

disordered state A state in which the positions and velocities of atoms have a random distribution. (page 242)

dispersion A phenomenon in which different wavelengths of an electromagnetic wave are refracted at different angles as they pass through a given medium. (page 357)

Doppler effect A phenomenon in which, due to the relative motion between the source and the observer, an observer perceives a frequency or wavelength coming from a source of wave motion that is different from what the source is actually emitting. (pages 262, 276)

Doppler shift A change in the wavelength or frequency of a source of wave motion produced as a result of relative motion between the source and the observer. (page 487)

efficiency The ratio of the useful work produced by a given machine to the total amount of energy required to operate it. (page 174)

elastic collision A collision in which both linear momentum and kinetic energy are conserved. (page 166)

elastic limit The point at which a given deformation causes the structure of a solid to be permanently altered. (page 212)

elastic modulus The ratio of the stress to the strain in a given solid; Young's modulus. (page 211)

elastic solid A solid that exhibits a restoring force when subject to a given deformation. (page 211)

electric charge A physical characteristic inherent to atomic and subatomic particles that is responsible for electrical and magnetic phenomena. (page 282)

electric current The flow of electric charges from one location to another. (page 286)

electric field The ratio of the total electrostatic force on an object to the total electric charge on that object; the electrostatic force per unit charge. (page 289)

electric potential The ratio of the total electrical potential energy to the total charge: and electrostatic potential energy per unit charge; voltage. (page 291)

electrical potential difference The difference in electrical potential energy between two points. (page 291)

electrical potential energy The energy associated with the position of a given electric charge in a given electric field. (page 291)

electromagnet A magnet produced from the magnetic field associated with the flow of electric current. (page 312)

electromagnetic radiation The radiant energy associated with electromagnetic waves. (page 344)

electromagnetic spectrum The sum total of the possible range of frequencies and wavelengths of electromagnetic waves. (page 353)

electromagnetic waves Waves produced by oscillating electric charges. (page 346)

electromagnetism The area of physics dealing with the various aspects of the relationship between electricity and magnetism. (page 312)

electron A negatively charged, comparatively low-mass particle generally found in the outer regions of atoms. (page 187)

electron gun A device for producing a beam of electrons in which electrons are ''boiled'' off a hot filament toward a positively charged grid or a plate having a small aperture and pass through the grid or aperture in the form of a collimated beam. (page 426)

electron microscope A type of microscope that employs the de Broglie wavelengths of electrons to resolve the images of very small objects. (page 448)

emitter The layer of a transistor from which electric charge carriers flow toward the collector. (page 452)

emp (electro**m**agnetic **p**ulse) A surge of electromagnetic energy associated with a nuclear detonation which has the effect of rendering any piece of unshielded electronic equipment virtually useless. (page 540)

encounter orbit An orbital trajectory in which one object makes a single encounter with a second object; a parabolic or hyperbolic orbit. (page 135)

endothermic reaction A chemical reaction in which outside energy must be supplied in order for the reaction to occur. (page 536)

energy The measure of the ability of an object, person, or system to deliver or perform work. (page 158)

entropy A measure of the state of disorder in a given system. (page 243)

era of decoupling A hypothesized stage in the Big Bang theory in which hydrogen and helium atoms were formed and the universe became transparent to photons; also called the era of recombination. (page 518)

escape velocity The minimum velocity required to escape the gravitational pull of a given object. (page 183)

event horizon The sphere surrounding a black hole whose radius is equal to the Schwartzchild radius and within which any form of particle or energy cannot escape the gravitational pull of the black hole. (page 526)

evoluation   A view of nature in which changes are thought to gradually occur over very long periods of time (page 503)

excited state   Any state of energy in an atom or molecule that is higher than the lowest energy state or ground state. (page 390)

exothermic reaction   A chemical reaction in which energy is given off. (page 536)

exponent   A number in algebraic notation placed above and to the right of a second number, called the base, which indicates how many times the base appears as a factor in the given algebraic expression. (page 38)

exponential growth   A type of growth in which the rate of increase can be mathematically described in the form of a power law, $N = N_0 A^t$, where N_0 is the initial number or population, N is the value after a time t and A is a constant larger than 1. (page 550)

extrapolation   A process in which results obtained over a given time period are projected beyond the given time period in an attempt to predict future behavior or results. (page 30)

Faculus   A bright region in the solar atmosphere that is characterized by higher temperatures and larger magnetic fields than the regions surrounding it; a white-light plage. (page 498)

Fahrenheit temperature scale   A temperature scale developed by Gabriel Fahrenheit in which 0°C is equal to 32°F and 100°C is equal to 212°F. (page 248)

Faraday's Law   The statement that the induced voltage is equal to the negative of the time rate of change of magnetic flux. (page 322)

fax machine   A device that can transmit photocopies of images and text over long distances; a facsimile machine. (page 434)

ferromagnetic material   A material such as iron, nickel, or cobalt which is capable of being magnetized. (page 303)

field lines   A ''picture'' representation of an electric, magnetic, or gravitational field by a set of lines and curves in which the concentration or crowding of the lines at a certain point represents the strength of the field and the arrows on the field lines the direction that a force would be exerted on a given particle by that field. (pages 289, 303)

filament   Jets of material ejected from the sun's surface that are silhouetted against the sun's disk. (page 498)

First Law of Thermodynamics   The statement that the total energy given off by a system cannot exceed the total energy added to the system. (page 241)

fission   A process in which the nucleus of a given atom is split into smaller fragments with an attendant release of energy. (page 539)

fission fragment   An isotope that is the product of a nuclear fission reaction. (page 539)

fissionable material   Material that is capable of sustaining a chain fission nuclear reaction. (page 402)

flare   An event on the sun or star in which a localized volume of atmosphere will brighten up over a period of a few minutes with the attendant release of a wide variety of high-energy particles and electromagnetic waves. (page 498)

fluid   Any substance that lacks the rigidity of the solid state; a substance that can flow and take on the shape of its container. (page 198)

fluidity   The ability of a substance to flow and take on the shape of its container. (page 198)

fluorescence   A process in which light is produced from a given substance by electrons making transitions from higher to lower energy states. (pages 198, 430)

focal point   The point through which passes all of the light rays gathered by a given lens, mirror, or optical instrument. (page 359)

force   Any agent in nature that can alter the state of an object's motion (page 53)

frame of reference   A backdrop or background that is assumed to be stationary and against which motion is measured; a reference frame. (page 50)

frequency   The number of cycles of a given vibration or wave motion which occurs per unit time. (page 256)

frequency modulation (FM)   A process in which a carrier wave frequency is modified by a signal wave. (page 428)

friction   A force between two surfaces sliding over each other in contact which acts in such a way as to oppose the motion of the object involved. (page 66)

fulcrum   The point about which all of the lever arms in a given lever operate; the pivot point of a lever. (page 180)

fusion   A nuclear process in which two or more nuclei collide under conditions of extraordinarily high temperatures and pressures to form a single nucleus having a larger mass number; also the process in which a solid melts into a liquid. (pages 228, 402)

galaxy   A vast system of stars, gas, and dust which is typically of the order of 50–100 thousand light years in diameter and typically contains about 100–200 billion stars. (page 514)

Galileo Trial   The trial in which Galileo was brought before the Roman Inquisition in the spring of 1633 and forced to renounce his belief in a heliocentric planetary system. (page 16)

galvanometer   A device that measures the amount of electric current flowing in a circuit through a coil placed in a magnetic field. (page 316)

gamma rays The radiant energy from the region of the electromagnetic spectrum having frequencies larger than about 10^{18} Hz and wavelengths smaller than about 3×10^{-10}m; the most energetic of the electromagnetic waves. (page 354)

gas The state of matter that has no definite form nor occupies a definite volume. (page 197)

general relativity theory That branch of relativity theory that deals with the behavior of quantities measured in accelerating reference frames, (page 463)

generator A device that converts rotational mechanical energy into electrical energy. (page 325)

geocentric theory A theory in which the earth is regarded as the center of the planetary system. (page 7)

globular cluster A large spherically-shaped star cluster that typically contains tens to hundreds of thousands of stars and lies in the outer or "halo" regions of a galaxy. (page 513)

globule A relatively small, irregular dark cloud of gas and dust about a half light year or so in diameter which is believed to be one of the early stages of star formation; a protostar. (page 505)

gram The approximate mass of one cubic centimeter of water; a mass equal to 1/1000 of a kilogram. (page 34)

gravitational potential energy The potential energy of a given object located at a given position in a given gravitational field. (page 159)

gravitational redshift A general relativistic effect in which the wavelength of a given photon is displaced to a larger value or "redshifted" as it leaves a gravitating mass. (page 473)

graviton A hypothesized quantum mechanical exchange particle that produces the gravitational interaction. (page 412)

gravity The attractive force between a given mass and any other mass in the universe. (page 123)

gravity waves The hypothesized waves in space-time produced by vibrating masses. (page 472)

Greek civilization The civilization that flourished in and around Greece and the Aegean Sea from about 600 B.C. to about 330 B.C. (page 6)

ground The position in an electrical circuit of zero or lowest electrical potential; the zero electrical potential of the earth's surface. (page 300)

ground state The lowest energy state for a given atom or molecule. (page 390)

hadron A nuclear particle such as a proton or neutron that is capable of exerting the strong nuclear interaction. (page 398)

harmonic law The statement that the ratio of the cube of the mean distance to the square of the sidereal period has the same value for each of the planets orbiting the sun. (page 122)

harmonic motion A back and forth or to and fro type of periodic motion. (page 255)

heat capacity The amount of thermal energy required to raise the temperature of a given object 1°C. (page 224)

heat engine A device that will convert thermal energy into mechanical energy (page 239)

heat of fusion The amount of energy required to convert one gram of a given solid into a liquid at a pressure of one atmosphere. (page 228)

heat of vaporization The amount of energy required to convert one gram of a given liquid into a gas at a pressure of one atmosphere. (page 231)

heliocentric theory A theory in which the sun is regarded as the center of the planetary system. (page 7)

Hellenistic civilization An amalgamation of Greek and Near Eastern cultures which developed in the eastern Mediterranean and the Near East in the wake of the conquests of Alexander the Great and lasted from about 330 B.C. until about 150 A.D. (page 6)

hertz A frequency of one cycle per second. (page 256)

hologram A seemingly three-dimensional image created from a laser-induced interference pattern. (page 444)

horsepower A unit of power equal to lifting a 550 pound weight at a rate of one foot per second; an amount of power equal to 746 watts. (page 172)

Huygens' Principle The concept that waves spreading out from a given source can be thought of as being composed of a set of much smaller-scale wavelets. (page 348)

hydraulic lift A device in which fluid pressure is employed to gain mechanical advantage. (page 214)

hydrostatic equilibrium A condition in which all of the forces on the interior of a given fluid object, most notably gravity and pressure, are in balance so that the size of the object neither increases nor decreases. (page 496)

impulse The product of the force on an object and the time interval over which the force is exerted; a change of momentum. (page 78)

incandescent light source A light source that shines by its own energy as opposed to shining by reflected energy. (page 486)

inclined plane A simple machine in which a reduced lift force can be exerted along a path slanted upward from the horizontal. (page 174)

index of refraction The ratio of the speed of light in free space to that in a given medium. (page 356)

inelastic collision A collision in which linear momentum is conserved but the total kinetic energy is not. (page 166)

inelastic solid   A solid that does not exhibit a restoring force when subject to a given deformation (page 211)

infrared rays   The radiant energy from the region of the electromagnetic spectrum having frequencies between about 4×10^{14} Hz and 10^{11} Hz and wavelengths between about 7.5×10^{-7} m and .003 m. (page 354)

infrasound   Sound waves whose frequencies are less than 20 Hz, the low-frequency limit audible to human hearing. (page 269)

insulator   A substance that can resist the flow of electric charges or thermal energy. (page 236)

integrated circuit   A circuit in which all of the component elements of a given electronic circuit which are required for the circuit to perform its desired function are properly arranged and mounted on a small single chip of semiconducting material, usually silicon. (page 453)

intensity   The amount of power per unit area delivered by a given wave. (page 266)

interference   The interaction between two waves in which the waves can reinforce one another, destroy one another, or exhibit some intermediate stage of interaction. (page 260)

internal combustion engine   A heat engine in which a work-delivering piston is driven by the pressure produced from a gas explosion. (page 240)

inverse proportion   A mathematical relationship between two quantities A and B which can be expressed in the form A = constant/B or B = constant/A; an inverse variation. (page 123)

inverse square law   A mathematical law in which the value for a given quantity varies inversely with the square of the distance. (page 123)

ion   Any atom or molecule with a net electric charge. (page 191)

ionic bond   A chemical bond created by the electrostatic attraction between two oppositely charged ions. (page 191)

ionization energy   The amount of energy required to cause an electron in a given atom to escape the atom and become a free electron. (page 392)

isotope   One of several possible forms of a given element that have the same number of protons in their nuclei but different numbers of neutrons. (page 189)

isotropic   The quality of having the same value or behavior regardless of direction. (page 347)

joule   The amount of work done when a force of one newton acts through a distance of one meter; an equivalent amount of energy. (page 154)

Kapteyn universe   An early 20th century view of the universe in which the entire universe was thought to be contained in an oblate spheroid 30,000 light years by 6,000 light years in size. (page 513)

kelvin   One degree of temperature change on the Kelvin temperature scale. (page 249)

Kelvin temperature scale   A temperature scale in which absolute zero at $-273.15°C$ is taken as the zero point and has a degree size equal to that of the Celsius temperature scale. (page 248)

kilogram   The mass of the standard platinum-iridium cylinder kept in France; the approximate mass of a one liter volume of water. (page 34)

kilowatt-hour   The expenditure of 1000 watts of power over a one hour interval of time; an amount of energy equal to about 3.6 million joules. (page 172)

kinetic energy   The energy associated with a moving object. (page 161)

laser   A device that produces a powerful, highly directional beam of monochromatic light through the process called *l*ight *a*mplification by *s*timulated *e*mission of *r*adiation. (page 441)

law of equal areas   The statement that the rate of orbital area swept out per unit time as one object orbits another has a constant value. (page 138)

law of fixed proportions   The statement that the component elements in a given chemical compound are always present in exactly the same fixed proportions. (page 186)

law of inertia   The statement that every body continues in its state of rest or of uniform motion in a straight line unless compelled to change that state by forces impressed upon it; Newton's First Law of Motion. (page 58)

law of reflection   The statement that the angle of reflection in a reflection process is equal to the angle of incidence. (page 356)

laws of exponents   The statements governing the algebraic operations and manipulations of exponents. (page 38)

length   A quantity that denotes the position and/or the extent of an object in space. (page 33)

length deformation   A type of deformation in which the length of a given object is changed due to the exertion of a stress or force per unit cross-sectional area. (page 211)

Lenz's Law   The statement that the polarity of an induced voltage and the direction of any current flow it produces will always be such as to oppose the change that initially produced them. (page 319)

lever   A device in which lever arms are employed to gain mechanical advantage. (page 180)

lever arm   The shortest distance between the axis of rotation of a given object and the line along which a given force acts. (page 95)

lift force   The difference between the buoyant force exerted on an object and the weight of the object. (page 204)

light year   The distance light travels in one year in a vacuum; a distance of about 9.5×10^{12} km. (page 463)

light-time effect   An effect in which a periodic event is observed sooner or later than its average period would predict due to the light travel time across the earth's orbit. (page 350)

line of apsides   A line joining the near and far points of a given elliptical orbit about a given object; the semimajor axis of an elliptical orbit. (page 457)

linear momentum   A "quantity" of motion equal to the product of an object's mass and its velocity. (page 75)

liquid   A state of matter that has a definite volume but cannot retain a definite shape. (page 195)

living matter   Any substance that can perform all of the life activities, including the ability to move, grow in size, take in food or other sustenance, respond to stimuli, and reproduce itself. (page 199)

logistic growth   A type of growth in which the total population asymptotically approaches the carrying capacity of the given environment. (page 550)

longitudinal wave   A type of wave in which the wave pattern is along the direction of the wave's motion. (page 258)

luminosity   The total amount of radiant energy produced by a given light source per unit time; the power of a light source. (page 379)

machine   A device that can convert or direct energy from a given source or supply into useful work or energy. (page 174)

magnetic core   A ring made of a magnetic material that is used to store binary information in older computer systems. (page 437)

magnetic disk   A disk coated with a layer of magnetic material used to store binary information in current computer systems. (page 438)

magnetic domain   An aligned group of atoms and molecules that possess a net magnetic field. (page 332)

magnetic flux   The product of a given surface area and the component of magnetic field strength that is perpendicular to that area. (page 319)

magnetism   A force associated with moving charges and certain materials such as iron or nickel. (page 302)

magnetosphere   The region of relatively high magnetic field intensity that surrounds a given planet and within which charged particles have been trapped. (page 320)

main sequence stage   A stage in the life cycle of a star during which time the star's physical conditions, most notably its self-gravitation and energy production are in equilibrium. (page 506)

many-body system   A mechanical system of three or more non-negligible gravitating bodies. (page 140)

mass   The amount of matter or material particles contained within a given object. (page 34)

mass number   The total number of protons and neutrons contained in a given nucleus. (page 187)

mass spectrometer   A device used to measure the actual masses of atoms and molecules. (page 420)

mechanical advantage   The ratio of the load force to the applied force for a given machine. (page 175)

melting point   The temperature at which a given substance is changed from its solid state into its liquid state at one atmosphere of pressure. (page 228)

meniscus   The concave surface of a liquid in a narrow tube produced by the adhesive or cohesive (self adhesive) forces in the liquid. (page 195)

meson   One of a number of subatomic particles having masses between that of a proton and that of an electron. (page 405)

metallic bond   A collective covalent bonding present in some substances that permits the valence electric charges to readily move throughout the assemblage of bonded atoms. (page 191)

metastable state   An excited energy state in an atom or molecule whose lifetime is longer by several orders of magnitude than that of a typical excited energy state before deexcitation occurs. (page 441)

meter   The distance travelled by a light ray in $1/299,792,458$ of a second. (page 33)

microscope   A device in which greatly enlarged images of very small objects are produced through the use of electromagnetic waves or electron de Broglie waves. (pages 359, 448)

microwaves   The radiant energy from the region of the electromagnetic spectrum having frequencies between about 10^9 Hz and 10^{11} Hz and wavelengths between about .003 m and 0.3 m. (page 354)

Milky Way   The vast, spiral-shaped system of stars, gas, and dust about 100,000 light years in diameter in which our sun is located about 2/3 of the way out from the center.

model interior   A theoretical portrait of the interior of a star based on the observed physical characteristics of the star and a set of mathematical equations which are used to describe the physical conditions for the regions inside of the star. (page 496)

molecule   The smallest division of a given substance composed of one or more elements which retains all of the properties of that substance. (page 189)

moment of inertia   The ability of an object to resist the effects of a given torque. (page 97)

monopole   A single, isolated magnetic pole that is not paired up with its opposite counterpart and whose existence has yet to be observed. (page 303)

motor   A device that converts electrical energy into mechanical energy. (page 326)

muon   A subatomic particle that has a mass about 210 times that of an electron and the electric charge of either a proton or an electron; a μ-meson. (page 405)

Murphy's Laws   A set of whimsical "laws" that summarize the frustrations and struggles of life's day-to-day operations. (page 23)

natural frequency   The frequency at which a given object will vibrate if it is struck by an outside blow. (page 270)

neap tide   The relatively small tide that occurs when the sun and moon are at right angles to each other as seen from the earth. (page 138)

neuron   A nerve cell. (page 295)

neutrino   A subatomic particle that has energy but little or no mass and no electric charge. (page 403)

neutron   An atomic particle generally found in the nuclei of atoms, which has a mass slightly larger than that of a proton and no electric charge. (page 397)

neutron star   A small, high-density, rapidly rotating star which is composed almost entirely of neutrons. (page 509)

neutronization   A process in the advanced stages of a massive star's life cycle in which protons and electrons in the center of the star are crushed into neutrons. (page 508)

newton   The force required to accelerate a mass of one kilogram at a rate of one meter/(second)^2. (page 59)

Newton's First Law of Motion   The statement that every body continues in its state of rest, or of uniform motion in a straight line, unless it is compelled to change that state by forces impressed upon it; the law of inertia. (page 58)

Newton's Second Law of Motion   The statement that whenever an object is subjected to a force, it will respond by accelerating in the direction of the force by an amount which is directly proportional to the force and inversely proportional to the object's mass; the statement that the force is equal to the product of the mass and the acceleration. (page 58)

Newton's Third Law of Motion   The statement that if an object is subjected to a force, it will respond by exerting an equal and opposite force on the agent exerting the force; the principle that action equals reaction.

(page 61)

nodal point   A point along a standing wave that does not experience any displacement from the wave's equilibrium point; a node. (page 260)

node   A position along a standing wave that does not experience any displacement from the wave's equilibrium point; a nodal point. Also, the spacings between segments of the myelin sheath in a human nerve cell. (pages 260, 295)

non-conservative force   A force, such as friction, in which the total amount of work done against the force to move from one position to another depends on the path taken between the two positions. (page 169)

normal force   A force between two objects that is exerted along a line perpendicular to the contact area between the two objects. (page 67)

normal line   A line perpendicular to a surface at a given point on the surface. (page 356)

npn transistor   A transistor which has a *p*-type semiconductor as the base and *n*-type semiconductors for the emitter and collector. (page 452)

n-type semiconductor   A semiconductor in which the charge carriers are negatively charged electrons. (page 451)

nuclear bounce effect   An event in the life cycle of a massive star in which the star's outer layers collapse and then recoil from a small, incompressible core with an attendant release of a considerable amount of energy; a core bounce. (page 508)

nuclear fission   A process in which the nucleus of a given atom is split into smaller fragments with an attendant release of energy. (page 539)

nuclear fragment   An isotope that is the product of a nuclear fission reaction. (page 539)

nuclear fusion   A process in which two or more nuclei collide under conditions of extraordinarily high temperatures and pressures to form a single nucleus having a larger mass number. (page 402)

nuclear winter   A hypothesized cooling of the earth resulting from the dust and debris released into the earth's atmosphere during a full-scale thermonuclear war. (page 541)

nucleosynthesis   A theory in which the chemical elements and their respective abundances are thought to be the result of nuclear reactions occurring for the most part at the centers of stars in advanced stages of their life cycles. (page 488)

oblate spheroid   A solid generated by rotating an ellipse about its minor axis. (page 151)

Occam's Razor   A philosophical view that if two competing theories are equally successful in describing a natural phenomenon, then the least complicated theory should be the preferred theory. (page 11)

ohm   The amount of electrical resistance that will permit a current of one ampere to flow when an electric potential difference of one volt is placed across the resistance. (page 299)

Ohm's Law   The statement that the current flowing in a given circuit is directly proportional to the voltage and inversely proportional to the resistance in the circuit. (page 299)

oil drop experiment   An experiment in which the value of the fundamental unit of electric charge is determined by observing the motions of charged oil drops in an electric field. (page 307)

optical activity   A property of certain transparent substances in which the plane of polarization of light passing through the substance will experience a certain amount of rotation. (page 365)

optical disk   A highly polished disk of metal onto which information is stored by means of laser-etched pits. (page 443)

orbital angular momentum   The product of an orbiting object's mass, velocity, and distance from the object about which it is rotating. (page 102)

ordered state   A state in which the positions and velocities of atoms have a non-random or ''ordered'' distribution. (page 242)

ordinate   The vertical or y-axis on a graph. (page 30)

overtone   One of a set of higher frequency, lower intensity sound waves produced when a given musical instrument sounds a particular musical note. (page 263)

ozone layer   The layer in the earth's atmosphere at an altitude of about 50 km which contains a relatively high percentage of ozone (O_3), a highly efficient absorber of high-energy ultraviolet rays from the sun. (page 541)

pair production   A process in which high-energy photons are relativistically converted into a particle-antiparticle pair. (page 517)

parallax angle   The angle subtended by the radius of the earth's orbit as seen from a star at a given distance. (page 490)

parallel circuit   An electrical circuit in which each resistance in the circuit is hooked up directly to the terminals of the circuit's voltage supply. (page 301)

parsec   The distance to a star that exhibits a parallax angle of one arcsecond; a distance of about 3.26 light years or about 3.09×10^{16} meters. (page 500)

particle asymmetry   The observed imbalance between dominant particles and almost non-existent antiparticles in the local region of the universe. (page 516)

pascal   The pressure exerted by a force of one newton over a surface area of one square meter. (page 201)

Pascal's Principle   The statement that any change of pressure on an enclosed fluid is transmitted undiminished to all parts of that fluid. (page 202)

period   The time for one complete cycle to occur in vibratory motion, wave motion, or orbital motion. (page 256)

periodic table   A table of chemical elements in which the various families of chemical elements are set up in columns of increasing atomic number for the elements in a given family. (page 189)

perturbation   The small-scale departures from the theoretical behavior of a two-body system due to other forces, especially the gravitational effects of other objects. (page 139)

photodisintegration   A process in which high-energy gamma rays cause a heavy element nucleus to be broken down into alpha particles. (page 508)

photoduplication   A process in which images are copied through the use of photoconductivity and static charges. (page 432)

photoelectric effect   An effect in which light incident on a given substance causes electrons to be emitted from that substance. (page 371)

photoelectrons   Electrons that are ejected from the surface of a given substance as a result of that substance being exposed to light. (page 371)

photon   A discreet unit of electromagnetic energy that exhibits particle-like behavior. A quantum of light. (page 372)

photosphere   The region of atmosphere of the sun or a star from which most of the object's energy is radiated into space. (page 493)

pi (π)   The ratio of the circumference of a circle to its diameter; a number with an approximate value of 3.1416. (page 6)

pion   A subatomic particle that has a mass about 273 times that of an electron and an electric charge of zero, that of a proton, or that of an electron; a π-meson. (page 405)

plage   A bright region in the solar atmosphere that is observable in the monochromatic light of some spectral line such as those of hydrogen or calcium, and which is characterized by higher temperatures and higher magnetic fields than the regions surrounding it. (page 498)

Planck length   The diameter of an expanding universe in the Big Bang theory when it reaches the point of being physically describable; a length of about 4.1×10^{-35} m. (page 515)

Planck time   The time required for light to traverse a Planck length; a time interval of about 1.4×10^{-43} seconds. (page 515)

Planck's constant   The constant ratio of the energy of a photon to its frequency; a value of about 6.6×10^{-34} J-s. (page 390)

planetary nebula   A glowing shell of gas ejected from a low-mass star which is in an advanced stage of its life cycle. (page 507)

plasma   A gaseous mixture of free electrons and ionized atoms. (page 197)

p–n junction   The interface boundary between a p-type semiconductor and an n-type semiconductor. (page 452)

pnp transistor   A transistor which has an n-type semiconductor as the base and p-type semiconductors for the emitter and the collector. (page 452)

polar molecule   A molecule that exhibits an asymmetric shape and charge distribution. (page 191)

polarization   A process in which the plane of vibration of a transverse wave motion is restricted to a single orientation. (page 364)

positron   A subatomic particle having the same mass as an electron and a positive electric charge; an antielectron. (page 404)

potential energy   The energy associated with the object's position in a given force field. (page 159)

power   The time rate at which work is performed or energy is released. (page 172)

pressure   The force per unit area exerted by a given fluid. (page 200)

Principle of Conservation of Energy   The statement that the total amount of energy in a closed, isolated system remains the same, even though the energy in the system may be converted from one form to another. (page 158)

Principle of Conservation of Momentum   The statement that in the absence of a net external force the total momentum of a given system is the same before and after a given interaction within the system. (page 76)

principle of equivalence   The statement that there is no way to observationally distinguish an accelerating reference frame from one that is inertial, but located in a constant gravitational field. (page 469)

principle of relativity   The statement that all laws of nature have exactly the same form in any reference frame that is moving at a constant velocity. (page 459)

principle of simplicity   The statement that the most acceptable explanation for a natural phenomenon must be the simplest explanation that is most consistent with the available facts; Occam's Razor. (page 11)

principle of universality   The statement that the behavior of the physical world does not change with time or one's location in the universe. (page 20)

projectile motion   A type of motion for a falling body that has both horizontal and vertical components. (page 70)

prominence   A flame-like jet of gas that appears at or near the visible edge of the sun. (page 498)

proton   An atomic particle, generally found in the nuclei of atoms, which has a mass of about 1.67×10^{-27} kg and a single unit of positive electric charge. (page 397)

protostar   A star that is still in the process of formation; a pre-main sequence star. (page 505)

p-type semiconductor   A semiconductor in which the charge carriers are positively charged electron holes. (page 452)

pulley   A type of simple machine consisting of a wheel mounted on a support frame. (page 174)

pulsars   Radio sources that emit radio pulses at very regular intervals ranging from 0.03 to 5 seconds and are believed to be rapidly rotating neutron stars. (page 509)

quantum theory   A theory that describes the behavior of matter and energy at the atomic and subatomic level; quantum mechanics. (page 407)

quark   One of six hypothesized particles that are believed to be the basic constituents of elementary particles and are characterized by a fractional electric charge of either 2/3 of the charge on a proton or 1/3 of the charge on an electron. (page 406)

quasar   One of a class of starlike objects that have very large redshifts in their spectral lines and are believed to be the active nuclei of remote galaxies. (page 527)

radial velocity   That component of an object's relative motion that is directed along the line of sight, either toward or away from the observer. (page 487)

radian   The central angle subtended by an arc length of a circle that is equal to its radius; an angle of about 57.3°. (page 115)

radian measure   The angular displacement expressed as a ratio of the arclength along the circumference of a circle subtended by the given angular displacement to the radius of the circle. (page 115)

radiative transport   A process in which energy is transported on electromagnetic waves. (page 238)

radio waves   The radiant energy from the region of the electromagnetic spectrum having frequencies smaller than about 10^9 Hz and wavelengths larger than about 0.3 m; the least energetic of the electromagnetic waves. (page 354)

radioactive dating   A technique for determining the age of an object by measuring isotope abundances in the object's composition. (page 400)

radioactive decay   A process in which an atomic nucleus spontaneously takes on a less energetic state with the attendant release of gamma rays and subatomic particles; radioactivity. (page 399)

radioactive half-life The time it takes for one-half of the initial number of atoms in a given sample of unstable nuclei to decay into a less energetic state. (page 399)

radioactivity The process in which atomic nuclei spontaneously decay to a less energetic state with the attendant release of gamma rays and subatomic particles. (page 384)

radiogenic An isotope produced from the radioactive decay of a ''parent'' isotope. (page 401)

RAM (Random Access Memory) chip A computer memory chip in which data are stored in the form of electrical charges located at cells consisting of a transistor and a capacitor. (page 439)

red giant stage The stage in a star's life cycle that follows its main sequence stage and in which the star is characterized by a relatively large size and luminosity, but a comparatively low surface temperature. (page 507)

reference frame A backdrop or background that is assumed to be stationary and against which motion is measured; a frame of reference. (page 50)

reflecting telescope A telescope whose principal light collecting element is a concave mirror. (page 360)

reflection A process in which a light ray ''bounces'' off a smooth surface. (page 355)

refracting telescope A telescope whose principal light collecting element is a lens. (page 359)

refraction A process in which the direction of a light ray is bent as it crosses a boundary between two different transparent media. (page 356)

relativity theory A theory that describes the behavior of quantities measured in reference frames having very large velocities and/or accelerations. (pages 459, 469)

Renaissance An era in Western European history that extended from about 1400 to 1600 and which was characterized, among other things, by a renewed interest in the workings of the physical universe. (page 12)

renewable energy source A source of energy that when used will be replaced in some fashion by natural processes. (page 535)

resistance The ability of a given circuit to resist the flow of electric charge. (page 298)

rest mass The mass observed for a particle that is not moving relative to an observer. (page 481)

restoring force A force that seeks to restore the shape of a deformed object. (page 211)

resultant vector The vector that results when two vectors are added together according to the rules of vector addition. (page 46)

reverberation The prolongation of a sound wave by multiple reflections; a multiple echo. (page 266)

revolution Circular motion by one body about a second, separate object or point. (page 88)

ROM (Read Only Memory) chip A permanent computer memory chip in which data are stored on a system of diodes and which can be read from but not written to. (page 439)

rotation Circular motion that occurs within an object about an imaginary line thrust through the object. (page 89)

rotational energy The energy associated with a rotating object; rotational kinetic energy (page 163)

rotational impulse The product of the torque on a given object and the time over which the torque is exerted; a change in angular momentum. (page 104)

rotational law of inertia The statement that every body continues in its state of rest or of uniform rotation, unless it is compelled to change that state by torques impressed upon it. (page 97)

RPM The angular velocity of an object expressed in *ro*tations *per* *m*inute. (page 93)

scanning electron microscope A type of electron microscope in which an image is obtained by analyzing the secondary electrons emanating from the surface of a given sample as it is scanned by a highly focused beam of primary electrons. (page 449)

scattering An interaction between electromagnetic waves and the atoms and molecules in a gas in which the waves are absorbed and then reradiated in random directions. (page 369)

scholasticism A medieval blend of faith and reason in which new ''knowledge'' was generated by rational appeals to revered sources of authority, most notably Aristotle and the Bible. (page 11)

Schwartzchild radius The radius to which a given object's mass must be reduced in order that its escape velocity is equal to the speed of light; the event horizon of a black hole. (page 525)

scientific hypothesis A proposed overall description of a given natural phenomenon based on scientific laws. (page 19)

scientific law A behavior pattern observed to occur in a given phenomenon which can be expressed as a definitive statement and/or a mathematical equation. (page 19)

scientific method A method of gathering knowledge about the physical world based on measurements, experiments, and observations. (page 18)

scientific notation A type of mathematical notation in which a given numerical value is expressed as the product of a number between 1 and 10 and then raised to some positive or negative integral power. (page 38)

scientific theory A scientific hypothesis that has been successful in predicting experimental and observational results. (page 20)

screw A type of simple machine in which an inclined plane is wrapped around a vertical shaft; a circular inclined plane. (page 174)

second The interval of time required for exactly 9,192,631,770 atomic vibrations of a cesium atom to occur. (page 32)

Second Law of Thermodynamics The statement that the total amount of work obtained from a given system is always less than the total amount of energy initially put into the system; also the statement that the entropy of a given system tends to increase in time. (page 242)

sedimentation A layering effect that occurs when mixtures of particles of various size or fluids of various densities are subjected to a centrifugal force. (page 90)

selective absorption A process in which only light of a certain color or colors is absorbed by a given object while all of the remaining colors are reflected. (page 363)

selective reflection A process in which only light of a certain color or colors is reflected by an object while all of the remaining colors are absorbed. (page 363)

semiconductor A substance whose electrical conductivity can be greatly altered by the addition of tiny amounts of impurities to its crystalline structure. (page 294)

series circuit An electrical circuit in which each resistance in the circuit is connected in such a way that the current flows along one single route through each of the resistors in the circuit. (page 300)

shear deformation A type of deformation in which the atoms and molecules in a given object are subjected to a twisting change of relative positioning. (page 211)

shear modulus The ratio of the shear stress on an object to the displacement per unit height experienced by the object as a result of the stress. (page 212)

shock wave The V-shaped or cone-shaped wave created by a source of wave motion moving through a fluid more rapidly than the waves it creates; a bow wave. (page 262)

short circuit A path of very low resistance in an electrical circuit, usually provided inadvertently, and which draws a relatively large current. (pages 299–300)

sidereal time A timekeeping system based on the rotational motion of the earth relative to the sun and the stars. (page 389)

simple harmonic motion A back and forth type of periodic motion. (page 255)

sine The ratio of the side opposite a given angle in a right triangle to the hypotenuse of that triangle. (page 256)

sine wave A wave-like figure obtained when the value of the sine (or cosine) of a given angle is plotted against the value of the angle itself. (page 256)

solar cell A device that uses solar energy to create an electric charge flow. (page 547)

solar constant The amount of power per unit area received from the sun as observed from the earth's distance; an energy intensity of about 1360 watts/m^2. (page 493)

solar activity cycle An eleven year cycle during which time the solar activity as evidenced by the numbers of flares, prominences, plages, and sunspots varies from a maximum level to a minimum level. (page 498)

solar granulation A mottled pattern of light and dark regions in the sun's photosphere which are the tops of convective cells extending deep into the solar interior. (page 497)

solar luminosity unit A unit of power equal to the luminosity or power output of the sun; a power output of about 3.8×10^{26} watts. (page 500)

solar mass unit A unit of mass equal to the mass of the sun; a mass of about 2.0×10^{30} kg. (page 500)

solar radius unit A unit of length equal to the radius of the sun; a length of about 6.9×10^8 meters. (page 500)

solar temperature unit A unit of temperature equal to the temperature at the sun's surface; a temperature of about 5800 K. (page 500)

solenoid A spring-like set of current loops used to create an enhanced and uniform magnetic field within the array; a coil. (page 327)

solid A state of matter that has a definite shape and a definite volume. (page 191)

space-time A four-dimensional geometry consisting of three spatial coordinates and one time coordinate used to represent our universe. (page 463)

special relativity theory That branch of relativity theory that deals with the behavior of quantities measured in reference frames moving at constant velocities relative to one another. (page 459)

specific heat The amount of thermal energy required to raise the temperature of one gram of a given substance one degree Celsius. (page 223)

spectral classification A system of classifying stars according to the appearance of the absorption lines present in their spectra. (page 387)

speed The magnitude of a velocity. (page 52)

spring tide The relatively large tide that occurs when the sun, earth, and moon are in a nearly colinear configuration. (page 138)

stable object An object that will not tip over due to the action of the earth's gravity. (page 108)

standing wave A type of wave disturbance that is produced by the interference of two sets of identical waves moving in opposite directions, resulting in a wave that does not appear to move. (page 260)

state variables A set of variables, usually temperature, pressure and density, that are used to specify the state of a given gas. (page 197)

statics That branch of physics that deals with the interplay of forces and torques when no motion is present. (page 107)

Stefan-Boltzmann Law The statement that the energy intensity at the surface of a radiating object is directly proportional to the fourth power of the Kelvin temperature. (page 486)

step-down transformer A device in which voltages can be decreased by electromagnetic induction. (page 329)

step-up transformer A device in which voltages can be increased by electromagnetic induction. (page 329)

stopping potential An electric potential employed to stop the flow of photoelectrons. (page 371)

strain The response of an object to a given stress force. (page 211)

stress A force that deforms an object in some way. (page 211)

strong nuclear interaction The fundamental interaction of nature that holds atomic nuclei together. (page 398)

subtractive color mixing The creation of a given color by the selected absorptions of the right proportions of the subtractive primary colors of cyan, magenta, and expanded yellow. (page 363)

subtractive primary colors The basic colors used in subtractive color mixing processes; the colors cyan, magenta, and expanded yellow. (page 363)

sunspots An irregular-shaped dark region in the solar photosphere which is characterized by a large magnetic field and a significantly lower temperature than its surrounding regions. (page 497)

sunspot cycle An eleven-year cycle during which time the average number of sunspots varies from a maximum value to a minimum value (page 498)

superconductivity A property of certain substances in which their ability to resist the flow of electric current essentially vanishes as their Kelvin temperatures become very low. (page 294)

surface gravity The acceleration of gravity at the surface of a given object. (page 130)

surface tension A contractive force within a liquid that causes the liquid's surface to contract itself into the smallest possible surface area under the given set of conditions. (page 195)

tachyons Hypothesized particles that can move only at speeds greater than that of light (page 464)

telemetry A process by which information is collected at a very distant location, such as from another planet, and then transmitted via electromagnetic waves to a receiving station for analysis and interpretation. (page 431)

telescope A device that produces an enlarged image of a distant object. (page 359)

temperature The measure of the ability of an object or system to transfer thermal energy. (page 248)

temperature inversion A condition in which warm, low density air is at a higher altitude than cool, high density air, thus preventing any convective air circulation from occurring. (page 237)

tesla A magnetic field strength equal to one newton per ampere-meter; a magnetic field strength of one newton per coulomb-meter/sec. (page 313)

thermal equilibrium A condition in which no net flow of thermal energy will occur between two objects that are brought into contact with each other. (page 241)

thermal excitation A process in which thermal energy excites one or more electrons in a given atom to higher energy states. (page 501)

thermal expansion The increase in an object's linear dimensions due to the addition of thermal energy to that object. (page 228)

thermal pollution The difference between the input energy and the work performed by a given heat engine which appears as unused thermal energy released into the environment. (page 242)

thermodynamics The branch of physics that deals with the transformation of thermal energy into mechanical energy. (page 239)

thermonuclear reaction A nuclear reaction in which two or more nuclei collide under conditions of extraordinarily high temperatures and pressures to form a single nucleus having a larger mass number; a nuclear fusion reaction. (page 402)

Third Law of Thermodynamics The statement that it is not possible for an object to have a temperature equal to 0 K or absolute zero. (page 234)

3 K microwave background A background of radiant energy intensity thought to be left over from the Big Bang and whose wavelength distribution is that of a blackbody having a temperature of 3 K. (page 515)

threshold frequency A frequency of monochromatic

light below which no electrons are ejected from a given photoemissive surface when light shines on that surface. (page 371)

tidal distortion The deformation of an orbiting body due to a differential gravitational force exerted across that body. (page 138)

time The duration between two observed events in nature. (page 32)

time dilation A relativistic effect in which time intervals observed in a moving reference frame by a stationary observer are perceived as being longer than those observed for an at rest reference frame. (page 481)

torque The product of a force and the lever arm of that force. (page 95)

total internal reflection A phenomenon in which all of the light in a medium having a higher index of refraction is reflected off a boundary with a medium of lower index of refraction. (page 357)

trajectory The path of an object whose motion has both vertical and horizontal components. (page 70)

transformer A device in which electromagnetic induction is employed to transfer electrical power from one coil at a certain voltage to a second coil at a higher or lower voltage. (page 329)

transistor A device consisting of a semiconducting emitter, base, and collector in which the current flowing from the emitter to the collector can be varied by placing a bias voltage at the base region of the transistor. (page 452)

translational motion A change of an object's position relative to a background frame of reference which is assumed to be motionless or at rest. (page 51)

transmission electron microscope A type of electron microscope in which an electron beam is directed through the sample to be analyzed and then onto a fluorescent screen or photographic plate where an image of the sample is recorded. (page 448)

transmutation Any process in which the nucleus of one element is transformed into the nucleus of a different element. (page 399)

transverse wave A type of wave in which the wave pattern is perpendicular to the direction of the wave's motion. (page 258)

triggering mechanism One of several possible processes in which star formation is induced in gas and dust clouds. (page 505)

triode A device that can amplify small-scale variations present in a signal current or signal voltage. (page 428)

two-body system A system in which two bodies are moving in each other's gravitational fields. (page 134)

ultrasound Sound waves having frequencies that are larger than about 20,000 Hz, the high frequency limit audible to human hearing. (page 268)

ultraviolet catastrophe The early 20th century discrepancy between the predicted behavior of blackbody radiation at smaller wavelengths with what was actually observed. (page 370)

ultraviolet rays The radiant energy from the region of the electromagnetic spectrum having frequencies between about 7×10^{14} Hz and 3×10^{15} Hz and wavelengths between about 4×10^{-7} m and 10^{-7} m. (page 354)

ultraviolet summer A hypothesized condition in the aftermath of a nuclear war in which high-energy ultraviolet rays from the sun would reach the earth's surface due to the damage inflicted on the earth's ozone layer. (page 541)

uncertainty principle The statement that there is a fundamental intrinsic limit on the accuracy to which one can make simultaneous measurements in the microscopic world. (page 410)

uniform circular motion Motion which occurs at a constant speed along a circular path. (page 88)

unit An indication of the measuring system employed to express the value of a given quantity. (page 35)

universal gravitation constant The constant of proportionality in the Universal Law of Gravitation, whose value is about 6.67×10^{-11} N-m^2/kg^2. (pages 126, 147)

Universal Law of Gravitation The statement that the gravitational force between two masses is directly proportional to the product of the masses and inversely proportional to the square of the distance between the masses. (page 126)

unstable object An object that will tip over due to the action of the earth's gravity. (page 108)

valence electron An electron in the outer regions of an atom which participates in chemical reactions. (page 189)

Van Allen belts A set of donut-shaped zones about the earth in which the earth's magnetic field has trapped high-energy charged particles coming from the sun and outer space. (page 320)

Van der Wall's bond A weak chemical bond that exists between polar molecules. (page 191)

vector quantity A quantity whose mathematical description requires that both its numerical value and the direction in which it operates be specified. (page 46)

velocity The distance an object moves in a given direction per unit time. (page 51)

vibration   Any type of repeating or periodic motion by a given object. (page 254)

vis viva   The 17th century term used in an attempt to describe the kinetic energy of an object. (page 161)

viscosity   The measure of the ability of a substance to resist fluid flow; the rigidity of a fluid. (page 200)

volt   A unit of electric potential that is equal to one joule/coulomb. (page 291)

voltage   The ratio of the total electrical potential energy to the total charge. (page 291)

waterdrop model   A view of the nucleus as a drop of incompressible nuclear fluid. (page 399)

watt   A unit of power equal to one joule/sec. (page 172)

wave barrier   A buildup of waves that occurs when the velocity of a source of waves approaches the velocity of the waves themselves. (page 262)

wave motion   An up and down or back and forth pattern of motion which propagates outward from a vibrating source. (page 258)

wavelength of maximum intensity   The wavelength at which the energy intensity of the radiation from an incandescent light source reaches its maximum value. (page 486)

wave-particle duality   A representation of light in which light can somehow behave as both a particle and as a wave, but not simultaneously. (page 373)

weak nuclear interaction   A fundamental interaction in nature which is thought to be responsible for beta decay in the nucleus. (page 403)

wedge   An inclined plane that is used to deliver a splitting force to a given object. (page 174)

weight   The gravitational force exerted on a given mass by the earth or other body. (page 59)

wheel and axle   A simple machine consisting of a shaft attached to a wheel and which allows one to apply a force through large distances by simply exerting the force in a circular path along the rim of the wheel. (page 174)

white dwarf   A star that can no longer generate nuclear energy and has collapsed down to an object roughly the size of the earth; a low-mass star that has burned out. (page 507)

Wien's Law   The statement that the product of the Kelvin temperature of a given incandescent light source and its wavelength of maximum energy intensity is equal to a constant value of 0.003 m-K. (page 486)

work   The product of a force on a given object and the distance the object moves as a result of the action of that force. (page 154)

work function   The energy barrier at the surface of a metal plate. (page 417)

work-energy theorem   The statement that the work done on an object by conservative forces is equal to the change in the object's kinetic energy plus the change in its potential energy. (page 170)

X-rays   The radiant energy from the region of the electromagnetic spectrum having frequencies between about 3×10^{15} Hz and 10^{18} Hz and wavelengths between about 10^{-7} m and 3×10^{-10} m. (page 354)

Young's modulus   The ratio of the stress to the strain in a length deformation process; the elastic modulus. (page 211)

Zeroth Law of Thermodynamics   The statement that if two bodies are each in thermal equilibrium with a third body, then they must be in thermal equilibrium with each other. (page 241)

Appendix A Selected Answers

CHAPTER 1

SECTION 1.1

Q2. Knowledge was a prerequisite to control and power over the natural world and other people(s). Those who learned and applied their knowledge survived and prospered more so than others. (Do you think things are very different now?)

Q4. A few of the most significant are:
   a. The Pythagorean idea that natural phenomena could be explained by relationships based on numbers.
   b. The Platonian idea that pure thought (alone) was necessary to discern the relationships that explained events in nature
   c. The Aristotelian notion that any valid explanation must be consistent with all known facts and observations. (This led to some rather ingenious rationalizations invented to resolve apparent contradictions between principles and observations.)

SECTION 1.2

Q2. Scholasticism can be regarded as a way of ''Saving the Phenomena'' in the sense alluded to in the answer to Question 4c in Section 1.1. It was (is) based on the assumption that all fundamental knowledge was already revealed by previous authorities or prophets. Clever arguments were (are) used to bridge the gaps in logical reasoning in order to derive explanations for all phenomena within the realm of human experience.

Q4. The principle that the simplest explanation for a class of phenomena that does not contradict any other observed facts is the best one. It serves as a guide to help avoid the introduction of unnecessary assumptions and complexities in the development of new theories.

SECTION 1.3

Q2. It was consistent with ''common sense,'' as well as church doctrine, much of which was based on the accepted authority of Aristotle. A change in this point of view would threaten their place in the scheme of things (and nobody likes that!).

SECTION 1.4

Q2. These three principles apply to different facets of our search for scientific truth. Universality has been verified by observation as well as taken on faith to extend our range of knowledge into new domains. Simplicity helps keep new theories from accumulating unnecessary complexities and makes it easier to relate them to older established theories. Correspondence ensures that the knowledge developed for the new domain is consistent with the knowledge that applies outside of the new domain. They are all important in different stages of the evolution of a new theory.

Q4. Without predictions, the laws could not be tested. If the laws cannot be tested, how do we know if they are true? This is as far as science can take us. Its truth is applicable only to what can be objectively demonstrated.

THOUGHT QUESTIONS

Special Note to the Reader:

These questions are meant to help stimulate *your* thinking about the ideas presented in the chapters. They are ''answered'' when you are able to incorporate the new ideas into your thinking process, and they became yours in the sense that they became part of your own ''mental map'' of the world.

My answers to a select group of these questions are presented for comparison and contrast with your own answers, not as *the* right answers. Some of my answers will occasionally be relevant to several questions, and will be referenced accordingly in those cases.

6. A real life example of this question was provided in 1991 by two scientists' claim that they had discovered ''cold fusion''. Their reports were not consistently verified by other researchers; if they continue to do basic research, it is likely that anything they publish will be met with considerable skepticism.

7. The fundamental issue here is the idea of what is ''self-evident''. Science takes nothing for granted, and holds only to those principles that have been supported by repeated observation and measurement. Since there are some aspects of human experience that cannot be measured or objectively verified, there will also be principles and explanations for things that are beyond the range of science. There can only be valid controversy between science and religion if they are talking about the same thing. In order to be as clear and specific as possible, and to avoid the ambiguity of the written or spoken word, science tends to rely more on logic and mathematics to keep its relations and conclusions precise and consistent. The strength of its conclusions stems from its limitation to objective phenomena, as suggested in Section 1.4, Review question 4.

CHAPTER 2

SECTION 2.1

Q2. To express relationships between measurable quantities. When symbols are used to represent these quantities, the rules of mathematics can then be used to make predictions and find new relationships between the quantities.

SECTION 2.2

Q2. A point in a plane has two numbers to specify its location with respect to the intersection of x and y axes. (This point of intersection is commonly referred to as the *origin*.) Find the *x* number on the *x* axis and imagine a line passing through the number, parallel to the *y* axis. Then go along this imaginary line until you are across from the *y* number on the *y* axis. You are now at the position of the point. "Plot" the point by marking the location with a small dot, circle, square, etc.

Q4. About $375.

P2. **a.** $x = 0$
    $y = 4$
  b. $x = 3$
    $y = 3$
  c. $x = 9$
    $y = 1$
  d. $x = 12$
    $y = 0$

SECTION 2.3

Q2. The second is currently defined as the time it takes a cesium atom to complete 9,192,631,770 vibration cycles.

  The meter is now defined as the distance a beam of light can travel during a time interval of 0.00000000333560952 second.

  The kilogram is defined as the mass of a particular platinum-iridium cylinder maintained in a vacuum chamber at constant temperature.

Q4. The meter is determined by the speed of light *c*, and a certain time interval *t*. Since the speed determines the distance traveled during any time interval, the meter is directly proportional to *c*, $1\ m = (c)(t)$.

SECTION 2.4

Q2. They are treated like numbers. The only difference is that they do not combine to produce a single value; they remain in their respective places, (numerator or denominator).

P2. 1 day = 24 hours $\times$ 60 min/hour $\times$ 60 sec/min = 86,400 seconds

SECTION 2.5

Q2. Remember, what we want to do here is change the way the number looks, not change its value. For example, one dollar is the same amount of money as 10 dimes or 100 pennies.

  a. If the number is larger than ten, move the decimal as many places to the left as necessary to make the apparent value between 1 and 10. This is the same as dividing the number by a power of 10; the power is the number of places you moved the decimal. To keep the number at the same value, you then multiply by 10 to the power.

    If the number is less than 1, you would have to multiply it by a power of 10 to make it come out between 1 and 10. The power is the number of places you would have to move the decimal to the right. You then have to divide by 10 to that power to keep the number value the same. Division is expressed as a negative power.

  b. This is easier. If the power is positive, just multiply the number by 10 that many times. If the power is negative, divide the number by 10 that many times.

Q4. Centi: one hundredth of; one penny is a centi-dollar (1 cent). Milli: one thousandth of; a milli-dollar would be 0.1 cent. Kilo: one thousand of; a kilo-dollar would be $1000.

P2. 96,000

CHAPTER REVIEW PROBLEMS

2. **a.** 3×10^3
   b. 7×10^{12}
   c. 4×10^{12}
   d. 6×10^9

4. 9.192631770×10^9

6. **a.** 0.133
   b. 2.5×10^5
   c. 2×10^{36}
   d. 1.29×10^{-30}
   e. 7.5×10^{-5}

8. Either way, you have to keep track of all the zeros. With scientific notation, you only have to count them all once

10. 4×10^{31} atoms

12. 2.74 billion

14. 10^{12}

16. 1609 km

18. The mile is 109 m longer (1 mi = 1609 m, or 1.61 km)

20. 1991

THOUGHT QUESTION

18. The prediction will be in error if any of the factors affecting the graphical relationship undergo any change outside the known range.

FORMULATION 2.1

Q2. Yes; suppose you were on vacation, driving along the interstate, and you were running low on gas. There is a station at the next exit but the price is sky high. The next town (with cheap gas) is 80 miles away. Can you make it? You need to know something about your gasoline consumption rate, either miles per gallon or miles per tank. If you knew you could drive 400 miles on one tank of gas, and your gas gauge read 1/8 full, then you could set up an equation like: $\dfrac{x}{1/8} = \dfrac{400}{1}$, where x is the number of miles you can go on 1/8 tank (which it is reasonable to assume will be the same ratio as 400 miles on a full tank. Solving for x you would get 50 miles, so you better stop and get some. What is the minimum amount you could buy to get to the next town? Another equation would answer that question too. Sometimes we do simple equations in our heads without even realizing it.

P2. $v = x/t => vt = x => t = x/v$

FORMUATION 2.2

Q2. Yes; consider an airplane, especially at takeoff. It starts out going in one direction along the runway, and without changing that direction it must begin to move upward as well. Once off the ground, to describe completely its direction in space would require three numbers, one for each dimension.

P2. **a.** 5, east
       b. 12, north
       c. 15, 37° south of east

CHAPTER 3

SECTION 3.1

Q2. The quantities that define what we mean by motion are velocity and acceleration. If an object moves (relative to a frame of reference), it has velocity. If its rate of movement changes, it has acceleration. The measured quantities are displacements and time intervals.

Q4. The ''definition'' given in Chapter 2 was based on a comparison to an arbitrary standard quantity of matter. In Newton's theory, mass can be defined as a measurable property of matter to resist changes in its motion.

P2. 12 miles/hour per second, or 17.6 feet/s/s, or 17.6 ft/s^2

SECTION 3.2

Q2. First law: People standing in a bus continue to move forward when the bus makes a sudden stop.
Second law: Throwing a ball.
Third law: The ''kick'' of a shotgun, rifle, or pistol when it is fired.

Q4. A very fundamental one. Force is the underlying concept in all three. In the first two, force is the agent of any change in motion. In the third, the idea that force is always an interaction between two or more objects is made explicit.

P2. $F = ma = 2000 \text{ kg} \times 0.5\text{m/s}^2 = 1000 \text{ kg-m/s}^2 = 1000 \text{ N}$

SECTION 3.3

Q2. The earth and moon are moving in their orbits as a result of a gravitational force that has a nearly constant magnitude (but a changing direction). Freely falling bodies are experiencing a constant net force until they are moving fast enough for air resistance to become significant. The particle beam in your TV tube is accelerated by a constant electrical force.

Q4. A number that specifies the amount of friction acting between two surfaces; it depends primarily on their texture and composition. It can be determined experimentally from the friction force and normal force.

P2. 0.510

SECTION 3.4

Q2.

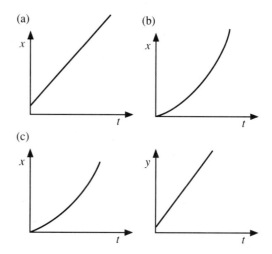

(a)

(b)

(c)

P2.

t (s)	x (m)	y (m)
1	0	25.1
2	0	40.4
3	0	45.9
4	0	41.6
5	0	27.5
6	0	3.6
7	0	−30.1
8	0	−73.6

SECTION 3.5

Q2. a. A Nolan Ryan fastball has about the same momentum as
    b. Nolan Ryan walking

Q4. Accident investigation: A model of the accident is produced from physical principles and data. The model can then be used to help determine how the accident happened, who was at fault, how it could be prevented, etc. A similar analysis is applied in the design and testing of seat belts, air bags, etc.

P2. To the left at 0.2 m/s.

CHAPTER REVIEW PROBLEMS

2. One minute and forty seconds

4. 4 m/s^2

6. 50 kg

8. 0.2 m/s/s

10. 12 m/s

12. 4 m/s^2

14. To the left at 0.1 m/s^2

16. 7.06 m/s^2

18. -25 kg-m/s, -25 kg-m/s, 250 N

20. To the right at 1.5 m/s

22. 4 km/s

24. 6.67 m/s

THOUGHT QUESTION

14. There are two reasons. First, the object starts with the same velocity as the earth. This part of the motion persists even after it is thrown. Second, because the earth's motion is not a straight line, the path of the object would eventually diverge from the earth's orbit if not for gravity.

FORMULATION 3.1

Q2. **a.** An object starting free fall from an initial height.
  b. A moving car suddenly locking up its brakes and skidding.
  c. A parachutist jumping out of a moving airplane.

P2. 42.5 s

CHAPTER 4

SECTION 4.1

Q2. Gravity; tension in a string, wire, or cable; cohesive forces in materials.

Q4. $kg \times \dfrac{m^2/s^2}{m} = kg\text{-}m/s^2 = N$

P2. 400 g

SECTION 4.2

Q2. In rotation, the body turns around a line that passes through the body, as in the spinning of a wheel on an axle. In revolution, the axis that the body turns around is outside the body, as in the motion of a satellite, or a car on a circular track.

Q4. **a.** A picture that is not hung straight has an angular displacement.
  b. A frisbee must have angular velocity to ''fly'' properly.
  c. A circular saw blade undergoes angular acceleration to reach operating speed.

P2. 300 degs/s or 0.833 rev/s

SECTION 4.3

Q2. Moment of inertia plays the same role in rotational motion that mass does in linear motion. It takes into account the way the mass is arranged around the axis of rotation.

Q4. A net torque on a body will cause it to change its rate of rotation. The larger the torque, the greater the change (angular acceleration). For a given amount of torque, the angular acceleration also depends on the mass and how spread out it is. More mass or bigger spread means less acceleration. Expressed more concisely as an equation, $\Gamma = I\alpha$, it allows us to analyze rotational motion in the same way as linear motion.

P2. 0.5 kg-m^2

SECTION 4.4

Q2. Since the rotational inertia takes into account the different distances of each part of the objects from the axis of rotation, it would naturally apply to a rotating object. If the whole object is going around the axis, the distribution of mass is not as important, so a different relation is more appropriate. The two are related, however, because the equation for the rotating object can be found by ''adding'' the motions of each individual fragment of the object as they revolve around the axis.

Q4. Our planet has a tremendous amount of angular momentum, but it takes a whole day just to make one rotation.

P2. 50 rad/s, if they have the same mass.

SECTION 4.5

Q2. The interplay of forces that result in equilibrium. The study of statics underlies the design and construction of buildings, bridges, dams, and utility lines.

Q4. When both the net force *and* net torque on the object are zero.

P2. 4 m

CHAPTER REVIEW PROBLEMS

2. 15 rad/s or 143 rpm

4. 200 rad/s or 1910 rpm

6. 140 rad/s^2

8. **a.** 0
   b. 40 rad/s
   c. 200 rad or 31.8 rev.

10. 3.2 m

12. Doubling mass doubles inertia; halving width cuts inertia to one-fourth of initial value.

14. 1 N, to the right

16. 2 s

18. 10 N·m

a. 25 N
b. 100 N
a. yes (more force if closer to hinge)

20. 33.3 mm

22. 1.98 m/s

THOUGHT QUESTIONS

8. Since both will experience the same torque, the disk will have a greater angular acceleration because it has less resistance to changes in its angular velocity (smaller rotational inertia).

FORMULATION 4.1

Q2. It allows the direct relation of angles and corresponding distances (from the center and around the circumference).

P2. 7.27×10^{-5} rad/s, 465 m/s, or 10,472 mph

FORMULATION 4.2

Q2. Yes! If there was no net force, it would travel in a straight line.

P2. 31.3 rad/s

CHAPTER 5

SECTION 5.1

Q2. 1. Since the planets are not going in straight lines, there must be a force acting on them.
  2. Since the orbits and periods are known, the acceleration is approximately centripetal, and this gives a relation between the force and other known quantities.
  3. Since forces act between objects, the gravitational force must depend on both masses.

Q4. As an intermediate step to get a relation between the force and distance that did not depend on time.

P2. Decreases by a factor of 400.

SECTION 5.2

Q2. The value of ''g'' would increase (at an increasing rate) as you approach the Earth's surface. It would peak at the surface (9.8 m/s^2) and then decrease at a constant rate as you approach the center. There it would be zero and you would be truly weightless!

Q4. By calculating the centripetal acceleration of the moon from his theory (law of gravitation and the second law), and then comparing his result with the acceleration found from measurements.

P2. 0.011 m/s^2

SECTION 5.3

Q2. Earth and sun, earth and moon, binary stars, hydrogen atoms.

Q4. a. The harmonic law says that if you take the mean orbital radius of a planet, raise it to the third power, and then divide that by the time for one revolution, raised to the second power, then your result will be the same for any of the nine planets.

  b. Imagine the orbit was a pie, with the big planet in the middle and the smaller one going around the edge. The size of a piece of the pie is then determined by how long it takes the smaller planet to traverse its outer edge. The law of equal areas says that if you take another piece of the pie and the planet takes the same time to traverse its outer edge, then the second piece will be exactly the same size as the first. (This may seem obvious if you are thinking of a circular pie, but remember Kepler's first law.)

P2.

Planet	Ratio
Venus	3.34×10^{18} m³/s²
Earth	3.32
Mars	3.37
Jupiter	3.36

These are all pretty close; the difference is only about 1%.

CHAPTER REVIEW PROBLEMS

2. 1.22×10^5 kg

4. a. x 3
  b. x 1/2
  c. x 1/16
  d. x 25
  e. x 3/32

6. 0.00578 m/s² or (1/1695) g

8. Mother: 2.50×10^{-7} N; Mars: 3.34×10^{-8} N. The gravitational influence of both are extremely small; the mother's gravitational influence is almost 10 times that of Mars.

10. $\sqrt{(1/6)}$ or 11/27

12. a. 4 m/s²
  b. 1.01×10^{23} kg

14. 1/1,000,000

15. g would be 39.2 m/s² at the surface.

16. 40 m

18. $$\frac{d_e}{d_m} = \frac{m_m}{m_e} => d_e = \frac{m_m d_m}{m_e} = \frac{m_m (384{,}000 \text{ km} - d_e)}{m_e}$$

$$=> d_e \left(1 + \frac{m_m}{m_e}\right) = \frac{m_m}{m_e}(384{,}000) => d_e = \frac{\dfrac{m_m}{m_e}(384{,}000 \text{ km})}{1 + \dfrac{m_m}{m_e}}$$

with $m_m = 7.36 \times 10^{22}$ and $m_e = 5.98 \times 10^{24}$, $d_e = 4.67 \times 10^3$ km, so $d_e - R_e = (4.67 - 6.37) \times 10^6$ m $= -1700$ km

20. 100,000 kg

22. 13×10^{16} m², five years

THOUGHT QUESTION

24. The discovery verified that the law was indeed universal and applied to all bodies, even those which hadn't been seen yet. It also confirmed very dramatically the predictive power of physical theory.

FORMULATION 5.1

Q2. No effect, for three reasons: (1) the earth's gravity is vertical and the forces measured were horizontal, (2) the force due to Earth's gravity was balanced by the tension in the supporting fiber, and (3) the earth's gravitational field is uniform throughout the space occupied by the apparatus.

P1. 7.68×10^{-11}

FORMULATION 5.2

Q2. No effect on results. Here again, the earth's gravity is vertical while the force between masses on opposite sides of the balance is horizontal. If you really wanted to get picky about it, you might want to consider the component of force between M and M' !

P1. 3.70×10^{17} kg

CHAPTER 6

SECTION 6.1

Q2. When you sit in a chair, you exert a force on it but it doesn't move; when you try to unscrew a jar lid that won't open.

Q4. Not the same! Torque is a vector; it has a direction. Work is a scalar quantity with no direction. Also, the distance involved in a torque is perpendicular to the force, while the distance involved in work is parallel to the force. A torque can be balanced by a counter torque; work is a *process* that transfers energy.

P2. 120 N

SECTION 6.2

Q2. A bird on the wing has translational kinetic energy (TKE) and gravitational potential energy (GPE). A moving car has TKE and rotational kinetic energy (RKE) in the spinning wheels, gears, etc. A speeding bullet has all three kinds of energy.

P2. 1.5 J

SECTION 6.3

Q2. The theorem is a mathematical expression of the energy conservation principle for situations involving conservative forces.

P2. 147 kJ

SECTION 6.4

Q2. A rate of doing work: one joule per second. The amount of work done in one hour by a thousand 1-W sources. A kWh is a quantity of energy. A watt is a rate of energy transformation; (watts) x (seconds) = joules.

P2. 12.5 seconds

CHAPTER REVIEW PROBLEMS

2. 125 N

4. 240 m, 60 kJ

6. 20.4 m

8. 60 kJ, 60 kJ

10. 0 m/s; KE = 0, PE = 0.375 J; KE + PE = 0.375 J

12. 40%; 1111 J

14. 504 J

16. 0.200 J

18. Same

20.
$$1500 \text{ kWh} \times 3.6 \times 10^6 \frac{J}{\text{kWh}} = 5.4 \text{ billion joules!}$$
(Mid-summer in Texas with low-efficiency AC!)

22. 0.54 seconds

24. 1.5 m/s (Do you think this collision could actually happen? *Hint:* Think about momentum...is it conserved?)

THOUGHT QUESTION

10. Of course! Can you see the wind? Can you see a force? No, but you can see their effects. It's the same with energy. These are all ideas we use to describe the physical world, and they have served us extremely well. The scientific method helps us find consistent relationships between any physical quantities that can be defined and measured.

FORMULATION 6.1

Q2. A bottle or can opener, wheelbarrow, car jack, pliers

P2. Not! A force of 392 N would be required to lift the stone. If this force was applied over 0.5 m, the stone would move 0.1 m.

FORMULATION 6.2

Q2. The minimum escape velocity is determined by giving a KE to the object that is large enough to do the work against gravity (PE) needed to escape. This KE is transformed to PE as the work is done.

P2. 11.2 km/s or about 25,000 mph

CHAPTER 7

SECTION 7.1

Q2. Atoms are unique to each element in that they all have the same number of protons; but they are not much like billiard balls. Atoms of different elements have different numbers of protons and different masses. The combination of atoms (elements) into molecules (compounds) has been verified and explained in terms of the interactions of the electrons and protons.

P2. 26.7 kg

SECTION 7.2

Q2. Retains shape and volume, even when subjected to forces; bricks, pencils, books, and boards.

P2. 66.7 kg, 3.45×10^{-3} m^3, or 3.45 liters

SECTION 7.3

Q2. The attractive force between neighboring molecules at the surface of a liquid. This ''pulling together'' of molecules results in the formation of spherical droplets instead of irregular blobs. You can feel it by placing an aluminum pie tin in a tub of still water. When you lift the pan slowly and evenly (keep it level!), you will feel the surface tension pulling against you.

SECTION 7.4

Q2. A gas whose molecules or atoms are so hot they won't stay together. It behaves like a gas in its mechanical properties, but is very different in its electrical properties. This is because the electrons are not bound to the nucleus as in a normal atom.

SECTION 7.5

Q2. The ratio of an applied force to its contact area. It describes how much the force is ''concentrated'' or spread out over its area of application.

P2. 1.02 cm, 2.55 cm

CHAPTER REVIEW PROBLEMS

2. **a.** 17
   b. 25% N atoms, 75% H atoms
   c. 82% N, 18% H (by mass)
4. **a.** 10^{24}
   b. 10^{30}

6. 1.88×10^{-3} m^3 or 1.88 liters

8. 3400 kg

10. 13 m/s^2

12. 3.06×10^6 N/m^2 or 30.3 atm

14. Yes, the density of the block is only 0.4 kg/m^3. This is much much less than the density of water, so it will float with only 0.04% submerged.

16. 1.02 kg, 3000 kg/m^3

18. 4000 N/m^2

20. 2.55 kg/m^3

22. **a.** 90.4 grams or 0.0904 kg
    b. 0.0784 N
    c. -0.8075 N
    cube sinks: (weight is greater than buoyant force)
             (negative "lift force")

24. 0.170 m

26. 4.42×10^{-10} m or 0.442 nm

THOUGHT QUESTION

4. No. For a similar reaction to occur, the element would have to have the same number of "outer" electrons. This would be indicated by the element being in the same column of the periodic table. Since chlorine and oxygen are not in the same column, they would react differently.

FORMULATION 7.1

Q2. They are all defined in terms of an applied force, contact area, and distortion. Since the distortion is relative (i.e., expressed as a fraction or percent of original size), the only units remaining are those of force and area.

P2. 1×10^9 N/m^2

FORMULATION 7.2

Q2. No. The reduced force will be balanced by increased distance of application. If the 100-N force is applied over 1 m, the 1000-N weight will only rise 1/10 m.

P2. 98 N

FORMULATION 7.3

Q2. The "bubble" in a convertible top as it speeds down the highway, the outflow of inside air at the rear of a vent window, the "floating" beach ball over the vacuum cleaner hose in a department store window, and the rise of liquid in a straw if you blow over the top of it.

P2. 7.39×10^4 N/m^2

CHAPTER 8

SECTION 8.1

Q2. The Farenheit scale, with 180 degrees between the freezing point and boiling point of water; the Celsius, or Centigrade, scale, with 100 degrees between the freezing and boiling points of water; and the Kelvin, or absolute scale, which also has 100 divisions between the freezing and boiling point of water, but has no negative values.

Q4. The specific heat depends only on what substance the object is made of. It indicates how much energy is needed to change the temperature of each gram (or kilogram) of the object one degree (Celsius). The heat capacity is the specific heat multiplied by the mass of the object. It indicates how much energy it takes to change the temperature of the object by 1 degree.

P2. 237 J/kg-°C, 474 J/°C

SECTION 8.2

Q2. The heat of fusion gives the atoms enough energy to move away from their fixed positions, but not very far. They are still very much under the influence of neighboring molecules. The heat of vaporization gives them enough energy to get far enough away from other molecules that the influence is reduced to almost none. Vaporization occurs at a higher temperature because it takes more work (energy) to separate the molecules than it does to shake them loose of their rigid framework.

Q4. a. Boiling occurs when energy is transferred to the liquid at a high temperature. This energy can then do the work needed to pull the molecules away from each other. Then the molecules take on the character of a vapor or gas.

   b. Evaporation is very similar, but much slower because it takes place at a lower temperature. Since some molecules are always moving faster than the average speed, occasionally some will escape from the surface of the liquid. With the faster molecules gone, the average speed of the molecules left in the liquid decreases, so the temperature decreases. If heat is absorbed to maintain a constant temperature, then the average speed goes back up and more molecules get away.

   c. Gas pressure is simply the net force exerted by the molecules when they hit something.

P2. a. 24 l

   b. 4 l

   c. 8 l

SECTION 8.3

Q2. Radiative transport is the way we get energy from the sun and how food gets energy in a microwave oven. The radiators in our cars use convection to transfer heat from the engine to the coolant and then from the coolant to the air. Conduction is how the heat gets through the bottom of the pan when you cook on a conventional stove.

Q4. Solids conduct better than they convect because the molecules stay in place. Since molecules of a gas or liquid can move around, fluids usually convect better than they conduct. Radiation depends more on temperature than whether the molecules are in a solid or fluid state.

SECTION 8.4

Q2. 0) If objects A and B are in thermal equilibrium, and objects B and C are in thermal equilibrium, then A and C are also in thermal equilibrium.

1) The net gain (or loss) of internal energy in a system is equal to the heat added, minus the work done *by* the system, plus any work done *on* the system.

2) No system can completely convert heat into work, thus the internal energy of a system will always increase if heat is applied to produce work.

Although all of the above relate to energy in some way, only the first law is directly related to the conservation of energy principle. Essentially they are equivalent: the first law is a specific application of the general case.

Q4. The heat left over after a system does work. There will always be heat left over (second law); so it can never be eliminated, only reduced to a minimum value that depends on temperature.

P2. 75%, 3000 K

CHAPTER REVIEW PROBLEMS

2. 88.9°C

4. 28.8 kcal or 121 kJ

6. 5.0005°C

8. 1600 cal or 6.69 kJ

10. 75 cal; yes, heat lost equals heat gained

12. 0.5 atm

14. 300 K

16. 2 kg/m^3

18. a. 200 N
b. 0.8 m^3
c. 80 J

20. 800 J, 80%

22. 3000 J, 2000 J

THOUGHT QUESTION

10. As a result of molecular collisions, the faster molecules are slowed down and the slower ones speeded up.

FORMULATION 8.1

Q2. On my scale 0°M will be the temperature at which the element mercury freezes, and the temperature at which the mercury boils will be 1000°M. To convert from the traditional scales to my new scale, use the following formulae:

$$°M = 2.5(°C + 39)$$
$$°M = 1.4(°F + 38)$$

P2. $-459°F$

FORMULATION 8.2

Q2. If there were another relation, it would have to give the same results as the first one. If it gives the same results, it can be shown to be mathematically equivalent.

P2. 0.028 l

CHAPTER 9

SECTION 9.1

Q2. A pendulum, a child on a swing, a vibrating tuning fork, vibrating quartz crystal, electric current (ac), and a car with bad shocks after it goes over a bump.

Q4. Because they describe the same thing in a different way. Frequency is cycles per second and the period is seconds per cycle; one is the inverse of the other.

P2. 0.0339 cycles/day or 1.03 cycles/month (1 month = 30.4 days)

SECTION 9.2

Q2. The difference is the direction the material particles move as the wave travels through them. Longitudinal waves shake the particles parallel to the wave direction. Transverse waves shake the particles perpendicular to the direction the wave is moving. In both cases, the particles move in simple harmonic motion as the wave goes by, and the wave speed is the product of frequency and wavelength.

Q4. Unlike particles, two or more waves can occupy the same place at the same time. The resulting wave motion is the sum of each individual wave motion. This is called *interference*. It is referred to as *constructive* when the individual wave motions are in the same direction at the same time, and *destructive* when the motions are in opposite directions.

P2. 75 cm

SECTION 9.3

Q2. Because the molecules are closer together and thus react sooner when the molecules nearby start moving.

Q4. The most common is an echo. This is essentially a mirror image of the original sound. A reflected sonar wave can be used to determine the distance and velocity of an underwater object. Bats use ultrasound in a similar way in the air, and the analysis of ultrasound reflections within the human body has contributed significantly to progress in the medical field. Reflected waves in ship channels must be damped out so they don't interfere and create disturbances for ships.

P2. The level of A is 30 dB less than the level of B.

CHAPTER REVIEW PROBLEMS

2. 4 Hz

4. 5 cm

6. 15 m/s

8. 200

10. 0.5 Hz (once every 2 seconds)

12. 5 Hz

14. 15 km

16. 1/100

18. 1/10,000

20.

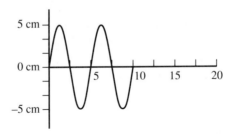

22. 110 Hz

THOUGHT QUESTION

4. There would be vibrations produced in the surrounding air and in the ground when the tree hit. These vibrations would propagate outward from the event whether or not there was an instrument (animate or inanimate) there to detect it.

FORMULATION 9.1

Q2. Yes. If only the observer is moving, the effect is produced by the observer moving through the wave crests, thus hitting them more or less often depending on whether the motion is toward or away from the source. If the source is moving, the distance between the crests is actually shortened or lengthened by its motion. Both the observer's and the source's motion can be included in the equations.

P2. 473 Hz; not likely at that speed.

FORMULATION 9.2

Q2. A similar scaling of frequencies is used in the analysis of electrical devices.

P2. 140 dB

CHAPTER 10

SECTION 10.1

Q2. It was found that rubber and glass rods could be charged. If any other object is charged and brought near the rubber rod and then the glass rod, it would always be attracted to one and be repelled by the other.

Q4. The two protons repel each other and the two electrons repel each other. Each proton is attracted to both electrons and each electron is attracted to both protons.

SECTION 10.2

Q2. Both laws describe a force that varies as the product of the entities involved (mass or charge), and decreases as the square of the distance between their centers. The differences are that Coulomb's law can allow for a repulsive force, and the constant in Coulomb's law is much much bigger than the constant in the law of gravitation.

Q4. Attractive forces pull the objects toward each other and repulsive forces push the objects directly away from each other.

P2. 12 m, attractive

SECTION 10.3

Q2. The field is stronger where the lines are closer together, and the direction of the field at a point is the direction the line is heading at that point.

Q4. The field is directly related to the force. The potential is directly related to the energy. The field is then related to the potential when the force does work on the charge. The potential is the amount of work the field can do on a unit of charge (1 coulomb) if it is allowed to move without restriction to a very great distance. The potential difference is the work done or energy transferred, per unit charge, by the field when the charge moves from one point to another; it is the same as the difference in the potential at each point.

P2. 4500 N

SECTION 10.4

Q2. Negative charge (electrons); much smaller mass and less restricted movement.

Q4. When the bulb is turned on, a potential difference occurs across the bulb. Energy is transferred to the charge. This kinetic energy is converted to thermal energy as the charges collide. Thermal energy is then radiated. Some of this radiation is visible light (most of it is infrared).

P2. 90,000 C; 5.62×10^{23} (almost 1 mole)

SECTION 10.5

Q2. For a large class of conducting materials, the current resulting from an applied potential difference is directly proportional to the potential difference. The current increases linearly as the potential difference is increased. The constant of proportionality is related to the resistance. The law allows us to predict the effects of potential differences and currents in these materials. Thanks to this we have stereos, TVs etc.

Q4. When one becomes part of a path between points of potential difference or when one becomes charged (by friction for example). The earth is normally considered to be at zero potential (ground). Any charge at a higher potential will move toward ground if there is a way for it to get there.

P2. 330 watts or 0.442 horsepower

SECTION 10.6

Q2. They exert attractive and repulsive forces on other magnets; they exert attractive forces only on certain kinds of materials, such as iron. These effects are distance dependant and are reduced when the magnets are heated. They also influence the motion of electric charges.

Q4. North and south magnetic poles exert attractive forces on one another.

P2. The force after distance is increased would be 1/25, or 4%, of the force before distance was increased.

CHAPTER REVIEW PROBLEMS

2. Force would be 1/9 of previous value.
4. $+20$ C
6. 5 A
8. 720 C
10. 30 V
12. 11 A
14. 128 W
16. 5 A, 20 V
18. 96 W
20. 48.4 ohms
22. 3162 A-m
24. 0.216

THOUGHT QUESTION

4. A negative electric charge will exist on an object if there are one or more

''extra'' electrons in the object. The net charge would be equal to the number of electrons minus the number of protons, multiplied by the charge per electron. We don't say the electron is the charge, we say it carries the charge, because it also has mass.

FORMULATION 10.1

Q2. Their weight (due to the gravitational field), the electric force, and the contact force of air resistance.

P1. 4.69×10^{-19} C, ($\approx 3e$)

FORMULATION 10.2

Q2. $P = I \times V = I \times (IR) = I^2R$
$P = I \times V = (V/R) \times V = V^2/R$

P1. **a.** 0.5 A
b. 240 Ω

CHAPTER 11

SECTION 11.1

Q2. This force arises when a charged particle moves across a magnetic field; the faster it moves, the greater the force. This force is different from the static forces in two important respects. One is the velocity and the other is the peculiar direction of the force. It is at right angles to both the velocity and the field!

Q4. This force is a good example of the reciprocity inherent in natural phenomena. Since moving charge (current) produces magnetic effects, two different currents can interact both electrically and magnetically. One way to describe this is that one current produces a magnetic field in the space around it. If another current passes through this space, there will be an electromagnetic force acting on it. Depending on the direction of the current, the force can be attractive or repulsive. It is also reciprocal because each current produces a magnetic field around it.

P2. 0.300 μN, repelling

SECTION 11.2

Q2. Lenz's law relates the direction of a magnetically induced current to the change in magnetic flux that produced it. Faraday's law relates the magnitude of the induced current to the change in flux.

Q4. Magnetic flux describes both the strength and extent of the field. It is equal to the product of the field strength and the area crossed by the field.

P2. 3.93 mV

SECTION 11.3

Q2. Both work on the same principle of magnetic induction. The motor uses electrical energy to produce mechanical motion. The generator uses mechanical motion to produce electrical energy.

Q4. As the current in the first coil changes, it sets up a changing magnetic field, which penetrates the second coil. This changing magnetic field then induces a current in the second coil. The magnitude of the second current depends on the primary current and the number of turns in each coil.

P2. 240

SECTION 11.4

Q2. Although all moving charges produce magnetic effects, and all substances are composed of these moving particles, in most substances these effects are evened out and there is no net effect. Only a few materials have atomic configurations in which these effects are not cancelled.

Q4. Since all magnetic effects arise from moving charges, any kind of magnet depends on the electromagnetic interaction for its properties.

CHAPTER REVIEW PROBLEMS

2. 0.012 N, to your right

4. 800 N

6. 0.16 N

8. 1 m; it would be possible to define the unit of magnetic field in terms of the force and the current. Historically, the current was already defined by the (magnetic) force between two wires, so the unit of magnetic field did not need to be defined separately; it followed from the definition of the ampere.

10. 4.8×10^{-13} N, downward

12. 224 A

14. **a.** 2×10^{-5} T
     b. 2×10^{-7} T
     c. 2×10^{-10} T

16. 1.25×10^{-3} T-m^2

18. 6.67×10^{-2} m^2

20. 30 V

22. **a.** $-BA$
     b. $-2BA$
     c. $-BA$
     d. 0

THOUGHT QUESTION

4. There will be forces where the wires cross, but they will be in opposite directions on opposite sides of the wires. The net force will be zero.

FORMULATION 11.1

Q2. So the magnetic fields due to the connecting wires do not interfere with those of the wire segments.

P2. 2.4×10^{-6} N

FORMULATION 11.2

Q2. No process has ever been found that violates the energy conservation principle. This is just another case of energy conversion, from lower voltage and higher current to higher voltage and lower current. Power in = power out + thermal loss.

P2. 200

CHAPTER 12

SECTION 12.1

Q2. He set up a circuit with a changing electric current and used a wire loop as a detector. When the circuit was activated, a corresponding electric current was observed in the loop.

Q4. Because the light energy spreads out as it travels through space.

P2. 24.9 m

SECTION 12.2

Q2. Yes, but the time interval he was attempting to measure was much smaller than human reaction time.

Q4. A number that is used to indicate the speed at which light travels through the substance.

P2. 10

SECTION 12.3

Q2. This varies a little from one person to another. The limits are about 360 to 720 nm. This corresponds to a frequency range of 8.33×10^{14} Hz to 4.17×10^{14} Hz, or about one octave.

Q4. **a.** Transmitting and receiving information quickly.
 b. Power transmission through space.
 c. Producing images of objects opaque to light.
 d. Provide information on the location and behavior of heat sources.
 e. Sophisticated space-based nuclear defense capability?

P2. 5000 km

SECTION 12.4

Q2. **a.** The angle at which a ray reflects from a surface has the same value as the angle at which it approaches the surface. This angle is always measured from the line perpendicular to the surface. This ⊥ line is called the *normal*.
 b. The angle (from the normal) at which a ray travels through a substance depends on the angle of incidence and the indices of refraction on both sides of the surface. If the angle of incidence is zero, the angle of refraction

will also be zero. If the incident angle is not zero, the refraction angle will be smaller or larger, depending on whether the index of the substance is larger or smaller than the index of the material the light is coming from.

Q4. Although it doesn't happen every day, my favorite example is a rainbow. More common examples are the spots of color produced by lenses, prisms, and crystals. If you worked in an optical research lab, you might be using dispersion to tune your laser to a particular wavelength.

SECTION 12.5

Q2. A white surface reflects all colors, and the combination of all colors appears as white light. Black is black because no light and thus no colors are reflected; all are absorbed.

Q4. Stage lighting uses both additive and subtractive color mixing to produce colors that enhance the tone of the performance. Color television produces a full range of colors by additive mixing. Subtractive color mixing can be employed in a film developing process to produce color photographs.

SECTION 12.6

Q2. Longitudinal waves cannot be polarized.

Q4. Selective absorbtion. The polaroid material only transmits light whose orientation is parallel to the polaroid axis. This axis is vertical in the sunglasses, so all light that has horizontal polarization will be absorbed.

SECTION 12.7

Q2. If a single beam of light waves is separated into different parts, and these parts travel different distances before recombining, then interference can occur. Some ways to separate the beam are by transmission through small apertures, reflection from small ridges or strips, and by reflection from the top and bottom surfaces of thin films.

Q4. Particles do not exhibit any interference effects. Only waves can occupy the same place at the same time, and thus interfere constructively or destructively.

SECTION 12.8

Q2. To see blue sky, you must look in some other direction than directly at the sun. Since the light you are looking at came from the sun as well, it had to change its direction of travel to reach you. This change of direction is called scattering, and blue light scatters best. At sundown, the light reaching you is mostly red because the blue has already been scattered in other directions (so someone else can see the blue sky). The sun appears yellow, or almost white, when viewed overhead because that is when the light passes through the least amount of atmosphere, so there is not as much blue scattered out.

Q4. More particles in the atmosphere are likely to scatter more light. That would mean less blue light reaching us, redder sunsets, and redder, irritated eyes from unfriendly chemicals.

SECTION 12.9

Q2. The process whereby electrons are emitted from the surface of a metal when it is exposed to light. According to the wave model, the kinetic energy of the emitted electrons should depend on the intensity of the incident light. But it doesn't; it depends only on the frequency. Also, if the frequency is too low, no electrons are emitted at all; the intensity doesn't make any difference.

Q4. A photon is the "particle of energy" that seems to be riding on the electromagnetic wave. When the wave interacts with an individual electron, the process appears to be localized in space rather than spread out over the wavelength. Planck's radiation law, the photoelectric effect, the Compton effect, and atomic spectra all confirm the existence of photons.

P2. 3.03×10^{13} Hz; not visible, lower frequency than red light.

CHAPTER REVIEW PROBLEMS

2. 20 m

4. **a.** 1.8×10^7 km,
b. 2.59×10^{10} km
c. 9.47×10^{12} km
d. 9.47×10^{13} km

6. 75,000 km/s

8. 30×10^6 km

10. 300,000 km/s

12. NOT! (Unless a new substance was also discovered with a refractive index of 100!) All electromagnetic energy travels at the speed of light, adjusted for index of refraction. A speed of 3000 km is much too slow for any type of radiation.

14.

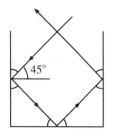

16. 2.31
18. 1.5×10^{-14} m, 1.33×10^{-11} J or 82.7 MeV
20. 17.3 km, 17.3 kHz
22. 6.27×10^{-34} J-s

THOUGHT QUESTION

2. The observed intensity would diminish more rapidly as the distance from the source increased. Some of the light would be absorbed instead of moving outward. If the intensity at 2 m was 10 W/m², the intensity at 4 m would be less than 2.5 W/m².

FORMULATION 12.1

Q2. The power is the rate at which radiant energy is leaving the source—in all directions. The intensity refers to how much of this power crosses one square meter of area.

P2. 0.446 m

FORMULATION 12.2

Q2. Instead of the ray passing through the cogwheel, it is reflected from one face toward a distant stationary mirror. After reflection from the fixed mirror, it returns to the rotating mirror and hits another face and is reflected to the observation point. Since these reflections can only occur if the angles between the mirror faces and light beam are just right, the rotational speed of the mirror is even more critical than it was for the cogwheel. Thus the speed of light can be measured more precisely by this method (see diagram).

P2. 5.77×10^8 m/s (not a very good experimental result!)

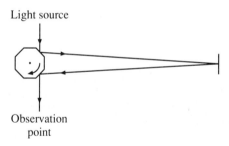

Light source

Observation point

CHAPTER 13

SECTION 13.1

Q2. If there were no other force acting than electrostatic, the nucleus would fly apart from the tremendous repulsive forces between the protons. Then the electrons would no longer be attracted to a central point, and each one would eventually pair off with a proton. Then there would be nothing in the universe but hydrogen!

Q4. By supposing that the angular momentum of the electron orbiting the nucleus was quantized in the same way energy was in the photon. This condition made only certain orbits possible, and the changes between these discrete orbits result in discrete spectral lines. When the atoms are crowded together, the orbits of neighboring atoms interact and the lines are smeared into a continuous spectrum.

P2. Wavelength must be longer than 2.43×10^{-12} m

SECTION 13.2

Q2. Three kinds of "rays" have been found to emanate spontaneously and randomly from the nuclei of certain atoms. Two of these rays are particles and one is electromagnetic radiation. Which types of rays are emitted, and their rates of emission, are unique to each element.

Q4. In this model, the protons in the nucleus behaved like water molecules in a drop of water. This explained the spherical shape of the nucleus, but could not explain the discrete energy levels within the nucleus, nor could it explain why only some atoms were radioactive.

P2. 9.5 minutes

SECTION 13.3

Q2. There are several "families" of particles produced in high-energy collisions. One family consists of μ and π-mesons which are many times more massive than electrons and have electric charges like electrons and protons. Electrons belong to the family of leptons, and protons are members of the baryon family. The building blocks of all of these families are thought to be the quarks, which have various masses and electric charges that are 1/3 or 2/3 of the electron and proton charges.

Q4. High-energy particles from outside our solar system. When they collide with atoms in the outer atmosphere, many of the elementary particles from different families are produced.

SECTION 13.4

Q2. A mathematical formula that describes the wave-like behavior of elementary particles and atoms. Perhaps the formulas used by insurance companies to predict the numbers of different kinds of claims they can expect during a given time interval.

Q4. An expression describing the limits of precision for simultateously measuring the position and speed of a particle. If we know one exactly, we have to sacrifice knowing the other. This idea is critical in any theory that attempts to describe atomic and subatomic processes because it tells us what the possible outcomes of our experiments will be.

P2. 2.1×10^{-25} m

CHAPTER REVIEW PROBLEMS

2. 3.31×10^{-19} J, 1.10×10^{-27} kg-m/s

4. 5000 km, 3.98×10^{-32} J, 1.33×10^{-40} kg-m/s
Wavelength is too long to be focused by any man-made object; it will just diffract around it.

6. 3.03×10^{-19} J or 1.90 eV

8. 6.63×10^{-33} m; too small to be measured

10. 1.32 km/s

12. 4.06×10^{17} kg/m^3

14. Nitrogen-14

16. 7.6 days

18. 1/16 or 6.25%

20. 3.9 billion years

22. At least 1.05×10^{-24} kg-m/s

24. let $\Delta x = v\Delta t$ and $\Delta p = h/\Delta\lambda$;
note that $\Delta\lambda = v/\Delta f$ so $p = h\Delta f/v$;
now $h\Delta f = \Delta E$, so $\Delta x\Delta p = v\Delta t \times \Delta E/v = \Delta E\Delta t$.
Therefore, if $\Delta x\Delta p = h$, so does $\Delta E\Delta t$.

THOUGHT QUESTION

8. The alphas and betas are electrically charged particles and will thus experience a force upon passing through an electric or magnetic field. The gammas are chargless photons, so they are not deflected.

FORMULATION 13.1

Q2. $h(c/\lambda_1 - c/\lambda_2) = e(V_1 - V_2)$

P1. 6.86×10^{-34} J·s

FORMULATION 13.2

Q2. If the substance is vaporized, ionized, accelerated by an electric field, and then directed into a magnetic field, the ions will turn in a circle determined by the mass and charge of the ions. The radius of the circle can be measured and thus the charge and mass of the ions can be calculated.

P2. 1.6×10^{-27} kg

CHAPTER 14

SECTION 14.1

Q2. Mostly Maxwell's equations! The electrical signals are converted to electromagnetic waves by the transmitter. The waves travel through space and interact with objects, etc., according to the principles of wave motion. Maxwell's equations are at work again to describe how the waves are changed back to electric current, and how this current is used to produce the picture on the screen. Throughout the entire process, the energy conservation principle determines the magnitudes of the currents and amplitudes of the waves. All four of Maxwell's equations are involved in almost every step in this process.

Q4. They could be considered ''mirror images'' of each other. The electrical components are almost the same except for the ''eye'' of the camera and the receiver screen. The main difference is that the camera changes the optical image to an electric signal and the receiver changes the signal back to an optical image. In between, one is operating like the other, only in reverse.

SECTION 14.2

Q2. Keeping accurate records and documentation (like your transcript of academic work). In general, preserving and transferring information. Like other modern conveniences, the time ''lost'' when they don't work is about the same as the time ''saved'' when they do. There are also hidden costs for the energy used and the environmental impact of chemicals used in the process.

Q4. The fax machine operates by the same basic principles as the television transmitter/receiver (see Section 14.1, Q2). However, in this case, there is only one image to transmit and it is transmitted electrically instead of by an electromagnetic wave. The photocopier relies more heavily on the behavior of static electric charge, the photoelectric effect, the mechanical interactions of rotating parts, and heat.

SECTION 14.3

Q2. These circuits represent logical operations. If 0/1 or -/+ stand for false/true, these circuits make logical comparisons between them. The ''and'' circuit will only send out a signal if it receives inputs that are both ''true.'' The ''or'' sends a signal if either input is ''true,'' and the ''not'' only sends a signal if both input signals are ''false.''

Q4. The on/off or +/- bits of information can be stored magnetically, through polarization of tiny magnets, or electrically, by charging tiny capacitors, or by configurations of solid-state diodes.

P2. 8 million

SECTION 14.4

Q2. They are used to establish and maintain alignment of components in large construction projects. The intense, narrow beam can be used in place of a surgeon's scalpel or a clothier's scissors. The monochromatic beam can be used to store, transmit, and retrieve information.

Q4. One method is *optical pumping*, in which the atoms are exposed to light energy, which can be absorbed by the atoms. Gas lasers are energized by thermal activity.

THOUGHT QUESTION

2. In principle, yes, but it would not work very well. The microphone is designed to accept and transmit only a small part of the sound energy reaching it. So if you ran it backward, you would get only a very small amount of sound energy out of it. The speaker is designed to transfer a large quantity of sound energy from a strong signal. It would therefore not be very responsive to the small signals corresponding to ordinary sound levels.

FORMULATION 14.1

Q2. Both instruments utilize a beam of electrons in place of a light beam. This allows better resolution of microscopic objects because of the shorter wavelength of the particles. Whereas the transmission microscope sends the beam completely through the sample and then refocuses it, the scanning microscope does not refocus the original beam. It gathers the electrons knocked off of the surface of the sample and focuses them to produce a picture of the surface.

P2. 1256 km/s

FORMULATION 14.2

Q2. Junction diodes act like one-way valves for electric current. Transistors are like diodes with one more layer of material that acts as a control valve. This addition can be used to increase or decrease the current going through the transistor.

Q4. Not likely! How would you prevent the different substances from mixing and eventually forming a homogeneous mixture? The semiconducting properties depend on a stable structure in each material that is different from the other.

CHAPTER 15

SECTION 15.1

Q2. Revolving charges in a rotating sun will produce electromagnetic radiation. Thermal energy generates radiation, as do some chemical reactions. The only type of reaction that releases as much energy as the sun does is nuclear, $E = mc^2$.

Q4. No material medium is necessary for it to travel through. The speed is the same for any observer moving at any constant velocity, and does not depend on any constant velocity the source may have. (However, there will be a Doppler effect.)

P2. 175×10^{12} s or 5.5 million years; at that rate it would have gone out long before now.

SECTION 15.2

Q2. In the Newtonian, or classical, view, all three were absolutes and would be the same for all observers. However, it has been verified that the motion of the observer affects the values that they can measure. Thus any set of values is relative to that observer.

Q4. Because there is no absolute or standard frame of reference in which to measure it. Everything is moving with respect to everything else. This would be in the same category as Archimedes' fulcrum for moving the earth.

P2. 11×10^{-15} kg

SECTION 15.3

Q2. When a proton is accelerated by an electric field in a high-energy particle accelerator, the increase in its speed is found to be less than it would be if the mass remained the same. Where did the rest of the energy go? A mass increase explains why the speed does not keep increasing.

Q4. No matter how much energy is transferred to the particles, their speeds never quite reach the value of *c* (but they have gotten pretty close!).

SECTION 15.4

Q2. Both state how the laws of physics and thus measurable quantities will vary between different observers. The special theory deals only with observers moving at constant velocity. The general theory includes observers whose velocity is not constant.

Q4. The irregularity in the orbit of Mercury, the ''bending'' of light rays passing near a large mass, and the small discrepancy between clocks in relative motion.

SECTION 15.5

Q2. The eclipse allows observers to detect light rays passing very close to the sun. These rays cannot be seen when the sun is at full brightness (just as the stars cannot be seen in the daytime).

Q4. This is similar to a Doppler effect for light leaving a source that is moving away from the observer. However, in this case the source is accelerating. The pulses will not ''arrive'' as rapidly because the source is ''moving away at increasing speed.'' (Note how the principle of equivalence is used here!)

CHAPTER REVIEW PROBLEMS

2. 9×10^{16} J

4. 9×10^{13} J

6. Same units as kinetic energy, $\frac{1}{2} mv^2$.

8. 5.49×10^{-14} J, 8.28×10^{19} Hz (would be 8.2×10^{-14} J, 1.24×10^{20} Hz, for $m = 9.1 \times 10^{-31}$ kg)

10. 4.42×10^{-13} m, 6.79×10^{20} Hz

12. 8.19×10^{-13} J (mostly mass energy) (would be 8.19×10^{-14} J for $m = 9.1 \times 10^{-31}$ kg)

14. 5×10^{10} s or 1584 years. Not even long enough for half of recorded history!

16. 4.5×10^{20} years

18. 1.64×10^{-9} J, 2.47×10^{24} Hz, 1.21×10^{-16} m (would be 1.64×10^{-13} J, 2.47×10^{20} Hz, 1.21×10^{-12} m for $m = 9.11 \times 10^{-31}$ kg)

20. 6.31×10^{12} km

22. 15.8 million years

24. Straight up at 9.8 m/s^2

THOUGHT QUESTION

2. This is not only possible, but likely! Suppose a star 1000 light-years away from us burns out. We won't know it for 1000 years. An observer halfway between us and the star would find out 500 years before us.

Formulation 15.1

Q2. I would certainly have to get moving! Or find something else moving toward or away from me at a speed of at least 30,000 km/s.

P2. 260,000 km/s

Chapter 16

Section 16.1

Q2. As an essential premise. If the laws of physics ''out there'' were any different than here, we would have very great difficulty constructing a consistent theory to explain our observations.

Q4. A velocity that is directly toward or away from the observer. It is determined by the shift of the pattern of spectral lines in the light emitted by the object.

P2. Intensity increases by a factor of 81. Maximum intensity wavelength is reduced by a factor of 3.

Section 16.2

Q2. It is the only process we know that is capable of sustaining the tremendous power output over so long a span of time. It is also consistent with almost all of our current knowledge of the sun.

Q4. It would be kind of like trying to hurry across campus in between classes on the first day of school! The atoms the photon has to travel through are moving so fast because of the high temperature that it doesn't get very far before it hits one. So it takes a very long time for a photon to finally get to the surface—and it may not even be the same photon that started out!

P2. 5.58×10^{26} watts

Section 16.3

Q2. The spectral lines depend on the energy levels in the atoms. As the temperature increases, the higher energy levels become activated so the line pattern shifts accordingly.

Q4. Its composition can be determined by its spectral lines because each element has a unique set. Its radial velocity can be found from the Doppler shift of the lines, and its temperature can be predicted from the wavelength at which the maximum intensity occurs.

P2. 11.5 solar radii

Section 16.4

Q2. From big clouds of hydrogen gas and cosmic dust. If some external event induces the cloud to contract or condense, then gravity takes over. The gravitational PE changes to KE and the cloud gets hotter as it contracts. The contraction continues until the pressure produced by gravity balances the gas pressure due to temperature.

Q4. The most "natural" is that where the star has exhausted its hydrogen fuel, cooled, and contracted to a cool dark dense cinder. If the star has sufficient mass, it may come to a more violent end. It can explode, leaving behind a superdense core of neutrons. This core usually rotates at a rapid rate until its energy is radiated away and it too finally finds repose as a cold dark very dense cinder. With more mass yet, the star may completely succumb to gravitational collapse and become what is known as a "black hole"—a very large mass compressed into such a small volume that its gravitational field will not even allow light to escape.

P2. 4×10^{49} m^3, 3.42×10^{16} m

SECTION 16.5

Q2. The theory that explains how the universe came to be as we observe it today. If it all began as a big "explosion of energy," then as it spread out and cooled it would give rise to the stars and galaxies, which would then continue spreading out. It would also leave a trace of residual radiation.

Q4. The first stage was all energy, or extremely high-temperature radiation. As the cooling took place, the fundamental particles and forces condensed into their current form. The gravitational interaction was the first to take form, followed by the nuclear or strong interaction. At a lower temperature, the weak and electromagnetic interactions took shape. Along with the latter came the protons and electrons, which then came together to form atoms as the cooling progressed further.

P2. 4.96×10^{17} s, 15.7 billion years

CHAPTER REVIEW PROBLEMS

2. 3×10^5 km/s $\times$ 3.16×10^7 s/yr $= 9.48 \times 10^{12}$ km/yr

4. 5.8×10^{-13} m, about ten times larger

6. 4.59 million watts per square meter

8. 750,000 km

10. **a.** 180×10^{45} J,
b. 450×10^{18} sec or 14.2×10^{12} years

12. 8500 W/m^2

14. 86.9×10^{18} m or 9165 light-years

16. 4.02×10^{26} watts

18. 24×10^{30} kg

20. $0.04\, R_s$

22. 8000 K

24. 3.02×10^{17} seconds or 9.56 billion years

THOUGHT QUESTION

2. Because stars go through stages of growth and decay. What we see now is the stage they were in when the light left them. Since the light may take hundreds to millions of years to reach us, they will have changed by the time their light gets here.

FORMULATION 16.1

Q2. It is assumed that the amount of radiation leaving the star's surface is related to its surface temperature in exactly the same way. Also, as is the case with most equations used in astronomy, the principle of universality is assumed from the start.

P2. One-half solar temperature

FORMULATION 16.2

Q2. The black holes themselves don't suggest anything, but we can make speculations based on their existence and physical characteristics. The idea is very interesting, but unless it can be tested by an experiment it cannot even be considered an hypothesis.

P2. Density $= \dfrac{3M}{4\pi R^3}$ and $c^2 = \dfrac{2GM}{R} => \dfrac{RC^2}{2G} = M$,

so density $= \dfrac{3}{4\pi R^3} \times \dfrac{RC^2}{2G} = \dfrac{3c^2}{8\pi GR^2}$.

CHAPTER 17

SECTION 17.1

Q2. **a.** Surface mining of coal has resulted in deforestation, erosion, and degradation of water quality in the surrounding watershed. In some cases run off from processing is acidic and poisons the soil and streams.

b. In addition to the spillage that occurs when transporting liquid and solid fuels, the transportation itself contributes to air contamination through unburned hydrocarbons and carbon dioxide. Even transporting electrical energy involves the use of potentially hazardous substances, especially in the cooling oils used in transformers.

c. This one depends a lot on the particular conversion process. Photovoltaic, wind turbines, and hydroelectric turbines are very clean. Burning coal or oil to produce electricity is not; in addition to air pollution, there is also thermal pollution produced in the water used for cooling.

Q4. The region at and around the earth's surface that includes all living creatures. There is almost no direct evidence for life existing elsewhere in the universe. However, since the universe is so vast the possibility that it does exist must be acknowledged.

SECTION 17.2

Q2. Some advantages are convenience, reliability, and portability. Some disadvantages are limited energy (small quantities), corrosive substances involved in manufacturing, and the disposal of nonrechargeables.

Q4. The most prevalent method to produce electrical energy is with a generator. The energy source utilized to turn the windings of the generator can be derived from falling water, wind, and heat from combustion or nuclear reactions. The battery involves a direct conversion from chemical (potential) energy to electrical energy.

SECTION 17.3

Q2. The wind turns a large fan blade, which then turns a set of coils in a magnetic field. An alternating current is induced in the coils as they cut across the field. This is an example of Faraday's law in action.

Q4. One that won't run out or be used up before more is available. Wood is a renewable energy source as long as we keep planting trees. Coal, oil, and natural gas are not renewable sources at the rate they are being consumed.

SECTION 17.4

Q2. The coal comes from plants. Plants store energy through a photochemical process. Burning releases this chemical potential energy. Glass is made of inorganic materials that are inert and do not have any stored energy.

Q4. These depend on which kind of chemical energy you choose. One example has already been discussed in Section 17.2, Q2. Some advantages of the energy produced by combustion are that the technology is relatively simple, the supply is easily accessible, and the rate of production is not limited. Disadvantages are that the supply itself is limited and that the acquisition and energy conversion process reduces the quality of the environment.

SECTION 17.5

Q2. In nuclear fission a large nucleus is split into two smaller nuclei. Fusion occurs when two or more small nuclei (hydrogen or dueterium) combine to form a single larger nucleus. Both processes release a tremendous amount of energy. The fusion process has less environmental impact because there are no radioactive by-products of the reaction.

Q4. Because the consequences of a mishap could be disastrous and the radio-active residue is very difficult to dispose of safely.

SECTION 17.6

Q2. Virtually unlimited supply, minimal environmental impact, and best of all, it's free!

Q4. The two-ply structure of the cell is designed to take advantage of the pho-toelectric effect. Electrons are liberated from one layer and captured in the other. This results in a potential difference (voltage) between the layers that can be used to produce an electric current.

THOUGHT QUESTION

4. No. The potential energy is already in the nucleus. The breeding process only makes it possible to release it by upsetting its stability.

Appendix B
Table of Fundamental Constants

Quantity	Symbol	Approximate Value
Speed of light	c	3.00×10^8 m/s
Gravitational constant	G	6.67×10^{-11} N-m^2/kg^2
Gravitational acceleration	g	9.81 m/s^2
Electron rest mass	m_e	9.11×10^{-31} kg
Proton rest mass	m_p	1.67×10^{-27} kg
Classical radius of proton	r_p	1.2×10^{-15} m
Proton/electron mass ratio	m_p/m_e	1836
Neutron rest mass	m_n	1.67×10^{-27} kg
Classical radius of neutron	r_n	1.2×10^{-15} m
Atomic mass unit	amu	1.66×10^{-27} kg
Avogadro's number	N_A	6.02×10^{-23}/mole
Bolzmann constant	k	1.38×10^{-23} J/K
Absolute zero	O K	-273 °C
Atmospheric pressure (sea level)	1 atm	1.01×10^5 N/m^2
Electron charge	e	1.60×10^{-19} C
Coulomb's law constant	K	8.99×10^9 N-m^2/C^2
Planck's constant	h	6.63×10^{-34} J·s
	$\hbar = h/2\pi$	1.05×10^{-34} J·s
Magnetic force law constant	K^{11}	1.00×10^{-7} N/ampere^2
Stefan-Bolzmann constant	σ	5.67×10^{-8} W/m^2·K^4
Bohr radius (radius of hydrogen atom)	r_o	529×10^{-11} m
Wien's law constant	w	.0029 m·k
Density of water	H_2O	1000 kg/m^3
Mass of sun	$M_\odot$	1.99×10^{30} kg
Radius of sun	$R_\odot$	6.96×10^8 m
Luminosity of sun	$L_\odot$	3.83×10^{26} w
Mass of earth	$M_\oplus$	5.98×10^{24} kg
Radius of earth	$R_\oplus$	6.37×10^6 m
Mean earth-sun distance (astronomical unit)	AU	1.50×10^{11} m
Light year	ly	9.46×10^{15} m
Parsec	pc	3.09×10^{16} m, 3.26 ly

Appendix C
The Periodic Table of the Elements*

Group I	Group II	Transition elements										Group III	Group IV	Group V	Group VI	Group VII	Group 0
H 1 1.0079 $1s^1$																	He 2 4.0026 $1s^2$
Li 3 6.941 $2s^1$	Be 4 9.01218 $2s^2$											B 5 10.81 $2p^1$	C 6 12.011 $2p^2$	N 7 14.0067 $2p^3$	O 8 15.9994 $2p^4$	F 9 18.9984 $2p^5$	Ne 10 20.179 $2p^6$
Na 11 22.9898 $3s^1$	Mg 12 24.305 $3s^2$											Al 13 26.9815 $3p^1$	Si 14 28.0855 $3p^2$	P 15 30.9738 $3p^3$	S 16 32.06 $3p^4$	Cl 17 35.453 $3p^5$	Ar 18 39.948 $3p^6$
K 19 39.0983 $4s^1$	Ca 20 40.08 $4s^2$	Sc 21 44.9559 $3d^14s^2$	Ti 22 47.88 $3d^24s^2$	V 23 50.9415 $3d^34s^2$	Cr 24 51.996 $3d^54s^1$	Mn 25 54.9380 $3d^54s^2$	Fe 26 55.847 $3d^64s^2$	Co 27 58.9332 $3d^74s^2$	Ni 28 58.69 $3d^84s^2$	Cu 29 63.546 $3d^{10}4s^1$	Zn 30 65.39 $3d^{10}4s^2$	Ga 31 69.72 $4p^1$	Ge 32 72.59 $4p^2$	As 33 74.9216 $4p^3$	Se 34 78.96 $4p^4$	Br 35 79.904 $4p^5$	Kr 36 83.80 $4p^6$
Rb 37 85.468 $5s^1$	Sr 38 87.62 $5s^2$	Y 39 88.9059 $4d^15s^2$	Zr 40 91.224 $4d^25s^2$	Nb 41 92.9064 $4d^45s^1$	Mo 42 95.94 $4d^55s^1$	Tc 43 (98) $4d^65s$	Ru 44 101.07 $4d^75s^1$	Rh 45 102.906 $4d^85s^1$	Pd 46 106.42 $4d^{10}5s^0$	Ag 47 107.868 $4d^{10}5s^1$	Cd 48 112.41 $4d^{10}5s^2$	In 49 114.82 $5p^1$	Sn 50 118.71 $5p^2$	Sb 51 121.75 $5p^3$	Te 52 127.60 $5p^4$	I 53 126.905 $5p^5$	Xe 54 131.29 $5p^6$
Cs 55 132.905 $6s^1$	Ba 56 137.33 $6s^2$	57–71‡	Hf 72 178.49 $5d^26s^2$	Ta 73 180.948 $5d^36s^2$	W 74 183.85 $5d^46s^2$	Re 75 186.207 $5d^56s^2$	Os 76 190.2 $5d^66s^2$	Ir 77 192.22 $5d^76s^2$	Pt 78 195.08 $5d^96s^1$	Au 79 196.967 $5d^{10}6s^1$	Hg 80 200.59 $5d^{10}6s^2$	Tl 81 204.383 $6p^1$	Pb 82 207.2 $6p^2$	Bi 83 208.980 $6p^3$	Po 84 (209) $6p^4$	At 85 (210) $6p^5$	Rn 86 (222) $6p^6$
Fr 87 (223) $7s^1$	Ra 88 226.025 $7s^2$	89–103‖	Rf 104 (261) $6d^27s^2$	Ha 105 (262) $6d^37s^2$	106 (263)	107 (262)		109 (266)									

Legend: Symbol—Cl 17 —Atomic number; Atomic mass**— 35.453; $3p^5$ —Electron configuration

Lanthanide series‡

La 57 138.906 $5d^16s^2$	Ce 58 140.12 $5d^04f^26s^2$	Pr 59 140.908 $4f^36s^2$	Nd 60 144.24 $4f^46s^2$	Pm 61 (145) $4f^56s^2$	Sm 62 150.36 $4f^66s^2$	Eu 63 151.96 $4f^76s^2$	Gd 64 157.25 $5d^14f^76s^2$	Tb 65 158.925 $5d^04f^96s^2$	Dy 66 162.50 $4f^{10}6s^2$	Ho 67 164.930 $4f^{11}6s^2$	Er 68 167.26 $4f^{12}6s^2$	Tm 69 168.934 $4f^{13}6s^2$	Yb 70 173.04 $4f^{14}6s^2$	Lu 71 174.967 $5d^14f^{14}6s^2$

Actinide series‖

Ac 89 227.028 $6d^17s^2$	Th 90 232.038 $6d^27s^2$	Pa 91 231.036 $5f^26d^17s^2$	U 92 238.029 $5f^36d^17s^2$	Np 93 237.048 $5f^46d^17s^2$	Pu 94 (244) $5f^66d^07s^2$	Am 95 (243) $5f^76d^07s^2$	Cm 96 (247) $5f^76d^17s^2$	Bk 97 (247) $5f^96d^07s^2$	Cf 98 (251) $5f^{10}6d^07s^2$	Es 99 (252) $5f^{11}6d^07s^2$	Fm 100 (257) $5f^{12}6d^07s^2$	Md 101 (258) $5f^{13}6d^07s^2$	No 102 (259) $6d^05f^{14}7s^2$	Lr 103 (260) $6d^15f^{14}7s^2$

*Atomic mass values averaged over isotopes in percentages as they occur on earth's surface.
**For many unstable elements, mass number of the most stable known isotope is given in parentheses.

Appendix D
Overview Questions

1. What is a conservation law or principle in physics? Which quantities in physics can be described by such a law or principle?

2. Discuss the physics that is involved in the functioning of each of the human senses, including hearing, sight, smell, taste, and touch.

3. What is meant by an inverse square law? Describe some examples of an inverse square law.

4. Identify and compare and contrast the four fundamental interactions of nature. Discuss the possible existence of additional interactions.

5. Select five individuals of your choice and describe the contributions that each person has made to the science of physics.

6. Describe the relationships between relativity theory, Newtonian mechanics, and quantum mechanics.

7. Discuss the impact that the Scientific Revolution has had on human history.

8. What are some of the things that the laws and principles of physics tell us we can never do? What is the justification for stating these limits?

9. Describe all the forms of energy that are currently known to exist in nature. Do you think there might be other, as yet undiscovered energy forms? Explain.

10. Discuss the relationship that exists between physics and each of the following fields of human endeavor:
    a. religion   **b.** history   **c.** politics   **d.** mathematics   **e.** the arts

11. What is meant by wave motion? To what extent has this concept been used to describe various phenomena in the physical world?

12. Discuss how physicists measure various lengths ranging from the diameter of an atom to the diameter of the universe.

13. What is meant by a fundamental constant in physics? Of what importance are these constants? Choose one such constant and describe how its value is measured.

14. List the five functions associated with living matter. Choose a living creature and discuss the physics involved in each of these functions.

15. Speculate on the nature of the physical world that might currently lie outside of our present limits of observation and detection.

16. Choose five devices important in our current technology and describe the physical principles upon which they operate.

17. Describe how each of the following is defined in terms of the fundamental quantities of the meter, the kilogram, and the second.
    a. force   **b.** energy   **c.** power   **d.** electric charge   **e.** work

18. Compare and contrast light and sound in terms of the way these phenomena are described in physics.

19. Pick five of the principles stated in the text and discuss their justification and importance in physics.

20. Describe five important discoveries in physics that have been made unexpectedly or by accident.

21. Compare and contrast circular motion translational motion, and wave motion.

22. List the sources of power currently available to human technology, along with the advantages and disadvantages associated with their use. Speculate on the possible existence of other power sources that may become available in the future.

23. Of what use is the study of the relatively ''inaccessible'' parts of the physical world such as outer space and subatomic space?

24. Describe a practical use human beings make of each of the following states of matter:
    a. solid   **b.** liquid   **c.** gas   **d.** plasma

25. What role, if any, do you think that physics will play in the coming years of human existence on the earth?

Index

PHOTO CREDITS *(Continued from page iv)*

7.13(c), page 198: National Solar Observatory/National Optical Astronomy Observatories.

Physics Facet 7.2(left), page 199: ©A. J. Copley/Visuals Unlimited.

Physics Facet 7.2(right), page 199: ©Biophoto Associates.

7.17, page 203: Author.

7.19, page 205: ©Scott Camazine/Photo Researchers, Inc.

8.3, page 228: Author.

Physics Facet 8.2(a), page 230: ©Joe McDonald/Visuals Unlimited.

Physics Facet 8.2(b), page 230: ©Richard C. Johnson/Visuals Unlimited.

Physics Facet 8.2(c), page 230: ©T. E. Adams/Visuals Unlimited.

Physics Facet 8.2(d), page 230: ©Dan Kline/Visuals Unlimited.

8.7 page 236: Author.

8.8, page 237: ©David R. Frazier/Science Source.

8.9, page 238: NASA/Science Source.

Physics Facet 9.1(a), page 263: UPI/Bettmann.

Physics Facet 9.1(b), page 263: Reuters/Bettmann.

Physics Facet 9.1(c), page 263: The Bettmann Archive.

9.16, page 270: Science VU/Visuals Unlimited.

9.18, page 271: UPI/Bettmann.

Physics Facet 10.1, page 291: ©Kent Wood/Photo Researchers, Inc.

Physics Facet 12.1, page 345: The Bettmann Archive.

12.12, page 358: Science VU/Visuals Unlimited.

12.23, page 365: Peter Aprahamian/Science Photo Library.

12.28, page 368: From LIFE SCIENCE LIBRARY: LIGHT AND VISION. Photograph by Ken Kay. ©1966 Time-Life Books Inc.

Physics Facet 13.1, page 385: The Bettmann Archive.

13.3, page 387: Illustration adapted from photo by Mt. Wilson and Campanus Observatories.

13.14, page 405: From *Physical Review,* Volume 45, 1934, p. 295 and Volume 50, 1936, p. 263, by Carl D. Anderson and S. Neddermeyer. Reprinted with permission of The American Physical Review.

13.15, page 406: Courtesy of Brookhaven National Laboratory.

Chapter 14 opener top left, page 423: NASA/Science Source.

Chapter 14 opener top right, page 423: Philippe Plailly/Science Photo Library.

Chapter 14 opener middle left, page 423: Bobbie Kingsley/Photo Researchers, Inc.

Chapter 14 opener bottom, page 423: Tim Davis/Photo Researchers, Inc.

14.8(a), page 431: NASA.

14.8(b), page 431: NASA.

14.8(c), page 431: NASA.

14.8(d), page 431: NASA.

Physics Facet, page 435: Imperial War Museum.

14.18(c), page 443: Dr. Jeremy Burgess/Science Photo Library.

14.19, page 444: ©1989 Tony Freeman/PhotoEdit.

14.24(a), page 450: Dr. Jeremy Burgess/Science Photo Library.

14.24(b), page 450: Science VU-IBMRL/Visuals Unlimited.

14.24(c), page 450: ©Don Fawcett/Visuals Unlimited.

Chapter 15 opener, page 455: The Bettmann Archive.

Physics Facet 15.1, page 460: The Bettmann Archive.

15.7, page 468: ©Seul/Photo Researchers, Inc.

15.13, page 475: University of Hawaii Institute for Astronomy from UH 88-inch telescope on Mauna Kea.

Chapter 16 opener, page 483: National Optical Astronomy Observatories.

16.4, page 487: Author.

16.8, page 497: National Solar Observatory/National Optical Astronomy Observatories.

16.9(a), page 497: National Solar Observatory/National Optical Astronomy Observatories.

16.9(b), page 497: NASA.

16.9(c), page 497: NASA.

16.11, page 501: Author.

16.13, page 504: National Optical Astronomy Observatories.

16.14(a), page 505: Lick Observatory Photograph.

16.14(b), page 505: Palomar Observatory, California Institute of Technology.

16.15, page 505: ©1984 Anglo-Australian Telescope Board.

16.18, page 507: ©Anglo-Australian Telescope Board 1979. Photograph by David Malin.

16.20, page 509: Lick Observatory Photograph.

16.21(a), page 509: Lick Observatory Photograph.

16.21(b), page 509: National Optical Astronomy Observatories.

Physics Facet 16.2, page 513: Palomar Observatory, California Institute of Technology.

16.23, page 514: Palomar Observatory, California Institute of Technology.

16.25, page 515: SOURCE: M. Seldner, B. Siebers, E. J. Groth, and P. J. E. Peebles, 1977, *Astronomical Journal,* 82:249.

17.2(left), page 531: ©Paul Conklin/PhotoEdit.

17.2(middle), page 531: ©J. Muir Hamilton/Stock Boston.

17.2(right), page 531: ©Richard Hansen/Photo Researchers, Inc.

Physics Facet 17.1, page 540: UPI/Bettmann.

The Metric System

Unit	Measure		Symbol	English Equivalent
Linear measure				
1 kilometer	= 1000 meters	10^3 m	km	0.62137 mile
1 meter		10^0 m	m	39.37 inches
1 decimeter	= 1/10 meter	10^{-1} m	dm	3.937 inches
1 centimeter	= 1/100 meter	10^{-2} m	cm	0.3937 inch
1 millimeter	= 1/1000 meter	10^{-3} m	mm	
1 micrometer (or micron)	= 1/1,000,000,000 meter	10^{-6} m	μm (or μ)	English equivalents
1 nanometer	= 1/1,000,000,000	10^{-9} m	nm	infrequently used
1 angstrom*	= 1/10,000,000,000	10^{-10} m	Å	
Measures of capacity (for fluids and gases)				
1 liter			L	1.0567 U.S. liquid quarts
1 milliliter	= 1/1000 liter		ml	
1 milliliter	= volume of 1 g of water at standard temperature and pressure (stp)			
Measures of volume				
1 cubic meter			m^3	
1 cubic decimeter	= 1/1000 cubic meter = 1 liter (L)		dm^3	
1 cubic centimeter	= 1/1,000,000 cubic meter = 1 milliliter (mL)		cm^3 = ml	
1 cubic millimeter	= 1/100,000,000 cubic meter		mm^3	
Measures of mass				
1 kilogram	= 1000 grams		kg	2.2046 pounds
1 gram			g	15.432 grains
1 milligram	= 1/1000 gram		mg	.01 grain (about)
1 microgram	= 1/1,000,000 gram		μg (or mcg)	

*The angstrom is not part of the metric system but is so frequently encountered in the literature that it is included.